Sustainable Crop Protection Strategies

Volume-2

Sustainable Crop Protection Strategies

Volume-2

H.R. Sardana
O.M. Bambawale
D. Prasad

2024

Daya Publishing House®
A Division of
Astral International Pvt. Ltd.
New Delhi – 110 002

First Published, 2010

Reprinted, 2024

ISBN: 978-81-7035-689-9 (HB)

Published by : **Daya Publishing House®**
A Division of
Astral International Pvt. Ltd.
– ISO 9001:2015 Certified Company –
4736/23, Ansari Road, Darya Ganj,
New Delhi-110 002
Ph. 011-43549197, 23278134
E-mail: info@astralint.com
Website: www.astralint.com

Printed at : **Replika Press Pvt. Ltd.**

प्रो. स्वपन कुमार दत्ता
उपमहानिदेशक (फसल विज्ञान)
Prof. Swapan Kumar Datta
Deputy Director General
(Crop Science)

भाकृ अनुप
ICAR

भारतीय कृषि अनुसंधान परिषद्
कृषि भवन, डा. राजेन्द्र प्रसाद मार्ग, नई दिल्ली - 110114
INDIAN COUNCIL OF AGRICULTURAL RESEARCH
KRISHI BHAWAN, DR. RAJENDRA PRASAD ROAD, NEW DELHI-110114

Foreword

Sustainable production of crops depends upon the several biotic and abiotic factors and particularly on the management practices. Among them, sound pest and disease management strategies contribute substantially to minimize losses caused by pests. Chemical method of control of pests is commonly adopted in many crops the world over. However, the indiscriminate use of pesticides has resulted in pest out breaks and pest resurgence on one hand and health hazards on the other hand. The importance of achieving sustainable food production through the use of eco-friendly sustainable pest management techniques is being realized more and more in the recent past. Thus, with the increasing importance of sustainable agriculture, the concept of integrated pest management (IPM) has emerged. To assess the present scenario in sustainable pest management, it becomes vital to systematically compile the recent advances in pest management to develop better crop protection programmes for future. The present book is an effort in this direction.

The editors of book '*Sustainable Crop Protection Strategies*' Drs. H.R. Sardana, O.M. Bambawale and D. Prasad have made a concerted effort and done a commendable job to compile the work done at various institutions under National Agricultural Research System (NARS) into 38 thought provoking chapters. Various aspects of pest management covered in the book on recent developments in the fields of pheromone technology, indigenous knowledge, biological control, biodiversity, simulation modeling and climate change, forecasting and geospatial technologies have been extensively and elegantly presented.

This book would not only serve as a reference source but also as a guide and inspiration for future research in the area of plant protection to achieve the most cherished goal of economically and ecologically sustainable agriculture. I hope the efforts of editors will certainly enriched the knowledge of the readers of crop protection. I congratulate the editors for this timely publication.

S. K. Datta
Swapan Kumar Datta

Preface

Over exploitation of natural resources and excessive chemicalisation of agriculture in the past few decades, have led to poor sustainability of farm production. Over dependency and indiscriminate use of agricultural chemicals have resulted in the problems of pest resurgence, secondary pest outbreaks, development of pest resistance, decimation of natural enemies on one hand and disturbances to environment and health hazards through residues in food chain on the other hand. These side effects have raised public concern about the safety of pesticides. Today we are confronted with the herculaean task of providing environmental and food security to our expanding population. This calls for reorientation of our strategies to minimize the use of chemical inputs in agriculture and depend more on eco-friendly sustainable crop protection approaches without causing disruption to fragile eco-system. Integrated pest management (IPM) is one such environmentally sensitive approach to pest management that relies on combination of common- sense approaches and fits into the concept of sustainable agriculture. IPM incorporates ecological and economic factors into agricultural ecosystem and decision making and addresses public concern about environmental quality and food safety. Today, IPM is considered to be the springboard to sustainable crop management and forms a focus of the present book.

The thirty eight chapters covered in these 2 volumes deal with various facets of pest management. The book begins with a chapter on maintenance of diversity in sugarcane ecosystem highlighting its importance from pest management perspective. The ramifications of introduction of new pests along with the movement of plant materials and germplasm are well known. The chapter 'Indian plant quarantine system' focuses on such sanitary and phytosanitory issues. The role of geospatial technologies, simulation modeling, entomopathogens and chemical pesticides in IPM programmes has been highlighted and discussed. There is need to integrate traditional knowledge of farmers with other advanced techniques in order to develop a profitable and sustainable approach to pest management. The chapter on 'ITK's' stresses upon the indigenous knowledge in cotton ecosystem.

Adequate focus is on pheromone technology which is a major tool for monitoring of pests for initiating pest control measures. The sustainable IPM technologies developed and practiced for major cropping systems *viz;* cotton, rice, vegetables, pulses, oilseeds, oilpalm and spices have been discussed in various chapters. In all, the book comprises of 38 chapters on various topics with recent technological advances such as maintenance of diversity in pest management, national plant quarantine system, geospatial technologies, simulation modeling and climate change, disease resistance, utilization of pheromones, diagnostics of viruses, eco-friendly pest management, mites and rodents pest management, bio-agents and entomopathogens, weed management and IPM in temperate horticultural crops.

The book '*Sustainable Crop Protection Strategies*' attempts to comprehend and consolidate the scattered knowledge on sustainable pest management strategies. The book would not only serve as a reference but also as a guide and inspiration for shaping future research on modern plant protection tactics. The book will also prove useful to all those interested in promoting the cause of IPM so that sustainability in agricultural system and environmental protection for future generations is achieved.

We received a tremendous response and support from all the contributors for preparing their chapters in tune with the theme of book, for which we express our sincere gratitude to them.

We are also grateful to Shri Anil Mittal of Daya Publishing House, New Delhi for timely taking up the task of publication of this book and providing us all technical help required during the preparation and finalization of the manuscript for the book.

H.R. Sardana

O.M. Bambawale

D. Prasad

Contents

— Volume 2 —

Chapter 20

Integrated Pest Management in Short Duration Pigeonpea

Nitin Joshi[1], M.K. Pandey[2], C.P. Srivastava[3] and Mukesh Sehgal[4]
[1]Bayer BioScience Pvt. Ltd., Delta Square, IFFCO Chowk, Gurgaon, Haryana
[2]Krishi Vigyan Kendra, Ujwa, New Delhi
[3]Department of Entomology, Institute of Agricultural Sciences, B.H.U., Varanasi - 221 005, U.P.
[4]National Centre for Integrated Pest Management, New Delhi - 110 012

Introduction

Pulses are among the ancient food crops with evidence of their cultivation for over 8000 years. They are rich in proteins, essential amino acids, particularly lysine, which is deficient in the cereals. A mixture of cereals and pulses along with some milk and eggs give a balanced diet. Besides, being a rich and the cheapest source of dietary protein and a valuable animal feed, they also play a key role in improving and sustaining soil productivity through biological nitrogen- fixation by bacteria, *Rhizobium* spp. prevalent in their root nodules. It is estimated that the country's population will touch nearly 135 crores by 2020 A.D. The country would then need a minimum of 30.3 million tonnes of pulses (Anonymous, 2002). The total area under pulses remained virtually stagnant (22–24 million hectares) with almost stable production (12–14 million tonnes) over the last four decades (Anonymous, 2003). These crops accounts for one–fifth of the total area under food grain crops and contribute a meager of one–twelfth of the total food grain production in the country. It was estimated that the annual requirement of pulses is likely to be around 19.0 million tonnes in 2010, 23 million tonnes in 2020 and 24–26 million tonnes in 2030. This will require about 75 per cent increase in the production of pulses by the year 2030 (Anonymous, 1999). Thus, the present productivity of 0.6 tonnes per hectare will have to be raised to 0.99 tonnes per hectare. The per capita availability of pulses also declined from 60.7 g/

capita/day in 1951 to less than 30 g/capita/day in 2001 as against the FAO/WHO's recommendation of 80 g/capita/day. As a result, Indian Government is importing pulses from other countries to meet the requirement. The import per cent of pulses has increased to a great extent as compared to the import of cereals in 21st Century. Therefore, we are paying a large sum of money to other countries for this purpose.

Role of Short Duration Pigeonpea in Cereal Based Cropping System

Pulses play a pivotal role in sustaining the cereal–based cropping systems in the country, where rice–wheat (9.7 million ha) and rice–rice (2.1 million ha) cropping systems are essential for food security. Pulses in rotation with cereals can effectively break disease cycles and inoculum built-up of major pathogens of both cereal and pulse crops, since both do not act as alternative host to harbour each others pathogens. Short duration pulse varieties developed for multiple cropping systems have tremendous scope in raising cropping intensity as well as improving soil health. Therefore, pulses have comparative advantage in contributing to crop diversification, rotation and mixed cropping.

The country contributes more than 75 per cent to the global production of chickpea and pigeonpea, the two major pulses of Indian sub–continent. Red gram or pigeonpea, *Cajanus cajan* (L.) Millsp. ranks second highest amongst the pulse crops in India, entering, as it does, into the daily food of considerable number of people. On an average it contains 22.3 per cent protein. It is largely eaten in the form of split pulse as 'dal', while its tender green pods constitute a very favourite vegetable in some parts of the country. The outer integument of its seed together with part of the kernel provides a valuable feed for the cattle. The stalks are utilized for roofing, wall sides of carts and basket making. In India, the total area under pigeonpea cultivation is 3.4 million hectares (*i.e.*, 15 per cent of total area under pulse production) with a production of 2.30 million tonnes (*i.e.*, 17 per cent of total pulse production). The average yield of pigeonpea is 600 kg/ha, inspite of its potential yield to 2.0–2.5 tonnes/ha. Between 1970 and 1990, though the area under pigeonpea cultivation increased at more than 2 per cent per annum, yield increased only 0.33 per cent annually during the same period (Ryan, 1994). At national level, the important states for pigeonpea cultivation and production are Uttar Pradesh, Maharashtra, Karnataka, Madhya Pradesh and Gujarat. The state of Uttar Pradesh has the distinction of contributing about 40 per cent of the total pigeonpea production in the country. Much of the area under pigeonpea is covered by medium and long duration regional and landrace cultivars, which are harvested after 200–280 days. Research efforts in recent years have led to the development of dwarf, short duration cultivars and hybrids (Nene and Sheila, 1990). These short duration cultivars mature between 120 to 150 days in North Indian conditions. Short duration pigeonpea are commonly grown as a monocrop, whereas medium and long duration pigeonpea are mostly grown as intercrops or mixed crops in rainfed or dryland agriculture. The newly developed short duration genotypes have a higher harvest index than currently grown cultivars, and have shown high productivity in sole cropping systems at high plant density (Ariyanayagam and Singh, 1994). Besides this, they are less sensitive to photoperiods and temperature and may be adopted to several new environments. However, it has not been possible to exploit the full genetic potential of high yielding pigeonpea cultivars because of extensive losses due to insects, diseases and moisture stress.

Crop Losses in Short Duration Pigeonpea

Pradhan (1971) had very rightly advocated the concept that in agriculture the production technology is inextricably interwoven in the protection technology and therefore, the production technology cannot succeed unless there is adequate progress in protection technology. The productivity of pigeonpea has not increased considerably during the last decade in our country. The damage caused by insect pests and diseases are the major reasons of low productivity.

Insect Pests

Sachan *et al.* (1994) have reported that in India pigeonpea is attacked by nearly 250 species of insects belonging to 8 order and 61 families, though relatively few cause serious yield losses. Pests that attack the reproductive structures cause the greatest yield losses (Rangaiah and Sehgal, 1984). The insect pest complex as well as their pattern varies according to the maturity groups of pigeonpea crop. Short duration pigeonpea provides less time for the plant to compensate the loss of first flush of flowers and pods, thus suffer comparatively greater losses. In contrast, medium and long duration pigeonpea genotypes produce additional flushes of flowers when the early flushes are lost and can tolerate early season insect attack. The insects which feed on medium and long duration cultivars, also attack short duration pigeonpea. The key pests of short duration pigeonpea include *Helicoverpa armigera* Hubner, *Maruca vitrata* Fabricius and *Melanagromyza obtusa* (Malloch). Others such as *Mylabris* spp., *Clavigralla gibbosa* Spinola, *Lampides boeticus* (L.) and *Exelastis atomosa* (Walsingham) are also potential pests and occasionally cause significant grain yield losses in short duration pigeonpea.

H. armigera is the major biotic constraint to increasing pigeonpea production throughout India. The four features of *H. armigera* life history which make it one of the most serious wide spread pest are; high fecundity, extensive polyphagy, strong flying ability and the facultative diapause. This insect pest has been recorded to feed on at least 181 plant species (Manjunath *et al.*, 1989). The larvae of the pest have preference for feeding on plant pods rich in nitrogen such as reproductive structures and growing tips (Fitt, 1989). In eastern U.P., pod damage by *H. armigera* to the short duration cultivar "UPAS-120" varied from 15.9 per cent to 32.5 per cent (Joshi and Srivastava, 2006). Sahoo and Senapati (2001) assessed that pod borer complex damaged 34.79 per cent seeds incurring the yield loss to the tune of 28.07 per cent in UPAS-120. However, in recent years, the podfly damage in short duration pigeonpea have increased considerably. Singh and Rai (1981) studied the losses caused to the early maturing pigeonpea variety T-21 by *Melanagromyza obtusa* (Mall.) in Uttar Pradesh. They observed that pod infestation varied from 35.83 per cent to 47.70 per cent and seed infestation from 24.27 per cent to 32.37 per cent. Kabaria *et al.* (1988) have reported 2.6 per cent to 11.2 per cent seed damage by *M. obtusa* in Gujarat. Lal and Yadav (1988) found that grain damage in treated plots ranged from 17.0 per cent to 30.2 per cent for *M. obtusa* and from 5.8 to 18.6 per cent for *H. armigera*. The annual loss of pigeonpea production due to podfly alone has been estimated to the extent of 25–30 per cent in Uttar Pradesh (Lal and Yadava, 1987), whereas the total seed loss due to pod borer damage has been worked out to the tune of 50,000 tonnes annually for Uttar Pradesh alone.

Diseases

Phytophthora blight caused by *Phytophthora drechsleri* Tucker f. sp. *cajani* has been reported as an important production constraint in Northeastern India particularly in low lying, poorly drained fields (Kannaiyan *et al.*, 1984; Mishra and Shukla, 1987 and Chauhan *et al.*, 2002). Cloudy weather accompanied by intermittent rains followed by mean temperatures 25±1°C favours the infection and development. Crop yield is also significantly reduced due to wilt disease caused by *Fusarium udum* Butler, with an estimated yield loss of US$36 million in India and $5 million in eastern Africa (Kannaiyan *et al.*, 1984). Like any other soil-borne diseases, the wilt disease of pigeonpea is difficult to control. Some pesticides and chemicals have been recommended for the management of the disease, but none have been proven to give the desired success in controlling the disease (Singh, 1998). Pesticides are reported to cause adverse effects on treated soil ecosystem because of their non-biodegradable nature and also because they induce resistance in pathogens (Cohen and Leavy, 1990). Sterility Mosaic Disease (SMD) is another important disease of pigeonpea, is so called because of the resulting lack of flower production and the mosaic symptoms that appear on leaves (Dharamraj *et al.*, 2004). The pathogen stimulates excessive vegetative growth, though infected plants appear to have a normal growth pattern. The leaf symptoms are often not conspicuous under field conditions and farmers are often unaware of the problem until the flowering stage, when the crop fails to produce flowers. Unaware of the nature of the disease the hapless farmers may resort to application of chemical sprays with the hope of inducing flowering, but this is a futile exercise and adds to their losses. SMD is spread rapidly by a tiny arthropod mite, *Aceria cajani*. Pigeonpea losses worth over US$300 million per annum are attributed to SMD in India and Nepal alone.

IPM Approaches

In recent years, pigeonpea production has shifted from a largely subsistence operation to more market oriented, commercial production. The substantial rise in the price of pigeonpea (Kelley and Parthasarathy Rao, 1993) has provided farmers an opportunity to consider pest management options which formerly would have been too costly. One option which farmers are increasingly choosing is the use of synthetic organic insecticides. There are three primary concern with a pest control technology based predominantly on the use of chemical pesticides; the effect on human health (Loevinsohn, 1987), the degradation of ecosystems (Pimental *et al.*, 1992) and the increasing amount of pesticides required to produce adequate levels of control due to the development of resistance in the pest insects (Armes *et al.*, 1992). This "pesticide treadmill" led ultimately to the collapse of pest control (Van den Bosch, 1978). Therefore, the basic concept of IPM approach is to be implemented which involves the containment of pests below economic thresholds, using an appropriate combination of control strategies. A more recent definition has highlighted the fact that farmers do IPM and that economic, social and environmental sustainability must be considered when pest management practices are being evaluated (Wightman, 1993). The insect pest management in short duration pigeonpea has been reviewed which involves the use of biorational insecticides to control pests (Shanower, 1996). Biorational insecticides include plant derived products such as neem and insect pathogens such as *Helicoverpa armigera* nuclear polyhedrosis virus (HaNPV) and *Bacillus thuringiensis* (Bt) based products. These are generally considered to be safer to the environment, have less impact on beneficial organisms and are safer for human beings. Based on these, many IPM modules have been developed and are being evaluated in short duration pigeonpea at different locations under All India Co–ordinated Research Project on Pigeonpea and by other scientists also on large plots in farmer's field. Integrated disease management is based on clear under standing of the genetic and biochemical bases of pathogenesis, ecology and other factors influencing the disease which may lead to the development of more diverse, effective long term, ecologically safe disease management strategies.

Keeping in view the above facts, the following IPM practices can be implemented by the farmers in short duration pigeonpea to enhance productivity.

1. Deep summer ploughing to expose the insects, pathogens and nematodes so that their immature stages or dormant stages get killed due to high temperature.
2. Selection of disease/insect resistant or tolerant variety based on the requirement of the local situation. For example, in NEPZ cultivation of variety UPAS-120 and Prabhat is recommended, which escapes the *H. armigera* damage if planted in middle or last week of June.

3. Seed treatment with *Trichoderma viride* and Vitavax @ 4g and 2 g/kg of seed respectively, provides protection against *Fusarium* wilt.
4. Sowing on ridges provides protection against water logging and *Phytophthora* blight.
5. Installation of sex pheromone traps @ 20/ha for monitoring of *H. armigera* at the initiation of flowering in north India is very effective to forewarn farmers for timely control of pest. The pheromone traps are set up at the height of 0.5 m above the crop canopy for trapping the male moths. The moths caught in these traps should be removed and killed at weekly intervals.
6. Erection of bird perches @ 20/ha at the time of 50 per cent flowering for sitting of insectivorous birds which will prey upon lepidopteron larvae (Gopali and Lingappa, 2001a). The farmers can easily make T–shaped bird perches with the help of bamboo sticks and these can be erected at 1.0 m above the crop canopy so that bird can easily locate and pick up the larvae from the pigeonpea plant.
7. Apply 5 per cent neem seed kernel extract (NSKE) or spray readymade Multineem 300ppm @ 1l/ha at the time of 50 per cent flowering phase which acts as ovipositional deterrent, repellant, antifeedant to insect pests. Sachan and Lal (1990) have reported 5 per cent NSKE to be highly effective and almost at par with the recommended insecticide endosulfan (0.07 per cent) against *H. armigera* on chickpea and pigeonpea.
8. In areas where *H. armigera* is the key pest, spray HaNPV @ 500 LE/ha at full flowering phase. It is a highly infective and selective microbial biopesticide for the management of *H. armigera*. Within 3–5 days of spray, the dead larva shows typical HaNPV symptoms, *i.e.*, hanging of diseased larvae upside down (Sarode *et al.*, 1994 and 1997; Gopali and Lingappa, 2001b). The spray of Bt formulations @ 1.0 kg/ha is also very effective against lepidopteran larvae (Santharam *et al.*, 1994; Sundara Babu and Muthuswami, 1995 and Reddy *et al.*, 2001). If these biorationals are not able to take care of *H. armigera* or pest load is high, Spinosad 45SC @73g a.i./ha can be sprayed for effective control of lepidopteran larvae (Haldar *et al.*, 2006).
9. In the regions where podfly is a problem, spray dimethoate @ 1l/ha or Imidacloprid 17.8SL @150ml/ha.

Preparation of Spray Materials

The spray mixture for each treatment should be prepared by mixing the required quantity of insecticides or biorationals according to required dose of application in clear water @ 200 litres (L)/ha to be sprayed by power sprayer with a cone type nozzle after thorough cleaning and calibration. Proper stirring is required after adding water for obtaining uniform emulsion. Spraying should be done in early hours of the day, when slow and stable wind velocity prevailed so as to avoid the drift of insecticidal fluid to the adjacent plots starting from full flowering and initial pod formation stage at 12–15 days interval. The spraying of Bt and HaNPV are suggested to be performed in evening hours to avoid the harmful effect of Ultraviolet (UV) radiation on Bt and HaNPV. Tinopal @ 0.1 per cent is mixed in spray solution for protection against UV radiation. Besides this, jaggery or sugar @ 0.5 per cent can be added in HaNPV as a phagostimulant. NSKE can be freshly prepared by the farmers itself through separation of seed and placing them in water. The neem seed covering, *i.e.*, kernel is extracted. Then, 5 kg neem seed kernel is grinded properly and soaked in 20 L of water in muslin cloth for overnight. Next morning, it is squeezed and filtered with the help of fine muslin cloth and finally the extract is made 100 L by adding additional water to obtain 5 per cent NSKE. The freshly prepared

NSKE is also mixed with soap solution @ 0.1 per cent, which serves as spreading agent by reducing the surface tension of the spray fluid. NSKE acts as antifeedant, ovipositional deterrent or repellant against lepidopteran larvae as well as checks the growth of fungal pathogens. In case neem seeds are not available, the neem formulations can be purchased from market.

Benefit : Cost Ratio

The Benefit : Cost ratio's are an important parameter to assess the returns obtained in terms of increased yield over control after deducting the total cost of production including amount spent in plant protection measures. The cost of grain prevailing at that time is also taken into account for evaluating B:C ratio. Studies conducted in eastern U.P. have found that short duration pigeonpea was an extremely remunerative crop as compared to any other crop. The highest B:C ratio's of 3.22:1 and 3.55:1 was recorded in module M2 (ridge sowing, pheromone traps, bird perches, methomyl 40SP, NSKE 5 per cent and Bt) during *kharif* 2001 and 2002, respectively (Joshi and Srivastava, 2006). Similarly, Siddhabhatti *et al.* (2003) found that module M4 (profenofos, NSKE, endosulfan and bird perches) and M5 (Bt, NSKE, endosulfan and bird perches) were most effective and recorded 20.38 per cent and 19.37 per cent pod damage, 6.45 per cent and 7.10 per cent grain damage, respectively. These IPM modules can be adopted by the farmers increase the productivity of short duration pigeonpea. This will ultimately improve the socio-economic status of the farmers.

Researchable Issues

1. Use of biotechnology for spreading transfer of genes, developing resistance to important diseases and key pests, *e.g.* transfer of Bt gene in pigeonpea for the control of pod borer.
2. Efforts should be made to improve ridger–cum–planter for hard soils.
3. Hybrids of short duration varieties should be developed with lesser maturity period.
4. There have to be integrated approaches for the management of diseases and pests *i.e.*, IPM as well as of nutrients *i.e.*, Integrated Nutrient Management (INM), depending upon the agro–climatic condition.
5. The forecasting and forewarning of the incidence of insect pests and diseases should be imitated and developed to keep their population below economic injury level (EIL).

Future Prospects

It is generally believed that there is little scope for area expansion for crop cultivation. However, there are still large tracts of land that remain fallow because of non–availability of irrigation water is not available. In some areas, the soil is able to hold moisture for a short span only. As a result, the presently available varieties cannot complete their life cycle on these lands. So, the short duration pigeonpea can work as an effective and economic substitute to combat this critical situation. Therefore, efforts should be made for area expansion and for cropping system manipulation through intercropping with short duration pigeonpea. The variety ICPL–87 has extended cultivation of the crop to drought prone areas in Maharashtra. Likewise, early maturing pigeonpea can be grown in irrigated tracts of Haryana, Punjab, North-west Rajasthan as well as in Western U.P. Besides this, to achieve higher economic returns, special emphasis should be given to crop protection (IPM approaches) as this will play a crucial role in enhancing productivity and ultimately uplift the socio-economic status of the Indian farmers. The integration of the seed production chain is of utmost importance in the case of pigeonpea. Proper planning and monitoring to meet the seed demand is lacking at the state level. So, there is an urgent need to revise the seed indents both quantitatively as well as qualitatively.

References

Anonymous, 1999. *The Hindu Survey of Indian Agriculture*, pp. 17–23.

Anonymous, 2002. *The Hindu Survey of Indian Agriculture*, pp. 63–69.

Anonymous, 2003. *The Economic Survey*, Ministry of Agriculture, Government of India, pp. 151–170.

Ariyanayagam, R.P. and Singh, N.B., 1994. Pigeonpea breeding : Accomplishment and challenges. *Plant Breeding Abstracts*, 64 : 773–782.

Armes, N.J., Jadhav, D.R., Bond, G.S. and King, A.B.S., 1992. Insecticide resistance in *Helicoverpa armigera* in South India. *Pesticide Science*, 34 : 355–364.

Chauhan VB, Singh, VB and Singh AK., 2002. Status of Phytophthora blight of pigeonpea in eastern Uttar Pradesh. *Annals of Plant Protection Sciences*, 10(2):402–404.

Cohen, Y. and Leavy, Y., 1990. Joint action of fungicides in mixtures: Theory and practice. *Phytoparasitica*, 18:159-169.

Dharamraj PS, Narayana YD, Lava Kumar P, Waliyar F and Jones AT., 2004. Pigeonpea Sterility Mosaic Disease: An Emerging Problem in Northern Karnataka, India. *International Chickpea and Pigeonpea Newsletter*, 11:47–49.

Fitt, G.P., 1989. The ecology of *Heliothis* in relation to agro-ecosystem. *Annual Review of Entomology*, 34 : 17- 52.

Gopali, J.B. and Lingappa, S., 2001a. Role of insectivorous birds in pigeonpea ecosystem. *Karnataka Journal of Agricultural Sciences*, 14 (4) : 955–960.

Gopali, J.B. and Lingappa, S., 2001b. Selection of virulent strain of NPV against *Helicoverpa armigera* in pigeonpea ecosystem. *Karnataka Journal of Agricultural Sciences*, 14 (4):1067–1071.

Haldar, B., Srivastava, C.P. and Joshi, N., 2006. Comparative performance of some newer insecticides against the major insect pests of short duration pigeonpea. *Pestology*, 30(9): 32-35.

Joshi, N. and Srivastava, C.P., 2006. Evaluation of integrated pest management modules for insect pests in short duration pigeonpea (*Cajanus cajan*) in eastern Uttar Pradesh. *Indian Journal of Agricultural Sciences*, 76(8): 481-484.

Kabaria, B., Goyal, S.N. and Shah, A.H., 1988. Surveillance of insect pests damage to pigeonpea at podding stage in Bharuch District, Gujarat, India. *International Pigeonpea Newsletter*, 8: 15–16.

Kannaiyan J., Nene Y.L., Reddy M.V., Ryan J.G. and Raju T.N., 1984. Prevalence of Pigeonpea diseases and associated crop losses in Asia, Africa and the Americas. Tropical Pest Management, 30:62–71.

Kelley, T.G. and Parthasarathy Rao, P., 1993. Chickpea: A preferred pulse ? *International Chickpea Newsletter*, 28: 6–7.

Lal, S.S. and Yadava, C.P., 1987. Estimates of crop losses in pigeonpea caused by the pod borer complex. *FAO Plant Protection Bulletin*, 35 (3): 93–98.

Lal, S.S. and Yadava, C.P., 1988. Efficacy of certain insecticides against pod borers infesting pigeonpea. *Pesticides*, 22 (12): 30–35.

Loevinsohn, M.E., 1987. Insecticide use and increased mortality in rural central Luzon, Phillippines. *The Lancet*, 13 June: 1359–1362.

Manjunath, T.M., Bhatnagar, V.S., Pawar, C.S. and Sithanantham, S., 1989. Economic importance of *Heliothis* spp. in India and an assessment of their natural enemies and host plants. In : *Proceeding of the Workshop on Biological Control of Heliothis : Increasing the Effectiveness of Natural Enemies*, 11-15 November 1985, New Delhi, India, pp., 197–228.

Mishra A.N. and Shukla P., 1987. Prevalence of Phytophthora blight of pigeonpea in Uttar Pradesh. *Indian Phytopatholgy,*. 40:56–58.

Nene, Y.L. and Sheila, V.K., 1990. Pigeonpea Geography and Importance. In : *The Pigeonpea* (eds. Y.L. Nene *et al.*) pp. 1-14. Wallingford, United Kingdom: CAB International and International Crops Research Institute for the Semi–Arid Tropics.

Pimentel, D. Acquay, A., Biltonen, M., Rice, P., Silva, M., Nelson, J., Lipner, V., Giordano, S., Horowitz, A. and D'Amore, M., 1992. Environmental and economic costs of pesticide use. *Bio Science*, 42 : 750–760.

Pradhan, S., 1971. In tropics protection research more needed than production research. *Indian Journal of Entomology*, 33 (3): 233–259.

Rangaiah, P.V. and Sehgal, V.K., 1984. Estimation of crop losses in pigeonpea due to insect pests associated within various stages of its growth. *Indian Journal of Plant Protection*, 12 (2):127–130.

Reddy, C.N., Singh, Yeshbir, Dureja, Prem, Singh, V.S., Singh, Y. and Dureja, P., 2001. Bioefficacy of insecticides, biopesticides and their combinations against pod borers in pigeonpea. *Indian Journal of Entomology*, 63 (2): 137–143.

Ryan, J.G., 1994. A global perspective on pigeonpea and chickpea sustainable production systems: Present status and future potential. *Paper presented at the International Symposium on Pulses Research*, 1–6 April 1994. Indian Council of Agriculture Research, New Delhi, India, 33 pp.

Sachan, J.N. and Lal, S.S., 1990. Role of botanical insecticides in *Helicoverpa armigera* management. *Paper presented in National Symposium on problems and prospects of botanical pesticides in Integrated Pest Management*, held at C.T.R.I., Rajamundary, 21–22 Jan., 1990.

Sachan, J.N., Yadava, C.P., Ahmad, R. and Katti, G., 1994. Insect Pest Management in Pulse Crops. In : *Trends in Agricultural Insect Pest Management* (eds. G.S. Dhaliwal and Ramesh Arora) pp. 308–344, New Delhi.

Sahoo, B.K. and Senapati, B., 2001. Extent of damage by different pod borer species in pigeonpea in coastal Orrisa. *Journal of Applied Zoological Researches*, 12 (1) : 19–22.

Santharam, G., Victoria, D.R., Rabindra, R.J. and Jayaraj, S., 1994. Studies on biological control of pigeonpea pod borers in India. *Anzeiger fur Schadlingskunde, Pflanzenschutz,Umweltschutz*, 67 (5): 103–106.

Sarode, S.V., Deotale, R.O., Jumde, Y.S. and Thakare, H.S., 1994. Field evaluation of *Heliothis* nuclear polyhedrosis virus (HaNPV) for the management of *Helicoverpa armigera* (Hübner) on pigeonpea. *Indian Journal of Entomology*, 56 (2):176–179.

Sarode, S.V., Jumade, Y.S. and Borkar, S.L., 1997. Efficacy of *Heliothis* nuclear polyhedrosis virus and neem seed kernel extract combinations against *Helicoverpa armigera* (Hb.) on pigeonpea. *PKV Research Journal*, 21 (2):227–229.

Shanower, T.G., 1996. Insect pest management in short duration pigeonpea status and needs, in *Plant Protection and Environments. Proccedings of the National Symposium on Plant protection and environment* (eds. D.V.R. Reddy, H.C. Sharma, T.B. Gour and B.J. Divakar). Plant protection association of India, Hyderabad, India : 125–136.

Sharma, V.K., Pandey, S.N. and Singh, R., 1991. Avoidable losses in pigeonpea (*Cajanus cajan* (L.) Millsp.) variety UPAS–120 due to insect pests. *Indian Journal of Entomology*, 53 (3): 511- 512.

Siddhabhatti, P.M., Khan, M.I., Satpute, N.S. and Mane, P.N., 2003. Evaluation of different integrated pest management (IPM) modules against pod borer complex of pigeonpea. *Crop Research*, 26 (1) :150–153.

Singh, N.N. and Rai, L., 1981. Assessment of damage by *Melanagromyza obtusa* Malloch to T 21 variety of pigeonpea. *Indian Journal of Entomology*, 43 (4): 445–447.

Singh, R., 1998. Chemical control of wilt disease of pigeonpea. *Korean Journal of Mycology*, 26:416-423.

Sundara Babu, P.C. and Muthuswami, M., 1995. Field efficacy of *Bacillus thuringiensis* var. *kurstaki* against *Helicoverpa armigera* on chickpea, pigeonpea and cotton. *Presented in Symposium on Integrated Pest Management and Environment* at Madras, Tamil Nadu, India, Feb. 2–4, 1995.

Van den Bosch, R., 1978. *The pesticide conspiracy*. University of California Press, Berkeley, USA, 226 p.

Wightman, J.A., 1993. Towards the rational management of insect pests of tropical legume crops in Asia: Review and remedy. *Crop protection and sustainable agriculture*, John Wiley and Sons, Chichester, UK. pp. 233–256.

Chapter 21

Pest Management in Changing Cotton Scenario

Rishi Kumar, Dilip Monga, S. L. Jat and K. C. Kumhar
Central Institute for Cotton Research,
Regional Station, Sirsa

Introduction

Cotton is a fibre crop of global importance, which is grown in more than 80 countries the world over. Out of about 42 species of the genus *Gossypium*, only four species, *viz. Gossypium arboreum*, *G. herbaceum*, *G. hirsutum* and *G. barbadense* are cultivated and rest are wild. The first two species are diploid (2n = 26) and are native of old world. Diploid cultivated species are also known as *Desi* cotton or Asiatic cotton because they are cultivated in Asian region. The last two of the above mentioned cultivated species are tetraploid (2n = 52) and are referred to as New world cotton. The *G. hirsutum* is also known as American cotton or Upland cotton and *G. barbadense* is also referred to as Sea Island cotton or Egyptian cotton or Tanguish cotton. The *G. hirsutum* is the predominant species, which alone contributes about 90 percent to the global production. The major cotton growing countries are USA, China, India, Pakistan, Uzbekistan, Egypt, Argentina, Australia, Greece, Brazil and Turkey, which contribute about 85 percent to the global cotton production. In India, cotton crop is grown throughout the country. However, there are nine major cotton producing states *viz.*, Punjab, Haryana, Rajasthan, Madhya Pradesh, Maharashtra, Gujarat, Andhra Pradesh, Karnataka and Tamil Nadu. These states contribute about 90 per cent to the national cotton production. Based on cotton cultivation, India is divided into three major zones *viz.*, north zone (Punjab, Haryana and Rajasthan), central zone (Madhya Pradesh, Maharashtra and Gujarat) and south zone (Andhra Pradesh, Karnataka and Tamil Nadu).

There has been a significant change in the cultivation of cotton in the countries by rapid growth of transgenic cotton hybrids with a myriad gene of *Bacillus thuringiensis* (Bt) Singly, (Cry1Ac- Mon 531 event and Cry1Ac event 1) and in combinations through stacking (Cry1Ac and Cry1Ab Mons. 15985 event) and fusion (Cry1Ac Cry1Ab GFM event) in India. An increase in the area under Bt-cotton from 0.038 m ha in 2002-03 to 5.5 m ha in 2006-07 has been observed. Simultaneously the number of approved hybrids and seed companies increased from 3 to 134 and 1 to 24 respectively during that period. In many states of north India like, Punjab and Haryana, around 80 percent of the area is under Bt-cotton. The total change in cultivation profile is evident with the Bt hybrids constituting 58 percent of the country 9.13 m ha cotton area during 2006-07. The increase in stagnated production and productivity of 158 lakh bales and 309 kg/ha during pre Bt benchmark year of 2001-02 to 310 lakh bales and 503 kg/ha, respectively in 2006-07 signify the notable contribution of Bt hybrids in India (Anonymous, 2007).

Cotton hybrids introduced with various endotoxins from *B. thuringiensis* imply genetic or repulsive and biological tools of integrated pest management which ultimately reduced the massive reliance on insecticides which is beneficial for cotton. In India for pest management 50 per cent of the pesticide was applied on cotton occupying on 5 percent of the total cropping area. As the Bt-cotton is effective against major bollworms of cotton *i.e. Helicoverpa armigera, Earias vitella, Earias insulana* and *Pectinophora gossypiella*, the crop is affected by sucking pests like jassid, *Amrasca bigutulla bigutulla*; whitefly, *Bemisia tabaci*; aphid, *Aphis gossypi*; thrip, *Thrips tabaci*; mired, *Creontiades biseratense* and mealy bug *Phenacoccus solenopsis* as a major loss causing entities.

Cotton eco-system throughout the world includes a wide variety of arthropods. Cotton being the perennial plant and well adapted to annual indeterminate growth characteristics provide favorable microclimate for sustaining a large number of insect-pests. The number of species may range from few to 162 species (Dhaliwal *et al.*, 2004). As many as 1326 species of insects has been listed on cotton crop world wide (Hargreaves, 1948), which also include those of little or no economic importance, sporadic, casual or accidental visitors. Dhawan (1999) observed so far 24 insect species have attained the pest status where as 9 could attain the status of key pests in different zones. According to Luttrell *et al.* (1994) significant crop damage is caused by 5-10 key pests in the most production systems. The major pests in the asian region include the pink bollworm, spotted bollworms, *Helicoverpa* spp., mites, thrips, and aphids and bug. In humid environment of Brazil and USA, the bollweevil *Anthonomus grandis* Bohemen is most destructive pests *Helicoverpa zea* (Boddie), *Helicoverpa inversencens* (Fab), plant bugs, *Lugens lineolenis* (Pohsot de Beaavois) and *Psallius seriatus* (Renter) also cause serious damage. Species of *Spodoptera, Earias, Alabana, Dysdercus*, jassid, aphid and whitefly are varying and of local importance in different region.

Cotton Pests Scenario: A change

Insects

Cotton crop has been found associated with a large number of insect-pests which cause economic damage either in quantitative or qualitative terms. The large number of insects though remained in cotton ecosystem throughout the season but only a few of them have been reported as major pests. There are many factors which have been found responsible for a change in scenario of cotton insect-pests.

- ☆ Cultivation pattern of diploid and tetraploid cotton.
- ☆ Adoption of Bt-cotton.
- ☆ Changing ecological scenario.
- ☆ Adoption of large number of cultivars with different habits.

- ☆ Use of unrecommended insecticides.
- ☆ Practice of mixing of synthetic pyrethroids with conventional insecticides.
- ☆ Faulty spray technology.

Table 21.1: Changing Status of Cotton Insect Pests in Irrigated Cotton

Name of Insect	*Up to 1970*	*1971-80*	*1981-90*	*1991-2000*	*2001-03*	*2004-08*
Jassid	Major	Major	Major	Major	Major	Major
Whitefly	Minor	Minor	Minor	Major	Major	Major
Aphid	Minor	Minor	Minor	Minor	Minor	Minor
Thrips	Minor	Minor	Minor	Minor	Minor	Minor
Mealy bug	-	-	-	-	-	New
Mirid bug	-	-	-	-	-	New (South India)
Tobacco caterpillar	-	-	New	Minor	Minor	Major (Sporadic)
Spotted bollworm	-	Major	Major	Major	Major	Minor
Pink bollworm	Major	Major	Major	Minor	Minor	Minor (Serious in later stage of Bt-cotton)
American bollworm	-	New	Minor	Major	Major	Minor (observed on Bt in Sirsa)

Source: Dhawan *et al.* (2008)

The indigenous cotton varieties of *Gossypium arboretum* were able to tolerate the attack of many insect species. With the introduction of high yielding varieties of *G hirsutum* receiving heavy dosages of chemical fertilizers (nitrogenous) and pesticides led to the attraction of many insect species changing the pest scenario of cotton. Competitive displacement among the fauna of cotton had been observed through epidemic outbreaks of many insect species resulting in the reduction in the population of another. The change in cultivation pattern of cotton from *desi* cotton to upland cotton (varieties to hybrids) and subsequently the adoption of Bt-cotton resulted in the new pest problem of sucking pest (jassid and whiteflies) to bollworms to foliage feeders and ultimately the bug's regime (mealy and mirid bugs). *Desi* cotton, *G. arboreum* sown in India in early fifties was fairly resistant to jassid and having lower preference for the pink bollworm incidence due to its short duration. After the introduction of *G. hirsutum* large area came under this upland cotton which attracted and helped in multiplication of pink bollworm (Singh and Khush, 1993). Jassid, *Amrasca biguttula biguttula* and pink bollworm, *Pectinophora gossypiella* were the serious pest of cotton in India till 1970. But the cultivation of susceptible, medium and long duration varieties of upland cotton allowed the multiplication of spotted bollworm, *Earias insulana* and *E. vittela* in attaining a status of major pest of cotton (Dhàliwal and Arora, 1996). The insect species which is responsible for devastation of cotton sustainability in India is american bollworm, *Helicoverpa armigera* since nineties though the pest appeared on cotton in 1977. The outbreaks of the pest have been noticed in Punjab by Dhawan *et al.* (1980). The attack of this pest has been reported in pockets but in 1990s it attained the status of a major pest.The changing agro climatic conditions since 1990s especially the duration and the magnitude of rainfall, cultivation of un-recommended and susceptible varieties as well as hybrids, indiscriminate use of pesticides (starting of era of over use of synthetic pyrethroids), development of resistance against most of the conventional insecticides led to the severe out breaks of *Helicoverpa armigera* in 1990, 1995, 1997, 1998 and 2001 rendering to huge loss to cotton growers in Punjab (Dhawan *et al.*, 2004) and other parts of north India including Haryana and Rajasthan. The another pest whitefly also emerged as a major pest because of the over reliance on synthetic pyrethroids for the management of cotton pests coupled with heavy use of nitrogenous fertilizers, closer spacing, availability of preferred and alternate host during in and off season, availability of weed and volunteer flora and elimination of natural enemies and improper spray technology (Dhawan *et al.*, 2007).The practice of seed treatment for jassid in 1998 (imidacloprid, gaucho 70 WS) and introduction of neonicotinoids after 1995 managed the sucking pest successfully. The pest status of bollworm complex declined with the introduction of new chemical insecticides like, spinosad and indoxacarb in 2003 and 2004.The introduction of Bt-cotton in 2005 was an additional advantage in the management of bollworms because of inherent resistance against the bollworm complex. The bollgard cotton having Cry1Ac gene is effective against the *Helicoverpa armigera, Earias insulana, E. vittella* and *Pectinophora gossypiella* but it is not effective against the sucking pests receiving 3-4 insecticidal sprays as management tactics but meanwhile tobacco caterpillar appeared as a pest of cotton. With the introduction of Bollgard-II, the tobacco caterpillar can also be managed effectively. But during 2006 mealy bug, *Phenacoccus solenopsis* Tinsley appeared in north India particularly in Punjab (Distt. Bathinda, Ferozpur and Mukatsar). The serious damage of this pest was reported in Punjab followed by Haryana and Rajasthan and now through out India Similarly in southern part of the country mirid bug appeared as a loss causing entity to Bt–cotton especially.

Diseases

The cotton disease scenario has shown a continuous change during the past sixty years. When mainly indigenous diploid cottons were being grown in fifties, the *Fusarium* wilt, root rot, seedling blight, anthracnose and grey mildew were the major problems. With the large scale cultivation of tetraploid upland cotton (*Gossypium hirsutum*), bacterial blight became the major problem to which indigenous cottons were highly resistant. *Fusarium* wilt has become less important as upland cotton now occupying above 80 percent area is immune to Indian races of the pathogen. *Verticillium* wilt which appeared in Tamil Nadu remained restricted mainly to that state only. The grey mildew, once a serious problem for diploid cottons especially in central India had almost disappeared with the large scale cultivation of tetraploid cottons. However, with the continued cultivation and imposed selection pressure the disease got adopted to tetraploid cotton and their hybrids as well. Presently, it is a major problem in central and south India in Bt-cotton hybrids. *Alternaria* blight and *Myrothecium* leaf spots are prevalent everywhere but are severe in the states of Karnataka and Haryana, respectively. Other diseases such as *Cercospora* and *Helminthosporium* leaf spots are sporadic only. Rust, although a minor disease may become a problem in southern states in near future. During late 1970's a new problem of wilt (new wilt) emerged and caused serious damage to some of the hybrids both intra *hirsutum* or inter specific (*hirsutum x barbadense*) in central and southern zones. However, its incidence later became negligible for the last one decade. In north India, the leaf curl disease caused by arbore virus and transmitted by whitefly, *Bemisia tabaci* has become a major threat to cotton cultivation since 1993.

After the introduction of Bt-cotton hybrids during 2002 onwards and continuous increase in area under these hybrids to more than two third of total cotton area till date, the disease scenario has also shown some change. The Bt-cotton hybrids have shown higher incidence of fungal foliar spots including grey mildews. The occurrence of para-wilt has also increased considerably in northern cotton zone. Para-wilt incidence has been reported in Ankur 2534 Bt and Ankur 651 Bt (Yadav, 2008).

Table 21.2: List of Important Cotton Diseases, their Causal Organism, Distribution and Status in India

Disease	*Pathogen*	*Distribution*	*Status*	*Affected Crop Stage*
Root rot	*Rhizoctonia solani, R. Bataticola*	All zones	Major in north zone	Seedling stage, vegetative stage
Fusarium wilt	*Fusarium oxysporum f. sp. Vasinfectum*	All zones	Affects *desi* (*G. rboretum* and *G. herbaceum*) cotton	At any growth stage
Verticillium wilt	*Verticillium dahlia*	South zone (mainly in Tamil Nadu)	Minor	Seedling stage
Alternaria blight	*Alternaria macrospora*	South zone (mainly in Karnataka)	Major	Vegetative and reproductive stage
Helminthosporium leaf spot	*Helminthosporium gossypii, Helminthosporium speciferum*	North zone	Minor	Vegetative and reproductive stage
Grey mildew	*Ramularia areola*	Central and south zone	Major	Late crop stage
Myrothecium leaf spot	*Myrothecium roridum*	North zone	Major	Vegetative and reproductive stage
Boll rot	Saprophytic fungi and some pathogenic fungi and bacteria	All zones	Major	Boll formation satge
Bacterial blight	*Xanthomonas axonopoides pv. Malvecearum*	All zones	Major	All crop stages
Cotton leaf curl	Incited by Gemini virus and transmitted by whitefly	North zone	Major	Vegetative stage
Para-wilt	Pathogen is unknown or may be physiological disorder	Central, south and north zone	Major in pockets	Late crop stage

Table 21.3: List of Insect Pests Attacking Cotton in India

Common Name	*Scientific Name*	*Crop Stage*	*Pests Status*	*Zone*
		Seedling stage		
Termite	*Odontotermus* spp.	Seedling stage	Non- pest	All zone
Cutworm	*Agrotis ipsilon* (Hufnagel)	Seedling stage	Non-pest	All zone
Field cricket	*Gryllus* spp.	Seedling stage	Non-pest	All zone
Surface grasshopper	*Chrotogonus brachypterus* Blanchard	Leaves	Non-pest	All zone
		Cotyledon to vegetative stage		
Serpentine leaf minor	*Liriomyza trifolii* Burgess	Leaves	Non-pest	All zone
Thrips	*Thrips tabaci* Lindeman	Leaves	Sporadic but serious after Bt-cotton	All zone, severe in CZ

Contd...

Table 21.3–Contd...

Common Name	Scientific Name	Crop Stage	Pests Status	Zone
		Vegetative to fruiting stage		
Jassid	*Amrasca bigutulla bigutulla* (Ishida)	Leaves	Major	All zone
Whitefly	*Bemisia tabaci* (Gennadius)	Leaves	Major	All zone, severe in NZ
Aphid	*Aphis gossypii* Glover	Leaves, flowers, buds	Major	All zone, less severe in NZ
Mealy bug	*Phenacoccus solenopsis* Tinsley	Twigs, leaves, fruiting bodies	Major	All zone severe after introduction of Bt
Brown mirid bug	*Creontiades biseratense* (Distat)	Square and small developing bolls	Major	Severe in South Zone only
Tobacco caterpillar	*Spodoptera litura* Fabricius	Leaves	Major	All zone, status can be change after introduction of Bollgard-II
Tobacco caterpillar	*Spodoptera exigua* Hubner	Leaves	Major	All zone
Leaf roller	*Sylepta derogate* Fabricius	Rolling and fastening of leaves	Minor	CZ
Shoot weevil	*Alcidodes affaber* (Aurivillius)	Bore into top shoots and leaf axils	Minor on Bt	SZ, getting serious
Stem weevil	*Pempherulus affinis* (Faust)	Feeding on soft tissues of cotton stem	Minor	SZ, getting serious on Bt
Grey weevil	*Myllocerus undecim-pustulatus* Faust	Leaves, flower, roots	Minor	All zone
Bihar hairy caterpillar	*Spilosoma oblique* Walker	Leaves	Sporadic	NZ
Spotted bollworm	*Earias vitella* (Fabricius) *Earias insulana* (Bousduvai)	Growing shoots, fruiting bodies	Major but after introduction of Bt least important	All zone
American bollworm	*Helicoverpa armigera* (Hubner)	Leaves, fruiting bodies	Major (till 2001)	All zone
		Fruiting stage		
Pink bollworm	*Pectinophora gossypiella* (Saunders)	Fruiting bodies	Major (Appearance recorded on Bt-cotton)	CZ, SZ and earlier major pest of north zone
Red Cotton Bug	*Dysdercus koenigii* (Fabricius)	Bolls	Non-pest	All zone
Dusky cotton Bug	*Oxycarenus laetus* Kirby	Open bolls	Non-pest	All zone
		Minor Pest		
Spider mite	*Tetranychus telarius* L.	Leaves	Minor	All zone
Cotton semilooper	*Anomis flava* Fabricius	Leaves	Minor	All zone
Cotton stem borer	*Sphenoptera gossypii* Cotes	Feeding on internal tissues of stem	Minor	CZ and SZ

Important Pests of Bt-Cotton: Diagnostic and Management Details

Insects

Cotton Whitefly

Scientific Name: *Bemisia tabaci* (Gennadius)

Family: Aleyrodidae

Order: Hemiptera

Description of the Pest

Whitefly is a major pest of cotton in India. Whitefly adults are tiny (0.06 inch or 1.5 mm long), yellowish insects with white wings. Their wings are held somewhat vertically tilted, or roof like, over the body and generally do not meet over the back but have a small space separating them. The louse

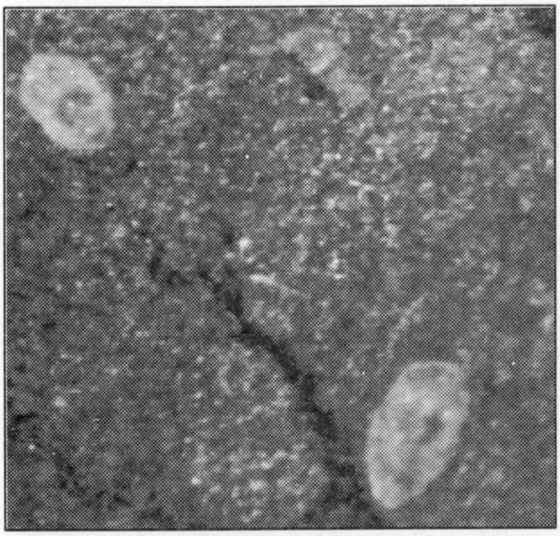

Eggs

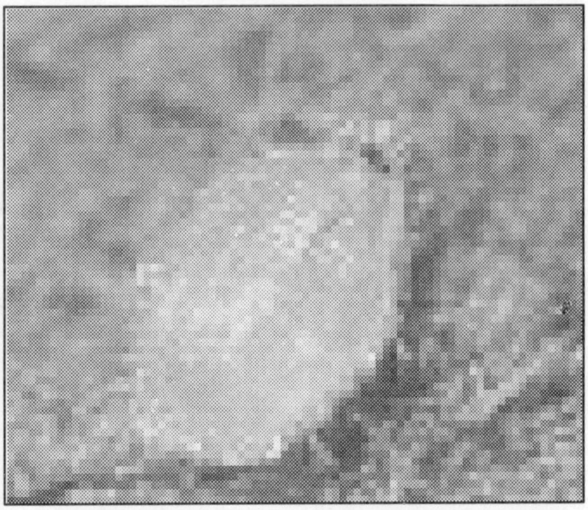

Nymph

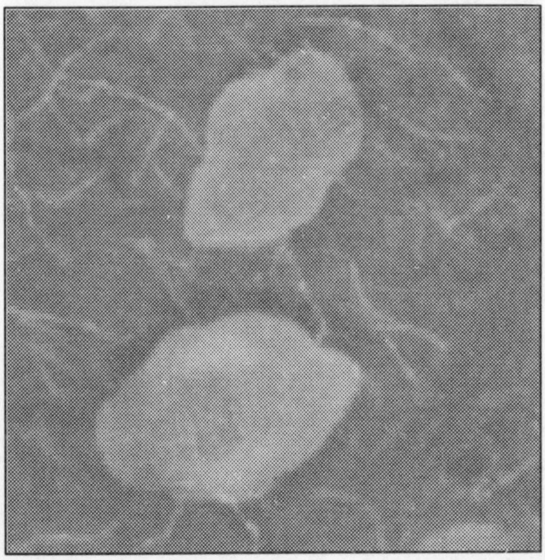

Pupa

Adult

like nymphs, which suck sap, are sluggish creatures, clustered together on the under surface of the leaves. Insect's breeds throughout the year and all the developmental stages are noticed although adults predominate in cold season. The eggs are laid singly on underside of the leaves. The female lays on an average 119 eggs which are stalked, sub elliptical and light yellow at first and turning brown later on. Hatching period is 3-5 days in April-September, 5-17 days in October-November and 33 days in December-January. Just after emergence, the nymphs look elliptical and start sucking the sap from plant tissues and passed through three stages for formation of pupae with in 9-14 days in April-September, 17-81 days in October-March. The pupal period is 2-8 days. Total duration of life cycle is of 14-22 days and 11 generations are completed.

White Fly Damage

Damage

Whiteflies are sucking insect and their feeding removes nutrients from the plant. Feeding by high populations may result in stunting, poor growth, defoliation, boll shed and reduced yields. As they feed, whiteflies produce large quantities of honeydew which, if deposited on fibers, will reduce cotton quality and may interfere with picking, ginning, and spinning. Honeydew also supports the growth of black sooty molds that stain lint, lowering its quality. The whitefly, vectors the cotton leaf curl virus in north India.

Management

Whiteflies are difficult to manage once their populations have reached high levels. Repeated exposure to insecticide treatments is very likely to lead to development of resistant strains. In general, the best approach is an integrated pest management strategy that relies first on cultural and biological control methods and uses chemical controls only when needed.

Cultural Control

Use resistant varieties or hybrids, if available. When possible, plant cotton away from other key whitefly hosts such as citrus, tomatoes, cole crops, brinjal and okra. Maintain good sanitation of areas from winter or spring host including weeds by destroying and removing all crop residues as soon as possible. Control weeds in non crop areas including head rows and fallow fields. Before destroying weeds, however, check them for whitefly predators and parasites because they can be an important source of these natural enemies. After harvesting, promptly destroy stalks to prevent regrowth and limit additional whitefly buildup.

Biological Control

Several wasps, including species in the genera *Encarsia* (5-10 per cent parasitization) and *Eretmocerus* (up to 33 per cent parasitization of nymphal stage) parasitize whiteflies. Whitefly nymphs are also preyed upon by big eyed bugs, seven spotted lady beetle, *Coccinella septempunctata*, *Chrysoperla* spp. and *Brumus* spp. preys upon nymphs and adult but do not keep it below damaging numbers.

Resistance

Some pyrethroid and organophosphate insecticides have lost their effectiveness for controlling whiteflies. Repeated applications of a product may build high resistance levels in whiteflies. To delay and manage resistance, do not treat successive generations of whiteflies with same product or with insecticides that have the same mode of action, rotate insecticides with a different mode of action during the season. Avoid the use of synthetic pyrethroid after 15th September on cotton.

Monitoring and Treatment Decisions

Routinely check field margins for whiteflies; these areas are usually infested first. Be especially alert for rapid population buildup when nearby host crops are in decline. During these critical periods, check cotton fields twice weekly. Whitefly adults and nymphs need to be monitored on under sides of leaves from early squaring to harvest. Check for whitefly adults on undersides of leaves, if 6 or more adults per leaf are found, rate the leaf as infested.

Chemical Control

When population level 6-8 adults/3 leaves (as mentioned in table-10) arrived, the insecticides recommended (table-11) can be applied as an action threshold.

Thrips

Scientific Names: *Thrips tabaci* Lindeman

Family: Thripidae

Order: Thysanoptera

Description of the Pests

Thrips are tiny, slender yellowish brown insects less than 0.06 inches (1.55 mm) long with bladder like legs. The tip of the abdomen is curved and the abdominal segments are transversally banded with dark lines. Males are wingless and females have a long narrow strap like wings, fringed with hairs.

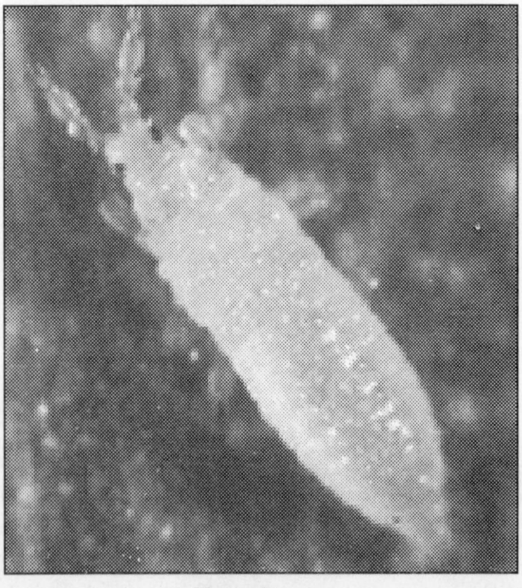

Nymph

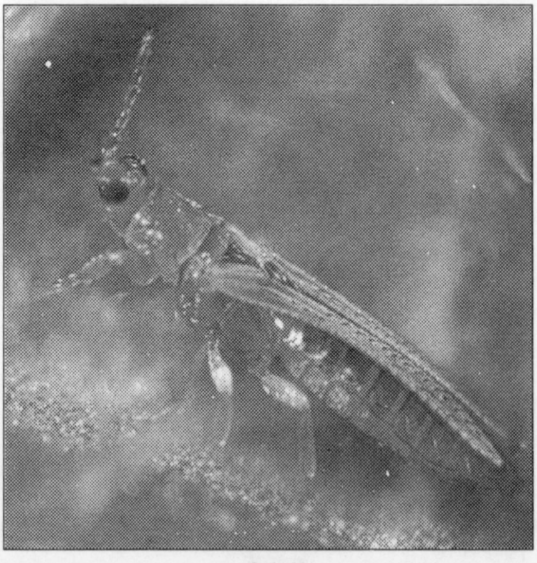

Adult

Nymphs are wingless and smaller in size and blackish in color. Kidney shaped eggs (50-60 per female) are laid in slits in leaf tissue. Eggs hatch in 5 days, nymphs and pupal period lasts for 5 and 4-6 days, respectively. Adults survive for 3-10 days.

Damage

The nymphs and adults suck cell sap from under surface of the leaves, lacerating the leaf tissue; such leaves shine in presence of sun light giving silvery appearance. Upper side of the affected leaves is leathery and hard. Leaves becomes curled, wrinkled and finely get dried. While infested seedlings may appear severely deformed, they will grow out of the damage rapidly with the onset of congenial weather.

Thrips Damage

Management

No treatment is generally recommended for thrips. Young plants will rapidly recover from injury. Some of the insecticides applied to control thrips are usually counterproductive, as they tend to promote outbreaks of mites so only in situations where a prolonged thrips infestation is destroying seedling terminals, the treatment is justified.

Cultural Control

- ☆ Use of resistant varieties.
- ☆ Balanced use of fertilizers.
- ☆ Use of yellow sticky trap.
- ☆ Timely irrigation.
- ☆ Remove the weeds from field.

Biological Control

Release of *Chrysoperla carnea* grub @ 50,000/ha or *Coccinella* spp. beetle @ 15,000/ha in the field.

Monitoring and Treatment Decisions

The critical time for monitoring thrips is from crop emergence through seedling stages. The efficiency of monitoring program can be increased after combining the sampling of thrips with other pests like jassids and mites.

Chemical Control

The pest is of miner importance and is becoming serious after the introduction of Bt-cotton but no chemical treatment is required. In north India no economic threshold level of thrips has been calculated but for action threshold level of thrips in cotton (as mentioned in table-10) arrived, the insecticide used like, acetamiprid (Pride) 50 g or thiomethoxam 100 g per acre can be applied as remedial measures.

Cotton Jassid

Scientific Name: *Amrasca biguttula biguttula* (Ishida)

Family: Cicadellidae

Order: Hemiptera

Description of the Pests

Most destructive insect of cotton. The adults are about 3 mm long, greenish yellow during summer and acquiring reddish tinge in winter. The wedge shaped winged adults having conspicuous black dot on vertex of forewings. They jump or fly at slight disturbance and are attracted towards light during night. Movement of jassid is in zigzag manner. The pest practically breeds throughout the year and during winter only adults are visible on potato, brinjal, tomato. The breeding is started in spring season on okra crop. Female lay about 15 eggs on the underside of the leaves embedded into leaf veins, which are elongate and yellowish in color. Eggs hatch within 4-11 days and wedge shaped highly active nymphs with wing pads covering first five abdominal segments, come out. While sucking cell sap from leaves the nymphs passes through 6 stages of growth in 7-21 days. Winged adults live for 5-7 weeks, feeding on plant sap and complete seven generations in a year.

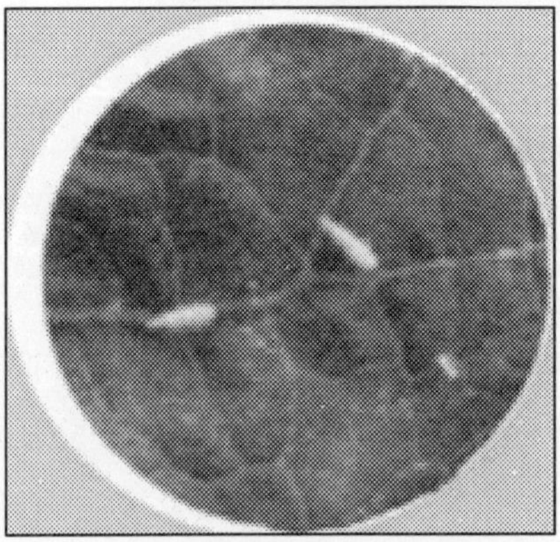

Nymph

Adult

Damage

The plants are injured due to loss of cell sap suck by adults and nymph from veins on underside of the leaves. The midrib veins become roughened. Affected leaves may become distorted, leathery, and develop yellow or red blotches. Though rare, severe infestation may cause plants to shed squares and small bolls. Larger bolls may turn soft and spongy, and fail to mature.

Jassid Damage

Management

Natural enemies usually keep leafhoppers from building up large populations in cotton. However, if large numbers migrate to cotton from other hosts, treatment may be needed if extensive symptoms appear.

Cultural Control

- ☆ Grow resistant varieties.
- ☆ Judicious use of nitrogenous fertilizers.

Biological Control

Chrysoperla spp. (Chrysopidae) and spiders like, *Distina albida* L. feed on nymphs and adults. No other parasitoids are known for this insect.

Monitoring and Treatment Decisions

Spray against jassid should be done at second grade injury level (yellowing and curling of leaves margins) or when population is 2 nymphs/leaf.

Chemical Control

The chemical treatment mentioned (Table 21.11) can be applied as action threshold.

Brown Mirid Bug

Scientific Name: *Creontiades biseratense* (Distat)

Family: Miridae

Order: Hemiptera

Description of the Pest

Adult of brown mired bug is elongated, about 7 to 9 mm in length, with long legs and antennae. Brown mired bug is a light yellow-brown color. Newly hatched nymphs of mirid bug are very pale green in color and about 1.4 to 1.6 mm in length with long antennae and legs. Antennae are four segmented; there is a distinctive red and green stripe. In brown mired bug, two reddish bands are present on the distal end (furthest away from the body) of the femur of the hind legs. As nymphs grow, their colour changes from light green to yellowish green. The wing pads start to develop at the 3rd instar. In the 4th instar, brown mired bug develop two black spots between the wing pads, the brightness of the stripe on the antennae diminishes to some extent and the reddish bands on the femur coalesce. In the 5th instar, the hind legs of brown mired bug are becomes a dark brown. Eggs are laid singly within plant tissue, making them virtually impossible to see. The only visible evidence is an oval shaped operculum (egg cap) used to 'breathe'. Immediately after being laid, the operculum is a light brown colour. A female can lay up to 80 eggs in her life time. Hatching period is 6-7 days. It has 5 nymphal stages which completed 12-14 days. Under summer conditions, one generation can be completed in about 3 weeks.

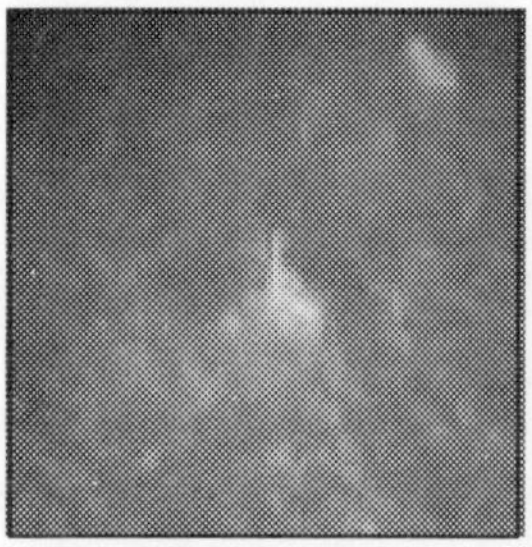
I-instar

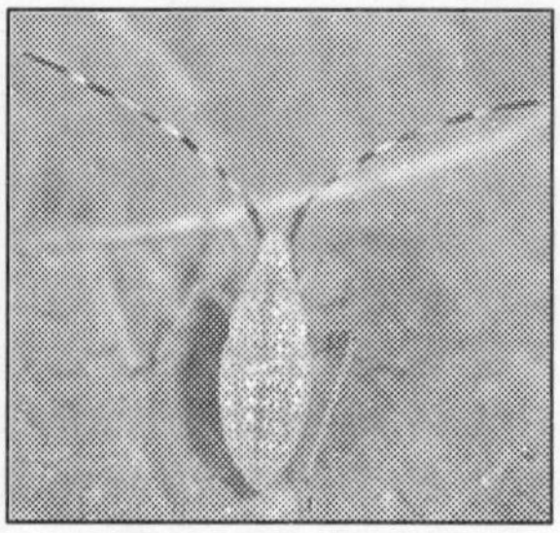
Eggs

V-instar

Adult

Damage

Nymphs and adults of both the species were observed to feed on squares and small developing bolls. During feeding the mirids pierce the plant tissues with their stylet. The affected area rapidly dulls in color, then blakens and ultimately resulting in death of the cells in the region. Feeding by these insects result in heavy shedding of medium sized squares and tiny bolls. Larger squares suffer damage that may cause development of deformed bolls which is often referred to as "parrot beaking". If the infestation is severe in older bolls, the damaged locks may not develop properly accompanied with the presence of stained lint. As the squares and bolls drop off, significant reduction in yield is noticed.

Brown Mirid Bug Damage

Management

Survival Strategies

Brown mirid bug are known to survive on weeds and native plant hosts surrounding cotton fields. They are also known to breed on native hosts in winter and can migrate to cotton growing areas in spring.

Alternate Hosts

Brown mired bug distinctly prefer lucerne to cotton. Lucerne strips or blocks can be used as trap crops to prevent the movement of mired bug into cotton crops. If using lucerne to manage brown mired bug, the lucerne should not be allowed to flower, seed or hay-off. Slashing half the lucerne at 4 weekly intervals and irrigating will ensure that fresh lucerne regrowth is constantly available for mirid bug feeding, thus preventing the movement into cotton. Other crop hosts include soybeans, moong beans, pigeon pea, safflower and sunflowers. It is assumed that mired bug migrate between these crops. Weeds hosts include turnip weed, noogoora burr, variegated thistle and volunteer sunflowers.

Biological Control

There are no beneficial species that are recognized to be regulators of brown mirid bug populations in cotton; however damsel bugs, big-eyed bugs, predatory shield bugs, as well as lynx, night stalker and jumping spiders are known to feed on mirid bug adults, nymphs and eggs.

Monitoring and Treatment Decisions

The action threshold for brown mirid bug is:

For irrigated cotton

- ✰ Squaring stage: 4 mirid bug/metre and/or <60 per cent retention.
- ✰ Early boll stage: 3 mirid bug/metre and/or 60-70 per cent retention.

For dryland cotton: 3 mirid bug/metre and/or <60-70 per cent retention throughout the season.

Chemical Control

Among the various treatments, acephate, imidacloprid and acetamiprid were superior in reducing the mirid bug infestation. Recent trials have shown that reduced rates of indoxocarb or fipronil combined with salt have given effective control of mirid bug population with reduced negative affects on natural enemies.

Cotton Mealy Bug

Scientific Name: *Phenacoccus solenopsis* Tinsley

Family: Pseudococcidae

Order: Hemiptera

Description of the Pests

Mealy bug is dark green in color when wax removed with 2 stripes on back. Short filaments are around body and caudal filaments about one-fourth length of body. Mealy bugs are generally oval, 3-4 mm long, and usually are covered with a white cottony or mealy wax coating. Ovisac regular remains beneath the body when squashed a pale greenish fluid observed. The female lay its egg in ovisac of cottony mass and pale greenish crawlers comes out and their development completed in 26-36 days. The crawlers not having wax on their body are very mobile. There are three nymphal instars in females and four in case of males. The total nymphal duration 13-19 days and last instar of male is an inactive stage with wing buds in cocoon of mealy wax.

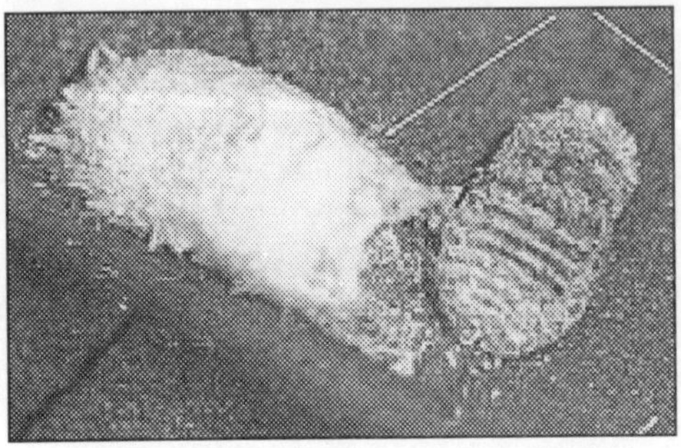

Ovisac

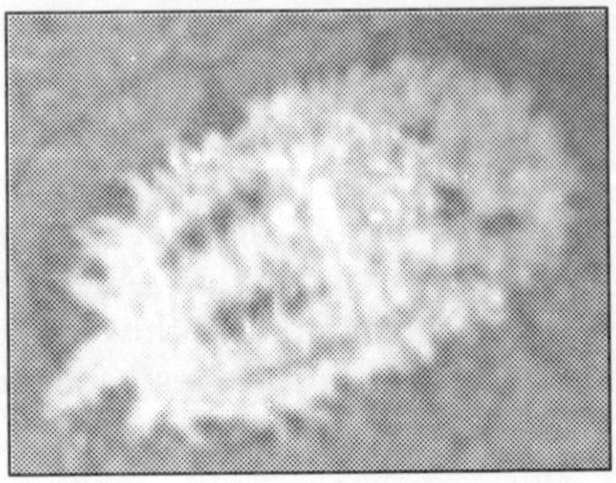

Female Adult

Damage

Both the adults and nymphs of this insect suck sap from the soft tissues of plant and while sucking they inject toxin in the plant system. The leaves of the attacked plants start curling and the plant remain stunted and give bushy appearance. Although mealy bug has a tendency to attack the stem portion of the cotton plant but in case of severe infestation, it is congregated on branches, leaves and fruiting bodies. The plant starts giving cottony appearance. There severe infestation led to premature flower dropping, deformities in the stem, reduction in size and number of bolls and delayed boll opening. The honeydew excreted by mealy bug attracts black sooty mould, which creates hindrance in photosynthesis activity. The severe attack of these mealy bugs led to the death of plants.

Mealy Bug Damage (Circle)

Factors Responsible for Emergence of Mealy Bug as Key Pest

- Before the introduction of Bt-cotton, the cotton crop was protected during vegetative phase from the sucking pests with insecticides. The removal of pesticides umbrella early in the season resulted in the build up of this pest, which was otherwise controlled with the insecticides used for the control of sucking pests.
- The practice of seed treatment of Bt-cotton further increased the insecticide free period to flowering phase and mealy bugs were able to do damage to the cotton crop.
- Mild winter season coupled with scanty rainfall.
- Non-adoption of non-Bt hybrids as refugia around Bt-cotton fields, which required insecticidal sprays for insect pests finally keeping the mealy bug population under check, which intend to appear around the field.
- Improper monitoring of pest population for timely management of pest in hot spot as the initial infestation always goes undetected.
- Use of unrecommended insecticides with improper sprays technology.
- The pest having wide host range and numerous source of survival during off season.
- Capacity of pest to breed with high reproductive rate, having natural protection and ability to hide during unfavorable environmental conditions.

Management

Cultural Measures

- Uproot and destroy the alternate host plants of mealy bug like *kanghi buti, peeli buti, congress grass, puthkanda, gutputna* and *tandla* growing on field bunds, water channels and wastelands in the area during the off-season of cotton.
- Do not throw the uprooted infested plants in cotton fields or water channels to check further spread of mealy bugs to far off areas.
- Follow cultural operations at monthly interval during off and in season.

Strategies During Off-Season for Reduction of Carryover

- Spray the infested plants or rows of cotton after last picking.
- Dislodge the mealy bugs by shredding the infested sticks against ground and destroy the mealy bugs by burying in the soil or use of insecticides.
- Stack the sticks from infested rows separately and use these sticks on priority for fuel up to end of February.
- Do not allow the movement of farm animals in infested fields.
- The cotton sticks are also a known source of spread of this pest: therefore, prevent the movement of sticks from the infested areas to the new areas.

Strategies for Alternate Hosts

- Regular monitoring of mealy bug incidence in major cotton growing areas on different crops, weeds and trees or fruit plants during *rabi* season. Identify the hot spots.
- Destruction of alternate hosts of mealy bugs to prevent carryover and spread of this pest at farmer's fields.

- ☆ Special efforts should be made to destroy the carryover sources of this pest through the eradication of weeds from wastelands, roadsides and irrigation channels or canals. The trees or fruit plants harboring mealy bug population should be sprayed with recommended insecticides.
- ☆ Do not throw the uprooted infested weeds in canal water or irrigation channels to prevent further spread of mealy bug.

Strategies During Crop Season

- ☆ Keep the fields free from weeds which are harbouring the pest populations.
- ☆ Sow only recommended hybrids or varieties.
- ☆ Cultivate non-Bt hybrids or varieties as refugia around Bt-cotton fields.
- ☆ Grow arhar or jowar or maize as barrier crops, being least preferred hosts.
- ☆ Do not grow guar, okra in or around the cotton crop, being most suitable hosts.
- ☆ Regular monitoring of the pest should be carried on after the sowing of cotton crop. Management of pest at early stage of infestation is a key to success.
- ☆ The pest initially restricted to a few plants in a row, adjoining source of infestation, thus spot treatment of pest with insecticides on cotton crop is advocated.
- ☆ Thorough coverage of plant is a must to check its multiplication.
- ☆ Restrict the movement of workers in the infested fields to prevent further spread of pest.
- ☆ Dissemination of information for timely action for management of this pest through mass media.

Chemical Measures

- ☆ Fine spray should be done with triple action nozzle with fine spray discharge.
- ☆ Spray thoroughly on stem or twigs along with the leaves of the plant.
- ☆ Spot treatment of recommended insecticides (table-11) to the plants in the affected border rows along with 4-5 adjacent rows.
- ☆ Do not use mixtures and un-recommended insecticides.

Pink Bollworm

Scientific Name: *Pectinophora gossypilla* (Saunders)

Family: Gelechidae

Order: Lepidoptera

Description of the Pests

Pink bollworm is a major pest of cotton but miner after 1991 because of the american bollworm appear as a major pest but now the appearance of this pest has been recorded on Bt-cotton. Adult are small, dark brown color, inconspicuous moths. When their wings are folded they have slender appearance. The wing tip is conspicuously fringed with blackish spots on forewings. Wing span is about 8-9 mm. Young larvae are tiny white or creamy caterpillars with dark brown heads. When mature, they are about 0.5 inch (12 mm) long and have transverse pink bands on back. Eggs are flat and small, slightly elongated and laid singly under the calyx of the green bolls, flower buds, leaves.

Eggs hatch in one week and larvae emerges which are white and turn pink as they grow older. Soon after emergence larvae enter into flower buds, flower or boll. After entering into boll it closes the hole and continuously feeds on seeds inside the boll. After full growth in about two weeks it comes out for pupation on the ground in fallen plant debris. After one week, the moth emerges and starts the life cycle again. By November it complete 4-6 generations. Fully grown larvae of last generation do not pupate, a few of them reach to the ground, but majority keeps feeding inside the bolls. They cut window holes inside the two adjoining seeds and joint together forming double seeds; finally they fall to the ground and form a major source of infestation for the next year. Some of them are picked with healthy cotton and reach to the ginning factories from where just a few return to the field along with the seed. The hibernating larvae lie curled in curled in the double seeds for many months and after passing winter, they emerged as moths. The last life cycle of the pest is very long covering 5-10 months, although life cycle is very short taking only 3-4 weeks.

Reason for Pink Bollworm Appearing as a Pest of Bt-cotton

1. Late season appearance: Pink bollworms appear late in the season and the concentration of cry1Ac toxin reduced after 95-110 days.
2. Feeding site and invariable expression of Cry1Ac in plant: the concentration of Cry1Ac is invariable and it is maximum in leaves followed by squares, bolls and flower, where as the toxin expression in the boll rind, squares, bud and ovary of flowers was clearly inadequate to confer full protection to the fruiting parts. Pink bollworm as a habit attack flowers buds, flowers and bolls (Kranthi *et al.*, 2005).

Eggs

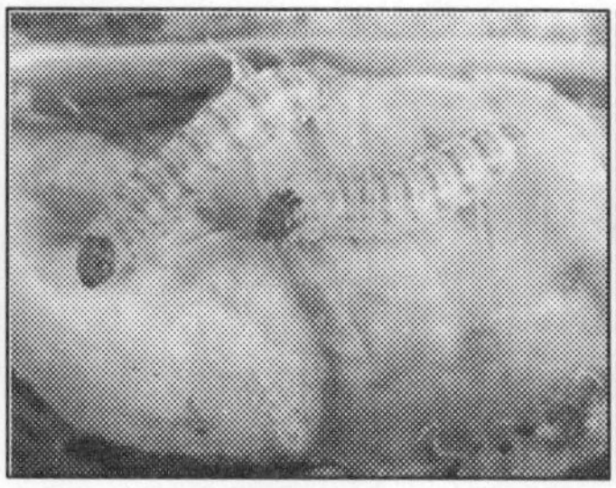
Larvae

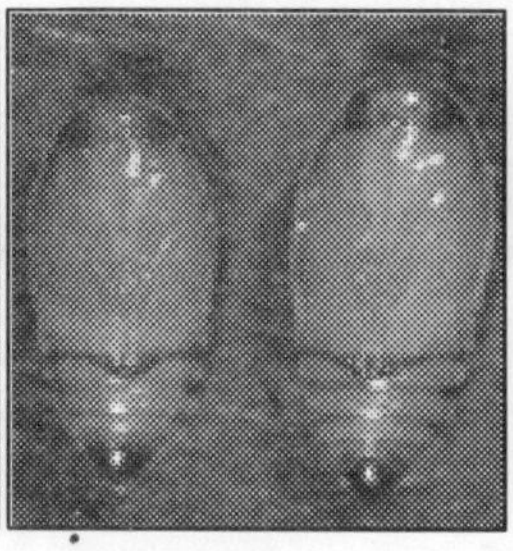
Pupae

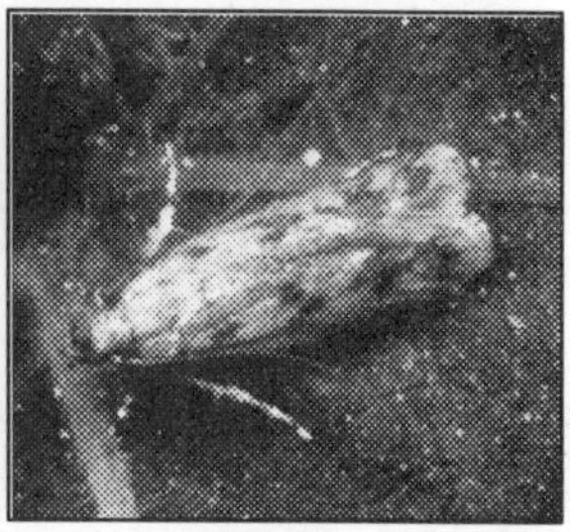
Adult

Damage

Moth emerges in the month of summer. The emergence take place at two distinct times *i.e.* May-June and then in July- August. Pink bollworms damage squares and bolls, the damage to bolls being the most serious. Larvae burrow into bolls, through the lint, to feed on seeds. As the larva burrows within a boll, lint is cut and stained, resulting in severe quality loss. Larvae cut two seeds and joint to both of them forming "double seed". Under dry conditions, yield and quality losses are directly related to the percentage of bolls infested and the numbers of larvae or boll. With high humidity, it only takes one or two larvae to destroy an entire boll because damaged bolls are vulnerable to infection by boll rot fungi.

Management

When high population levels of pink bollworm occur, the objectives of management are to keep infestations below damaging levels in the current season without creating secondary outbreaks of other pests and to reduce the over wintering population that will threaten the following season's crop.

Damage Cotton Bolls

Rosette Flower

The main control tools are:

- Observance of host-free period.
- The judicious use of insecticides.
- Timely crop termination and harvest.
- Rapid crop destruction.
- Cleaning of field before sowing.
- Deep ploughing.
- Encourage bioagents.
- Use of pheromone traps for monitoring and moth killing.

It is rarely necessary to apply insecticides against moths from the over wintered population of pink bollworm and, often treatments are not needed against the first generation of moths that develop from larvae within squares.

Cultural Control

The use of Bt-cotton will help prevent damage by pink bollworm during early part of season. Recently developed transgenic cotton, Bollguard-II, offers suppression of cotton bollworms, along with pink bollworm. Eliminate the food supply for pink bollworm by cutting off irrigation early enough to stop production of green bolls at the end season. Regardless of when the crop is terminated, immediately shred the cotton plants following harvest. Shredding destroys some larvae directly and promotes rapid drying of unharvested bolls. Though winter irrigations can reduce populations of over wintering pink bollworms by as much as 50 to 70 per cent. If cotton is being followed with cotton, pre irrigate in and plant as early as possible.

Organically Acceptable Methods

Cultural controls, with the exception of the use of Bt-cotton but when the pest attack of the later stages of crop growth after fourth week of October, pink bollworm can afford to even survive and damage to Bt-cotton and the use of mating disruption and sprays of the Entrust formulation of spinosad are acceptable to use on organically grown cotton.

Biological Control

Table 21.4: Parasitoids of Pink Bollworm

Parasitoids	*Stage of Pest Attacked*	*Parasitization (Per cent)*
Trichogramma	Egg	1-10
Trichogramatoidea	Egg	21-25
Apanteles angaleti	Larva	20-25
Bracon greeni	Larva	2-6
Rogas algarhensis	Larva	33
Elasmus johnstoni	Larva	10-12
Goniozus spp.	Larva	0.3-17
Chelonus blackburni	Egg-Larva	5

Monitoring and Treatment Decisions

Pink bollworm is primarily managed with a host-free period, sampling bolls is the most reliable way to monitor pink bollworm populations. 5 per cent bollworm incidence or 8 male moths per trap per night are the appropriate indicators to start the action threshold.

Chemical Control

Insecticides mentioned in Table 21.11 can be applied simultaneously; the use of gossyplure, a sex attractant that disrupts mating when distributed throughout the field, may be effective against pink bollworm when it is supplemented with cultural control practices that minimize the number of over wintering bollworms.

Tobacco Caterpillar

Scientific Name: *Spodoptera litura* (Fabricius)

Family: Noctuidae

Order: Lepidoptera

Description of the Pest

The adult moths possess brownish forewing and whitish hind wings. The forewing is speckled with beautiful, golden and grayish brown pattern. Moth is about 22 mm long and measures 40 mm

Eggs

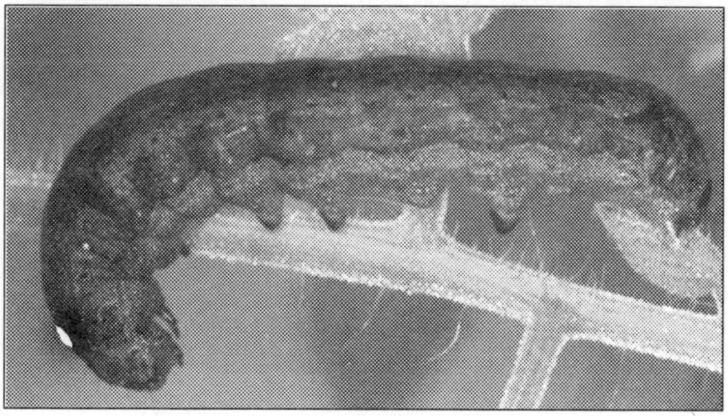

Larva

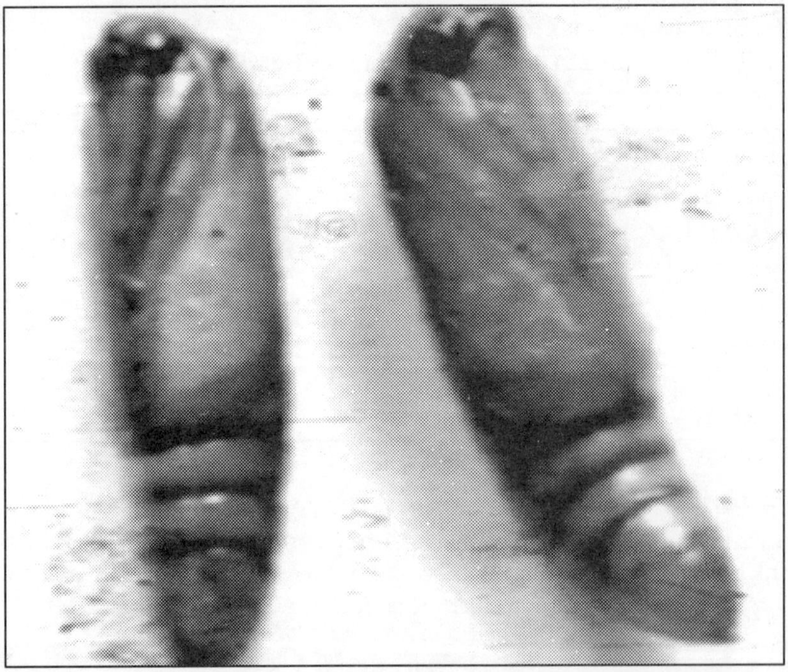

Pupae

Adult

wing span. Adult moths are active during night; they mate with female, female lays about 300 eggs in cluster. These clusters are covered with brown hairs and eggs hatch within 3-5 days. The larvae feed gregariously for first few days and then disperse to feed individually. They pass through 6 instars and are fully developed in 15-30 days. The fully grown larvae enter the soil where they pupate. The pupal stage lasts for 7-15 days and moths on emergence survive for 7-10 days. The life cycle is completed in 50-60 days.

Damage

Tobacco caterpillar is a polyphagous pest and active from August to October. As the insect lay eggs on the underside of mature leaves its initial instar feed gregariously on the underside of leaves leaving below only the veins. Later on the larvae continue feeding by spreading to the whole plant and make irregular holes in the leaves. In case of severe infestation the plants are capably defoiled leaving behind only the net of veins.

Tobacco Caterpillar Damage

Management

The impact of a tobacco caterpillar infestation depends on the number of larvae present, the age of the larvae and the timing of damage relative to the crops. Monitoring and control must therefore be aimed at the collection of eggs (laid in groups on the underside of the leaves) and small larvae (feeding on underside of the leaves gregariously). Natural enemies are very important in managing populations of tobacco caterpillar. Insecticides are needed only if the population exceeds the treatment threshold level.

Cultural Control

- ☆ Plough deeply to expose the pupae and hibernating larvae and complete the sowing operations at earliest possible.
- ☆ Avoid growing castor, moong, *dhaincha* and *bhindi* in and around the Bt-cotton. These are the most preferred host of tobacco caterpillar, helps the pest to multiply and shift to cotton.

☆ Collection and destruction of egg masses and young larvae.

☆ Recently developed transgenic cotton, Bollgard-II, offers suppression of tobacco caterpillar along with the cotton bollworms.

Biological Control

Many predators and parasites combine to substantially maintain tobacco caterpillar populations at low levels. Insecticide sprays for other pests will disrupt this natural control.

Monitoring and Treatment Decisions

Monitoring of tobacco caterpillar can be done by using sleeve or moth catch trap. Presence of 4 eggs masses per 50 metre row length are the indicator of action threshold.

Chemical Control

Insecticides mentioned in Table 21.11 can be applied simultaneously and cultural practices can be adopted.

Integrated Pest Management Strategies for Insects of Bt-cotton

1. Varietal resistance
2. Cultural and mechanical control
3. Biological control
4. Microbial control
5. Chemical control (ETL)
6. Insect resistant management (IRM) strategies

Varietal Resistance

Varietal resistance being eco-friendly is a key factor in pest management programme. Though the source of resistance in case of bollworms were not effective but after the commercial utilization of transgenic cotton as a component of IPM, the cotton cultivation has been revived throughout the country, similarly after the introduction of second generation of genetically modified cotton *i.e.*, Bollgard-II (Cry1Ac and Cry2Ab) has provided an additional advantage of repulsion of tobacco caterpillar, a sporadic pest which otherwise has emerged as a major pests of Bt-cotton in north India after the introduction of Bollgard-I (Cry1Ac).

Table 21.5: Resistance/Tolerant Sources Identified Against Insect Pests of Cotton in India

Sl.No.	*Common Name*	*Scientific Name*	*Cultivar (s)*
1.	Jassid	*Amrasca bigutulla bigutulla*	Khandwa 2, Badnawar, MCU 5*, Krishna*, Mahalaxmi*, Sujay*, Sanguioneum*, Eknath*
2.	Whitefly	*Bemisia tabaci*	LK 861*, Amravathi*, Kanchan*, Supriya*, LPS 141*
3.	Pink bollworm	*Pectinophora gossypiella* (Saunders)	G 27, LD 135*, Lohit Adadhita, MCU 7, Sujata, Digvijay, Sanguineum*
4.	Spotted bollworm	*Earias vitella* (Fabricius)	L 1245*, JK 119-25-54, BCS 10, BCS 10-75, FBRN 2-6, HAO 66-107-1/1, G 27, Sanguineum*

Source: Dhaliwal and Arora (2005). *: Released for cultivation.

Cultural and Mechanical Control

Cultural, mechanical and physical control practices are historically the most important methods of preventing crop losses. As early as 1911, removal of cotton sticks by 1st August every year was made compulsory by law to minimize the incidence of pink bollworms on cotton in Madras state (Dhaliwal and Arora, 1993). The cultural practices like, planting time, plant geometry, tillage, crop sanitation, fertilizer application, irrigation, intercropping, crop rotation, manipulation of carryover sources, barrier or trap crop etc. have been found useful in managing pest population in the simplest and cheapest way and are still popular and important even after the introduction of Bt-cotton.

Table 21.6: Recommended/Approved, Cultural and Mechanical Control Strategies for Insect Pests in Cotton

Sl.No.	*Physical/Cultural/ Mechanical Practices*	*Target Pests*	*Impact*
1.	Uprooting and destruction of weeds during off season	Mealy bug, whitefly and CLcuV	Reduces carry over infestation
2.	Deep ploughing for soil exposure before sowing and after harvest	American boll worm, tobacco caterpillar	Pupal mortality through birds and other factors
3.	Early sowing (Haryana, Punjab and Rajasthan)	American boll worm	Escape mechanism
4.	Proper plant spacing and maintenance of plant population	Sucking pests, bollworms and foliage feeders	Reduces incidence, proper application of insecticides through foliar application
5.	Growing bajra/sorghum/maize as barrier crop	Mealy bug	Act as a physical barrier and reduces population build up
6.	Growing of (refugea) non Bt around Bt-cotton field	Boll worm complex	Delay of resistance to Bt mechanism
7.	Over dose of nitrogenous fertilizer	Jassids, whitefly and bollworms	Congenial for population build up
8.	Non judicious use of irrigation water	Entire pest complex	Good for population build up
9.	Cultivation of undescript material	Entire pest complex	More damage
	Cultivation of Adjoining Crops in and Around Cotton Crop		
10.	Okra, moong and arhar	Jassid, whitefly, spotted and American bollworm	Most preferred increase in population build up and incidence on cotton crop
11.	Okra, moong, castor and dhaincha	Tobacco caterpillar	
12.	Okra, moong and guar	Mealy bug	
	Uprooting and Destruction of Weeds During the Season		
13.	Kanghi buti, Peeli buti	Whitefly	Reduced carry over and population uild up
14.	Itsit	Tobacco caterpillar	
15.	Congress grass	Mealy bug	
16.	Do not throw uprooted infested plant/weeds in cotton plant/water channel/common place	Mealy bug	Check further spread

Contd...

Table 21.6–Contd...

Sl.No.	Physical/Cultural/ Mechanical Practices	Target Pests	Impact
17.	Collection and destruction of egg masses and early instar feeding gregariously along with leaves	Tobacco caterpillar	Reduces population build up and incidence
18.	Removal of infested terminal shoots	Spotted bollworm	Reduces incidence
19.	Detopping *arboreum* cotton plants height beyond 1.5 m	Pink bollworm and Spotted bollworm	Reduces incidence increase seed cotton and easy in picking operations
20.	Early termination of crop	Bollworms	Reduces carry over
21.	Grazing of cattle, sheep, goat after last pick to freedom plant debris and unopened bolls	Pink bollworm	Reduces carry over
22.	Before stacking dislodge and burn the burs and unopened bolls	Pink bollworm	Reduces carry over
23.	Stacking of cotton from stalks in open in villages	Pink bollworm and mealy bug	Reduces carry over
24.	After last pick, dislodging of mealy bug	Mealy bug	Reduces carry over
25.	Do not allow the movement of farm animals	Mealy bug	Reduces carry over
26.	Restrict the movement of farm workers in infested fields	Mealy bug	Reduces carry over
27.	Prevent the movement of sticks from the infested areas to the new area	Mealy bug	Reduces carry over

Source: Dhawan *et al.* (2008).

Biological Control

Biological control agents have played a pivotal role in regulating pest population. A rich fauna of naturally occurring bioagents have been reported on insect pests of cotton crop. To conserve natural enemies or approach of no spraying during first three months and if required use of relatively safe herbal insecticides like neem oil and prevention of broad spectrum organophosphate as early season sprays have been advocated as a stratigies under IRM project in north India (Kranthi, 2005). The requirement of the exploitation of natural regulators for managing the population of sucking pests before embarking on recommended curative measures in Bt-cotton is important.

Table 21.7: Predators of Major Sucking Insect Pests of Cotton

Sl.No.	Name of Natural Enemies	Pest	Stage of the Pest
1.	*Coccinella septum punctata*	Sucking pests	Eggs, nymph, young larvae
2.	*Chrysoperla carnea*	Sucking pests	Eggs, nymph, young larvae
3.	*Cryptolaemous montrozieri*	Sucking pests	Eggs, nymph, young larvae
4.	*Cheilomenes sexmaculatus*	Sucking pests	Eggs, nymph, young larvae
5.	*Brumoides suturalis*	Sucking pests	Eggs, nymph, young larvae

Important Natural Enemies (Predators and Parasitoids) of Emerging Pests of Bt Cotton

Whitefly Parasitoid (*Encarsia formosa*)

Tobacco Caterpillar Parasitoid (*Telenomus remus*)

Mealy Bug Parasitoid (*Aenasius bambawalei*)

Pink Bollworm Parasitoid (*Trichogramma* spp.)

Pink Bollworm Parasitoid (*Apanteles* spp.)

Pink Bollworm Parasitoid (*Bracon* spp.)

Preying Mantid Feed on Insects

Spider Feed on Cotton Insects

Grub of *Chrysoperla carnea*

***C. montrozieri* Adult Feed on Mealy Bug**

***C. sexmaculatus* Adult Feed on Mealy Bug**

***C. septumpunctata* Adult Praying on Mealy Bug**

***C. montrozieri* Grub Feed on Mealy Bug**

***C. septumpunctata* Grub Praying on Mealy Bug**

***B. suturalis* Adult Praying on Mealy Bug**

Table 21.8: Parasitoids of Major Insect Pests of Cotton

Parasitoid	*Host*	*Parasitization (per cent)*	*Period of activity*
		A. Egg Parasitoids	
Trichogramma achaeae	P	5.9-27.2	September-October
T. chilotraeae Nagaraja	P	30-35	September-November
Telenomus remus Nixon	S	-	
Trichogrammatoidea spp. near *guamensis* Nagaraja	P	21.4-25.0	October-November
		B. Larval Parasitoids	
Agathis fabiae	P	2	September-October
Apanteles angaleti Muesebeck	P	2.5	January-December
Bracon chinensis	P	5.0	September-November
Bracon greeni	P	7.0-10.0	October-January
Campotolithlipsis gossypiella	P	5	September-November
Rogas aligarhensis Quadri	P	3.3	August-December
	P	59.07	-
Scambus limepes Morley	P	Negligible	December
Goniozus spp.	P	0.3-16.7	
Elasmus johnstoni Ferriere	P	10-12	August-November
E. philippinensis	S	1.7	October
Cremastus spp.	S	1.0	September-October
Pyemotes ventricous (Newport)	P	26.0-39.0	August-November

Contd...

Table 21.8–Contd...

Parasitoid	Host	Parasitization (per cent)	Period of activity
		C. Egg-larval parasitoids	
Chelonus spp.	P	0.5-3.6	September-November
C. blackburni Cameron	P	5	September-November
		D. Pupal parasitoids	
Xanthopimpla punctata	S	5.5	October
Brachymeria spp. n. *euploeae*	S	16.5	October

Source: Dhawan (1999).

A number of promising natural enemies have been reported attacking cotton pests in the country. However, large scale use of insecticides has reduced the population of most of these natural enemies's to insignificant level. In the field where biocontrol was practiced, 25 natural enemies's were recorded as compare to only 2 in the field sprayed with insecticides. A number of exotic parasitoids like *Chelomenes blackburni* Cameron from spotted bollworm and *Bracon kukpatricki* (Walkinson) and *Trichogramma brasiliense* (Ashmned) have been recorded from pink bollworm in the states of Maharastra, Karnatka, Tamil Nadu, Gujarat, Haryana and Punjab. Release of *Trichogramma* spp. at 1, 50,000 parasitized eggs/ha at weekly interval have proved promising for bollworm control. Sucking pests may be checked by releasing *Chrysoperlacanea* at 1 lakh/ha at fortnightly intervals. Eight releases of *T. chilonis* @ 1,5,0000/ha at weekly interval during July- August to October + one release of *C. carnea* @ 10,000/ha + need based application of insecticides for control of sucking pests (3 sprays) and for bollworms (3 sprays) during Sept- Oct was found effective and at par with spray schedule (3 sprays for sucking pests and 6 sprays for bollworms) in reducing the bollworms damage and increasing the yield (Brar *et al.*, 2001). Eleven to thirteen releases of multiple insecticide tolerant strain of *T. chilonis* @ 1,50,000/ha at weekly interval during July to October along with 6-9 insecticidal sprays proved significantly better for the control of bollworms and increasing egg parasitization and seed cotton yield (Brar *et al.*, 2007). After the introduction of Bt-cotton, the reduction in use of broad spectrum insecticides against bollworm, the other insects both harmful and beneficial are now increasing in abundance.

Microbial Control

Various diseases causing organisms perform an important role in dynamics and natural regulation of insect population. Among different microbial agents, the utilization of nuclear polyhedrosis virus (NPV) and bacteria *Bacillus thuringiensis* Berlinzer (Bt) formulations have been tested extensively for the control of cotton pests. Dhawan and Simwat (1998) observed that biolep, dipel and bioasp formulations of Bt @ 1.5 kg or litre/ha were effective as insecticides quinalphos against young larvae of *Helicoverpa armigera*. However, they were significantly inferior against grown up larvae. Dhawan *et al.* (2001) recorded that NPV sprays applied alone resulted in significantly less seed cotton yield and higher larval population and extent of damage as compared to NPV in combination with insecticides or NPV alternated with insecticides which were at par with insecticidal sprays. But after the introduction of Bt-cotton and as a part of insect resistant management (IRM) strategy avoid in use of Bt based biopesticides that may contribute to selection of a broad spectrum resistance bollworm to several Bt gene of interest is advocated.

Chemical Control

The use of insecticides is indispensable tool of pest management. Pesticide poisoning, insecticide resistance, resurgence of pests (Table 21.9), effect on non target organisms and pesticide residues are some of the factors to be kept in mind while selecting and using the pesticide. To use the insecticide judiciously the economic threshold (Table 21.10) values established for individual insect should be followed. Different group of insecticides *viz.*, nitroguadine, organophosphate, synthetic pyrethroids, carbamates, organophosphates, synthetic pyrethroids, carbamates, organochlorines, naturalyte, oxadizine, miscellaneous and insect growth regulators have been recommended against different insect pest of cotton in high of their safety to natural enemies.

Table 21.9: Insecticides Causing Resurgence of Insect and Mite Pests of Crops

Pest	*Scientific Name*	*Insecticide (s)*	*Mode of Application*
Jassid	*Amrasca bigutulla bigutulla* (Ishida)	Chlopyriphos, Quinalphos	Foliar
Aphid	*Aphis gossypii* Glover	Carbaryl,	Dust,
		DDT,	Foliar,
		Monocrotophos,	Foliar,
		Phorate, Disulfoton, Dimethoate, Cypermethrin, Deltamethrin, Fenvalerate, Flucythrinate, Permethrin, S-524	Soil, Foliar
Whitefly	*Bemisia tabaci* (Gennadius)	DDT, Dimethoate,	Foliar,
		Monocrotophos,	Foliar,
		Cypermethrin, Deltamethrin, Fenvalerate,	Foliar,
		PP-321,	Foliar,
		Toxaphene	Foliar
Spider mite	*Tetranychus cinnabarinus*	DDT, Carbaryl, Cypermethrin, Fenvalerate, Fluvalinate	Foliar
Mealy bug	*Ferrisia virgata* (Cockerell)	Cypermethrin, Deltamethrin, Fenvalerate, Permethrin	Foliar

Source: Dhawan *et al.* (2000).

Table 21.10: Economic Threshold levels (ETLs) of Major Insect Pests of Cotton at Different Locations in India

Pest Species	*ETLs*	*Location (State)*
Jassid	1 nymph/leaf	Ludhiana (Punjab)
	II grade injury symptoms	Faridkot (Punjab)
	2 nymphs/leaf	Coimbatore (Tamil Nadu)
	1-2 nymphs/leaf	Coimbatore (Tamil Nadu)
	3.5-5.1 nymphs/leaf	Surat (Gujarat)
	50-100 nymphs or adults/50 terminals	Coimbatore (Tamil Nadu)
	10 nymphs/leaf	Surat (Gujarat)

Contd...

Table 21.10–Contd...

Pest Species	*ETLs*	*Location (State)*
Whitefly	6-8 adults/leaf	Ludhiana (Punjab)
	20 nymphs/leaf	Ludhiana (Punjab)
	Stickiness due to honeydew excreted on leaves in 50 per cent of plants	Ludhiana (Punjab) Coimbatore (Tamil Nadu)
	8-10 adults or 20 nymphs/leaf	Coimbatore (Tamil Nadu)
	5-10 adults/leaf	Bapatla (A.P.)
Thrips	10 thrips/leaf	Coimbatore (Tamil Nadu)
	1-2 nymphs or adults/leaf	Coimbatore (Tamil Nadu)
Aphids	15-20 per cent affected plants	Coimbatore (Tamil Nadu)
Mites	10 mites/cm^2	Coimbatore (Tamil Nadu)
Pink bollworm	8 male moths/trap/night	Hisar (Haryana)
	4-6 moths/trap/night	Ludhiana (Punjab)
	4 per cent bollworm incidence in bolls on plants	Ludhiana (Punjab)
	5 per cent bollworm incidence	Coimbatore (Tamil Nadu)
	10 per cent bollworm incidence	Ludhiana (Punjab)
	Spray initiation at 75 days on LH 900 variety	Ludhiana (Punjab)
Spotted bollworm	1 larva/10 plants	Surat (Gujarat)
	10 per cent incidence in squares	Coimbatore (Tamil Nadu)
	10 per cent incidence shoots or reproductive parts	Coimbatore (Tamil Nadu)
	5 larval units/10 plants	Surat (Gujarat)
	10 larval units/10 plants	Anand (Gujarat)
American bollworm	25 adults/trap/night or 1 larva/ 10 plants or 20 bolls	Bapatla (A.P.) Coimbatore (Tamil Nadu)
	1 larva/plant	Coimbatore (Tamil Nadu)
	10 per cent incidence in reproductive parts	Bathinda (Punjab)
	2 larval units/10 plants	Surat (Gujarat)
	7 larval units/10 plants	Surat (Gujarat)
American bollworm (Bt-cotton)	Experiment in progress	CICR, Sirsa;RS, Faridkot
Bollworm complex	5 per cent bollworm damage	Akola (Gujarat)
	10 per cent bollworm incidence	Bhubaneswar (Orissa)
	5 per cent bollworm incidence in freshly shed fruiting shed fruiting bodies	Coimbatore (Tamil Nadu) Bathinda (Punjab)
	Spray initiation at squaring 25 per cent of plants in cotton	Ludhiana (Punjab)
Stem weevil	10 per cent plant infestation	Coimbatore (Tamil Nadu)
Tobacco cutworm	4 egg masses/50 m row length	Coimbatore (Tamil Nadu)

Source: Simwat (1994).

Table 21.11: Insecticides Recommended for the Management of Sucking Pests, Foliage Feeder and Pink Bollworm and their Safety to Natural Enemies of Cotton

Sl.No.	Insecticides	Doses/ha	Natural Enemies
		Jassids	
		Nitroguadine group	
1.	Imidacloprid 200SL	100ml	Toxic
2.	Imidacloprid 555	100 ml	Toxic
3.	Imidacloprid 17.8	100 ml	Toxic
4.	Acetamiprid 20SP	50 gm	Toxic
5.	Thiomethoxam 25 WG	100 gm	Toxic
		Whitefly	
1.	Triazophos40EC	1.5 litre	Toxic
2.	Ethion 50 EC	2.0 litre	Toxic
		Mealy Bug	
		Carbamate group	
1	Carbaryl 50WP	2.5 Kg	Toxic
2	Thiodicarb75 WP	625 g	Toxic
		Organophosphate	
3.	Profenohos50 EC	1.25 litre	Toxic
4.	Quinalphos 25 EC	2.0 litre	Toxic
5.	Acephate 75 SP	2.0 Kg	Toxic
6.	Chlorpyriphos 20 EC	2.5 litre	Toxic
		Pink bollworm	
		Carbamate group	
1.	Carbaryl 50 WP	2.5 kg	Toxic
2.	Thiodicarb 75 WP	625 gm	Toxic
		Organochlorin group	
3.	Endosulfan 35 EC	2.5 litre	Safe
		Organophosphate group	
4.	Profenophos 50 EC	1.25 litre	Toxic
5.	Quinalphos 25 EC	2.0 litre	Toxic
6.	Monocrotophos 36 SL	1.25 litre	Toxic
7.	Triazophos 40 EC	1.5 litre	Toxic
		Synthetic pyrethroids	
8.	Alphamethrin 10 EC	250 ml	Toxic
9.	Cypermethrin 10 EC	500 ml	Toxic
10.	Cypermethrin 25 EC	200 ml	Toxic
11.			
	Deltamethrin 2.8 EC	375 ml	Toxic
12.	Fenvalerate 20 EC	750 ml	Toxic

Table 21.11–Contd...

Sl.No.	Insecticides	Doses/ha	Natural Enemies
		Tobacco caterpillar	
		Carbamate group	
1.	Thiodicarb 75 WP	625 gm	Toxic
		Organochlorin group	
2.	Endosulfan 35 EC	2.5 litre	Safe
		Organophosphate group	
3.	Quinalphos 25 EC	2.0 litre	Toxic
4.	Acephate 75 SP	2.0 kg	Toxic
5.	Quinalphos 25 EC	2.0 litre	Toxic
		Insect growth regulator	
6.	Novaluron 10 EC	375 ml	Safe

Insect Resistance Management Strategies for Bt-cotton in India

Insecticide resistance management strategies are based on use of a rational and sensible sequence of insecticides that are effective on the target species, cause less disturbance to the beneficial fauna and minimize selection pressure and rotation of insecticide group based on unrelated resistance mechanisms. Based on the status and performance of insecticides, appearance of insect pests, windows have been developed for dissemination of the technology among the farmers of the country to delay in insecticide resistance and increase the functional life of the insecticides. With the introduction of noval and eco-friendly insecticides, pest management now appears to be very promising. The chloronicotinyls (imidacloprid, acetameprid and thiomethoxam) and the growth regulators are comparatively more effective against sucking pests and less toxic to natural enemies as compared to all conventional insecticides. Apart from the introduction of cry toxins in the form of transgenic technology, chemicals such as spinosad, indoxicarb, emmamectin benzoate, novaluron provided effective control of *H. armigera*, while being less toxic to beneficial and the ecosystems. However, as a rule of IRM, the over use of these molecules can lead to the development of pest resistance to the insecticides. The Bt-cotton technology is a powerful, eco-friendly tool introduced for cotton pest management. The transgenic crop now popularly called as Bt-cotton hold great promise on controlling cotton bollworm, *H. armigera*, which has developed resistance to all commonly used insecticides in the country. The technology has provided cotton growers with significant ecological and economical advantage and acknowledged that the benefits accrued from Bt-cotton outweigh risks substantially. But the durability of Bt-cotton to insect resistance is a matter of concern and need strategies of insect resistance management to Bt-cotton. Through regular monitoring and the use of effective pest management measures in Bt-cotton fields will reduce insect populations that survive Cry1Ac toxin and thus represent resistant genotype. Hence, it is important to use various pest management strategies including a few selected insecticides to ensure reduction in bollworm population or per the IRM windows:

Sucking Pest Control in Bt-cotton

1. Intercropping with cowpea, soybean and black gram was found to encourage natural enemies.

2. Chemical seed treatment (imidacloprid 70 WS or thiomethoxam 70 WS @ 5-7 g/kg seed) followed by 40 DAS. Stem application of acetamiprid or thiomethoxam or imidacloprid (Confidor) and ETL based spray of endosulfan or diafenthiuron (Polo) against jassid or whitefly or aphids.

Window 1: 60-90 DAS

Use eco-friendly methods such as cultural control or hand-pinking of surviving bollworms in Bt-cotton fields.

Window 2: 90-120 DAS

Conventional insecticides such as endosulfan, thiodicarb, quinalphos and chlorpyriphos or new molecules such as spinosad, indoxicarb, emmamectin benzoate, novaluron or lufenuron can be used on 90 and 120 days old crop to reduce populations of resistant genotypes.

Other Useful Strategies to Mitigate Resistant in Bt-cotton

1. Identify and use attractive synchronous alternate host crops for *H. armigera*, which could be used as intercrop or trap crop refuges.
2. Avoid use of Bt based biopesticide that may contribute to selection of a broad spectrum resistance to several useful Bt gene of interest.
3. Use alternate genes that do not share common resistance mechanisms as that of Cry1Ac, in transgenic plants either in rotation or alternation or mixtures.
4. Finally, resistance development would be slower if the Bt-cotton technology is targeted more for areas, which are identified as hot-spot problematic zones for *H. armigera* management. Though this is difficult to implement without a stringent policy restriction, it makes meaningful sense to focus efforts properly in addressing an intractable problem in problematic zones using a powerful eco-friendly technology such as Bt-cotton.

The resistance management strategies proposed herein have been validated in farmer participatory programmes, for large scale use to mitigate the problem. IRM strategies were designed based on results of the network project on 'Insecticide resistance management of *Helicoverpa armigera*' carried out over a period of ten years, at six different research institutions in India. The strategies place emphasis on efficient use of insecticides to conserve the ecosystem for better pest management. After the introduction and large scale cultivation of Bt transgenic cotton it is reasonably certain that bollworms particularly *H. armigera* will respond to intense selection pressure through decline in susceptibility to cry-toxins. Continuous selection pressure with the toxicant eventually lead to increase in numbers of tolerant or resistant individuals in populations. So, the strategies to reduce the rate of resistance development should be developed.

Resistance Management Strategies

Refugia

Refuge is one of the best favored resistance management options where an appropriate area of non Bt is cultivated in the vicinity of the Bt transgenic crop to ensure the survival of susceptible insect genotype. Though the concept of refuge was discovered for the mono-cultivation scenario having very remote possibilities of alternate or natural host available for target pest. The refuge strategies will help in mating of rare resistant adults emerging from Bt-cotton with more abundant susceptible adults from refuges. In India the adoption of refugia is not been consistently followed but the diverse cropping systems and wide variety of natural hosts certainly provide an alternate approach to IRM for Bt cotton. The target pests are infesting these hosts and are not being controlled using Bt on these other hosts, so therefore structured refuges for Bt-cotton in India is may not be necessary but in some of the states of North India (Punjab and Haryana) where more than 80 per cent area under Bt cotton and the availability of alternate host is limited, in these situations, the structured refugia is a necessasity as IRM strategies.

Regulating Expression

(Optimum dose strategies): Bt-cotton produces Cry1Ac protein throughout the season but an adequate level of expression of Cry1Ac throughout the critical part of the cotton growing season must be ensured. Bollworms, as a habit prefer fruiting parts and rarely feed on leaves. So, the option of tissue specific promoters can be used to ensure only fruiting parts produce toxin and are protected while the insect pests is encouraged to feed on leaves without causing any economic damage. But conversely it is possible that young larvae survive on leaves, grow into older instar capable of surviving the toxins levels in fruiting parts and thus cause yield loss. But the toxin expression in almost all commercial Bt-cotton crop known to decline after a certain stage so the use of Late Embryogenesis Abundant- LEA promoters to ensure the toxin at late stages in the crop life can be considered. Reduction in selection pressure can be brought about by either a temporal regulation of toxins expression as an insect inducible response on a tissue specific expression.

Gene Pyramiding

Gene stacking offer an alternate option of delaying resistance especially if two or more genes are available with high toxicity levels having independent modes of action and are do not have cross resistance. Stacking or pyramiding is an important IRM strategy and involves more than one insecticidal gene into the same plant. Commercial introduction of two genes Bt-cotton (Bollgard-II) in India was considered at the earliest opportunity being an essential part of the IRM strategies. Through many field and laboratory studies indicated that the Cry2Ab protein is expressed in Bollgard-II cotton at high dose for control of key target pests where as the Cry1Ac protein is expressed at high dose for both the pink bollworm and a moderate dose for cotton bollworm. The Bollgard-II containing Cry2Ab as event 15985 in addition to Cry1Ac has been released in India. The stacked genes are expected to increase the spectrum of toxicity to wide range of lepidopterous insects include, *Spodoptera litura* known to survive Cry1Ac. It has been summarized through several experiments that Cry1Ac and Cry2Ab proteins compliments each other are resistance management programme for all three cotton bollworm species. Therefore if one of the target pests should develop resistance to one of the proteins the other protein will provide control of that resistant pest. Spatial and temporal transgenic development where the selection of two or three different transgenic crops alternatively one after the other each year would cause a reduction in selection pressure due to each of the single toxins and thus delay resistance development.

Resistance Management Strategies: New Insight

Use of alternate gene that do not share common resistance mechanisms as that of Cry1Ac, in transgenic plants either in rotation or alteration or mixtures. Use eco-friendly methods such as cultural control or handpicking of surviving bollworms in Bt-cotton fields. Biopesticides that are neem based or HaNPV would be useful to manage younger larvae on 60-90 days old crop. Alternatively, conventional insecticides such as endosulfan, thiodicarb, quinaphos, and chlopyriphos, or new molecules such as spinosad, emamectin benzoate, novaluran or indoxicarb can be used on 90 and 120 days old crop to reduce populations of resistant genotypes. Identify and use attractive synnchronus alternate host crops for *H. armigera* which could be used as intercrop or trap crop refuges. Avoid use of Bt based biopesticides that may contribute to selection of a broad-spectrum resistance to several useful Bt genes for interest.

Diseases

Bacterial Blight

Pathogen

Xanthomonas axonopopodis pv. *malvacearum* (Smith) Vauterin, Hoste, Kersters and Swings

Symptoms

The disease affects all the aerial parts of the plant. The main symptoms are seedling blight, black arm, angular leaf spot and boll lesions. The cotyledons are the first to show the symptoms in young seedlings. Such infection comes from the seed which may carry the inoculum on the fuzz or less often internally. Small water soaked, circular or often irregular lesions appear on the margins of the cotyledons. Such lesions are not angular in shape like those on the true leaves. The lesions turn black or brown and cotyledons shrivel. From the cotyledons the expanding lesion passes on to the young stem and from this point, the seedling may be affected by the secondary infection up to the apex and may often die. Such extensive damage constitutes "seedling blight"

The most common and striking symptoms appears on the leaves where a few to numerous, minute, water soaked lesions appear scattered on the ventral surface of the young leaves but are later seen on both surfaces of the lamina. The lesions may increase in size and individually may be 5 mm in

Small Lesions

Vein Blight

Angular Lesion

Lesion on Boll

diameter, bounded by the smaller veins, giving them an angular shape, therefore its naming as "angular leaf spot". In severe infection, several spots may coalesce to form large dead areas. Bacterial blight lesions on the leaf often induce the formation of an abscission layer at the base of the petiole and premature shedding of leaves take place.

Blackarm Lesions

The lesion may extend to the mesophyll on both sides of the vein forming a black border with an irregular margin. The vein lesion may start from the point of attachment of the lamina to the petiole; such symptoms is called 'black vein' or 'vein blight'. Vein blight and angular leaf spot may appear together or independently of each other. Lesions on the stems, petioles and sympodial and monopodial branches are elongated, grayish to black. The stem may be girdled and not only the bark but the woody part of the stem also brittle and weak. Cracks may develop and productive branches may break off during strong winds. This phase is called 'blackarm'. The bolls are often often affected under hot and humid conditions. Buds and young bolls affected in their stalks or basal region may undergo abscission and drop. Spots on the bracts and calyx may extend to the developing boll. On older bolls small, circular, raised, water soaked spots appear. They may enlarge, become irregular in outline and spread through the greater part of, or even, the entire b ll. o

Epidemiology

Primary, infections from seed is responsible for dissemination of disease for almost all the cotton growing areas. The irrigated cotton (*G. hirsutum*) of Punjab, Haryana, Rajasthan and western U.P. suffer heavy loss due to this disease. The pathogen is capable of surviving on the fuzz and inside the seed for couple of years (Hunter and Brinkerhoff, 1964). The fuzz can get infested while ginning. Pathogen enter interior of the seed while it is developing inside the boll through chalasa (Tennyson,

1936). Seed borne infection important in north and central India but not to the same extent in south India.

Management

Cultural Control

Destruction of infected plant parts and debris after picking. Sowing of crop at recommended time minimizes the disease while more disease occurs in late sown crop (Rao, 1957).

Chemical Control

Use of disease free seed or treated seed: Seed treatment can be done by soaking the seed in solution of fungicides (emisan 0.5 g and streptocycline 100 mg in one litre water). One litre solution is sufficient for 500 to 800 g seed. Spray the crop with streptocycline 6-8 g/acre + copper oxychloride 600–800 g/acre in 200–250 litre of water. Spray should be initiated before the onset of monsoon or by the end of June or 6 weeks after sowing. 3–4 sprays should be given at 10–12 days interval.

Cotton Leaf Curl Virus Disease

Pathogen

Gemini virus of Begomo virus group

Vector

Whitefly (*Bemisia tabaci*)

Cotton leaf curl disease, earlier known as african leaf curl of cotton was first noticed in Nigeria on *Gossypium peruvianum* and *G. vitifolia* (Farquharson, 1912). In India, the disease was first noticed on few *G. barbadense* plants in 1989 and later on *G. hirsutum* near Sriganganagar in Rajasthan in 1993 (Ajmera, 1994)

Symptoms

The initiation of the disease is characterized by small vein thickening (SVT) type symptoms on young upper leaves of plants. Dark green bead like thickening of small veins can be clearly seen by holding the leaf against sunlight and observing it from lower side. These irregular thickenings gradually extend and coalesce to form continuous reticulation of small veins. The disease is further characterized

Whitefly

Small Vein Thickening

Upward Curling

by upward curling of the leaves which occurs because of the uneven growth of veinal tissues on the abaxial side of the leaves. Veins of the leaves become thickened which are more pronounced on the underside. Affected become thick, leathery, brittle and more green than healthy leaves. Similar symptoms are also observed on bracts. Later, formation of cup shaped or leaf laminar out growth called "enation" appear on the underside of the leaf. In severe cases and in plants affected at early age, reduction of internodal length leading to stunting and reduced flowering or fruiting and ultimately reduction in yield and quality.

Epidemiology

A number of weeds and other hosts *viz.*, *Sida species*, *Abutilon indicum*, *Hibiscus rosa sinensis*, *Althea rosea*, *Phaseolus vulgaris*, pepper, tobacco, tomato, okra, hollyhock, *Physalis floridana*, *Nocotiana benthamiana*, french bean, *Sida ageratum*, china-rose *Tribulus terrestris* and *Cucumus* spp harbour this virus in the off season and serve as main source of primary inoculum for disease development on cotton. The disease transmitted from these plants to cotton crop through its carrier whitefly, *Bemisia tabaci*. Weeds and hosts from other crops are also present during the season alongside the crop on which the disease keeps on multiplying (Table 21.12).

Table 21.12: CLCuV Disease Hosts Reported from India

Name of Host	*Type of Test*	*Reference*
Sida spp, *Abutilon indicum, Hibiscus rosa sinensis, Althea rosea*	Based on visual symptoms	Singh *et al.,* 1994
Phaseolus vulgaris, pepper, tobacco, tomato	Transmission studies and ELISA	Nateshan *et al.,* 1996
Okra, hollyhock, tomato, *Physalis floridana, Nocotiana benthamiana* and french bean	Transmission studies	Radhakrishnan *et al.,* 2001
Hollyhock, *Sida ageratum,* china-rose	DNA-A probe hybridization	Sharma, 2002
Tribulus terrestris and *Cucumus* spp.	Based on the presence of DNA- A and Beta DNA	Sivalingam *et al.,* 2004
Chorchorus acutangularis, Melilotus indica and *Ageratum conyzoides*	Radiolabelled probes for DNA-A and Beta DNA	Radhakrishnan *et al.,* 2004
Tomato, tobacco, zinnia, jungli pudina, chilli, China rose, okra and *Sida alba*	CP primers	Kang *et al.,* 2004
Sida spp., *Achyranthus* spp., *Clearodeadron* spp.	CP gene amplification	Monga *et al.,* 2005

Management

Cultural Control

1. Use of CLCuV resistant varieties/hybrids. H-1117, F-1861, RS-875, RS-810, RS-2013, LHH-144, CSSH-198, CSHH-238 and CSHH-243 are resistant to cotton leaf curl disease for north zone of India (Table 13).
2. Removal and destruction of weeds and other host plants around the cotton fields, which serve as hosts of virus during off season and on which whitefly continue to survive and multiply.

Table: 13. Promising CLCuV resistant varieties/hybrids in north zone

Name of Variety/Hybrid	*Source*
H-1117	CCS HAU, Hisaar
F-1861	PAU, Ludhiana
RS-875	RAU, Sriganganagar
RS-810	RAU, Sriganganagar
RS-2013	RAU, Srignganagar
LHH-144(hybrid)	PAU, Ludhiana
CSSH-198, CSHH- 238, CSHH-243	CICR,RS, Sirsa

Chemical Control

Population of vector can be minimized by judicious use of plant protection chemicals and neem products. Cotton seed oil, *Beauvaria bassiana* and insecticides like difenthuron and triazophos are effective for the control of whitefly.

Grey mildew

Pathogen

Ramularia areola Atk

Perfect Stage

Mycosphaerella areola Ehrlich and Wolf

Symptoms

Generally, disease appears on the older leaves as the plants reach maturity, in the form of irregularly angular, pale translucent spots, 3-4 mm in diameter and with a definite and irregular margin formed by the veins of the leaf. The lesions are light to yellowish green on the upper surface. As the spots become older, the leaf tissue turn yellowish brown while a whitish frosty growth appears mainly on the under surface. Affected leaves are ultimately dropped. Lesions occur on the brackets subtending the bolls. They also appear on the cotyledons. As the leaf becomes chlorotic, the lesions turn reddish brown and defoliation takes place. Early and severe defoliation leads to premature boll opening and immature lint.

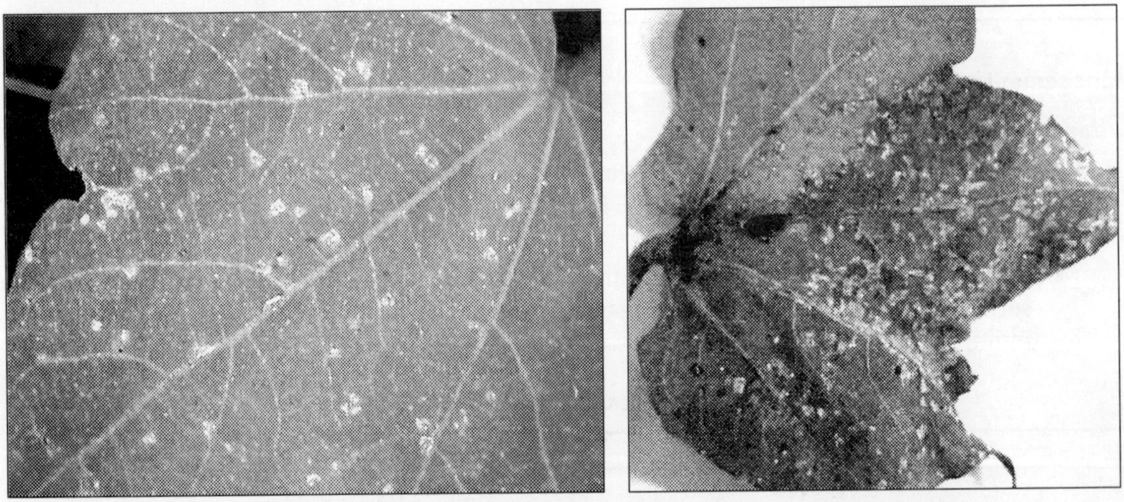

Grey Mildew Spots

Epidemiology

The disease earlier used to affect the *Gossypium arboreum* and *G. herbaceum* in India but now its attack on *G. hirsutum* and Bt cotton hybrid is quite severe and wide spread. This disease occurs in Madhya Pradesh, Maharashtra, Karnataka, Andhra Pradesh, Tamil Nadu and Punjab. Dead leaves act as primary source of inoculum for the new crop. The disease is spread by wind, running water and movement of persons and animals. The pathogen is confined in its host range to the genus *Gossypium*.

Management

Cultural Control

No appropriate method of cultural practices is efficient to control this disease.

Chemical Control

Dusting the crop with sulphur @ 16.8 to 22.4 kg per ha or spraying it with copper fungicide or spray with 0.1 per cent carbendazim 50 WP may control the disease to the great extent.

Leaf Spots

Pathogen (s)

Alternaria gossypina, A. macrospora, A. alternate, Myrothecium roridum, Helminthosporium gossypii, H. spiciferum, Colletotrichum capsici, C. gossypii, Cercospora gossypina and *Curvularia lunata*

Leaf spots caused by various fungal pathogens interfere with the photosynthesis of the plants which result in to lowering down the production of the cotton and quality of fibre. Mainly, leaf spots incited by species of *Alternaria* and *Myrothecium* are of great importance while remaining is of minor importance.

Alternaria Leaf Spot

Symptoms

Small, dull to dark brown, circular or irregular shaped spots appear on the leaves, varying in diameter from 0.5 to 10 mm. They often develop concentric ridges and present a target board appearance which is clearly visible on the upper surface. When mature they have dry, grey centers which may crack and even drop. The spots may coalesce and occupy large area of the leaf.

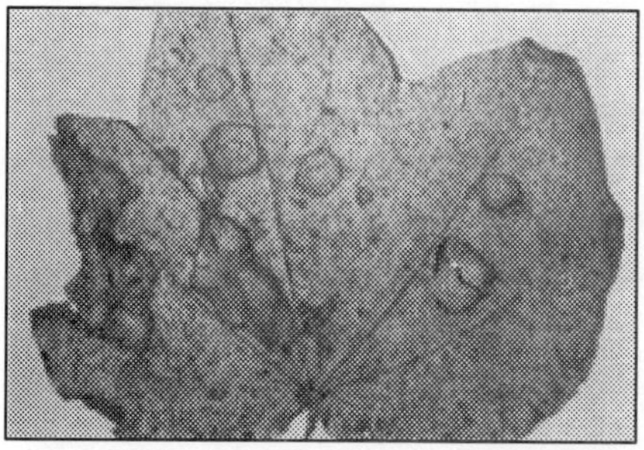

Alternaria Leaf Spot

Epidemiology

The pathogen lives on infected plant debris during off season. Padaganur (1980) has found that in infected bolls, A. macrospore becomes seed borne both internally and externally to the extent of 50 and 100 per cent, respectively.

Myrothecium Leaf Spot

Symptoms

Leaf spots circular or irregular in shape, brown in the centre with a darker brown or purple margin, 2 to 10 mm in diameter. Often the spots coalesce and form large patches. The centre may drop leaving a shot hole. There is a conspicuous ring or more than one concentric ring of sporodochia on the spots. The sporodochia may be of the size of a pin head and dark green or greenish in color. Lesions may appear on the petiole and stem. In severe attack the stem may break. Lesions may also appear on the bracts and bolls. Boll lesions can lead to damage to the lint which may become brittle and discolored.

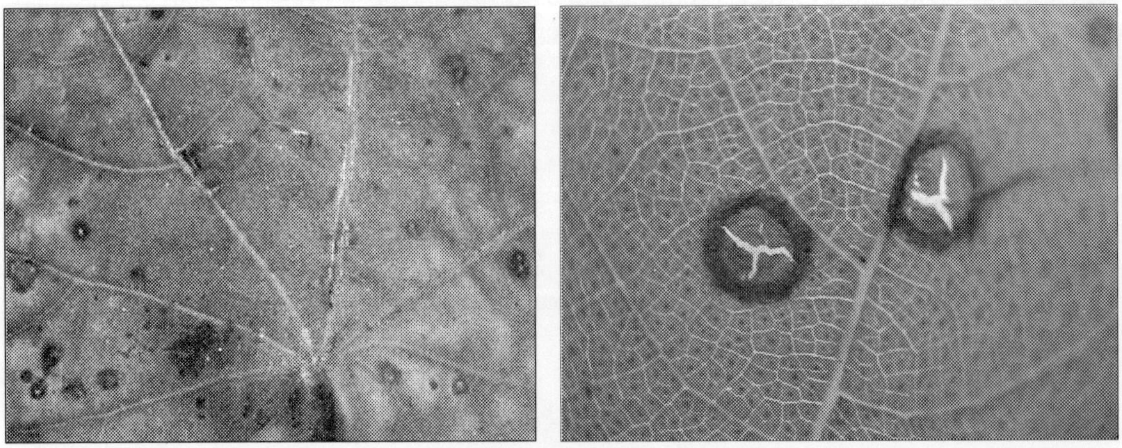

Myrothecium Leaf Spots

Myrothecium Spots on Boll

Epidemiology

The pathogen is seed borne in nature (Srinivasan and Kannan, 1974) and reduces the germination of several cotton varieties.

Management

Cultural Control

Avoidance of dense crop and paired sowing reduces the disease intensity.

Chemical Control

Seed treatment with carbendazim @ 2g/kg seed or emisan @ 0.05 per cent + streptomycin sulphate 0.001 per cent is helpful in reducing the leaf spots. Spray the crop with copper oxychloride (500 g/200 litre water) at fortnightly interval is recommended. It is advisable to start spray just starting of first shower of rains.

Root Rot

Pathogen(s)

Rhizoctonia solani Kuhn (perfect stage-*Pellicularia filamentosa* (Pat.) Rogers, known as *Corticium vagum* var. *solani* Burt ex Rolfs), *R. bataticola* (Taub.) Butler (pycnidial stage-*Macrophomina phaseoli* (Maubl.) Ashby

Symptoms

This disease occurs in circular patches. It appears either at seedling stage or at vegetative stage. In young seedlings when the stem is still soft, a yellow patch appears on the lower part of stem and if the attack is severe the seedling may collapse. In older seedlings which have developed some wood, the patch becomes blackened at ground level and the seedlings dry up and wither. Affected plants are easily pulled out of the ground. The tap root and a few secondary roots are decayed. The tips of the roots are moist, discolored yellow and sticky with soil adhering with them. The bark is shredded on the roots and basal part of the stem

Epidemiology

Maximum sclerotia of *Rhizoctonia bataticola* are found at a soil depth of 6-12 inches (15-30 cm) and decline at a depth of 15-18 inches (37.5-45 cm). Sclerotial number remains low during monsoon. It causes highest cotton plant mortality at 39 °C. This fungus is pathogenic in a range of 25 °C to 39 °C. High humidity and low temperature is unfavorable. The disease appears in June and is favored by high temperatures prevailing in June in the Punjab and declines during September. It is most severe at 28 °C and 20 per cent soil moisture in Maharastra. At Coimbatore, the disease develops well at 28 °C which prevails in September-October.

Management

Cultural Control

Successful resistance is not possible and can not be expected against polyphagous pathogen having wide host range. Higher resistance to *R. bataticola* in older seedlings than in younger ones has been attributed to higher contents of gossypol and phenols and low levels of sugars. Adjustment of sowing date and mixed cropping with *Phaseolus aconitifolius* are the simplest and most effextive methods for the control of the disease. Crop sowing in April or June instead of May also reduces the disease

Root Rot Affected Plants in Patch

Affected Root

incidence in the Punjab (Vasudeva, 1960). Heavy applications of FYM and cellulose containing composts which increases microbial activities, reduces the occurrence of the disease.

Chemical Control

Seed treatment with *Trichoderma viride* and *T. harzianum* is an effective means of control (Monga and Raj, 1996)). These antagonists establish itself in the soil and provide protection to the cotton plant for longer time.

Fusarium Wilt

Pathogen

Fusarium oxysporum f. sp. *vasinfectum* (Atk.) Snyder and Hansen.

Symptoms

Disease symptoms may appear at any growth stage. In early stage of seedling, plant mortality may be sudden and confusing to be due to damping- off. Cotyledons turn chlorotic, then yellow and brown and may drop. At this stage, vascular browning may be observed but there is no discoloration or necrosis of the roots. In older plants, though the disease may appear at any stage but it is most prominent at square and flower formation. The margins of the leaves wilt and leaves may turn chlorotic and brown, followed by epinasty and they may ultimately be shredded down. The lowest leaves show symptoms first then upper leaves. Plant may be affected partially and or completely. In partially and completely affected plants, bark of the stem turn brown. Vascular browning occurs continuously from the main stem into branches and even into leaves. The brown discoloration is confined to the vascular elements and does not spread to intervascular tissue or cortex. In heavily affected fields, the plants may be extensively defoliated exposing bare stems. Many plants die and dry up.

Wilted Crop

Epidemiology

Wilt of cotton, is believed to have originated in Maxico or Central America. In India it was first reported in Nagpur in 1908. The disease has now spread to almost all the black cotton soil tracts of the country. The pathogen causes extensive damage in the heavy soils. Pathogen is soil borne, leaving on host plant residues and later as resting structure as chlamydospore and sclerotium. It spread by irrigation and flood water and by the presence of debris containing fungal inoculum on seed. Cotton wilt is a single cycle disease and has only one disease cycle in every crop season.

Management

Cultural Control

1. Potassic fertilizers improve the resistance of the plants. Nitrogen and phosphorus are likely to increase the severity of the disease, unless balanced with appropriate potash applications. Micronutrients like zinc can mitigate the incidence and severity of the disease.
2. Resistant varieties like LD-327, LD-694 are the most effective and economical means of control of the disease.

Verticillium Wilt

Pathogen (s)

Verticillium dahliae Klebahn, *Verticillium albo-atrum*

Symptoms

Symptoms of *Verticillium* wilt at seedling stage includes cholorosis of cotyledons, yellowing and then quick drying up. However, the earliest diagnostic symptoms under south Indian conditions is a bronzing of the smaller veins of the true leaves starting from the lower ones. This is followed by an interveinal foliar chlorosis which may start from the margins, yellowing and later by a scorched appearance of the leaves. Often in the early stage, the leaves turn purple particularly on the lower surface, lose turgor, droop and become softer than healthy leaves. The interveinal yellowing and drying take on a characteristic appearance of "tiger strips". *Verticillium* wilt, in India was first observed in 1967 around Coimbatore in Tamil Nadu (Natarajan *et al.*, 1968).The disease reduces the yield, while the bolls form on the plants showing symptoms at late stage develop immature fibre which may give rise to a considerable number of neps in the yarn. It has also been reported that it affects mean fibre length, fibre weight, maturity coefficient, bundle strength, seed index and seed germination.

Epidemiology

The pathogen has poor competitive saprophytic ability and does not grow through soil. The microsclerotia are embedded in the leaves, stems and roots. Cotton roots can penetrate deep in to the soil and microsclerotia may found at depth of even 20 cm or more. After decay of the host tissue, they lie in the soil embedded in small pieces of host tissue, and forced into dormancy by soil fungistasis. They can survive in the soil for 14 years. The microsclerotia lying in the soil, enforced dormancy, germinate when the soil fungistasis is relieved by root exudates which, however, are, in general, nonspecific. This ability to germinate in the presence of non-host can be exploited for reducing the population of propagules

Management

Cultural Control

1. Cultivation of resistant cultivars (Sujata, Suvin and MCU 5 VT).
2. Growing cotton during the summer in south India gives a disease free crop and healthy seed.
3. A two–three year rotation involving lucern (*Medicago sativa*) or *Chrysanthemum* reduces wilt by 65-76 and 54-72 per cent, respectively.

Chemical Control

1. Debris adhering to the fuzz can carry microsclerotia, therefore to get rid off microsclerotia seed should be acid delinted.
2. Carban disulphide, vapam (nemagon), if applied three weeks before sowing, disease may control by 90-100 percent. Although these chemicals are toxic and expensive but benefits last for over three seasons.

Para-wilt

Pathogen (s)

Unknown or physiological disorder.

This wilt was first noticed on intra- hirsutum hybrid JKHy.1 at Adilabad in Andhra Pradesh in 1978. It is known by various names like Adilabad wilt (Srinivasan, 1984), new wilt (Raj, *et al.*, 1988), para-wilt (Mayee *et al.*, 1991) and others.

Symptoms

Affected plants show moderate stunted growth with dull green appearance, which turns yellow and then red. There is partial epinasty, drooping of leaves, without chlorosis, necrosis and loss of turgidity. Leaves fall down later, denuding the plant. Regeneration of leaves may take place and some leaves may die with 10-15 days. Bad boll opening takes place on affected plants. There is no vascular discoloration and rotting of roots has been seen. The fibre obtained from affected plants is of poor quality.

Epidemiology

Para-wilt is a physiological disorder which appears at flowering stage to fruiting stage. It does not appear to be associated with any pathogenic fungus, bacterium, virus, MLO, nematode or insect and studies for satisfying Koch's postulates have not yet been reported. Sahay *et al.* (1984) suggested that ethylene could produce similar symptoms and showed that foliar sprays of silver nitrate solution which suppresses the effect of ethylene prevented the expression of symptoms. All the four cultivated species of *Gossypium* are susceptible but varieties of *G. hirsutum* are more susceptible. Shastry (1993) reported that in the leaves of affected plants nitrogen, potassium, phosphorus, magnesium, sulphur, iron, copper and zinc content increases while calcium content decreases.

Management

Cultural Control

Cultural method of controlling this disease is not clearly known so far. Derivatives of Bikaneri narma do not affected by this disorder (Narayanan *et al.*, 1984).

Para_wilted Plants

Bad Boll Opening

Chemical Control

Spray of cobalt chloride @ 10 ppm (1g/100 litres of water) is helpful to some extent. It should be sprayed just after appearance of the disease.

References

Ajmera, B. D., 1994. Occurrence of leaf curl virus on american cotton (*G. hirsutum*) in north Rajasthan. Poster presentation, National Seminar on cotton production challenges in 21st century, April 18-20, Hisar, India.

Anonymous, 2007. Annual Rport of All India Coordinated Cotton Research Project, 2006-07 Dineen Jacqueline (1988). *Cotton and Silk.*

Brar, K. S., Sekhan, B. S., Singh, J., Snahmer, M and Bakhetia, D. R. C., 2001. Evolution of biocontrol based IPM modules for the control of bollworm complex on cotton. *J. Insect Science*, 14 (1 and 2): 19-22.

Brar, K. S., Virk, J. S., Jalali, S. K., Sohi, A. S. and Shenhmar, M., 2007. Evolution of multiple insectide tolerant strain of *Trichogramma chilonis* Ishii for the management of cotton bollworms. *J. Insect Sci.*, 20 (2): 208-211.

Dhaliwal, G. S. and Arora, R., 1993. *Changing status of insect pests and their management strategies.* In: K. S. Gill, G. S. Dhaliwal and B. S. Haursa (eds). Changing scenario of Indian Agricultures, Commonwealth Publishers, New Delhi, India, pp 98-145.

Dhaliwal, G. S. and Arora, R., 1996. *Principles of insect pest management.* Kalyani publishers, New Delhi.

Dhaliwal, G. S. and Arora, R., 2005. *Integrated Pest Management: Concepts and Approaches.* Kalyani Publications, New Delhi.

Dhaliwal, G. S., Arora, R. and Dhawan, A. K., 2004. Crop losses due to insect pests in Indian agriculture. *Indian J. Ecology*, 31 (1): 1-7.

Dhawan, A. K., 1999. Major insect pests of cotton and their integrated management. In : R. K. Upadhyay, K. G. Mukerji and O. P. Dubey (eds). *IPM System in Agriculture.* Vol. 6: *Cash Crops.* Aditya Books Pvt. Ltd., New Delhi, pp. 165-225.

Dhawan, A. K. and Simwat, G. S., 1998. Evolution of different biopesticides against cotton bollworm, *Heliverpa armigera* (Hubner). In Dhaliwal, G. S., randhawa, N. S., Arora, R. and dhawan, A. K. (Eds). *Ecological agriculture and sustainable development.* IES and CRD, Chandigarh. pp. 274-280.

Dhawan, A. K., Shera, P. S., Jindal, V. and Agarwal, N., 2008. Changing scenario of cotton insect pests and their management strategies in the Punjab. In: *Cotton Research in Punjab*, AICCIP, Ludhiana, april, 9-11, 2008. pp. 81-99.

Dhawan, A. K., Singh, J. and Sindhu, A. S., 1980. *Maconellicoccus* spp. attacking arborium cotton in Punjab. *Science and culture*, 46: 258.

Dhawan, A. K., Singh, K., Saini, S., Mohindru, B., Kaur, A., Singh, G. and Singh, S., 2007. Incidence and damage potential of mealy bug, *Phenacoccus solenopsis* Tinsley on cotton in Punjab. *Indian J. Ecol.*, 34 (2): 166-172.

Dhawan, A. K., Sithanantham, Baumgartner, J. and Rao Ranga, G. V., 2001. Evolution of nuclear polyhedrosis virus against *Helicoverpa armigera* on uplant cotton, *Gossypium hirsutum* Linnaeus. *Conference on Biopesticides: Emerging trends.* February 7-9, 2001, Chandigarh. pp 119.

Dhawan, A. K., Sohi, A. S., Sharma, M., Kumar, T. and Brar, K. S., 2004. *Insecticide Resistance Management in Cotton*. Department of Entomology, Punjab Agricultural University, Ludhiana, India. pp 56.

Dhawan, V. K., Dhaliwal, G. S. and Chelliah, S., 2000. Insecticide-induced resurgence of insect pests in crop plants. In: G. S. Dhaliwal and B. Singh (eds). *Pesticides and Environment*. Commonwealth Publishers, New Delhi, pp. 86-127.

Farquharson, C.O., 1912. A report of the mycologist. A report Agric. Deptt. Nigeria. In M.A. Siddique and L.C. Humgus (Eds). Cotton growth in Gezira environment. W. Haffer and Sons Ltd. Cambridge, England. pp 106.

Hargreares, H., 1948. *List of recorded cotton insects of the world*. Commonwealth Institute of Entomology, London. pp. 50.

Hunter, R. E. and Brinkerhoff, L. A., 1964. Longivity of *Xanthomonas malvacearum* on and in cotton seed. *Phytopathology*, 54: 617.

Kang, S. S., Athar, M., Cheema, S. S., Malathi, V. G. and Radhakrishnan, G., 2004. Quick detection of cotton leaf curl virus. *Indian Phytopathology*, 57: 45-46.

Kranthi, K. R., 2005. *Insecticide Resistance: Monitoring, Mechanisms and Management Manual*. Central Institute for Cotton Research, PB. No.2, Shankarnagar P.O. Nagpur.

Kranthi, K. R., Dhawad, C. S., Naidu, S., Mate, K., Patil, E. and Kranthi, S., 2005. Bt-cotton seed as a source of *Bacillus thuringiensis* insecticidal Cry1Ac toxin for bioassay to detect and monitor bollworm resistance to Bt-cotton. *Curr. Sci.*, 88: 796-800.

Lutrell, R. G., 1994. *Cotton pest management, part-II*. Alls perspective. A. Lev. Ent. 39: 527-542.

Mayee, C. D., Chaulwar, A. B. and Quadri, S. M., 1991. *Etiology and Epidemiology of Parawilt of Cotton*, Final Rep., Marathwada agric. Univ., Parbhani, pp. 99.

Monga, D, and Raj, S., 1996. Biological control of root rots of cotton. *J. Indian Soc. Cotton Improvement*, 21: 58-64.

Monga, D., Kumar, M. and Kumar, R., 2005. Detection of DNA-A and satellite (DNA-beta) in cotton leaf curl virus (CLCuV) infected weeds and cotton plants using PCR technique. *J. Cotton Res. Dev.*, 19: 105-108.

Narayanan, S. S., Singh, V. V. and Singh, P., 1984. *Proc. Group Discussion on New Wilt in Cotton*, Cent. Inst. Cott. Research, Nagpur, pp. 56-57.

Natarajan, M.K., Sivaprakasan, K. and Ramakrishanan, K., 1968. Record of Verticillium wilt of cotton in Madras state. *Madras agric. J.*, 55: 455-457.

Nateshan, H. M., Muniyappa, V., Swanson, M. M. and Harrison, B. D., 1996. Host range, vector relations and serological relationships of cotton leaf curl virus from southern India. *Ann. appl. Biol.*, 128: 233-244.

Padaganur, G. M., 1980. *Alternaria macrospore* Zimm. cause of Alternaria blight of cotton becomes seed borne. *J. Indian Soc. Cott. Improve.*, 5: 32-33.

Radhakrishnan, G., Malathi, V. G. and Verma, A., 2004. Detection of DNA–A and DNA beta associated with cotton leaf curl and some other plant diseases caused by white fly transmitted Gemini viruses. *Indian Phytopathology*, 57: 53-60.

Radhakrishnan, G., Malathi, V. G. and Verma, A., 2001. Novel features of cotton leaf curl citrus disease in India. *In 3rd international Gemini virus Symposium*. July 24-28, 2001, John Irr. Centre, Narwich, Noafolk, U.K., pp. 53.

Raj, S., Chakraborty, P. K. and Meshram, M. K., 1988. *Proc. Biennial Workshop, All India Coord. Cott. Improvement Project*, Hyderabad, 1987.

Rao, T. A., 1957. Comparative morphology and ontogeny of foliar sclereids in seed plants–1. Memecylon L. *Phytomorphology*, 7: 306-330.

Sahay, R. K., Pundarikakshudu, R., and Gawande, M. C., 1984. *Proc. Discussion Group on New Wilt in Cotton*, Cent. Inst. Cott. Research, Nagpur, pp. 33-34.

Sharma, P., 2002. Molecular approaches for detection and diagnosis of cotton leaf curl Gemini virus and its mode of disseminationin the field. Ph.D. thesis, Department of Plant Pathology, College of Agriculture, CCS HAU, Hisar.

Shastry, P. P., 1993. *Studies on the New Wilt of Cotton*, Doctoral Thesis, JNKVV, Jabalpur, India, pp. 124.

Simwat, G. S., 1994. *Modern concepts in insect pest management in cotton*. In: G. S. Dhaliwal and Ramesh Arora (eds.) *Trends in Agricultural Insect Pest Management*. Commonwealth Publishers, New Delhi, pp. 186-237.

Singh, J., Mann, H. S and Kapoor S. P., 1994. Studies on whitefly *Bemisia tabaci* (Genn.) transmitted cotton leaf curl virus disease in Punjab. *J. insect Sci.*, 7: 194-98.

Singh, T. H. and Khush, G. S., 1993. *Resume of cotton research in Punjab silver jubilee 1967-1992*. All India coordinated cotton improvement project (ICAR), Punjab Agricultural University, Ludhiana.

Sivalingam, P. N., Padmalatha, K. V., Mandal, B., Monga, D., Ajmera, B. D. and Malathi, V. G., 2004. Detection of begomoviruses in weeds and crop plants in and around cotton field infected with cotton leaf curl disease. Poster presented in National symposium on Crop surveillance: Disease forecasting and management held at IARI, New Delhi, February 19-21,2004. pp. 36.

Srinivasan, K. V., 1984. Proc. *Group Discussion on New Wilt in Cotton*, Cent. Inst. Cott. Research, Nagpur, pp. 9-13.

Srinivasan, K. V. and Kannan, A., 1974. Myrothecium and Alternaria Leaf Spots of Cotton in South India *Curr. Sci.*, 43: 489-490.

Tennyson, G., 1936. Invasion of cotton seed by *Bacterium malvacearum*. *Phytopathology*, 26: 1083-1085.

Vasudeva, R. S., 1960. Diseases. In R. H. Dastur, R.D. Asana, K. Sawhney, S. M. Sikka, R. S. Vasudeva, Q. Khan, V. P. Rao and B. L. Sethi (eds), *Cotton in India, A monograph*, Vol. II, pp. 164-216.

Yadav, M. S., 2008. Disease reaction of newly introduced Bt transgenic cotton hybrids in south–west region of Punjab. *Indian Phytopathology*, 61 (2): 141-145.

Chapter 22

Rodent Pest Mangement in Arid Land Farming System

R.S. Tripathi

Central Arid Zone Research Institute, Jodhpur – 342 003, Rajasthan

Introduction

The hot arid zone forms the eastern side of Thar Desert in India and is one of the most densely populated deserts of the world. Human and livestock population is consistently increasing. During the last 20 years, human population has increased by 50 per cent and that of livestock by 25 per cent. Such an increase in population is leading to greater exploitation of natural resources, which is otherwise quite meager in the region. Agriculture and Animal husbandry are the two main occupations of desert people. To meet the food and fodder requirements, farmers have adopted rainfed farming, development of range/gochar lands, arid horticulture and tree plantations. Thus, arable farming, horticulture and management of range/gochar lands for livestock production, and Silviculture/forestry provide sustainability to the region and form the back bone of rural economy. Arid zone occupying nearly 3,17,090 sq.km. area covers both cold arid in Jammu and Kashmir State and hot arid regions in north wester parts of India Most of hot arid region (about 62 per cent) falls in 12 districts of western Rajasthan only followed by Gujarat (19.6 per cent), Haryana (4.0 per cent), Punjab (4.6 per cent), Andhra Pradesh (6.9 per cent) Karnataka (2.7 per cent) and Maharastra (0.4 per cent) (Rao and Singh, 1998). The ecosystem is hyperthermic and is characterized by low and erratic rainfall, poor soil health and frequent droughts. With the canalization of river waters and large-scale exploration of ground water irrigated farming is also practiced in some parts. Besides the in built natural vagaries in arid agro-ecosystems, the arid crops and other productive vegetations are always under threat by native rodent, which are primarily herbivorous and constitute one of the largest mammalian group in arid region. Rodents aggravate the problem of desertification by depredating upon the sparse vegetation and disturbing the soil through their extensive fossorial activities.

Rodent Diversity in Arid Zone

Rodents form a very diverse group of mammals in arid regions comprising porcupines, squirrels, rats, gerbils, mice, mole rats etc. Extensive surveys in the region revealed occurrence of 18 rodent species belonging to three families *viz.*, Sciuridae, Hystricidae and Muridae. Of these, former two families are monotypic *i.e.*, represented by one species only (Tripathi *et al.*, 1992). The species biodiversity in the region is presented in Table 22.1.

Table 22.1: Rodent Species Occurring in Arid Zone (Tripathi *et al.*, 1992)

Family/Sub Family	*Species*
(a) Family Hystricidae :	(1) *Hystrix indica* Kerr
(b) Family Sciuridae :	(2) *Funambulus pennanti* Wroughton
(c) Family Muridae	
(i) Sub-family Gerbillinae	(3) *Tatera indica* (Hardwicke)
	(4) *Meriones hurrianae* (Jerdon)
	(5) *Gerbillus nanus indus* Thomas
	(6) *Gerbillus gleadowi* Murray
(ii) Sub-family-Murinae	(7) *Vandeleuria oleracea* (Bennett)
	(8) *Rattus rattus* (Linn.)
	(9) *R. cutchicus* (Wroughton)
	(10) R. (=*Millardia*) *meltada pallidior* (Gray)
	(11) *R. gleadowi* Gray
	(12) *Mus musculus* Linn.
	(13) *M. cervicolor* Hodgson
	(14) *M. platythrix* Bennett
	(15) *M. booduga* Gray
	(16) *Golunda ellioti* Gray
	(17) *Nesokia indica* (Gray and Hardwicke)
	(18) *Bandicota bengalensis* (Gray)

Distribution of rodent fauna in various habitats of desert biome of Rajasthan has been well documented by Prakash (1975) and Tripathi *et al.* (1992). Four types of rodents' habitats (sandy, rocky, gravel and ruderal) in Thar Desert have been identified by Prakash, 1975. Most of the species are found in more than one habitat, but a few *viz.*, *Rattus cutchicus* and *Mus cervicolor* are found in rocky habitat and *Gerbillus nanus* occurs in sandy habitat only. Similarly *M. musculus* and *M. booduga* inhabit ruderal habitats *i.e.* crop fields and/or residential areas. Other species *viz.*, *Funambulus pennanti*, *Meriones hurrianae*, *Tatera indica*, *Millardia meltada*, *Golunda ellioti*, *Gerbillus gleadowi*, *Mus platythrix* etc are abundant in sandy to ruderal habitats. Following sequence of rodent abundance in different habitats has been recognized;

- ✰ Sandy habitat: *M. hurrianae* > *T. indica* > *G. gleadowi* > *M. meltada*
- ✰ Gravel habitat: *M. hurrianae* > *T. i. indica* > *M. platythrix sadhu.*
- ✰ Rocky habitat: *R. c. cutchicus* > *M. p. sadhu* > *M.cervicolor i*>*F. pennanti.*
- ✰ Ruderal habitat: *T. indica* >*R. meltada*> *M.hurrianae*>*G.gleadowi*> *F. pennanti*

Why Rodents are a Serious Problem

Pest status of any species is a combination of cropping pattern of the region *vis-à-vis* eco-biology of rodent species. Generally crop fields are homogeneous and man made ecosystem with a plant community dominated by one or two crop species. Most of the rodent species of the region breed round the year with certain peak periods in breeding, commonly referred as breeding seasons, which largely depends on availability of nutritious food in plenty. The desert region show rich floral diversity, particularly in monsoon season (July-September) with lush green vegetation all over containing an inexhaustible source of foraging material for the very large number of herbivores, domestic as well as wild ones including rodents. Availability of green food in plenty during this season triggers the breeding activity of native rodents. In winters very limited area is under cropping due to limitations of irrigation facilities therefore the rodents are observed congregated in and around crop fields. This leads to extensive damage to standing *rabi* crops. This period also records increased breeding activity in certain rodent species. Summers are really a tough time for desert rodents when they thrive on roots, rhizomes of desert vegetation and their breeding activity attains lowest ebb (Tripathi 2005). Therefore the desert rodents are adopted to take advantage of the temporarily abundant food and at the same time overcome the crisis periods make rodents' a good candidate as agricultural pests.

Rodents are characterized by their "chisel" shaped ever growing incisors, which is a real weapon of destruction. It grows at a rate of 0.4 mm a day. In order to maintain the size of the incisors they gnaw and nibble whatever hard substance they find in their habitat causing extensive damage to fields crops, stored grains and several other household articles. Besides this, their fast breeding potential makes them to maintain high population levels in any habitat. Some characteristics features of breeding biology of rodents is given in Table 22.2.

Table 22.2: Breeding Biology of Rodents

Sl.No.	Parameters	Period
1.	Age at puberty	6-16 weeks
2	Estrous cycle	3-7 days
3.	Duration of heat	9-24 hours
4.	Mating habit	Promiscuous
5.	Gestation period	18-30 days
6.	Breeding season	year round
7.	Litter size	1-22
8.	Post partum heat	4-96 hours
9.	Breeding potential	800-1200/pair/year
10	Annual Productivity in arid zone	20-52/female/year

Cropping Patterns and Rodent Pests

The most predominant cropping pattern of the arid region in rajasthan are: (*i*) pearl millet-green gram-moth bean-sesame-cluster bean (as rainfed) and cotton and groundnut (irrigated) in *kharif* season and (*ii*) wheat-mustard-cumin-chickpea, (irrigated) in *rabi* season. Among vegetable crops, solanacious vegetables (tomato, brinjal and chillies) and cucurbits are most commonly grown in the region. Jujube is the most important fruit crop of the region, however, in recent years farmers have also started growing pomegranate, kinnow, amla, datepalm etc. underlimited irrigation. Generally the agro-economy of the arid region is mainly livestock based farming system. Thus a vast area is under perennial forage grasses or gochar lands. Major grass species in the region are *Cenchrus ciliaris, C. setigerus* and *Lasiurus sindicus*. Besides these grasses, foliage of several native bushes and trees (*Ziziphus nummularia* and *Prosopis cineraria*) are used as good source of animal feed. Besides, ber, pomegranate, date palm etc) are major arid horticultural crops. All these cropping systems are vulnerable to rodents attack which inflict serious pre harvest losses.

Of the eighteen species, only 7-8 species may be considered as pests and a complex of 3-4 species are reported to inflict economic damage in any particular cropping system/habitat. Both the gerbils, *viz., M. hurrianae* and *T. indica* are regarded as most predominant pests in rain fed as well as irrigated arable cropping systems and grasslands. Besides these, *R. meltada and M. booduga* prefer to inhabit irrigated croplands and *G. gleadowi* damages rainfed crops grown on dunes/interdunal spaces (Rana and Tripathi, 1999). The squirrels, *F. pennanti* occupy the trees scattered in the crop fields and *G. ellioti* prefers the thickets of the bushes commonly found on the periphery of the crop fields. In Indira Gandhi Canal command areas of Sriganganagar and Bikaner Districts, the crop fields are mainly infested with *T. indica, R. meltada, M. booduga, B. bengalensis*, and *Nesokia indica*. Since rodents are omnivorous. The may not show specific preference for any crop, however a complex of 3-4 rodent species are reported to have some association with cropping systems in arid ecosystem (Table 22.3)

Table 22.3: Pest Species Composition in Dry Land Cropping Systems

Cropping systems	*Rodent Pest Species*
Rain fed crops	M. hurrianae–T. indica–G. gleadowi
Irrigated crops (arid region)	*T. indica–M. hurrianae–M. meltada*
IGNP areas (arid region)	*T. indica–M. meltada–N. indica–Mus sp.*
Semiarid areas	*B.bengalensis–T. indica–M.meltada*
Grasslands	*M. hurrianae–G. gleadowi–T. indica*
Horticultural crops	*F. pennanti–M. hurrianae–T. indica*
Forest/silva fields	*M. hurrianae–T. indica–M. meltada–N. indica* (sandy areas)
Forest and fruit nurseries	*F. pennanti–T. indica–M. hurrianae*

The crop fields under farming systems with perrinneal components like, grasses, trees or fruit crops, besides fossorial field rodents, the field is also inhabited by another arboreal rodent *i.e.*, squirrel, *Funambulus pennanti* which is an additional menace for the system. therefore arid horticultural system harbours maximum 40 per cent rodent abundance as compared to silvipature or crop fields sharing around 30 per cent abundance.

Major Rodent Pest Species

The Five Striped or Northern Palm Squirrel, *Funambulus pennanti* (Family Sciuridae)

It is a medium sized rodent (90g) with a bushy tail. The dorsal side is greyish brown with five distinctly white stripes separated by four off white bands. It is diurnal and peaks of activity occur in the morning and early evenings. The squirrel nests in holes in tree trunks and crevices in the walls of buildings, windowsills, compounds and is pest of vegetable and fruit crops in nurseries fields and kitchen gardens. It breeds round the year with peaks during March–April and July-September in Rajasthan with a litter size of 1-5 (Plate 22.1).

The Indian Gerbil, *Tatera indica* (Family Muridae, Subfamily Gerbilliane)

It is a medium sized rodent weighing 70 to 150g. Its tail is covered with hair, which terminates at the tip as a tuft, a characteristic feature of all gerbils. The hind feet are longer than fore feet. Eyes are large and ears round. It is a nocturnal animal and inhabits open plains, loose sandy soils of the desert, and is usually found at the edges of cultivation. Burrows are dug near hedges, thickets or under bushes, sometimes inside the field also when conditions are dry. *T.indica* breeds throughout the year in arid Rajasthan with maximum littering in the month of August and a minor peak in February with a litter sizi of 1-9 (Plate 22.1).

The Desert Gerbil, *Meriones hurrianae* (Family Muridae, Sub family Gerbillinae)

It is a true desertic fauna and is is restricted to arid tracts of Rajasthan and adjoining regions of Haryana, Punjab and North West Gujarat. In Rajasthan it is widely distributed in sandy areas. These are and well adopted to survive the extremes of arid climates. Its body colour is sandy grey to brownish grey dorsally and white to off-white ventrally. Tail is pale with dark brown tussle of hair at the tip. These gerbils are diurnal and inhabit cropfields, grasslands, waste lands, thorny forests etc.They are far more gregarious than Indian gerbils but live in smaller colonies. Females breed throughout the year with two peaks in February and July but a third peak is also observed during September-November (Plate 22.1).

The Hairy Footed Gerbil, *Gerbillus gleadowii* and *G. nanus indus* (Family, Muridae, Sub family Gerbillinae)

Both the species of Genus *Gerrbillus* occurring in Rajasthan are true xeric fauna and therefore occur only in extreme desertic tracts in Jaisalmer, Jodhpur, Barmer, Bikaner, Churu, Sikar and Jhunjhunu districts in Rajasthan. These are small sized rodents and inhabit dunny, hummocky or interdunal plains in sandy habitats. Both the species have two breeding seasons in a year *i.e.*, winter and summer months. It is regarded as pest of rainfed crops (Plate 22.1).

The Short Tailed Mole Rat, *Nesokia indica* (Family Muridae Sub family Murinae)

The species is relatively large sized (>200g) with heavily built body, which is grayish brown on the dorsum and has a lighter ventral side. Very short tail of the animal and soil molehills present near its burrow openings gives it the name as short tailed mole rat. It prefers bunds in cultivated fields along water channels but also occurs in natural vegetationIt is a very shy rodent with nocturnal and fossorial habits. In orchards and afforestation areas in Rajasthan its burrow openings plugged with excavated soil mole hills can be seen near base of fruit or forestry trees and all the openings are interlinked under ground. Although it breeds throughout the year under laboratory conditions, in nature breeding occurs only during winter with a litter size of 1-6 (Plate 22.1).

The Soft Turred Field Rat, *Millardia meltada palliodor* (Family Muridae Sub family Murinae)

It is a medium sized (weighs 40-60g) field rodent. Fur is very soft. Body color light to dark grey dorsally with foot and belly being off white. Tail is naked with annular rings similar to body color with dark grey above and pale below. It is distributed through out Rajasthan in irrigated crop fields, scrub grassland, sandy plains and is therefore regarded as sub-mesic rodent fauna. In Rajasthan this

Plate 22.1: Common Rodent Pest Species

Tatera indica

Meriones hurrinae

Mellardia meltada

Contd...

Plate 22.1–Contd...

Nesokia indica

Bandicoota bengalensis

Funambulus pennanti

species breeds throughout the year with peak reproduction occurring in monsoon and spring season (Plate 22.1).

The Indian Field Mouse, *Mus booduga* (Family Muridae Sub family Murinae)

Mus booduga is a tiny mouse weighing 10-12g. Dorsal fur on the body including tail varies in colour from pale sandy or dark brown to greyish and the ventral side is pure white, thus the tail looks bicolored. It is a ruderal species and occurs in and around irrigated crop fields throughout India. It breeds throughout the year except during very cold months.

The Lesser Bandicoot Rat, *Bandicota bengalensis* (Family Muridae, Subfamily Murinae)

The lesser bandicoot is robust with a round head and a broad muzzle weighing 200-350g. The body is covered with coarse fur, which forms black-tipped piles on the dorsal side. Tail is completely dark and paler below occasionally. Belly grey or light grey and rarely whitish. It is the No one rodent pest of the country, but being a mesic species isnot reported in extreme desert regions. However, due to changed land use pattern and urbanization it is being reported in irrigated croplands and in urban locales of Jodhpur and Bikaner cities. Breeding occurs throughout the year. Definite peaks occur during different seasons across the country. It is a fossorial rodent and burrows are elaborate with several openings characteristically covered with a heap of dug out soil. Considerable amount of grain is hoarded inside the burrows.

The House Rat, *Rattus rattus* (Family Muridae, Subfamily, Murinae)

It is a medium sized rat weighing 100-150g. It is the most abundant and widely distributed rodent species in the world. *Rattus rattus* is characterised by long tail, slender body and pointed snout as well as the belly. The dorsal fur is mostly blackish in commensal forms. It is a commensal species living in houses, godowns, stores, poultry farms, crop fields, adjacent to villages and causes seriouslosses to stored commodities. It breeds round the year.

The House Mouse, *Mus musculus* (Family Muridae)

The house mouse, *Mus musculus* is a tiny animal weighing 15-20g. Tail is naked and longer than body length. Dorsally, the color varies from brown to light brown with belly being whitish to light gray. It is mainly a commensal species but in recent years it is being reported from fields also. The mouse also breeds round the year with a litter size between 1-8.

Losses Due to Rodents

The rodent population being high in the desert and their demand for food being insatiable, these pests maintain an appreciable pressure on the desert croplands, grasslands and afforestation plantations causing serious losses (Table 22.4 and Plate 22.2). A single individual may typically cause damage to several plants in one night. Pearl millet, the staple food of desert people suffers to the tune of 2-5 per cent rodent damage. The crop experiences maximum loss when harvested crop is heaped in the fields and the gerbils follow the cobs by digging tunnels under the heap and feed on the cobs. The arid pulses *viz.*, moong and moth bean too experience around 3 per cent pod damage (Tripathi *et al.*, 2004). In chillies these pests damaged 15-20 kg fruits per ha at a population density of 142 burrows/ha. Similarly wheat crop experienced 18.7-21.3 per cent damage due to rodents Advani, *et al.* (1982). Mustard crop suffers about 22.9-43.5 per cent reduction in plant stand in the peripheral

Plate 22.2: Rodent Problem in Arid Crops

Rodent Damage in Pulse Crop

Rodent Damage in Groundnut

Rodent Damage in Cumin

Rodent Damage in Sesame

Contd...

Plate 22.2–Contd...

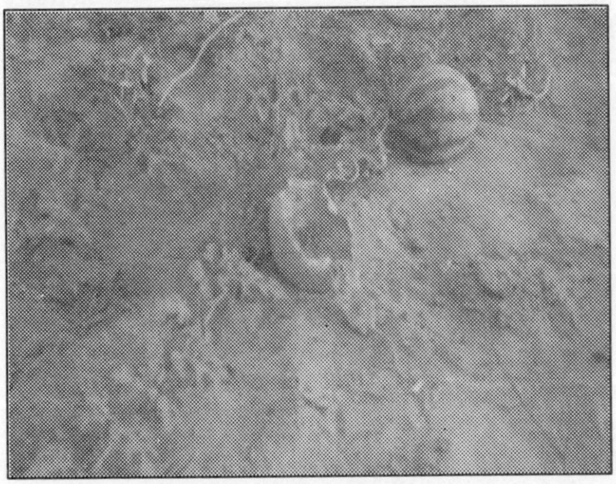

Rodent Damage in Cucurbits

Rodent Problem in Fruit Orchards

areas of the crop fields. Groundnut cultivation has suffered a major setback due to rodent devastations (30-50 per cent) during the last 4-5 years.

Table 22.4: Magnitude of Losses Caused by Rodents in Arid Agro-ecosystems

Crops	*Losses*
Wheat	11.1-11.9 per cent
Pearl millet	107.7 kg/ha
Groundnut	4.1-25.8 per cent
Bengal gram	2.5 per cent
Vegetables	8.7-10.0 per cent
Pomegranate	17-22 per cent
Date palm	18-20 per cent
Grasses	1040 kg/ha
Tree plantations	4.0-10 per cent

Among vegetables *viz.*, tomato, carrot and raddish were reported to experience 15.8-29.9 per cent damage by rodents (Advani and Mathur1982). Cucurbits recorded 4.1 per cent fruit damage resulting in a loss in yield to the tune of 6.5 q/ha. Similarly, the fruit crops *viz.*, ber, rodents also extensively damage pomegranate and date palm. In orchards *F. pennanti* is the predominant pest whereas in nurseries, a complex of *F. pennanti, T. indica, M. hurrianae* and *R. rattus* are quite commonly encountered. Squirrels were reported to cause a loss of 29 per cent of ripe fruits of pomegranate (Patel *et al.*, 1995). Studies on susceptibility of different pomegranate cultivars against squirrel attack revealed that Jalore Seedless and Khog cultivars experience least squirrel damage. Similarly, date palm at ripening stage suffers 60-80 per cent squirrel damage.

In the rangelands fodder grasses are damaged by gerbils causing immense losses to these perennial cropping system. They dig out the seeds of *Cenchrus ciliaris, C. setigerus* and *Lasiurus sindicus* and feed on them almost to the roots of the fodder. Similarly in afforestation plantations, particularly young saplings, two types of rodent attack *i.e.* (i) debarking and (ii) slicing are noticed. The debarking activity is done mainly by *M. hurrianae, T. indica and R. meltada.* Observations indicated that these rodents debarked *Acacia tortilis* and *Parkinsonia aculeata* plantation up to a height of 13.6 and 11.6 cm, respectively. Slicing activity by *Nesokia indica* from underground was particularly responsible for 4.4, 10.0 and 10.0 per cent dfamage in *A. tortilis, Prosopis juliflora, Acacia nilotica,* respectively (Tripathi and Jain, 1990).

Principles of Rodent Management

The primary and basic objective of a rodent pest management is to suppress rodent numbers to a very low level on a sustainable basis with out any effect on non-targets and the environment. Rodent Management therefore is applied ecology where the principles of population growth, limitations and regulation are utilized to maintain minimal numbers of a pest population. Normally the population fluctuates seasonally with alternating periods of high and low density, active and inactive periods, breeding and lean periods as well as periodical dispersals. Several environmental factors of the habitat like food, shelter, predators, disease and inter and intra specific competition limit the size of rodent populations. Since these factors change with seasons and locations managing rodent pests becomes a highly specialized strategy. Manipulation of all the factors that provide food, shelter and protection from predators, a integration of different techniques viz,. habitat manipulation, mechanical, physical and chemical methods yield better results (Jain, and Tripathi, 2000).

Most of the rodent pest species breed round the year but register peak breeding activities for some periods (Table 22.5). The seasonal fluctuations of rodent population in Indian desert revealed that *F. pennanti* shows gradual decrease of its population from April to October (Prakash and Kametkar, 1969) whereas *T. indica,* registered with two minor peaks in the population, the first one during March-April and the other during August-September (Jain, 1970) The desert gerbil, *M. hur*rianae has been reported to show a population buildup during winters, which continues till spring and then their number declines during summer (Prakash *et al.,* 1971). An overall comparison of seasonal fluctuations of predominant species revealed that (i) their number is lowest during the hot summer months due to hostile living conditions in the desert and (ii) the population buildup is during spring and soon after the monsoon. These two population build up periods coincide with that of highest breeding activity among desert rodents (Prakash and Mathur, 1987). Bimodal breeding patterns and corresponding increase in rodent population coincides with the cropping patterns of the region and exhibits serious economic concerns. The young ones borne during both the peak breeding periods (monsoon and spring seasons) become mature when the respective *kharif* and *rabi* crops are at flowering/ maturity stage which is considered as highly vulnerable stage to rodent attack in large numbers.

Since rodents are highly evolved mammals and its long association with the wisest mammal *i.e.,* man, has made them wiser than man by learning various tactics to avoid many harmful activities performed by man against them. Trap avoidance, bait shyness/aversion to acute poisons, resistance/ cross resistance to anticoagulants are only few examples of rodents behaviour, which very often questions the man's endeavors' for their control. Therefore an IPM approach based on sound eco-biology and ethology of pest species vis-à-vis population reduction technologies having economic viability and sociological acceptance needs to be evolved. Thus IPM is a bio-ecological system and is therefore dynamic one.

Table 22.5: Peak Breeding Season and Litter Size of Rodent Pests (Adapted from Tripathi *et al.*, 1992)

Rodent Species	*Peak Breeding Season*	*Litter Size*
F. pennanti	March to September March-April and July-September	1-5
G. gleadowi	May, June and October to January	2-5 (summer) 5-6 (winter)
T. indica	February, July- August and November	1-9
M. hurrianae	February to April	2-7
	July and September to November	2-7
R. meltada pallidor	Monsoon and Spring	3-9
M. musculus	All the year round	1-8
M. booduga	September-October and February and June	6-13
N. indica	January-March and August-October	2-5

Rodent Management Techniques

Habitat Manipulation

Habitat includes all aspects of physical environment including food. Destruction or disturbance of habitat significantly affects the population. This technique is primarily based on ecological concepts wherein the rodents' habitat is manipulated in such a way that creates stress among native rodent pests. The methods are low cost treatments and involve little modification in crop husbandry practices, like deep ploughing, removal of wild vegetation and refuge of previous crops, reduction in bund size etc. It helps in migration of pest rodents from crop fields and falling prey to predators.

Mechanical Control

Mechanical removal of rodent population from any habitat is mainly done by use of different types of traps. Trapping can no doubt provide, the information on species composition and density of pest population, hence can be successfully used for survey and monitoring purpose In indoor habitats, this method is quite effective, however for fields this method may be used as a follow up measure after bringing down the rodent pest population by poison baiting. The traps may be of single (sherman/ snap traps) or multiple catch (wonder/glue traps) type (Jain *et al.*, 1992 and Jain *et al.*, 1994). Rodents use visual and olfactory cues towards traps, therefore camouflaging the traps with bushes, grasses etc. in fields enhances the trap success considerably (Jain and Tripathi, 2000)

Use of Predators

Several vertebrates, mainly birds and mammals are listed as natural predators of rodents in literature (Deoras, 1964; Prakash and Mathur, 1987, Tripathi *et al.*, 1992; Neelanarayanan, 1997; Chaudhary *et al.*, 2001). However cats in domestic situation and snakes, owls, mongoose and varanids are predominant vertebrates predators of rodents. Studies on barn owl as potential bio control agent of bandicoots revealed a predation rate of 1-6 (Av 1.58) rodents/night. *B. bengalensis* (40 per cent) and *M. musculus* (33 per cent) constituted the major prey items of barn owls (Neelanarayanan, 1997). Success rate of biological control of rodents through other predators is not very encouraging because rodents do not constitute the sole diet of most predators and exert relatively lower predatory pressure as compared to faster turnover of rodents.

Rodenticides

Present day rodent management technology is mainly dependent on use of rodenticides because it is expedient and humane method to control pest rodents. Rodenticides have greater scope in large-scale control operations, since mixed population of several species are encountered in fields. Zinc phosphide is most widely used rodenticide in India. It is recommended at 2 per cent concentration in cereal baits. It yields around 60 per cent control success. It is highly toxic to a variety of rodent species. Major limitation in its frequent use is its high toxicity to non-target species and development of bait shyness/poison aversion in the target species after sub-lethal consumption (Prakash and Mathur, 1987). Bait shyness may persist for more than 2-3 months in different rodent species (Table 22.6). Its bait can be prepared by mixing the poison (2 per cent) in oil smeared cereals *viz.*, pearl millet, wheat, jowar, etc. For controlling the residual rodents (surviving after zinc phosphide treatment), another rodenticide *viz.*, aluminium phosphide (a fumigant) or second-generation anticoagulant rodenticide, bromadiolone (0.005 per cent) in baits has been advocated (Rana and Tripathi, 1999).

Table 22.6: Persistence of Zinc Phosphide-Induced Bait Shyness among Desert Rodents

Rodent Species	*Persistence of Bait Shyness (Days)*
F. pennanti	30
G. gleadowi	10-15
T. indica	115
M. hurrianae	35
R. rattus	75
R. meltada	135
Mus musculus	20
Mus booduga	95

Among second generation anticoagulants several products *viz.*, bromadiolone (Anonymous, 1986), brodifacoum (Jain and Tripathi, 1988), flocoumafen (Jain *et al.*, 1992) and difethialone (Chaudhary *et al.*, 2002 and Chaudhary and Tripathi, 2003) have shown their potency against Indian rodent species (Table 22.7), however only bromadiolone (0.005 per cent) is registered with Central Insecticide Board since 1988 for rodent control in fields and also in commensal situations (Rao, 2003). Coumatetralyl, also an anticoagulant is used in cereal mixed baits at 0.0375 per cent for indoor habitats (Rana and Tripathi, 1999). The second-generation anticoagulants are effective as single dose rodenticide and have very potent antidote in vitamin K_1 (Tripathi and Chaudhary 2006) Moreover they are highly effective at a very small concentration in baits (0.0025-0.005 per cent), thus are considered relatively safer to non-targets. They can be used for rodent management not only in fields but also in domestic situations, stores, godowns, poultry farm etc. (Anonymous, 1986 and Parshad, 1999).

Rodenticial Applications

Rodent pest management is mainly depends on the use of rodenticides, timing and placement of poison baits plays large scale rodenticidal baiting is advocated during summer months, when most of the rodent species maintains lower population and the acceptability of poison bait is very high due to less/non-availability of green food. In fact the acceptance of the technology by farming community is also quite encouraging as they are relatively free during summers and can undertake rodent control campaign on cooperative basis. Similarly second operation should be undertaken in October-November *i.e.*, before sowing of rabi crops. Two methods of bait placement is recommended for fields *i.e.* burrow baiting and use of bait stations. The former method provides better opportunity to access fresh baits by target rodents and protects the non-targets from its exposure.

Table 22.7: Single Dose Toxicity of Grain Baits of Second Generation Anticoagulant Rodenticides Against Common Rodent Pests

Rodent Species	*Per cent Mortality and (Mean days to death) with*			
	Bromadiolone (0.005 per cent)	*Brodifacoum (0.005 per cent)*	*Flocoumafen (0.005 per cent)*	*Difethialone (0.0025 per cent)*
B. bengalensis	100	100 (6.6)	100	100 (5.8)
T. indica	100 (9.5)	90 (6.9)	90 (8.2)	100 (7.9)
R. meltada	100 (6.4)	80-100 (6.6-7.1)	100 (5.7)	100
M. hurrinae	88	83-100 (4.7-8.0)	100 (7.0)	100 (5.9)
R. rattus	100 (7.5)	83-100 (8.0-10.8)	100 (8.6))	100 (5.9)
M. musculus	91 (8.1)	60-100 (4.6-8.4)	100 (6.8)	-
M. booduga	83	83 (9.4)	-	-
F. pennanti	-	66 (7.0)	100 (8.1)	100 (6.1)

Bait Preparations

It is an important aspect of rodenticidal application technique, which is often overlooked. The proportion of toxicant to bait should be maintained properly. Too heavy dosages may repel the pest and with too light dosages rodents may stop eating before consuming the lethal dose resulting into development of bait shyness. It is, therefore, important that the toxicant is uniformally distributed through the bait mixture rather than left as distasteful clumps. Since most of the rodneticides are available in powder from, the oil component of the bait is most important. AICRP, on Rodent Control Jodhpur has developed a very effective, easy and economic technique of bait preparation, which is as follows:

For Zinc Phosphide (2.0 per cent)

(*a*) *Pre-bait material :* (for one kg of bait)

- (*i*) Take 960g of locally grown food grains (wheat/bajra/rice/ragi/jowar).
- (*ii*) Mix 20g vegetable oil in food grain with bare hands.

(*b*) *Poison bait :* for one kg bait.

- (*i*) Prepare pre-bait as suggested above.
- (*ii*) Sprinkle 20g of zinc phoshpide and stir with wooden stick till uniform mixing is achieved. (no house hold utensil be used for this purpose).
- (*i*) Use any plant leaf as applicator for bait placement inside the burrows.

For Bromadiolone (0.005 per cent)

1. Take 960 g food grain of the area (Rice/wheat/ragi/bajra).
2. Mix 20 g vegetable oil.
3. Sprinkle 20 g bromadiolone concentrate powder (0.25 per cent).
4. Mix these ingredients in a container with wooden stick(no house hold utensil be used for this purpose).
5. Use any plant leaf as applicator for bait placement inside the burrows.

Burrow Fumigation

Aluminium phosphide, a burrow fumigant is available in tablet and pellet form. For fumigation, all the existing burrow openings are plugged with wet mud and 1.5 g aluminium phosphide is to be inserted in the active burrows, which should also be plugged with mud to check the escape of lethal gas. The fumigant is recommended after zinc phosphide baitng to control the residual rodent population. The fumigant is more effective in humid zone and irrigated fields with heavy soil.

Considering the havoc caused by rodents to agriculture and options available to farmers, it is advisable to integrate these methods in proper scheme of things. Cultural control methods coupled with poison baiting and trapping at proper time and location is being recommended for long lasting solutions to rodent problem.

Community Participation

Farmers are not well aware of the rodent problem and the management technologies therefore there remain a major gap in technology development and its popularization. More over, rodent problem is not the problem of one farmer but of whole village. Besides there are many common property resources like panchayat lands, gochars, rail and roadsides, which act as regular breeding and safe shelters for native rodents. Therefore rodent management activities should be undertaken on community basis in campaign form at least once in a year (preferably in summers) and regular then follow up can be done by farmers. By education and training one can increase the knowledge and awareness of end users and also change the attitude of the local people for rodent management. Therefore, we have to infuse in students and youths, the idea, that rodents are one of our greatest enemies and need to be dealt tactfully. For preparation of the community for rodent management various media may be approached for effective education, however, farmers trainings, field demonstrations etc. may also be arranged by Government departments, NGOs and or voluntary agencies.

Two pronged strategy is urgently required for involving the common man to undertake rodent management practices on a larger scale *viz.*, whole village/community basis (Jain *et al.*, 1993). First strategy should be to create a nucleus of trained personnel engaged in extension activities. This is most important, because there is paucity of trained manpower in this field. The trained official would then act as trainers for farmers and other end users. Rodent management technology is easy to operate, quick, result oriented and cost effective. All the strata dealing with extension and education need to be invariably trained. Different government agencies like KVKs, ATCs, States Extension Departments, State Agril. Universities (SAU's), ICAR Institutes should organize regular training programmes for farmers and farmwomen. These training programmes must include rodent management as a major input. There are several other agencies working among farmers, which may prove highly beneficial in creating awareness among farmers, if trainings are organized at their doorsteps. Rural youths and students may be trained and 'Rodent management Squads' can be formed at village levels. When these squads are trained rigorously, the media may be utilized to popularize the rodent control campaign on a whole village/community basis.

References

Advani, R. and Mathur, R.P., 1982. Experimental reduction of rodent damage to vegetable crops in Indian villages. *Agroecosystems* 8 : 39-45.

Advani, R., Prakash, I. and Mathur, R.P., 1982. Assessment of rodent damage and yield losses in standing wheat crop in a desert village complex. *Z. Angew. Zool.* 99 : 257-266.

Anonymous, 1986. Evaluation of second-generation anticoagulant rodenticides in India I. Bromadiolone. AICRP on Rodent Control, CAZRI, Jodhpur (Memio) 21 p.

Chaudhary, V. and Tripathi, R.S., 2002. Evaluation of toxicity and acceptability of difethialone treated baits in Indian desert gerbil, *Meriones hurrianae* Jerdon. *Curr. Agric.* 26 (1-2) : 31-34.

Chaudhary, V. and Tripathi, R.S., 2003. Susceptibility of house rat, *Rattus rattus* Linneaus against difethialone treated baits in laboratory. *Pestology* 27 (4): 38-41.

Chaudhary, V. Tripathi, R.S. and Rana, B.D., 2001. Obeservation on predation of *Tatera indica* by varanids in arid ecosystem. *Rodent Newsletter* 25: 5-6.

Chaudhary. V, Tripathi, R.S. and Poonia, F.S., 2002. Evaluation of toxicity and acceptability of difethialone treated baits against Indian gerbil, *Meriones hurrianae* Jerdon. *Current Agriculture* 26: 31-34.

Deoras, P.J., 1964. Rats and their control. A review of an account of work done on rats in India. *Ind. J. Ent.* 26: 407-408.

Jain, A.P. and Tripathi, R.S., 1988. Evaluation of second generation anticoagulant rodenticides in India. I. Brodifacoum. AICRP on Rodent Control, CAZRI, Jodhpur. Pp. 1-21.

Jain, A.P. and Tripathi, R.S., 2000. An Integrated approach to rodent pest management. In *IPM System in Agriculture*, Vol. 7 (Eds. R.K. Upadhyay, K.G. Mukerjee and O.P.Dubey). Aditya Books Pvt. Ltd., New Delhi. pp. 275-307.

Jain, A.P., Mathur, M. and Tripathi, R.S., 1992. Bio-efficacy of flocoumafen against major desert rodent pests. *Indian J. Pl. Prot.* 20 : 81-85.

Jain, A.P., Mathur, M. and Tripathi, R.S., 1992. Traps and Trapping. In *Rodents in Indian Agriculture* (Eds. I. Prakash and P.K. Ghosh). Scientific Publishers, Jodhpur. Pp. 585-894.

Jain, A.P., Tripathi, R.S. and Rana, B.D., 1993. Rodent Management : The State of Art. AICRP on Rodent Control, Central Arid Zone Research Institute, Jodhpur. 138 p (Mimeo).

Jan, A.P., 1970. Body weight, sex ratio, age structure and some aspect of reproduction in the Indian gerbil, *Tatera indica indica* Hardwike in the Rajasthan desert, India. *Mammalia* 34:415-432.

Neelanarayanan, P., 1997. Predatory pressure of barn owl (*Tyto alba stertens* Hartert, 1928) on rodent pest- a field evaluation. Ph. D. Thesis, Mayiladuthurai, Bharthidasan University, Thiruchirapalli 92p.

Parshad, V.R., 1999. Rodent control in India. *Integrated Pest Management Reviews* 4: 97-126.

Patel. N., Jain, A.P., Tripathi, R.S., Parveen, F. and Mathur, M., 1995. Losses caused to pomegranate fruits by rodents in arid horticulture. *Geobios New Reports* 14 : 37-38.

Prakash, I., 1975. The population ecology of the rodents of the Rajasthan desert, India. In *Rodents in desert environment* (Eds. Prakash, I. and Ghosh, P.K.). Dr. W. Junk b.v. Publishers, The Hague. Pp. 75-116.

Prakash, I. and Kamethar, L.R., 1969. Body weight, sex and age factors in population of the northern palm squirrel, *Funambulus pennanti* Wroughton. *J. Bombay Nat. Hist. Soc.* 66: 99-115.

Prakash, I. and Mathur, R.P., 1987. Management of rodent pests, Indian Council of Agricultural Research, New Delhi. pp. 1-133.

Prakash, I., Taneja, G.C. and Purohit, K.G., 1971. Ecotoxicology and control of Indian desert gerbil, *Meriones hurrianae* (Jerdon). VII. Relative numbers in relation to ecological factors. *J. Bombay nat. Hist. Soc.* 68 : 86-93.

Rana, B.D. and Tripathi, R.S., 1999. *Recent trends in Coordinated Research on Rodent Control.* Central Arid Zone Research Institute, Jodhpur. 48 p.

Rao, A.M.K.M., 1992. Integrated rodent management. In *Rodents in Indian Agriculture,* Vol. I (Eds.Prakash, I. and Ghosh, P.K.). Scientific Publishers, Jodhpur, India. pp. 651-668.

Rao, A.M.K.M., 2003. Rodent problem in India and Stratigies for their management. In *Rats Mice and People: Rodent Biology and Management.* (Eds G.R. Singleton, L.A. Hinds, C.J. Krebs and D.M. Spratt) Australian Centre for International Agricultural Research, Canberra. pp 203-212.

Rao, A.S. and Singh, R.S., 1998. Climatic feature and crop production. In *Fifty Years of Arid Zone Research in India* (Eds. A.S. Faroda and M. Singh). Central Arid Zone Research Institute, Jodhpur. Pp. 17-18.

Tripathi, R. S., 2005. Reproduction in Desert Rodents. In *Changing Faunal Ecology in the Thar Desert.* (Eds Tyagi, B.K. and Baqri, Q.H) Scientific Publishers, Jodhpur pp289-304.

Tripathi, R.S and Chaudhary, V., 2006. Antidoting effect of vitamin K_1 supplementation on difethialone poisoned house rat, *Rattus rattus* Linneaus. *Annals of Arid Zone* 45(2):203-210

Tripathi, R.S, Mathur, M., Jain, A.P. and Patel, N., 1994. Relative efficacy of glue and other traps for commensal rodent management. *Ann. Arid Zone* 33 : 143-45.

Tripathi, R.S. and Chaudhary, V., 2004. Principles and Practices of Rodent Pest Management. In *Advances in Plant Protection Sciences* (Eds. Prasad, D. and Singh, A.) Akansha Publishing House, New Delhi, India. pp 245-259.

Tripathi, R.S., and Jain, A.P., 1990. Rodent damage to desert afforestation plantations. *Rodent Newsletter* 14 : 7-8.

Tripathi, R.S., Chaudhary, V. and Idris, M., 2004. Incidence of rodent pests and their management in pulse crops under arid agro-ecosystem. *Pestology* 28 (6) : 67-70.

Tripathi, R.S., Jain, A.P., Kashyap, N., Rana, B.D. and Prakash, I., 1992. North-western Desert: Special Problem Areas. In *Rodents in Indian Agriculture,* Vol. I (Eds. I. Prakash and P.K. Ghosh). Scientific Publishers, Jodhpur. Pp. 357-396.

Chapter 23

Pheromones in Indian Agriculture and Future Scope

V. Nandagopal
Central Rice Research Institute (ICAR),
Cuttack – 753 006, Orissa

Introduction

Chemicals used for communication between the organisms are known as semio-chemicals or info-chemicals. These chemicals are also called "assembling scents" by Kettlewell (1942, 1943, 1946, 1955). In Greek *Pherein* means to carry and hormone (to excite, stimulate) as described by Karlson and Butenandt (1959). Further Kirschenblatt (1957) proposed the term "*telegones*". Micklem (1959) termed it as "*pherormone*" instead of pheromone. Karlson and Luscher (1959) have given a more scientific clarification for the use of the term *pheromone*.

Pheromone has been in use in India over more than 30 years as a monitoring tool under All India Coordinating programme for some of the crops likes cotton, groundnut and pigeon pea. The monitoring technology is well knitted in different centers in these crops. The monitoring helps in developing forewarning systems and plan management strategies. Unfortunately, as such efforts have not been made even at research station; the farmers are not convinced about practical utility of pheromone technology. Some farmers use pheromone traps especially in cotton and pigeon pea ecosystem, as a mass-trapping tool. Surprisingly, in a recent study at the NRCG; a maximum of 1470 males/day/trap in case of *Spodoptera litura* was recorded showing its potentiality for mass trapping.

Even after 30 years of introduction of pheromone technology in Indian Agriculture, it remains still in its infancy. This may be attributed to, high cost of pheromone chemicals, defects in manufacture of the lures and traps, insect related issues, difficulties in installation of traps at farmers level etc. An analysis of these issues has been made and suggestions are given to overcome the same. Then only there be actual fillip to the novel, eco-friendly pheromone technology in Indian agriculture (Nandagopal, 2004). There are, at present ten different types of pheromones identified world wide (Nandagopal *et al.*, 2008).

In India, monitoring of the key insect pests such as *H. armigera, S. litura, P. gossypiella,* were undertaken as early as 1985. Intensive work has been done on these insects spreading over Karnataka, Haryana, Madhya Pradesh, Tamil Nadu, Andhra Pradesh, Maharashtra, Himachal Pradesh, Punjab and Gujarat. Pheromone as a mass trapping tool have also been utilized in a few insect such as *Pthorimoaea opercullela, S. litura, Chilo sacchariphagus indicus, H. armigera, A. modicella, Lymantria obfuscata* Wlk (Porthetria dispar), Tortricid *Cydia pomonella* (L.), Codling moth, *Cydia pomonella,* and *Pectinophora gossypiella* (Saund.). The communication disruption as a tool has been tried in *S. litura, A. modicella, Peripleneta americana* and *Chilo auricilius* in India with limited success. There are attempt made to design the trap for efficient trapping of the target insect, some of them were successful in the case of cotton, sugarcane and groundnut ecosystem. Interestingly, lot of work has been done on the basic aspects of pheromone since 1975.

Basic Studies

Intensive work has been done on the basic aspects of the sex pheromones in various insects and mammals in India since 1980. Majority of the information are on the location of sex pheromone glands, occurrence of sex pheromones etc. These basic studies have given a fillip to the further studies on the pheromone work.

Isolation and Synthesis of Pheromones

The Indian Institute of Chemical Technology (CSIR), Hyderabad is the only full pledged institute having maximum facilities for undertaking synthesis of the components of the pheromone. Many of ICAR institute have been in collaboration with IICT for synthesis of the pheromones.

Whitegrub is very serious phytophagous pest of rainy season crops of many states of India including Rajasthan. Its larval stage, grub feeds on roots of wide variety of cultivated as well as uncultivated plants and beetles on foliage of tree plants. The major whitegrub species in the country are *Holotrichia consanguinea, H. serrata, H. reynaudi, Anomola dimidiata, H. seticollis, H. longipennis. Holotrichia consanguinea* is the most serious scarab in India and is dominant whitegrub species in Rajasthan, Gujarat, Haryana, Punjab. It feeds on all kharif crops but the damage is more evident on plants with tap roots like groundnut, its beetles feed on foliage of host trees preferably Jujube (Ber), Neem, Khejari, Guava.

Intially research work on whitegrub was undertaken to study their biology and behaviour (Yadava *et al.*, 1978, Yadava, 1981). Since 1980, more systematic efforts have been made to survey various species of whitegrubs in India through an All India Coordinated Research Project of ICAR, New Delhi. Later on focus was directed on integrated management of whitegrub using ecofriendly means. Recently few strains of *Metarhizium anisopliae* have been identified which are very effective against whitegrub as revealed through laboratory bioassays (Sharma *et al.*, 1999). With *Beauveria bassiana* about 40 per cent mortality of *Holotrichia consanguinea* grubs has been reported in the laboratory (Anonymous, 1990). Entomogenous nematodes *Steinernema glaseri* has been found effective against first instar grubs of *H. consanguinea* (Bareth *et al.*, 1995). Although field trials of these bioagents are in progress, their field efficacy is not as convincing as in the laboratory. This might be due to non-availability of required conditions for their growth and multiplication in the field. As an alternative management strategy, they have isolated and identified aggregation pheromone of *H. consanguinea*

(Leal *et al.*, 1996) and put to field trials. Subsequently, they have demonstrated kairomone activity for *H. consanguinea* in the leaf extracts from Neem *(Azadirachta indica)* and Khejari, *Prosopis cinerarium* (Yadav and Yadava 2000). They have found that green leaf volatiles (GLVs) of Neem (*Azadirachta indica*) had higher kairomone activity. One of the fractions of *Azadirachta indica* GLVs obtained using reverse phase HPLC showed maximum kairomone activity (our unpublished results). Subsequently they have also identified the active chemical component of kairomone from Neem, which is under field trials.

Pheromones have been identified and put to field application for the management of pests of many crops (Ganeswara Rao *et al.*, 1995, Ramchandran *et al.*, 1990). Following the identification and synthesis of female sex pheromone of the three main pests of cotton *viz. Earias insulana* (Hall *et al.*, 1980), *E. vittella* (Cork *et al.*, 1988) and *Pectinophora gossypiella*, field trials indicated that all the three could be controlled by mating disruption by a single, early season application of their combined pheromones (Chamberlain *et al.*, 1992). The groundnut leafminer, *Aproerema modicella* is the most important insect pest of groundnut in India. The sex pheromone in this pest was demonstrated by Nandgopal (1995) and identified as blend of three compouunds. Likewise, chemical components of the pheromone of *Diacrisia obliquq* and *Amsacta albistriga* the major insect pests of groundnut and castor were identified by Persoons *et al.* (1993). Most of the sex pheromones of the main lepidopteran pests of sugarcane were identified in early 1980 (Nisbitt *et al.*, 1986, David *et al.*, 1985) but little work has been undertaken to develop their use in pest management. Pheromone binding proteins (PBPs) present in the antennae of several species of scarab beetles have been characterized (Leal *et al.*, 1998). In most case only one class of PBP was expressed in both sexes. Ishida *et al.* (2002) have identified and cloned a cDNA encoding odorant-binding protein isolated from mosquitoes. In this case the protein was detected only in male antennae of *Culex quinquefasciatus*

Whitegrubs (Coleoptera : Scarabaeidae) are among the five most important 'national pests' in India with over 45 spices reported as serious pests of a diverse range of crops (Kumar and Veeresh, 1995). However, despite the seriousness of the damage they cause, very little work has been conducted on the pheromones of these species. Recently, identification of sex pheromone of *Holotricha consanguinea* (Leal *et al.*, 1998) and *H. reynaudi* has been reported (Ward *et al.*, 2002). Since many species of *Holotrichia* produce aggregation pheromone, they would be potentially valuable tools for mass trapping (Ganeshiah and Kumar 1993; Kumar and Veeresh 1994).

Biocontrol Research Laboratories of Pest Control India (PCI) played a pioneering role in promotion of pheromone technology in India by carrying out studies on mating disruption of pink bollworm of cotton in 1979, jointly with the Directorate of Plant Protection, Quarantine and Storage, Govt. of India; Directorate of Agricultural Aviation; Indian Cotton Mills Federation, Mumbai and Conrel of USA. These studies carried out over 290 hectares at Bhatinda in Punjab, by aerial application of pheromone using hollow fiber controlled release technology, demonstrated the efficacy of the method with 34 per cent increase in yield (Paharia, 1981).

PCI introduced sex pheromone lures for monitoring of pests in IPM in 1987 and set up a pilot scale pheromone synthesis facility in 1999. Investigations were carried out to develop protocols indigenously for production of sex pheromones for *H. armigera*. Based on the success of these efforts, processes were developed for up-scaling production of sex pheromones of various insect pests. The major challenge was synthesis of insect sex pheromones on a commercial scale, which was achieved for the first time in India by PCI. These initiatives were followed up by introduction of sex pheromone lures for mass trapping of sugarcane borers in 2003.

Several traps such as bucket trap, delta trap, fligh-T and funnel trap were designed for trapping various insects. PCI also successfully designed and developed the first portable, easy to use and cost effective water trap for use with sex pheromone lures for trapping sugarcane borers (Jayanth and Bhanu, 2004). The portable water traps are proving to be effective for management of other crop pests such as brinjal shoot and fruit borer, diamond back moth, etc.

Work pertaining to pheromone synthesis is still in a primitive level in India and the organizations, which are showing interest in this discipline are dependant on international collaboration. Most of the applications and trials with pheromone lures are being attempted with the products developed from outside India. No single company in India is producing the pheromones lures with indigenous technologies. Hence, there is definitely a need to develop technologies to produce the pheromones of important crop pests in the most economical and feasible way for regular supply. In addition there is a need to develop and establish expertise in isolation and identification of hitherto unknown pheromones of insect pests in India.

In India considerable work has been done in different crops using pheromone for different purposes.

Field Application of the Pheromones

The presence of sex pheromones of leaf miner has been identified for the first time (Nandagopal and Reddy, 1991) by the National Research Centre for Groundnut (NRCG) in collaboration with NRI, U.K. and field evaluation of the components of the sex pheromones have been also done by the NRCG. Experiments were also conducted by NRCG for standardizing the number of traps per unit area and use of the sex pheromone as one of the components of Integrated Pest Management (IPM) in groundnut where leaf miner damage is extensive under Saurashtra conditions. Nandagopal and Soni (1993) have evaluated a number of traps for the groundnut leaf miner (*Aproaerema modiceela* Deventer. Hall *et al.* (1993) have identified the total components of the pheromone compounds of leaf miner.

At Junagadh, a square glue trap developed at NRCG (Nandagopal and Soni, 1993) was used, while in other centres water trap designed by ICRISAT was used. Different types of traps were designed and tried in the trapping of males of *A. modicella*. The results indicated that a square glue trap designed

GLM Damage

Leaf Miner Moth

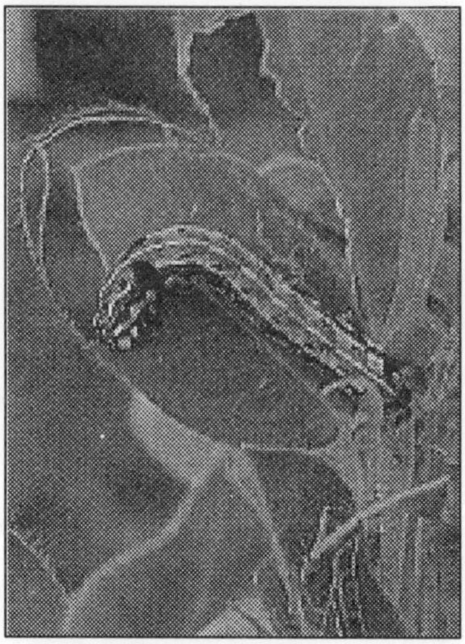

Larva of *S. litura*

Egg Mass of *S. litura*

with 22 cm side and with 4 entry holes (Nandagopal and Soni, 1993) was the most efficient in trapping the males. The pheromone was also used for monitoring the activity of the moths in ground nut (Yadav *et al.,* 2004). A trap efficient to a tune of 60 to 90 per cent has been developed for Spodoptera litura in groundnut. Nandagopal *et al.* (1995) have reported on the effects of certain components of IPM on the damage and yields in groundnut. Nandagopal (1998) explained about hove the eco-friendly management of Groundnut Leaf miner in India can be achieved. Nandagopal *et al.* (1998) reported on the development of an efficient pheromone trap for *Aproaerema modicella* Dev. (Gelechiidae: Lepidoptera) population in groundnut crops. Further, Nandagopal *et al.* (2006) have developed a viable efficient sex pheromone trap for *Aproaerema modicell.* In *Spodoptera,* Nandagopal *et al.*(2003) reported on the ppulation dynamics of *Spodoptera litura* (F.) in relation to weather parameters in groundnut in India using pheromone trap catch. Nandagopal *et al.*(2004) have developed an efficient pheromone trap for field catch of *Spodoptera litura* Fabricius in groundnut and castor ecosystem. Based on the pheromone trap catches the prediction of *Spodoptera litura* (F.) males in relation to key weather factors in groundnut have been reported. A technical bulletin has been published on the holistic IPM wherein the use of pheromone has been explained (Ghewande *et al.,* 2002).

Central Institute of Cotton Research, Regional Station, Coimbatore tried three types of pheromone traps, *viz.,* sticky, liquid and dry are used in cotton pest management. The mostly used trap in cotton is the delta trap. This trap is however, has limited trapping capacity, and is not adequate for large insects like *Helicoverpa* and the scale of trapped insects and dust affect its trapping efficacy.

Monitoring

Several studies have been conducted on cotton pests to establish a quantitative relationship between pheromone trap catches of particular pest and the damage caused with an aim to develop relationship would help to forewarn.

Mating Disruption

This technique was used in Punjab during 1992-1993 resulting decrease in boll damage from a pre-programme level of 23.3 per cent to 0.0 per cent requiring only an occasional post programme with use of insecticides.

Larva Damaging Cotton Boll

Adult of *H.armigera*

Leaf Folder

At Directorate of Rice Research, Hyderabad, the Rice leaf folder *Cnaphlocrocis medinalis* is a serious pest in India. Ganeswara Rao *et al.*, 1993 carried out investigations on the pheromone of *C. medinalis* using insects of Indian origin and evaluated blends of the compounds identified in the field in India.

The authors identified the presence of three electrophysiologically (EAG) active compounds *viz.*, Z 13-18:Ac (major component), Z 11-16:Ac and Z 13-18:OH (minor components) on the basis of GC retention times. The minor components were present at less than 10 per cent relative to the major component. Trace quantities of octadecyl acetate were also identified but no EAG response was observed.

The field experiments conducted (Ganeswara Rao *et al.*, 1995) with the above three blends showed that Z 13-18:Ac alone was not significantly attractive to male leaf folder *C. medinalis* in contrast to the findings from Kudagamage (1983). However, the addition of 1-20 per cent of Z 11-16:Ac to Z 13-18:Ac significantly increased attractiveness. The addition of third EAG active component in 1-50 per cent of Z 13-18:OH to Z 13-18:Ac significantly reduced attractiveness to male moths. As a result of this work, a lure with 1mg of a blend of Z 13-18:Ac with 5-10 per cent Z 11-16:Ac impregnated in a polythene vial has been recommended for monitoring *C. medinalis* in India (Ganeswara Rao, 1996).

The subsequent field trials conducted by Directorate of Rice Research, with lures impregnated with Z 13-18:Ac and Z 11-16:Ac showed that the trap catches were inconsistent with wide variations even within the same season. This might be due to the missing of some of the minor components in the synthetic blend supposed to be present in natural blend/ weather factors which needs to be investigated. The yellow stem borer is a very serious pest on rice cultuvation.

Kawazu and Tatsuki (2002) showed that calling activity in *C. medinalis* a distinct diel periodicity. Moths showed high activity during 5-7 h into the scotophase. To clarify the temporal change of pheromone titres in the pheromone gland within 1 day, 4-day old females were used. The changes in the calling activity and titre of the four sex pheromone components, (Z)-11-octadecenal (Z11-18:Ald), (Z)-13-octadecenal (Z13-18:Ald), (Z)-11-octadecen-1-ol (Z11-18:OH) and (Z)-13-octadecen-1-ol (Z13-18:OH), in the pheromone glands were roughly synchronous. At age 4, however, the pheromone components could already be detected at the onset of scotophase, while calling began at 3 h into the scotophase. Furthermore, calling was most active between age 3 and 7, whereas the highest level of pheromone titre at 6 h after light off was observed from age 3 to 5. These findings suggest the presence of different controlling mechanisms between calling behaviour and pheromone production. The ratio of the four sex pheromone components remained relatively constant at all times of the day and all ages.

At the Sugarcane Breeding Institute, Coimbatore, All the work carried out in India on synthetic pheromones of sugarcane moth borers originates only from Sugarcane Breeding Institute, Coimbatore. The work was initiated in 1981 and continued up to 1990. Now the work has been revived from 2000 and is being continued under the sponsorship of three sugar factories.. The work on synthetic pheromones of different sugarcane moth borers by Sugarcane Breeding Institute is presented under the national status below.

Internode Borer

Internode Borer (INB), *Chilo sacchariphagus indicus* K.

The pest is distributed all over India. It is major pest in Addhra Pradesh, Tamil Nadu and Karnataka and minor in North India. The pest inflicts serious damage in the soft joints within the leaf tissues. Severe attack causes rotting and death of whole stalk. Weakening of internode results in breakage.

The sex pheromones of INB were identified as (z)-13-octadecenyl acetate and (z)-13-octadecen-1-ol with the help of Overseas Development and Natural Resource Institute (ODNRI), London (David *et al.*, 1981) The lures consisting of the two components were tested in five different combinations in field trial and a ratio of 7:1 was found optimum. There was no significant variation in moth catches between the traps set at 16.5m and 24.8m distances (Anonymous., 1983). Of the varying heights of traps tested, traps set at 60cm from the ground level were found to attract significantly more number of moths per water trap (Anonymous., 1984). The efficacies of different types of sticky and funnel traps over the water traps were tested for seven days. The water traps were significantly more effective than the other traps (Anonymous, 1986).

In trials on mass trapping of male moths with pheromone lures using water traps, a total of 80,603 moths were caught during May – February from a 10.0 ha plot in Pugalur,Tamil Nadu. The intensity of attack was significantly less and the cane yield and CCS /plot were significantly higher in treated plots compared to the check plot (David *et al.*, 1985). Indigenisation of the lures was attempted with the association of the Bhabha Atomic Research Centre (BARC), Mumbai. The BARC pheromone formulation was found to be only 50 per cent effective as that of ODNRI's (Anon., 1990). Now, the work on indegenisation of pheromone lure has been revived in Sugarcane Breeding Institute, Coimbatore as INB problem has become severe in Co 86032, a ruling variety of Peninsular India. The studies indicate that some of the indigenous pheromone lures were on par in efficacy with the imported lures. As the present water trap used for trapping moths with pheromone lure is expensive and cumbersome to operate, new designs are being developed.

Stalk Borer, *Chilo auricilius* Dudgeon

Stalk borer is a serious pest in India. Sugarcane Breeding Institute in collaboration with ODNRI, London, initiated Stalk borer pheromone identification work in 1981. From the female pupae of the stalk borer, four compounds *viz.* (z)-7-dodecenyl acetate, (z)-8-tridecenyl acetate, (z)-9-tetradecenyl acetate, (z)-10-pentadecenyl acetate were identified (David *et al.*, 1985). Their combination at 4:8:4:1 ratio at the dose of one mg per vial was found effective through successive field tests at Shamli in U.P and Yamunanagar at Haryana. The highest number of male moths attracted to a trap in a day was 259.

While testing the four compounds individually, (z)-7-dodecenyl acetate was found to be an inhibitor of pheromone (David *et al.*, 1985). Using the inhibitor and the attractants, mating disruption by disrupting the communication between male and female moths, and confusing male moths in differentiating the virgin females from the synthetic lure were respectively attempted. Micro encapsulated spray formulations of the inhibitor and the attractant were sprayed at 0.0, 2.5, 5.0, 10.0 and 20.0 ml per $10m^2$ plot to saturate the atmosphere with the odour of the respective compounds (Anonymous., 1986). The pheromone vial in the water trap simulated the virgin female and was deployed in the centre of the treated plots. The per cent reduction in the number of moths caught in the traps of treated plots over the control plot determined the efficacy of the communication disruption or confusion technique. The results indicated that both the inhibitor and the attractants were equally effective in preventing 46.8 per cent of moth from reaching the vials simulating the virgin females. To further improve the efficacy and to have a long lasting effect of the inhibitor or the attractant in the field and due to the difficulty in spraying the microencapsulated formulations, adhesive flakes of the inhibitor and attractants were tried. The flakes each of approximately one cm^2 were distributed at the rate of one flake per clump. The flakes had improved efficacy with 76 per cent reduction in moth catches which however was below 99 per cent efficacy required for mating disruption technique (Curtis *et al.*, 1987).

Indigenisation of pheromone lures of stalk borer were also attempted in association with BARC, Bombay. The field trials on comparison of the lures with that of ODNRI indicated that the formulation of the BARC lure is required to be improved 20 times (Anonymous, 1990).

ESB Adult

Shoot Borer, *Chilo infuscatellus* Snellen

Shoot borer *Chilo infuscatellus* Snellan is generally destructive in hot dry months. In North India it is more destructive during pre-monsoon period and in south India on eksali crop during April to August and on adsali crop during September to January. The larvae feed on growing plants resulting in dead heart at shoot stage which come out easily when pulled and emit foul odour and sometimes canes are also damaged. Young shoots up to 100 days and during drought the internodes are affected. 60 per cent shoots are damaged due to this pest and 3.5 per cent loss in yield for every 5 per cent increase in incidence of the pest in cane. (David *et al.*, 1991).

The pheromone work on shoot borer was initiated in 1986 at Sugarcane Breeding Institute, Coimbatore with the collaboration of ODNRI, to dispense with pesticide application which is popular for early shoot borer control and to make management more easy and economical.

The pheromone, identified in China (Wu,1984) as (z)-11-hexadecenol, was tested in the field in 229 traps for 42 days in 15 ha area with high incidence of shoot borer at the endemic location *viz.*, Thiru Arooran Sugars Ltd., Vadapathimangalam, Tamil Nadu. But the mean moth catch was less than one/trap/day and there was no difference between treated and control plots in shoot borer incidence at the end of the experiment. So the pheromone was reidentified at ODNRI from the pupae sent from Coimbatore and confirmed to be the same compound. To improve the efficacy, different concentrations *viz.*, 0.1,1.0 and 5.0mg were tested in two types of dispensers *viz.*, vials and septa. Five mg though not economically viable, in order to assess the efficacy was also included in the study. The highest mean number of moths caught was 1.32 /trap/day in 5.0 mg septa (Anon., 1989). In Taiwan, it has been reported that the pheromone is effective at 13mg concentration (Chen *et al.*, 1994). In their experiment in a maturing cane field, traps placed at the height of 0.2 m attracted a daily average of 1.6 males per trap. This was significantly greater than the average number of males caught at heights of 1 and 2 m (0.2 and 0.4 males daily per trap, respectively). In a seedling cane field, the average number of males caught at each height differed significantly with 0.9, 0.5 and 0.1 males caught daily per trap at 0.2, 1 and 2 m, respectively.

To observe the peak attraction period, if any, of the male moths to the lures, observations were made at bihourly intervals from 18.00h to 6.00h on full moon and new noon nights. Male moths were attracted only after mid night and peak attraction was from 0.00 to 4.00 h.(Anonymous,1989).

Southern Petrochemical Industries Corporation, Chennai came out with lures for confusion technique using the same compound (z)-11-hexadecenol. A collaborative field evaluation study was taken up in two locations in Tamil Nadu. They were tested at the rate of 250 dispensers /ha in a two ha plot in each location with check plots. The dispensers were dispersed in three week old sugarcane crop. The shoot borer incidence did not differ between treated and control plots indicating that the pheromone is ineffective (Anonymous, 1987).

Adult Male

Top Borer, *Scirpophaga excerptalis* Walk.

Top borer *Scirpophaga excerptalis* is a serious pest in both Northern and Southern India The pheromones of top borer have been identified in China as (E)-11-hexadecenal and (z)-11-hexadecenal. They have been reported effective against top borer at 78:22 (Wu,1990) and 70:30 (Lin,1992) ratios. So the efficacy of both the ratios was tested by Sugarcane Breeding Institute. The Plant Research International, The Netherlands readily supplied free samples of lures of both the formulations impregnated at one mg concentration in rubber septa as per our request. The preliminary field studies at Harinagar Sugar Mills Ltd., Bihar, have revealed that 70:30 ratio is significantly more effective ['t' value at 0.05 for 66 df = 2.462] trapping an average of 10.29 / day. The work is in progress to indigenise the lures, to improve the trap catch by increasing the dose per septum and to evaluate the efficacy of different dispensers.

At the Indian Institute of Vegetable Research, Varanasi *Leucinodes orbonalis* is a serious pest on brinjal and several control measures are being attempted. Under Eu-sponsored research project on sustainable pest management in brinjal is being carried out in collaboration with the Natural Research Institute (NRI), UK. Trials with Pheromone lures of *L. orbonalis* supplied from NRI, UK conducted at Coimbatore in 2000 using white rubber septa revealed its effectiveness in trapping *L. orbonalis*, to a maximum of 117 adults / trap / week but could not correlate with the field damage level.

Fruits Damaged

The brinjal shoot and fruit borer, *L. orbonalis* (Lepidoptera: Pyralidae) is the major insect pest of eggplant throughout Asia. The larvae bore into the shoots during the vegetative growth stage and later into flowers and fruits rendering them unfit for consumption. Even small attempt of the insect on fruits will result in loss of market value further in terms of yield. Fruit losses in excess of 50 per cent are commonly reported due to the boring activity of larvae.

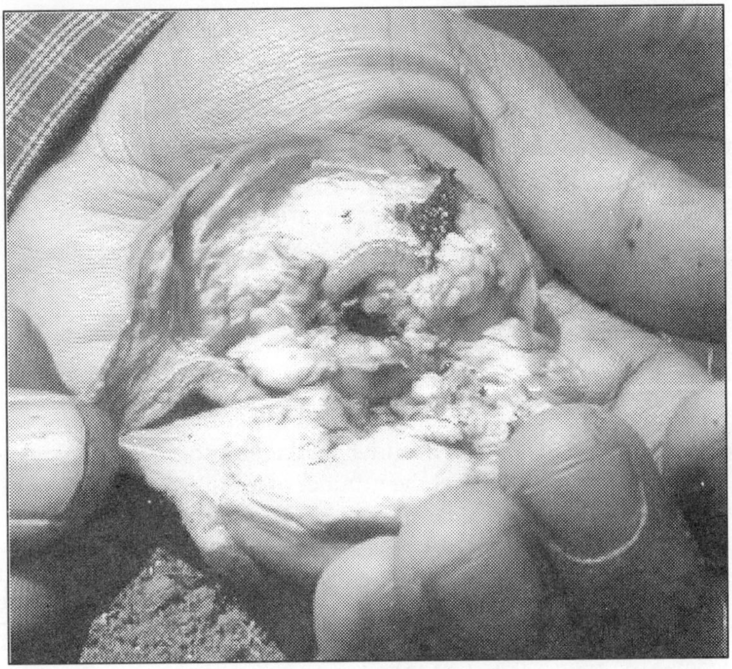

Damaged Fruit with Larva

Many of the insecticides used against this pest have become ineffective due to development of resistance. Alternative control strategies are needed but, to be effective, they require sound understanding of the biology and ecology of the pest. One of such methods is use of pheromones for monitoring and mass trapping in IPM. Integrated management of *L. orbonalis* was undertaken in South Asia involving India, Sri Lanka, Bangladesh and Taiwan (AVRDC, 2003). In this network programme, the sex pheromone against the target pest was evaluated. Different types of traps and field installation were reported to be effective at various test locations.

Insects were also trapped using yellow plastic funnel traps, white sticky delta traps at NARDI farm, Andhra Pradesh, West Bengal and Bangladesh. These studies concluded the effective ratio of pheromone components and dose per dispenser. White rubber septa and polyethylene vials were reported to be effective at a dose of 1-3 mg per lure. Field studies conducted at West Bengal to study the release rate indicated that white rubber septa loaded with 100, 300 and 1000 µg of E11-16:Ac and 1 per cent E11-16:OH approximately 86, 89 and 100 per cent of E11-16:Ac remained after 40 days (Cork *et al.*, 2001).

The studies undertaken to optimize the trap design showed varied results as delta trap was found to effective at GAU, Anand while funnel traps at IIVR, Varansi and a locally designed water trap is being recommended by scientists of BARI, Bangaldesh. Considering the trap height, the trap catch was high when placed 0.5 m above soil surface at GAU while 0.25m above the crop canopy at IIVR while 1.5 m above the soil surface at BARI, Bangladesh. The traps are to be placed at a distance of 10 to 15 m. (AVRDC, 2003). Field studies on the effect of dispenser type on *L. orbonalis* male moths attraction with three different types namely polyethylene vials, white and black rubber septa showed that polyethylene vials caught more male moths (3.38/trap/night) followed by white septa (1.42/trap/night) or black septa (0.79/trap/night).

Plutella xylostella is an important pest of crucifers. It has countrywide distribution. The attacked leaves are sometimes badly defoliated, and the attacked plants look pale sick, show larval feed holes and are soiled with excreta of the larvae. Such cabbage heads do not fetch any value in the market and are discarded. Abraham and Padmanapan (1968) reported 31 per cent yield loss in cabbage due to *P. xylostella*. The crop loss is estimated to vary from 52 per cent (Anuraha, 1997) to 100 per cent (Carleson and Hare, 1986). Chandrasekharan and Ragupathy (1996) reported that the level of resistance varied from 8.2 to 85.7 per cent to different insecticides in various locations. Again it was reported that the resistance level varied from 66.7-100 per cent for fenvalerate, 45.50 to 92.3 per cent for quinalphos. 32.6 to 85.7 per cent for monocrotophos. 14.3 to 55.2 per cent for carbofuran and 17.9 to 52.4 per cent for cartap hydrochlorite. DBM has developed resistance to a wide variety of insectides (Sannaveerappanavar, 1995).

Pheromone compounds *P. xylostella* was identified as (cz)-11-hexa decanal, (Z)C-11-hexdecenge acetate, Z-11-hexa decen-1 ol. A blend consisting the above compounds along with antioxidant in the ratio 50:50:1:5 at 1 mg/septum was tested under field conditions. Various types of traps are being used (Plate 15d). The results indicated that the number of moths caught was significantly higher than other treatments (Reddy & Urs, 1997). Placement of the pheromones at 30 cm level also-captured significant number of adults (Reddy and Urs, 1996). Chandra Mohan (1999) observed the seasonal variation in pheromone catches, with maximum catches being recorded during April to July. Different types of catches were evaluated and stick-a-trap captured maximum number of adults (Kumar, 2000). Bio rational IPM based on pheromone catches recorded least pest incidence and higher profit (Reddy and Guerrero, 2000). Nandagopal *et al.* (2004) developed an efficient pheromone trap for catching Brinjal shoot and fruit borer, *Leucinodes orbonalis*.

Mass Trapping

With the commercial availability of insect sex pheromones for several agricultural pests in the 1970s, scientists and entrepreneur turned their attention to push mass trapping as a tool for insect control. Due to want of clear directives on the number of traps to be used/unit area, the use of pheromone as mass trapping tool could not take off.

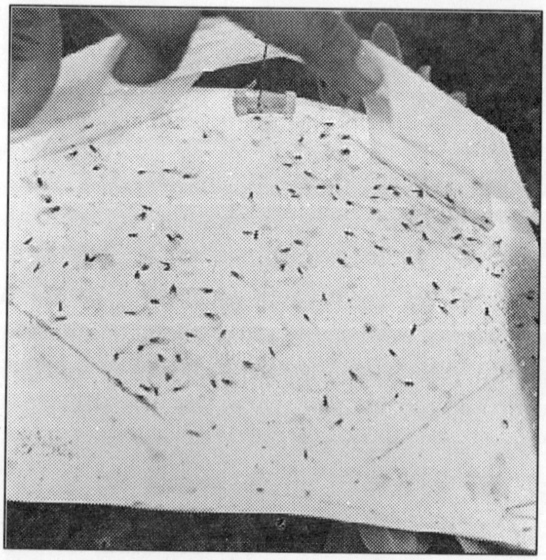

Mass Trapping of GLM

Mating Disruption Technique in Indian Agriculture

In a very few cases mating disruption technique was successful in India. Based on the information's available, communication disruption

may be successful in crops where dense foliage results in reduced wind velocity and this pheromone persists for a few hours. Some examples are *Pectinophora gossypiella* in cotton ecosystem which has resulted a grand success at Punjab during 1979. *Chilo auricilius* Dudgeon (Crambidae: Lepidoptera) in sugarcane in UP. (Nesbitt *et al.*, 1980), *Aproaerema modicella* (Dev.) (Gelechiidae: Lepidoptera) in pigeon pea intercropped with groundnut in Tamil Nadu. (Nandagopal *et al.*, 2001), *S. litura* (Fabricius) (Noctuidae: Lepidoptera) in groundnut in A.P. (Srinias and Rao, 1993; Rao *et al.*, 1989), *Scirpophaga incertulas* Walk (Pyralidae: Lepidoptera) in rice ecosystem in A.P. (Pasalu *et al.*, 2004).

Mating Disruption in Rice

Delivery System of the Pheromone

The delivery system is very import for release of pheromone for longer period to avoid lure replacement with in a short period (now it is within 21 days). There are several delivery system available for the mating disruption. Some of them are rubber septa, plastic laminate, hollow fiber, polyethylene tube, PVC resins, sprayable polymer beads, sprayable hollow fibers, spray able micro-encapsulations etc.

The microencapsulated pheromone offers the advantage of both utilizing the conventional pesticides application equipments and of distributing pheromones in a relatively uniform manner. The other option is to use micro-dispersible technique, hollow fiber dispensers applied with a modified spray procedure, appears to obviate the difficulty of maintaining the required emission rate, since longevity of emission depends on the length of the fiber.

Trap Design

Limited information is available on the development of traps for various insects in Indian agriculture. Efficient traps were design for groundnut leaf miner (Nandagopal and Soni,1993); *H armigera* (Pawar *et al.*, 1988 and Nandagopal *et al.*, 2003); Sugarcane internode borer, *Chilo sachariphagus indicus* (David *et al.*, 1986); for *Spodoptera litura*, (Prasad *et al.*, 1985; Krishnananda and Satyanarayana, 1985). The efficiency of these traps were between 60 to 90 per cent. The trap and the lures used for *H. armigera* attracted the *Pectinophora* males in pigeon pea ecosystem (Nandagopal *et al.*, 2003). Reed *et al.* (1975) and Satpute *et al.* (1985) reported an efficient trap for *P. gossypiella*. Various types of traps are available.

Monitoring

Yadav *et al.* (2004) suggested that timely and efficient monitoring should be the foundation of sound IPM programs, without which no proper pest management decision can be made.

The occurrence of the major insect pests is difficult to forecast unless it is noticed in a damaging proportion. Hence monitoring of the insect pests using various gadgets becomes imperative. There are various traps used in various situations and based on the behaviour of the target insects. Light trap for adults of Lepidoptera, Coleoptera; sticky trap for aphids, in some of the cases laser light traps are also used such as house fly; micro, Lepidoptera, thrips etc. Now the emphasis is on the use of pheromones. Pheromone traps are being used for all the major Lepidoptera insects, Coleoptera, Diptera etc. In the monitoring programme it is very important to recommend the required number of the traps to be used per unit area. Based on the trapping efficiency of the trap, the number of insects trapped, the management options have to be decided. Careful consideration is to be given to the availability of the host plant, the mode of pupation, occurrence of natural enemies of the target insects etc. Lot of research has to be done before recommending the number of traps to be used / unit area.

Pest monitoring using sex pheromones facilitates early detection of the pest occurrence, through moth population dynamics during crop season and this information helps in making decisions on the choice of pressing control tactics in other words precise timing of insecticide or bio-agents applications. The Table 23.1 gives the details of the pheromones used in Indian agriculture (Natarajan,2004).

In India, pheromones are used for monitoring of some important insects affecting various crops (Table 23.1).

Table 23.1: Pheromones Used for Monitoring Purposes in India

Pest Species	*Pheromone Components*
	CEREALS
Yellow stem borer, *Scirphophaga incertulas*	Hexadecenal (Z)-9-Hexadecenal (Z)-11-Hexadecenal (Z)-11-Hexadecen-1-ol (Z)-9-Octadecenal
Leaf folder, *Cnaphalocrocis medinalis*	Hexadecyl acetate (Z)-11-Hexadecenyl acetate Octadecyl acetate (Z)-13-Octadecenyl acetate
Leaf folder, *Marasmia patnalis*	(Z)-13-Octadecenyl acetate (Z)-11-Hexadecenyl acetate
Maize stalk borer, *Chilo partellus*	(Z)-11-Hexadecenal (Z)-11-Hexadecen-1-ol
Striped stem borer, *Chilo suppressalis*	Hexadecenal (Z)-9-Hexadecenal (Z)-11-Hexadecenal (Z)-11-Hexadecen-1-ol Octadecan-1-ol (Z)-11-Octadecenal
Climbing cutworm, *Mythimna separata*	(Z)-11-Hexadecenal (Z)-11-Hexadecen-1-ol (Z)-11-Hexadecenyl acetate
Purple stem borer, *Sesamia inferens*	(Z)-11-Hexadecenyl acetate (Z)-11-Hexadecen-1-ol (Z)-11-Hexadecenal

Contd...

Table 23.1–Contd...

Pest Species	*Pheromone Components*
Rice green caterpillar, *Naranga aenescens*	(Z)-9-Tetradecenyl acetate (Z)-9-Hexadecenyl acetate (Z)-11-Hexadecenyl acetate
COTTON	
Pink bollworm, *Pectinophora gossypiella*	(Z,Z)-7,11-Hexadecadienyl acetate (Z,E)-7,11-Hexadecadienyl acetate
Spotted bollworm, *Earias vittella*	(E,E)-10,12-Hexadecadienal (E,E)-10,12-Hexadecen-1-ol Z11-Hexadecenal Z11-Octadecenal
Spiny bollworm, *Earias insulana*	(E,E)-10,12-Hexadecadienal
Gram pod borer, *Helicoverpa armigera*	(Z)-9-Hexadecenal (Z)-11-Hexadecenal Hexadecenal (Z)-11-Hexadecen-1-ol Hexadecan-1-ol
Oriental tobacco budworm, *Helicoverpa assulta*	Hexadecenal (Z)-9-Hexadecenal (Z)-11-Hexadecenal Hexadecencyl acetate (Z)-9-Hexadecencyl acetate (Z)-11-Hexadecencyl acetate Hexadecen-1-ol (Z)-9-Hexadecen-1-ol (Z)-11-Hexadecen-1-ol
Tobacco caterpillar, *Spodoptera litura*	(Z)-9(E)-11-Tetradecadienyl acetate (Z)-9(E)-12-Tetradecadienyl acetate
OILSEEDS	
Groundnut leaf miner, *Aproaerema modicella*	(Z)-7-Decenyl acetate (E)-7-Decenyl acetate (Z)-7,9-Decadienyl acetate
Red palm weevil, *Rhynchophorus ferrugineus*	(4S,5S)-4-methyl-5-nonanol (4S)-4-methyl-5-nonanone
Rhinocerous beetle, *Orcytes rhinoceros*	Ethyl-4-methyloctanoate Ethyl-4-methylheptanoate 4-Methyloctanoic acid
VEGETABLES	
Diamond-back moth, *Plutella xylostella*	(Z)-11-Hexadecenal (Z)-11-Hexadecenyl acetate (Z)-11-Hexadecen-1-ol
Cabbage stem borer, *Hellula undalis*	(E, E)-11, 13-Hexadecadienal
Brinjal borer, *Leucinodes orbonalis*	(E)-11-Hexadecenyl acetate (E)-11-Hexadecen-1-ol

Contd...

Table 23.1–Contd...

Pest Species	*Pheromone Components*
Cutworm, *Spodoptera exigua*	Tetradecyl acetate (Z)-9-Tetradecenyl acetate (Z)-11-Tetradecenyl acetate (Z,E)-9,12-Tetradecadienyl acetate (Z,Z)-9,12-Tetradecadienyl acetate (Z)-9-Tetradecen-1-ol
Tobacco caterpillar, *Spodoptera litura*	See under item 14
Gram pod borer, *Helicoverpa armigera*	See under item 12
Earias sp.	See under item 10 and 11
Potato tuber moth, *Phthorimaea operculella*	(E, Z)-4,7-Tridecadienyl acetate (E, Z, Z)-4,7,10-Tridecatrienyl acetate
Sweet potato weevil, *Cylas formicarius*	(Z)-3-Dodecenol-(E)-2-butenoate
SUGARCANE	
Sugarcane stalk borer, *Chilo auricilius*	(Z)-7-Dodecenyl acetate (Z)-8-Tridecenyl acetate (Z)-9-Tetradecenyl acetate (Z)-10-Pentadecenyl acetate
Early shoot borer, *Chilo infuscatellus*	(Z)-11-Octadecen-1-ol
Internode borer, *Chilo sacchariphagus indicus*	(Z)-13-Octadecenyl acetate (Z)-13-Octadecen-1-ol
Armyworm, *Mythimna separata*	See under item 6
Purple stem borer, *Sesamia inferens*	See under item 7
TREE CROPS	
Shoot and panicle borer of cardamom, turmeric and ginger, *Dichocrocis punctiferalis*	(E)-10-Hexadecenal (Z)-10-Hexadecenal

Krishnaiah (1995) reported several parameters that would enable efficient pheromone based monitoring for insect pests of rice crop. Krishnaiah *et al,* (1998) optimized three traps per acre as trap density for monitoring of rice yellow stem borer. Various parameters of an efficient pest monitoring system for rice pests are presented in Table 23.2.

Monitoring System in Cotton

Several studies have been conducted on cotton pests to establish a quantitative relationship between pheromone trap catches of particular pest and the damage caused with an aim to develop relationship would help to forewarn. Utilization of pheromones of various cotton pests as follows (Table 23.3).

Monitoring System in Groundnut

The presence of sex pheromones of leaf miner has been identified for the first time (Nandagopal and Reddy, 1991) by the National Research Centre for Groundnut (NRCG) in collaboration with NRI,

U.K. and field evaluation of the components of the sex pheromones have been also done by the NRCG. Experiments were also conducted by NRCG for standardizing the number of traps per unit area and use of the sex pheromone as one of the components of Integrated Pest Management (IPM) in groundnut where leaf miner damage is extensive under Saurashtra conditions. Nandagopal and Soni (1993) have evaluated a number of traps for the groundnut leaf miner (*Aproaerema modiceela* Deventer. Hall *et al.* (1993) have identified the total components of the pheromone compounds of leaf miner.

Table 23.2: Pheromone Based Monitoring Systems of Rice Insect Pests

Insect	*S. incertulas*	*C. suppressalis*	*S. inferens*	*C. medinalis*
Components	(Z) 11-16: Al (Z) 9-16: Al	(Z) 11-16: Al (Z) 9-16: Al (Z) 13-18: Al	(Z) 11-16: Ac (Z) 11-16: OH (Z) 11-16: Al	(Z) 13-18: Ac (Z) 11-16: Ac
Best blend	3:1	10:1:1	40:10:1	10:1
Dose	2 to 5 mg	0.3 to 0.6 mg	2mg	1mg
Dispenser	rubber	rubber	rubber	polythene vial
Replacement period	3-4 weeks	4 weeks	3-4 weeks	4 weeks
Trap design	sleeve	plastic funnel	sleeve	delta sticky
Sleeve color	white	yellow	–	–
Trap elevation	0.5-1 m	0.5 m	1.0m	canopy level

Table 23.3: Utilization of Pheromones of Various Cotton Pests

Bollworm	*Trap HT*	*Predictability of Darval Damage*	*Catch/Night/Trap*
Bollworm		For 3 nights	Expect egg laying in the next 5-8 days
Pink bollworm	0.03m	8-10males/trap/ night for 4 nights	Expect egg laying in in the next 10-15 days
Spotted bollworm	0.03m	1 males/night for 3 nights	Expect egg laying in in the next 3-5 days
Spiny bollworm	0.03m	1 males/night for 3 nights	Expect egg laying in in the next 3-5 days

At Junagadh, a square glue trap developed at NRCG (Nandagopal and Soni, 1993) was used, while in other centres water trap designed by ICRISAT was used. Different types of traps were designed and tried in the trapping of males of *A. modicella*. The results indicated that a square glue trap designed with 22 cm side and with 4 entry holes (Nandagopal and Soni, 1993) was the most efficient in trapping the males. The pheromone was also used for monitoring the activity of the moths in ground nut (Yadav *et al.*, 2004). A trap efficient to a tune of 60 to 90 per cent has been developed for *Spodoptera litura* in groundnut. Nandagopal *et al.* (1995) have reported on the effects of certain components of IPM on the damage and yields in groundnut. Nandagopal (1998) explained about hove the eco-friendly management of Groundnut Leafminer in India can be achieved. Nandagopal *et al.* (1998) reported on the development of an efficient pheromone trap for *Aproaerema modicella* Dev. (Gelechiidae: Lepidoptera) population in groundnut crops. Further, Nandagopal *et al.* (2006) have developed a viable efficient sex pheromone trap for *Aproaerema modicell.* In *Spodoptera,* Nandagopal *et al.*(2003) reported on the population dynamics of *Spodoptera litura* (F.) in relation to weather parameters in groundnut in India using pheromone trap catch. Nandagopal *et al.*(2004) have developed an efficient pheromone trap for field catch of *Spodoptera litura* Fabricius in groundnut and castor ecosystem. Based on the pheromone trap catches the prediction of *Spodoptera litura* (F.) males in relation to key weather factors in groundnut have been reported. Holistic IPM was developed wherein the use of pheromone has been explained (Ghewande *et al.*, 2002).

Monitoring System in Red Palm Weevil *Rhynchophorus ferrugineus* Oliv.

Attraction of red palm weevil *Rhynchophorus ferrugineus* to ferrugineol based pheromone lures in coconut. The trapped weevils with different lures are given in Table 23.4 (Faleiro, 2004).

Table 23.4: Weevils Trapped with Different Lures

Sl.No.	Treatments	Strength of Lure	Mean Weevil Catch per Trap			
			Trial I (23/12/02 to 24/1/03)	Trial II (25/1/03 to 26/2/03)	Cumulative	Sex ratio
1.	Pherobank RPW lure	400mg	3.52 (12.0)	2.90 (8.3)	3.26 (10.1)	1: 2.05
2.	Pherobank RPW lure	700 mg	2.70 (7.0)	2.41 (5.6)	2.57 (6.3)	1: 2.8
3.	Pherobank RPW lure	1000 mg	3.02 (8.7)	2.41 (5.6)	2.76 (7.1)	1: 1.39
4.	Ferrolure+	800 mg	2.88 (8.0)	2.66 (7.6)	2.79 (7.8)	1: 1.35
5.	ISCA Technology	900 mg	2.76 (7.3)	2.22 (4.6)	2.55 (6.0)	1: 1.25
6.	CPCRI lure	0.157 g	1.56 (2.0)	1.26 (1.3)	1.47 (1.6)	1: 2.00
7.	Ferrolure+ only (no food)	800 mg	2.18 (4.3)	1.74 (2.6)	2.00 (3.5)	1: 1.63
8.	Food only (coconut petiole)	–	0.88 (0.3)	0.71 (0.0)	0.81(0.1)	1: 0.00
	CD (P = 0.05)		0.49	NS	0.46	–

Monitoring of Coffee White Stem Borer: *Xylotrechus quadripes* (Coleoptera: Cerambycidae)

Coffee is one of the major plantation crops grown in India and occupies a place of pride among the plantation crops. Arabica and Robusta are the two types of coffee cultivated on commercial scale. Arabica is known for its quality and has high international demand. Coffee Arabica in India, China, Srilanka, Vietnam and East African Countries however, has been threatened by the presence of a dreaded pest called Coffee White Stem Borer, *Xylotrechus quadripes*, Chevrolat (Coleoptera: cerambicidae) that has the potential to devastate the entire plantation if uncontrolled. Chemical methods developed to control the pest have their own disadvantages, as they are harmful to the environment. Biological methods of control proved to be a failure so far. Therefore, an alternate eco-friendly technique like use of sex pheromone has been developed after taking a clue that these beetles communicate each other through chemical medium. Studies showed that male borers attract the females by secreting sex pheromone. The sex pheromone components from the males have been successfully isolated, chemically characterized and synthesized in the laboratory. The components were identified to be (*S*)-2-hydroxy-3-decanone, (R)-3-hydroxy-2-decanone and 2,3-dihydroxy octane. Laboratory bioassay conducted using the live beetles confirmed that (*S*)-2-hydroxy-3-decanone component is very active followed by (R)-3-hydroxy-2-decanone and the third component did not elicit any response from the females. Large-scale field trials were conducted for over 8 seasons (4 years) using the synthetic pheromone in sticky cross vane traps. Field trials showed that the male pheromone of coffee WSB could be effectively used to trap the female beetles in the field (Jayaramma *et al.*, 2004).

Table 23.5: Field Trap Catches of Beetles in Pheromone Baited Traps (1997-98 season)

Pheromone Source	*March–June (62 days Observation)*		*Sept–December (102 days Observation)*		*Total*	
	Female	*Male*	*Female*	*Male*	*Female*	*Male*
R	2	3	8	1	10	4
RS	7	4	15	2	22	6
S	13	2	44	2	57	4
Male	-	-	6	-	6	-
Control	2	6	3	4	5	10
Total	24	15	76	9	100	24

Locations 4, Trap height 1-4 meters, 8 traps/location.

Table 23.6: Field Trap Catches of Beetles in Pheromone Baited Traps (1998-99 season)*

Pheromone Source	*March–June (81 days Observation)*		*Sept–December (70 days Observation)*		*Total*	
	Female	*Male*	*Female*	*Male*	*Female*	*Male*
R	1	4	n.t	n.t	1	4
RS	11	3	n.t	n.t	11	3
S	57	2	45	5	102	7
S+ss	n.t	n.t	6	2	6	2
Control	n.t	n.t	1	0	1	0
Total	69	9	52	7	121	16

n.t: Not tested during the period; *: 16 locations during Mach-June period & 22 locations during Sept-Dec with 8 traps per location.Trap height 1.5 meters, Trap: Sticky cross vane.

Table 23.7: Field Trap Catches of Beetles in Pheromone Baited Traps (1999-2000 season)*

Pheromone Source	*March–June (87 days Observation)*		*Sept–December (39 days Observation)*		*Total*	
	Female	*Male*	*Female*	*Male*	*Female*	*Male*
S	112	11	126	24	238	35
Control	1	6	31	13	32	19
Total	113	17	157	37	270	54

n.t: Not tested during the period; *: 12 locations during Mach-June period, 8 traps/location & 76 locations during Sept-Dec with 4 traps/location, Trap height 1.5 meters; Trap: Sticky cross vane.

Table 23.8: Field Trap Catches of Beetles in Pheromone Baited Traps at Chinese Fields (2000-01)

Pheromone Source	*May–June*				*Total*	
	Gonglang 2 Coffee Base		*Daheishan Coffee Base*			
	Female	*Male*	*Female*	*Male*	*Female*	*Male*
S	16	21	7	7	23	28
Control	-	-	-	-	-	-
Total	16	21	7	7	23	28

Number of baited traps 8 and control 8 in Gonglang while in Daheishan 3 baited and 3 control.
Trap= sticky cross vane and dispenser = vial

Monitoring in Banana

Banana Corm Weevil (*Cosmopolites sordidus*)

The release of a volatile aggregation pheromone by male *Cosmopolites sordidus* was first reported in 1993 by Buddenberg *et al.* (1993) and isolated the pheromone component and identified as sordidin (Beauhaire *et al.* (1995). Nakayama & Mori (1997), Mori *et al.* (1996) reported the synthesis and stereo chemistry of sordidin. Weevil trapping using banana pseudostem trap is the commonly advocated component of the IPM options (Gold, 1998). Yadav (1999) emphasised the use of Pheromone as a tool in Integrated Pest Management. Pheromone studies on banana corm weevil has been reviewed (Padmanaban, *et al.*, 2001). Alpizar and Fallas (1997) have studied the use of the Pheromone traps and indicated that the the pheromones found to reduce the damage due to *Cosmopolites sordidus* and increase the yield in bananas and plantains. 5-10 times efficient than the stem traps in Costa rica. Cosmolure baited buried pitfall trap containing 3 per cent laundry detergent in water was 2.5 times more effective than Cosmolure baited stem traps, 20 per cent more capture of weevil was noticed when plastic gallons with a ramp was used (Alpizar and Fallas, 1997). Pitfall Cosmolure + traps caught 18 times the number of weevils as compared to pseudostem traps (Tinsara *et al.*, 1999). It was reported that 25 pheromone traps/hectare without changing the location of traps in the field was more effective than 4 traps/hectare (Alpizar and Fallas 1992). Compared to pseudostem trapping, the pheromone traps had a cost advantage, *Cosmopolites sordidus* capture rates in banana and plantains declined by 75 per cent over 10-12 months. Corm damage decreased by 61.64 per cent over the same period. Fruit bunch weight increased by 27 per cent in plantain and 23 per cent in bananas (Padmanaban *et al.*, 2004).

Traps made of cooking type bananas are more attractive to the weevil than traps made of dessert type of bananas (Seshu Reddy *et al.*, 1993). The 47 per cent increase in banana yield was reported from Kenya over 2 year period.

Banana Stem Weevil (BSW)

Banana stem weevil is an emerging pest problem in many banana growing areas. Chemical control methods are available (Mathew *et al.*, 1997). Pseudostem trapping of banana weevil is recently reported by Padmanaban, *et al.*(2001). Pheromone studies conducted jointly by NRCB, Trichy and IICT, Hyderabad indicated the presence of pheromone component(s) from the volatile and body washes.

Electroantenno Gram and Gas chromotography-Electroantennographic detector studies revealed that the influence of host plant (plant kairomones) in the banana stem weevil attraction (Padmanaban *et al.*, 2004).

Year and Month	*Number of Beetles Trapped*
January, 2001	21
February	11.68
March	4
April	4.5
May	4.92
June	1.87
July	4.5
August	9.35
September	22.72
October	20.25
November	19.6
December, 2001	18.8

Monitoring in Oilpalm

Managing with Aggregation pheromone of the coconut rhinoceros beetle, *Oryctes rhinoceros* L. (Coleoptera: Scarabaeidae as one of the tools of IPM is commonly practiced in all the Oil Palm growing areas. The aggregating pheromone, *oryctalure* (Sime RB pheromone) is responsible for effective trapping of beetles. The total beetles trapped are a function of the longevity of the pheromone sachet and the emergence of beetles (depending on amount of breeding material available). Though the pheromone and the trap were effective in trapping the beetles, but high temperatures could reduce the effectiveness. Cost benefit ratio for the pheromone trap and the number of beetles trapped was found uneconomical in the coastal areas of Andhra Pradesh under the existing atmospheric conditions. However, these can be utilized for dissemination of the other biological agents particularly the virus, *Baculovirus oryctes* (Kalidas, 2004).

Mass Trapping

This is a direct method of managing the insects. Based on the number of insects trapped, a decision can be taken to continue the recommended number of traps/unit area for mass trapping or remove all the traps and use only for monitoring. Because, if the anticipated insects are going to occur only below ET levels, there is no need for continuing the traps meant for mass trapping which will lead to unnecessary waste of resources. If the population is of moderate level, mass trapping followed by application of bio-pesticides would be sufficient to contain the insect before reaching economically important level. If the insect is occurring in a endemic way, necessarily, we have to think of going for too many methods of available tactics to manage the insects. The following are the insects for which mass trapping was tried (Table 23.9) (Nandagopal and Prasad, 2004).

Table 23.9: Mass trapping were Tried in India

Species	*Location*	*Reference*
Pthorimoaea opercullela	Karnataka and Tamil Nadu	Trivedi *et al.*, 1994
	Karnataka	Nandihalli *et al.*, 1993
	New Delhi	Gupta *et al.*, 1990a,b
	Uttar Pradesh	Siddiqi, 1989
	Punjab	Anon,1981
Chilo auricilius	Uttar Pradesh, Hariyana	David *et al.*, 1986
S. litura	Uttar Pradesh	Singh and Sachan,1991
	Andhra Pradesh	Krishnaiah,K., 1986; Lalith kumari *et al.*, 1992
	Tamil Nadu	Dhandapani,1985
	Gujarat	Nandagopal *et al.*, 2004b
H. armigera	Andhra Pradesh	Lalith kumari *et al.*, 1992
	Haryana	Pawar *et al.*, 1984.
	Gujarat	Patel *et al.*, 1985
	Tamil Nadu	Balakrishnann, 2004a
	Tamil Nadu	Balakrishnann *et al.*, 2004b
	Gujarat	Nandagopal *et al.*, 2004b
	Gujarat	Patel *et al.*, 2004f
Aproaerema modicella	Gujarat	Nandagopal,1992; Nandagopal *et al.*, 2004a; Nandagopal *et al.*, 2004b
	Andhra Pradesh	Yadav *et al.*, 2004
Lymantria americana (=*Porthetria dispar*)	Jammu Kashmir	Masoodi *et al.*, 1990
Cydia pomonella	Jammu Kashmir	Pawar *and* Tuhan, 1985
	Ladakh	Pawar *et al.*, 1982; Pawar *et al.*, 1984
Pectinophora gossypiella	Tamil Nadu	Balasubramanian *et al.*, 1979; Satpute *et al.*, 1985; Balakrishnann *et al.*, 2004a
	Gujarat	Patel *et al.*, 2004e
Periplaneta mericana	Karnataka	Urs *et al.*, 1989
Orcytes rhinoceros	Andhra Pradesh	Kalidas *et al.*, 2004
Rhynchophorus ferrugineus	Goa	Faleiro *et al.*, 2004
	Karnataka	Kalleshwara swamy *et al.*, 2004b
Scirpophaga incertulas	Andhra Pradesh	Yadav *et al.*, 2004; Pasalu *et al., 2004*
Xylotrechus quadripes	Karnataka	Jayarama *et al., 2004*
Chilo sacchariphagus indicus	Karnataka	Jayanth and Bhanu, 2004
Scirpophaga excerptalis	Karnataka	Jayanth and Bhanu, 2004
Chilo infuscatellus	Karnataka	Jayanth and Bhanu, 2004

Contd...

Table 23.9–Contd...

Species	Location	Reference
Batocera dorsalis	Gujarat	Patel, 2004b
	Gujarat	Patel, 2004c
Odoiporus longicollis	Tamil Nadu	Padmanaban *et al., 2004b*;
Padmanaban *et al., 2004c*		
Earias spp.	Tamil Nadu	Balakrishnann *et al.*, 2004a,2004b
	Gujarat	Patel *et al.*, 2004f
Leucinodes orbonalis	Gujarat	Jhala, 2004
Cylas formicarius	Kerala	Pillai *et al.*, 1996
Plutella xylostella	Karnataka	Reddy and Urs,1997

Mating Disruption

Male confusion is thought to be the result of ambient pheromone concentrations sufficient to hide the trails of the calling female to the large doses from diffuse sources such as microcapsules or larger doses of pheromone at the point source dispensers such as tie-on polyethylene rope. Added to the effect, is the adaptation of antennal receptor sites and for habituation of the insect's central nervous system. Specific receptor sites on the antennae respond to only the pheromone molecules. When a receptor site is continually activated by high ambient concentrations of the pheromone, the resulting electrical signal diminishes (measured by an electroantennogram). The receptor site becomes unresponsive and the insect becomes navigationally blink. When the insect's central nervous system is inundated with signals from the receptor sites, it becomes habituated and no longer remains cabable to provide the directed behaviour. The net result of confusion is that the male is unable to orient to any pheromone source and follow the upwind trail to a mate (Flint and Doane, 1996).

Some possibilities of the mechanism of the confusion approach as put forth by Pedigo (2002) include:

1. Camouflage or covering up the natural pheromone scent of females
2. Misdirection of males to scents from multiple point sources of synthetic pheromone, and
3. Adaptation/habituation by desensitizing male antennal receptors through constant pheromone-like exposure

Bartell (1982) delineated five mechanisms by which mating disruption could be successful as below:

1. Sensory adaptation of the pheromone receptors and central nerve habitauation,
2. False trial following caused by a multiplicity of the trials produced by synthetic odour source,
3. Inability of insects to discriminate odour trials from background,
4. Imbalance of sensory input caused by the release of an unnatural ratio of pheromone components,
5. Peripheral sensilla blocked by pheromone analogues

Mating Disruption Technique in the Key Insect Pests in India

In a very few cases, mating disruption technique was successful in India. Based on the information available, communication disruption may be successful in crops where dense foliage results in reduced wind velocity and thus pheromone persists for a few hours. Some examples are *Chilo auricilius* in sugarcane in Uttar Pradesh (Nesbitt *et al.*, 1980), *Spodoptera litura* in groundnut in Andhra Pradesh (Rao *et al.*, 1989), Hall *et al.* (1994) used the pheromones for mating disruption of cotton bollworms and rice stemborer in developing countries. Slow-release pheromone formulations have been used for control by mating disruption of cotton bollworms (*Pectinophora gossypiella, Earias vittella* and *E. insulana*) in Pakistan and yellow rice stem borer (*Scirpophaga incertulas*) in India. Single applications of pheromone formulations gave season-long control and final yields at least as good as those achieved with conventional insecticides, and savings of up to five applications of insecticides against cotton bollworms and two against rice stem borer.

Mating disruptions in *Aproaerema modicella* in pigeon pea intercropped with groundnut in Tamil Nadu (our unpublished work) reported to be successful. Similarly, Pasalu *et al.* (2004) and Cork *et al.* (1996) have used unnatural blend of pheromones for the successful mating disruption of rice yellow stem borer, *Scirpophaga incertulas*. The first trial compared the efficacy of 2 formulations containing 1:10:1 and 1:10:0 ratios of (Z)-9-hexadecenal, (Z)-11-hexadecenal and (Z)-9-octadecenal, components of the sex pheromone of yellow stem borer and a commercially-available formulation, Selibate CS, containing the related pheromone of *Chilo suppressalis*, a 1:10:1 blend of (Z)-9-hexadecenal, (Z)-11-hexadecenal and (Z)-13-octadecenal, with farmers' practice plots treated with insecticides. Pheromone trap catches indicated that in each of the pheromone-treated plots the catches of adult males were reduced by up to 98 per cent compared with catches in the insecticide-treated plots, suggesting that pheromone-mediated communication was disrupted. Larval damage ranged from 5.7 to 8.1 per cent white heads (WH) in the insecticide-treated plots compared to a significantly reduced range of 2.1 to 2.4 per cent WH in the pheromone-treated plots.

Cork and Basu (1996) also attempted to manage rice yellow stem borer by mating disruption with a PVC resin formulation of the sex pheromone of *C.suppressalis* in West Bengal.

Dhawan and Sidhu (1978) have also reported on the effect of location of gossyplure traps on catches of the pink bollworm, *Pectinophora gossypiella* males. They conducted field studies in 1979-80 in India to observe the effect of position of gossyplure-baited traps on the capture of males of this gelechiid pest on the cotton. Traps placed at 25 cm above the crop canopy or near to stacked cotton stalks had the highest catch. Before the start of flowering, catches in traps inside and outside the cotton field were similar, but during flowering catches were lower in traps placed outside the cotton fields. Taneja and Jayaswal (1986) worked out the population dynamics of *P. gossypiella* on upland cotton in Haryana. The number of adult males caught in traps baited with the sex pheromone gossyplure increased from mid-July to a peak in September-November, and declined thereafter. The incidence of larvae on cotton flowers peaked in mid-to late August, whereas the incidence of larvae in bolls increased during the cropping season. Using regression techniques, a linear relationship was established between the per cent incidence of larvae in flowers and bolls and the number of males caught/ trap/night. Damage caused by the larvae was positively correlated with trap catches during the consequative crop seasons.

Rup and Sharma (1978) studied the role of the sex pheromone in the mating behaviour of pulse beetle, *Callosobruchus maculates* (F.), with a simple olfactometer in the laboratory. Observations confirmed an earlier finding that only the females of this species produce an attractant. The insignificant attraction of the females to the males, males to males and females to females confirmed the absence of an aggregation pheromone in this bruchid. Kanaujia and Sidhu (1979) were the pioneers who worked on pheromones of storage insects in India. The effects of age, mating, time of day and larval diet on the production of sex pheromone by adult females of the Angoumois grain moth *Sitotroga cerealella* were investigated in the populations reared on maize, wheat and rice grains in the laboratory at Ludhiana. Pheromone production commenced in female pupae 1.5 days before adult emergence, reached a peak intensity 1-2 days after emergence and declined very slightly thereafter, persisting even in dead females. After mating, it declined to a level below that of virgin females but did not cease altogether. Pheromone production occurred throughout the diel but was highest at 24.00-04.00 h. There was no significant difference between the pheromone content of females that developed from larvae reared on the 3 different kinds of grain. The same authors have also studied on the factors affecting the responsiveness of males to female sex pheromone in *S. cerealella* (Kanaujia and Sidhu, 1980). The laboratory studies were carried out in India on the response of males to the female sex pheromone, with regard to age, mating, pheromone concentration and time of day. Mating response of males increased with increase pheromone concentration. Males of all the age-groups tested (0.25-9 days) showed similar responses to the pheromone. Mating induced a very high reversible inhibition of response that was independent of pheromone concentration and persisted for 14 h after mating. Although males responded to the pheromone at all hours of the day and night, a peak was observed from 24.00 to 05.00 h with a gradual increase during the dark hours and an abrupt decrease during the light hours.

Sharma and Deora (1980) reported that females of rice weevil, *Sitophilus oryzae* produced and released a sex pheromone on the day after adult emergence but the optimum level of response by males occurred on the fifth day. Pheromone responsiveness of males varied with the time of day showing the maximum response between 10.00 to 14.00 h and the minimum at 02.00 h. A normal photoperiod (LD 14:10) was required for the full male response to the pheromone. Mated and unmated males responded similarly. Starvation of males and females enhanced male responsiveness.

Mating Disruption

Nesbitt *et al.* (1986) have reported the first ever work on the sugarcane internode borer in India in collaboration with Sugarcane Breeding Institute, Coimbatore. Four pheromone components were detected in ovipositor washings and volatiles from females of the pyralid *Chilo auricilius* using combined gas chromatography and electroantennography. The components were identified as (I) (Z)-7-dodecenyl acetate [looplure], (II) (Z)-8-tridecenyl acetate, (III) (Z)-9-tetradecenyl acetate and (IV) (Z)-10-pentadecenyl acetate by comparison of their gas chromatographic behaviour with that of synthetic standards. In field tests carried out in northern India during 1982-84, a combination of II, III and IV components in their naturally occurring ratio (8:4:1) had shown a highly attractive synthetic source of pheromones. Looplure was found to reduce catches of males of *C. auricilius*, both when dispensed with the other 3 components and when released from dispensers surrounding a trap baited with the other 3 components.

Studies on mate seeking behaviour and anemotactic response of tropical warehouse moth, *Ephestia cautella* to (Z,E)-9,12-tetradecadienyl acetate were made by Singh and Majumder (1983). Probably, this is a first attempt on the synthesis of a pheromone in the India by Majee *et al.* (1983), who have synthesed Queenbee pheromone. Chattopadhyay *et al.* (1983) also synthesized the pheromones of the queen bee and cabbage looper from aleuritic acid. The effects of various concentrations of (9Z,12E)-9,12-tetradecadienyl acetate on the mate-seeking and upwind anemotactic behaviour of *E. cautella* were studied in the laboratory in India. The mating rate was inversely related to anemotactic flight at all pheromone concentrations tested. Extensive mating was observed during periods of low flight activity.

The pheromone had no effect on calling females. Much anemotactic zig-zagging flight activity was recorded 5-15 min after exposure to the pheromone at a rate of 10-3 ng, while there was little straight flight activity in comparison with zig-zagging flight at all pheromone dosages. The highest upwind anemotactic response was observed 5-15 min after exposure to 10-4 ng of the pheromone. Olfactory studies indicated decreased response to increasing dosages of the pheromone.

Siddiqi and Khan (1983) have established the presence of a female sex pheromone which was detected for the first time in the grass hopper, *Hieroglyphus nigrorepletus* (or indeed in any acridid) in the laboratory at Aligarh Muslim University, Aligarh, India, by means of a glass olfactometer. The pheromone was secreted in the digestive tract and expelled with the excreta, and it was perceived by the male through the antennae. No pheromone was secreted by newly emerged females, but secretion increased from 72 h to 8-9 days after emergence, when the first mating usually occurred; older females continued to release the pheromone, although in reduced quantities up to the age of 29 days.

Chari *et al.* (1985) have studied the populations of the noctuid *H. armigera* and monitored with pheromone and light traps in field studies in bidi tobacco in Gujarat, India, during 1981-83. The pheromone traps were baited with 1 mg of a mixture of the synthetic pheromones (Z)-11-hexadecenal and (Z)-9-hexadecenal in the ratio 97:3. Catches in the pheromone traps peaked in December-April while catches in the light traps fluctuated. Light-trap catches were found to be significantly influenced by weather parameters and the moon phase. The study indicated that pheromone traps were more reliable for monitoring populations of *H. armigera* than light traps.

Sain and Kalode (1985) have used the virgin female of the rice gall midge, *Orseolia oryzae* to attract males of the same species using various types of traps was investigated in field studies in India in 1982. Of three types of traps tested, only the delta trap was found suitable. When traps containing 3 and 6 virgin females placed in the middle of rice fields, 24 and 125 males were caught/night, respectively. No males were trapped in the absence of virgin females, indicating the presence of strong female pheromone. The population of *O. oryzae* was monitored weekly with the delta traps. The average catches varied from 2.3 males/trap/night in December to 120 in October, while there were no catches in February and May-July. The trap had several advantages over the light trap presently used for monitoring. A technique for catching live males of *O. oryzae* for laboratory cultures using an inverted funnel trap was described.

A good amount of information has been brought out by Sahu and Hameed (1983) on the pheromones in insect control. After five years of the earlier report on the Induction of oestrous cycle irregularities in mice: presence of pheromonal stimulus in urine and excreta of females was reported by Gangrade and Dominic (1983). They have reported that Oestrous cycle irregularities and prolongation of the duration of vaginal cornification were induced in individually housed female by exposure to the urine of female housed in single-sex groups, or to bedding soiled by the excreta of such female. The results suggest the involvement of a non-volatile urinary contact pheromone in the incidence of oestrous cycle irregularities in unisexually grouped mice. Chandla *et al.* (1987) worked on monitoring of adult potato tuber moth, *Phthorimaea operculella*, with sex pheromone by conducting in potato fields at Shimla, India, during 1983-84. Sex-pheromone-baited water traps were found to be effective as a means of detecting and monitoring the occurrence of adults of *P. operculella* at an early stage, so that control measures could be applied at the most appropriate time. The traps were baited with (4E,7Z)-4,7-tridecadienyl acetate and (4E,7Z,10Z)-4,7,10-tridecatrien-1-ol acetate. A total of 575 and 631 adult males were caught during May to August in 1983 and 1984, respectively.

Trap Design

Limited information is available on the development of traps for various insects in Indian agriculture. There has been efficient trap design for groundnut leaf miner (Nandagopal and Soni,1993) in groundnut ecosystem, for *H armigera* (Pawar *et al.*, 1988 and Nandagopal *et al.*, 2003), For sugarcane internode borer, *Chilo sachariphagus indicus* Kapur (David *et al.*, 1986), for *S. litura*, (Prasad *et al.*, 1985;Krishnananda *and* Satyanarayana, 1985). The efficiency of these traps were between 60 to 90 per cent. In the case of *Pectinophora gossypiella* (Saunders) in cotton ecosystem an interesting observation was reported (Nandagopal *et al.*, 2003). The trap and the lures used for *H. armigera* attracted the *Pectinophora* males. Reed *et al.* (1975) and Satpute *et al.* (1985) reported a efficient trap for *P. gossypiella*.

Trap Design Developed

Information available on the development of traps for various insects in Indian agriculture is limited. Efficient traps were designed for *H. armigera* (Tamhanker *et al.*, 1993; Pawar *et al.*, 1988 and Nandagopal *et al.*, 2003); groundnut leaf miner (Nandagopal and Soni,1993; Nandagopal *et al.*, 2006); Sugarcane internode borer, *Chilo sachariphagus indicus* (David *et al.*, 1986); and *Spodoptera litura*, (Prasad *et al.*, 1985; Krishnananda *and* Satyanarayana, 1985; Ranga Rao *et al.*, 1993). The efficiency of these traps were between 60 to 90 per cent. The trap and the lures used for *H. armigera* attracted the *Pectinophora* males in pigeon pea ecosystem (Nandagopal *et al.*, 2003). Reed *et al.* (1975) and Satpute *et al.* (1985) reported an efficient trap for *P. gossypiella*. The efficacy of 4 types of pheromone traps for catching adult males of *Spodoptera litura* in groundnut fields was evaluated in studies in India in 1988 (Table 23.10). There was no significant difference in the performance of single-and double-funnel traps, and the single-funnel trap (20 cm in diameter) captured more moths than any other type of trap. Septa of 4 weeks or less exposure attracted most moths. One and 2 traps per ha caught significantly fewer moths than 4 and more traps per ha; however, there was no significant improvement in capture when 4 or more traps per ha were installed. Night observations indicated that many moths escaped from sleeve traps. Single plastic funnel trap found suitable when being used in monitoring *S. litura* in the national monitoring network in India (Raa *et al.*, 1991) (Nandagopal and Prasad, 2004).

Table 23.10: Trap Designs Developed in India

Species	*Reference*
Pectinophora gossypiella (Saunders) (Gelechiidae:Lepidoptera)	Reed *et al.*, 1975 Nandagopal *et al.*, 2003
Helicoverpa (Hub.) (Noctuidae:Lepidoptera)	Pawar *et al.*, 1988; Tamhankar *et al.*, 1993; Nandagopal *et al.*, 2003
Sitotroga cerealella (Olivier) (Gelechiidae: Lepidoptera)	Kanaujia and Sidhu, 1981.
S. litura (Noctuidae:Lepidoptera)	Prasad *et al.*, 1985; Krishnananda and Satyanarayana, 1985
Chilo sachariphagus indicus Kapur (Pyralidae:Lepidoptera)	David *et al.*, 1986; PCI (pvt) Ltd,Bngalore
Aproaerema modicella (Dev.) (Gelechiidae:Lepidoptera)	Nandagopal and Soni,1993, Nandagopal *et al.*, 2004; 2006
Spodoptera litura	Ranga Rao *et al.*, 1991
Bactrocera dorasalis walker	Patel *et al.*, 2004

Traps Available and Improved

There are many kinds of traps available for various important insect and mite pests. These include Sleeve trap, Water trap made of Galvonized Iron trays or Plastic Watta –Trap, Sticky trap such as Delta trap, Diomond trap, Panel trap, Unit trap, Rhegalatis trap.

There are particular trap for each of the species of insects belonging to a family/order to trap a specific pest. For example water traps are mostly used for Pyralid moths of the order Lepidoptera. When it was modified using a plastic, the trap is used for catching males of brinjal fruit and shoot borer and Groundnut leaf miner a Gelechiidae of order Lepidopera.

A typiocal example of how a available trap can be improved has been demonstrated by Nandagopal *et al.* (2006) which has been presented below:

Groundnut leaf miner (GLM), *Aproaerema modicella* Dev. (Gelechiidae: Lepidoptera) is a serious oligophagus insect pest of groundnut, soybean and other few legume crops in south and south East Asia (Mohammad, 1981). GLM populations fluctuate widely between seasons. Population peaks have been reported in July and August in Thailand (Campbell, 1983), at the end of the rainy season (September and October) specially in drought or low rain fall years (Amin, 1983), in November and December (Mohammad, 1981) as well as at the end of the post-rainy season *i.e.* March and April in Bangladesh and India (Islam *et al.*, 1983).

The identification of sex pheromone of GLM (Nandagopal and Reddy, 1990; Hall *et al.*, 1994) has enabled to monitor this pest with pheromone trap fabricated based on the behavior of this species (Nandagopal and Soni, 1993). Male moth catches of pheromone traps at ICRISAT, Hyderabad showed that two highest peak populations of this pest appear in 30-33 standard weeks (July-August) with 4-5 generations during rainy season and 6-11 standard weeks (February-March) with 6 generations during post-rainy season. Several years of pheromone trap data is required to indicate population fluctuations of GLM over seasons.

To mitigate the hazardous consequences due to indiscriminate use of chemical pesticides and reduce the cost of cultivation, use of sex pheromones for monitoring, mass trapping and mating disruption has become essential tool in IPM programme. Experiments were carried out on developing the efficient pheromone trap based on the behavior of this pest. Among various designs of traps such as delta, funnel, sticky and water trap evaluated against GLM, water trap was found effective (Ranga Rao *et al.*, 1993). Based on the trapping efficiency of the trap, number of insects trapped, the management options have to be decided. Hence the present investigation was taken up to evaluate different types of traps available in market and make necessary modifications so as to increase the trapping efficiency.

Randomized field experiment was conducted at National Research Centre for Groundnut (NRCG), Junagadh, Gujarat during rainy season of 2004. Pheromone lures of polythene vials impregnated with the blend of (Z,Z) 7-9-Decadienyl acetate (1) + (E) 7-Decenyl acetate (II) + (Z) 7-Decenyl acetate (III) in the ratio of 10:2:1.4 manufactured by Pheromone Research Group, Indian Institute of Chemical Technology (IICT), Hyderabad were used in the experiment. Each trap was fitted with lures loaded with 3 mg pheromone. The traps were placed at 0.5 m above the ground in the groundnut crop using the inverted ' L' shaped bamboo sticks (Ranga Rao *et al.*, 1993) and with a distance of 10 m apart on all the sides.

Four types of traps namely Acrylic Sticka trap (small), Acrylic Sticka trap (big), plastic tray trap, Wota-T trap were used in the study. Wota – T trap from Pest Control India (PCI), Bangalore was used with castor oil and water+ kerosene, while castor oil was used in the remaining traps as trapping material. Castor oil was smeared on the inside and sides of the trap as trapping material. All the traps with three replications were installed in groundnut fields where GLM infestation was more than 60 per cent. The positions of pheromone traps were changed clockwise on alternate day to eliminate locational/positional effect, if any on the trap catches. The traps were arranged following the procedure of Nandagopal (1992). Each trap was checked everyday for male moths trapped and observations were recorded for five consecutive days. The data obtained was statistically analyzed using two way analysis of variance.

The results of the present study showed that the difference in the male moths trapped in various traps were highly significant. Out of the five different types of traps evaluated against *A. modicella*, catches of male moths were significantly higher in Wota –T trap compared to other traps tested. Wota –T trap with castor oil was found superior (196.8 male moths /trap/day) which was on par with Wota –T trap having water and kerosene as trapping material (194.8 males/trap/ day).

The traps with two and four openings differed significantly, where 4 entry opening trap caught more male moths compared to two opening trap. Even the bigger acrylic trap which was found to be superior in our previous trials found to trap only 66 male moths / day however, the number of moths trapped in bigger acrylic trap with four openings were more than the smaller acrylic trap with two openings. The probable reason could be four opening of the bigger acrylic trap facilitate more number of male moths to enter into the trap compared to trap with two openings.

Since GLM is a passive flier with limited distance by itself and mostly carried by the downward wind whose direction may change. It is this which makes the four opening trap ideal as there is an opening at each of the four corners of the compass to facilitate the catching of the pest. The low population of males trapped in small acrylic sticka trap clearly indicates the difficulty that the male pest have when trying to negotiate entry to the trap when the wind is blowing in the wrong direction.

It is interesting to note that though the trapping area is more in Plastic tray trap (2040 cm^2) compared to Wota-T trap (1584 cm^2) the number of male moths trapped are less in plastic tray trap (108.5 male moths /trap/day) than in Wota –T trap (196.8 male moths /trap/day) indicating that the efficiency of the trap depends upon the design of the trap rather than the total trapping area. It was observed that the males when visited the lures run all around the lure, probably in search of the females, when found otherwise, they tried to move out. This behaviour of moving out is very important in designing the trap. When we measured the angles of movement and the site of the catch, it was below 45° angle and those which were caught on the sides were of more than 45° angle. However, there may be certain per cent of males moths might have escaped by flying above 45 ° angle (ACD) up to just below 180° angle BDE. Probably the increased angle might have facilitated for catching those males which may even move above 45 ° angle.

Ranga Rao *et al.* (1993) reported open water trap having 28cm x 22cm with a gap of 11 cm clearance was effective against GLM. The water trap used by the Ranga Rao *et al.* (1993) was made up GI sheets fitted with a rod. The cost of the water trap was estimated to be Rs. 200 per trap as per the local manufactures. Whereas the Wota –T trap used in this trial is about Rs. 39 per trap. There was no significant difference between castor oil and water + kerosene as trapping material in Wota –T trap indicating that these can be efficiently used for trapping of GLM however, we recommend to use castor oil in the Wota –T trap for easy and economic considerations (Nandagopal *et al.*, 2006).

References

Anonymous, 1981. An operational field trial project in India for suppression of the cotton pink bollworm, *Pectinophora gossypiella* (Saunders) (Gelechiidae: Lepidoptera) employing 'gossyplure hollow fiber' controlled release sex pheromone formulation. Technical Report Plant Protection Adviser to the Government of India. 1981, recd. 1983, No. 1, [7+] 44 pp.

Balakrishnan, N., Baskaran, R.K. Murali and Mahadevan, N.R., 2004. Pheromone trap catches of bollworms in different IPM modules in rainfed cotton ecosystem p. 16. In: *National Seminar on Trends in Pheromone Research and Technology*, 6-7 February. National Research Centre for Groundnut, Junagadh, Gujarat.

Balasubramanian, M., Murugesan, S. and Parameswaran, S., 1979. Trapping of the cotton pink bollworm *Pectinophora gossypiella* (Saunders) Gelechiidae, Lepidoptera moths using synthetic gossyplure. *Pesticides*, 13(6): 49, 51.

Bartell, R.J., 1982. Mechanism of communication disruption by pheromone in the control of Lepidoptera: a review. *Physiological Entomology*, 7: 353-354.

Campbell, W.V., 1983. Management of arthropods on peanut in southeast Asia. *Ann. Rep. Peanut Collab. Res. Support Prog.* (CRSP), University of Georgia, Georgia Experiment Station, USA.

Chandla, V.K., Bhalla, O.P., and Dogra, G.S., 1987. Monitoring of adult potato tubermoth, *Phthorimaea operculella* Zeller, with sex pheromone. *National Academy Science Letters*, 10(11): 397-399.

Chari, M.S., Patel, A.R., Rao, B.S., Bharpoda, T.M., and Patel, N.M., 1985.Population studies on tobacco capsule borer *Heliothis armigera* Hubner. *Tobacco Research*, 11(2): 98-104.

Chattopadhyay, A., Mamdapur, V.R., and Chadha, M.S., 1983. Synthesis of the queen bee and cabbage looper the pheromones of from aleuritic acid. *Indian Journal of Chemistry*, B. 22: 158-159.

Cork, A. and Basu, S.K., 1996.Control of the yellow stem borer, *Scirpophaga incertulas* by mating disruption with a PVC resin formulation of the sex pheromone of *Chilo suppressalis* (Lepidoptera: Pyralidae) in India. *Bull. Entomol. Res.*, 86(1): 1-9.

Cork, A, Souza-Kde, Krishnaiah, K. Kumar, D.V.S.S.R., Reddy, A.A., and Casagrande, E., 1996. Control of yellow stem borer, *Scirpophaga incertulas* (Lepidoptera: Pyralidae) by mating disruption on rice in India: effect of unnatural pheromone blends and application time on efficacy. *Bulletin of Entomological Research*, 86(5): 515-524.

David, H., Nesbitt, B.P., Easwaramurthi, S. and Nandagopal, V., 1985. Application of sex pheromone in sugarcane pest management. *Proceedings of the Indian Academy of Science*, 94: 333-340.

Dhandapani, N., 1985. Trapping of *Spodoptera litura* (Fb) moths using sex pheromone. *Madras Agricultural Journal*, Recd., 1987, 72(4): 236, 238.

Dhawan, A.K., and Sidhu, A.S., 1978. Effect of position of gossyplure-baited traps on the capture of males of the gelechiid *Pectinophora gossypiella* on cotton. *Journal of Insect Science*, 1(2): 136-141.

Faleiro, J.R., 2004. Pheromone trapping technology for the management of red palm weevil: An overview. *National Seminar on Trends in Pheromone Research and Technology*, 6-7 February. National Research Centre for Groundnut, Junagadh, Gujarat.

Faleiro, J.R., Ashok Kumar, J. and Rangnekar, P.A., 2002. *Crop Protection*, 21: 171-176.

Flint, H.M. and Doanne, C.C., 1996. Understanding semiochemicals with emphasis on insect sex pheromones in Integrated Pest Management Programs. http://www.epa.gov/

Gangrade, B.K. and Dominic, C.J., 1983. Induction of oestrous cycle irregularities in mice: presence of pheromonal stimulus in urine and excreta of females. *Archives-de-Biologie*, 94(3): 239-246.

Ghewande, M.P., Nandagopal, V., Desai, S. and Basu, M.S., 2002. Integrated pest management in groundnut. *NRCG Technical Bulletin*, pp. 19.

Gupta, G.P., Katiyar, K.N. and Vashisht, A.K., 1990a. Behaviour of male pink bollworm (*Pectinophora gossypiella*) toward gossyplure and its relationship with larval population and weather condition. *Indian Journal of Agricultural Sciences*, 60(6): 411, 416.

Gupta, G.P., Kishore, P. and Vashisht, A.K., 1990b. Monitoring of pink bollworm, *Pectinophora gossypiella* (Saunders), males through pheromone traps and weather parameters affecting population build. *U.P. Journal of Entomological Research*, 14(1): 21, 29.

Hall, D.R., Peter S. Beevor, Derek G. Campion, David J. Chamberlain, Alan Cork, Rosemary White, Aurelio Almetre, Thomus J. Henneberry, Nandagopal, V., Wightman, J. and Ranga Rao, G.V., 1993. Identification and synthesis of new pheromones. *IOBC Wprs Bulletin*, 16: 1-10.

Hall, D.R., Chamberlain, D.J., Cork, A., Desouza, K., McVeigh, L.J., Ahmad, Z., Krishnaiah, K., Brown, N.J., Casagrande, E. and Jones, O.T., 1994. The use of pheromones for mating disruption of cotton bollworms and rice stemborer in developing countries. *Proceedings Brighton Crop Protection Conference, Pests and Diseases*, Vol. 3, p. 1231-1238, BCPC Publications.

Islam, W., Ahmed, K.N., Nargis, A. and Islam, U., 1983. Occurance, abundance and extent of damage caused by insect pests of groundnut (*Arachis hypogaea*). *Malaysia Agri. J.*, 54: 18-24.

Jayarama, Vinod, Kumar, P.K., D'souza, M.Violet and Hall, David R., 2004. Identification, Characterization, Synthesis and Field application of Male Sex Pheromone of Coffee White Stem Borer (*Xylotrechus quadripes*).p. 14-25. In: *National Seminar on Trends in Pheromone Research and Technology*, 6-7 February, National Research Centre for Groundnut, Junagadh, Gujarat.

Kanaujia, K.R. and Sidhu, H.S., 1979. Factors affecting production of sex pheromone in females of Angoumois grain moth, *Sitotroga cerealella* (Olivier) (Lepidoptera: Gelechiidae). *Current Science*, 48(9): 420, 422.

Kanaujia, K.R. and Sidhu, H.S., 1980. Factors affecting the responsiveness of males to female sex pheromone in Angoumois grain moth, *Sitotroga cerealella* (Olivier). *Zeitschrift-fur-Angewandte-Entomologie.*, 89(4): 387-390.

Kanaujia, K.R. and Sidhu, H.S., 1981. Influence of trap design, pheromone source and release on moth trapping of Angoumois grain moth, *Sitotroga cerealella* (Olivier). *Current Science*, 50(11): 512, 514.

Karlson, P. and Butenandt, A., 1959. Pheromones (Ectohormones) in Insects. *Ann. Rev. Entomol.*, 4: 49-58.

Karlson, P. and Luscher, M., 1959. Pheromones : A new term for a class of biologically active substances. *Nature*, 183 : 55-56.

Kettlewell, H.B.D., 1942. The assembly scent of *Arctia villica* and *parasemia plantaginis*. Entomol. Res., 54: 62.

Kettlewell, H.B.D., 1943. Further observation on non-specific assembling scents in macro-lepidoptera. *Entomol. Res.*, 55: 107.

Kettlewell, H.B.D., 1946. Female assembly scents with reference to an important paper on the subject. *Entomologist*, 79: 8.

Kettlewell, H.B.D., 1955. A further case of similar assembling scents. *Entomologist*,. 88: 19.

Kirschenblatt, J.D., 1952.Telergones and their biological significance. *USP. Soverem. Biol.*, 46: 322.

Krishnaiah, K., 1995. Insect pheromones in integrated pest management of cereals. In: *Pheromones in Integrated Pest Management* (Eds. Krishnaiah K and Divakar B J). Plant Protection Association of India, pp. 33-39.

Krishnaiah, K., Rama, Gopala , Varma, N., Pasalu, I.C. and Zainulabeuddin, S., 1998. Improvement in the efficacy of sex pheromone of yellow stem borer, *Scirpophaga incertulas* Walker with co-attractants and plant extracts. *Indian Journal of Plant Protection*, 26(2): 131-137.

Krishnaiah, K., 1986. Studies on the use of pheromones for the control of *Spodoptera litura* Fab. on black gram grown in rice fallows. *Indian Journal of Plant Protection*, 14(2): 43-46.

Krishnananda, N. and Satyanarayana, S.V.V., 1985. Effect of height of pheromone trap on the capture of *Spodoptera litura* moths in tobacco nurseries. *Phytoparasitica.*, 13(1): 59, 62.

Lalithakumari, V.L. and Reddy, D.D.R., 1991. Sex pheromone in groundnut leafminer, *Aproaerema modicella. International Arachis Newsletter*, 9: 15-16.

Mani, E., Schwaller, F. and Riggenbach, W., 1978. Trap catches as indicators of disruption efficiency and uniformity of pheromone dispersal in *Cydia pomonella* trials. *Bulletin-SROP.*, 10(3): 17. In Working Group "Use of pheromones and other semiochemicals in integrated control", Neustadt, 8-12 September, 1986.

Masoodi, M.A., Trale, A.R. and Bhat, A.M., 1990. Suppression of *Lymantria obfuscata* Walker by sex pheromone trapping of males. *Indian Journal of Entomology*, 52(3): 414, 417.

Micklem, H.S., 1959. The proposed biological term Pheromone. *Nature (London)*, 83: 1935.

Mohammad, A., 1981. The groundnut leafminer (*Aproaerema modicella* Deventer) (*Stomopteryx subsecivella* Devventer) (Lepidoptera: Gelechiidae) : A review of world literature, 33 pp. occasional paper 3, Groundnut Improvement Program, International Crops Research Institute for Semi-arid Tropics, Patancheru.

Nandagopal, V., Prakash, A. and Rao, J., 2006. Status of pheromone research, application and future requirements in Indian agriculture.*J. Appl. Zool. Res.*, 17(2): 121-129.

Nandagopal, V. and Soni, V.C., 1993. Evaluation of pheromone traps for groundnut leafminer (*Proaerema modicella*). *Indian Journal of Agril. Sciences*, 63(1): 64-67.

Nandagopal, V. T.V., Prasad, Rajiv Rathod and Pradip Fulmali, 2004. Compatibility of sex pheromones of *Spodoptera litura* (F.), *Helicoverpa armigera* (Hub.) and *Aproaerema modicella* Dev. in cotton and groundnut ecosystem. In: *Abstract, National Seminar on Trends in Pheromone Research and Technology*, held on 6th and 7th of February, NRCG (ICAR), Junagadh, pp. 18.

Nandagopal, V., Prakash, Anand, Rao, J., Yadav, J.S. and Prasad, A.R., 2007. *Pheromones: Principles and Practices.* Central Rice Research Institute, Cuttcak, Indian Institute of Chemical Technology, Hyderabad, Applied Zoologist Research Association, CRRI, Cuttack, pp. 333.

Nandagopal, V., Fulmali, Pradip, Gedia, M.V.and Rathod, Rajiv, 2003. Development of an efficient pheromone trap for catching *Helicoverpa armigera* in pigeonpea ecosystem in the field. *International Pest Control* (London), 45(1): 32, 36.

Nandagopal, V., Yadav, J.S., Prasad, A.R., Fulmali, Pradip and Razak, Umashanker, 2004. Development of efficient pheromone trap for *Aproaerema modicella* Dev. (Gelechiidae: Lepidoptera) population in groundnut crops. In: *Souvenir, National Seminar on Trends in Pheromone Research and Technology*, held on 6th and 7th of February, NRCG (ICAR), Junagadh, pp. 96-103.

Nandagopal, V., 1992. Studies on integrated pest management in groundnut in Saurashtra. *Ph.D. Thesis*, Saurashtra University, Rajkot, pp. 247.

Nandagopal, V. and Reddy, P.S., 1990. Report of sex pheromone in groundnut leafminer, *Aproaerema modicella* Deventer. *International Arachis News Letter*, 7: 5-6.

Nandagopal, V., and Prasad, T.V., 2004. Why no fillip for pheromone technology in India, pp. 29-34, *National Seminar on Trends in Pheromone Research and Technology*, 6-7 February, National Research Centre for Groundnut, Junagadh, Gujarat.

Nandagopal, V., Prasad, T.V., Makwana, A.D. and Rathod, Rajiv, 2004. Pheromone in Indian Agriculture: A review nad requirements. In: *Souvenir, National Seminar on Trends in Pheromone Research and Technology*, held on 6th and 7th of February, NRCG (ICAR), Junagadh, pp. 96-103

Nandagopal, V., Soni, V.C., Hall, David R. and Gedia, M.V., 1995. Effects of certain components of IPM on the damage and yields in groundnut. *Research Presented in the National Seminar on IPM in Agriculture*, held at the College of Agriculture, Nagpur on 29 and 30th of December.

Nandagopal, V., 1998. IPM approach on the insect Pest Management in Groundnut Lecture delevered at the UNDP Trainers Training Programme on Advanced Technology in Groundnut Processing and Aflatoxin management held at NRCG, 11-13, November.

Nandihalli, B.S., Patil, B.V. and Hugar, P., 1993. Optimisation of pheromone traps for cotton pink bollworm, *Pectinophora gossypiella* Saunders. *Karnataka Journal of Agricultural Sciences*, 6(3): 300, 301

Natarajan, L., 2004. Isolation and identification of insect pheromones. pp. 144-159. *National Seminar on Trends in Pheromone Research and Technology*, 6-7 February, National Research Centre for Groundnut, Junagadh, Gujarat.

Nesbitt, B.F., Beevor, P.S., Cork, A., Hall, D.R., David, H., and Nandagopal, V., 1986. The female sex pheromone of sugarcane stalk borer, *Chilo auricilius*: identification of four components and field tests. *Journal of Chemical Ecology*, 12(6): 1377-1388.

Nesbitt, B.F., Beever, P.S., Hall, D.R., and Lesta, R., 1980. (Z)-9-hexadecenal, a minor component of the female sex pheromone of *Helicoverpa armigera* Hubner). (Lepidoptera : Noctuidae). *Entomologia Experimenta Applicata*, 27: 306-308

Padmanaban, B., Sundarajan, P. and Sathiamoorthy, S., 2004. PheromoneL Its scope in IPM in Banan. 121-127. In: *National Seminar on Trends in Pheromone Research and Technology*, 6-7 February, National Research Centre for Groundnut, Junagadh, Gujarat.

Pasalu, I.C., Varma, N R.G., Krishnaiah, K. and Katti, G., 2004. Pheromones: Its scope in IPM in rice. 56-71. In *National Seminar on Trends in Pheromone Research and Technology*, 6-7 February, National Research Centre for Groundnut, Junagadh, Gujarat.

Patel, Z.P., 2004. Potentiality of methyl eugenol trap against fruit-fly in sapota orchard of south Gujarat. (abstract) p. 8. In: *National Seminar on Trends in Pheromone Research and Technology*, 6-7 February, National Research Centre for Groundnut, Junagadh, Gujarat.

Patel, R.P., 2004. Fruit-fly management by using fruit-fly trap. (abstract) p. 9. In: *National Seminar on Trends in Pheromone Research and Technology*, 6-7 February, National Research Centre for Groundnut, Junagadh, Gujarat.

Patel, V.N., 2004. Population dynamics of pink bollworm, *Pectniphora gossypiella* Soaund based on sex pheromone trap under dry land condition. (abstract) p. 19. In: *National Seminar on Trends in Pheromone Research and Technology*, 6-7 February, National Research Centre for Groundnut, Junagadh, Gujarat.

Patel, V.N., Kabaria, B.B., Patel, G.M. and Chaudary, F.K., 2004. Population dynamics of pink bollworm, *Pectniphora gossypiella* Saund based on sex pheromone trap under dry land condition (Abstract). p. 19. In: *National Seminar on Trends in Pheromone Research and Technology*, 6-7 February, National Research Centre for Groundnut, Junagadh, Gujarat.

Patel, S., Patel, G.M.and Chaudary, F.K., 2004. Monitoring of male moths of *Earias vitella* and *Helicoverpa armigera*Hub. Through sex pheromone trap in cotton crop in North Gujarat (abstract) p. 2. In *National Seminar on Trends in Pheromone Research and Technology*, 6-7 February, National Research Centre for Groundnut, Junagadh, Gujarat.

Patel, R.C., Yadav, D.N., Joshi, A.D., Patel, A.K. and Regupathy, A. (ed.), Jayaraj, S (ed.)., 1985. Impact of mass trapping of *Heliothis armigera* and *Spodoptera litura* from cotton with sex pheromones. *Behavioural and Physiological Approaches in Pest Management*, p. 120, 122.

Pawar, C.S., Sithanantham, S., Bhatnagar, V.S., Srivastav, C.P. and Reed, W., 1988. The development of sex pheromone trapping of *Heliothis armigera* at ICRISAT. *Tropical Pest Management*, 34: 39-43.

Pawar, A.D. and Kapil, N., 1982, Publ., 1983. Monitoring of codling moth *Cydia pomonella* (Linnaeus) in Kinnaur district (Himachal Pradesh) by sticky traps baited with codling moth sex pheromone. *Plant Protection, Bulletin*, India, 34: 3, 4, 13, 14.

Pawar, A.D. and Tuhan, N.C., 1985. Codling moth (Lepidoptera: Olethreutidae), suppression by male removal with sex pheromone traps in Ladakh, Jammu and Kashmir. *Indian Journal of Entomology*, 47(2): 226, 229.

Pawar, A.D., Prasad, J., Singh, Raj and Singh, R., 1984. Field evaluation of traps and sex pheromones of *Heliothis armigera* Hb. (Noctuidae: Lepidoptera) and effect of releases of parasites on male moth catches in pheromone traps. *Plant Protection Bulletin*, India, 36(4): 29, 30.

Pedigo, L.P., 2002. Management by modifying insect development and behaviour. In: *Entomology and Pest Management*. Prentice Hall of India Private Limited, New Delhi, 457 p.

Pillai, K.S., Palaniswami, M.S., Rajamma, P., Ravindran, C.S. and Premkumar, T., 1996. An IPM approach for sweet potato weevil. In: *Tropical Tuber Crops: Problems, Prospects and Future Strategies*, p. 329-339.

Prasad, N.K.K., Mallikarjunaiah, H., Nandihalli, B.S., Regupathy, A. and Jayaraj, S., 1985. Efficiency of different types of traps in monitoring *Spodoptera litura* (F.) populations with synthetic sex pheromones. In: *Behavioural and Physiological Approaches in Pest Management*, p. 53-55.

Prasad, N.K.K., Mallikarjunaiah, H., Nandihalli, B.S. and Regupathy, A. (ed.) and Jayaraj, S (ed.), 1985. Efficiency of different types of traps in monitoring *Spodoptera litura* (F.) populations with synthetic sex pheromones. In: *Behavioural and Physiological Approaches in Pest Management*, p. 53, 55.

Rangạ Rao, G.V., Wightman, J.A. and Ranga Rao, D.V., 1991a. Monitoring *Spodoptera litura* (F) (Lepidoptera: Noctuidae) using sex attractant traps: Effect of trap height and time of the night on moth catch. *Insect Sci. and Appl.*, 12: 443-447.

Ranga Rao, G.V., Wightman, J.A. and Ranga Rao, D.V., 1991b. The development of a standard pheromone trapping procedure for *Spodoptera litura* (F) (Lepidoptera: Noctuidae) population in groundnut (*Arachis hypogaea* L.) *Crop. Trop. Pest Management*, 37: 37-40.

Ranga Rao, G.V., Wightman, J.A., Ranga Rao, D.V., Rameshwar Rao, V., and Hall, D.R., 1993. Optimization of pheromone traps for the groundnut leaf miner *Aproaerema modicella* (Deventer) (Lepidoptera: Gelechiidae). *Journal of Plant Protection in the Tropics*, 10(3): 260-266.

Rao, D.V.S. and Subbaratnam, G.V., 1998. Sex pheromone monitoring of the ragi cutworm, *Spodoptera exigua* (Hubner) in onion: Effect of trap height and pheromone dose on moth catch. *Pest Management and, Economic Zoology*, 6(1): 21, 25.

Rao, G.V.R., Wightman, J.A. and Rao, D.V.R., 1991. The development of a standard pheromone trapping procedure for *Spodoptera litura* (F) (Lepidoptera: Noctuidae) population in groundnut (*Arachis hyopogaea* L) crops. *Tropical Pest Management*, 37(1): 37-40.

Reddy, G.V.P. and Urs, K.C.D., 1997. Mass trapping of diamondback moth *Plutella xylostella* in cabbage fields using synthetic sex pheromones. *International Pest Control*, 39(4): 125-126.

Reed, W., Vendamoorthy, G., Raghavan, M.V., Rajan, M.P. and Vijaya, Raghavan, M., 1975. Pink bollworm moths: Catches in sex, attractant traps and nocturnal behaviour in South India. *Cotton Growing Review*, 52(4): 350, 359.

Rup, P.J. and Sharma, S.P., 1978. Behavioural response of males and females of *Callosobruchus maculatus* (F.) to the sex pheromones. *Journal of Ecology*, 5(1): 72-76.

Sahu, R.C. and Hameed, S.F., 1983. Pheromones in insect control: A review. *Pesticides*, 17: 12, 18-20.

Sain, M. and Kalode, M.B., 1985. Traps to monitor gall midge population in rice. *Current Science, India.*, 54(17): 876-877.

Satpute, U.S., Taley, Y.M. and Bhalerao, P.D., 1985. Male catches of pink bollworm with various concentrations of gossyplure. *Punjabrao Krishi Vidyapeeth Research Journal*, 9(1): 85, 88.

Sharma, S.P. and Deora, R.K., 1980. Factors affecting production, release and response to female sex pheromones in *Sitophilus oryzae* (L.) (Coleoptera: Curculionidae). *Indian Journal of Experimental Biology*, 18(5): 463-465.

Siddiqi, J., 1988, Publ., 1989. Trapping of pink bollworm, *Pectinophora gossypiella* Saunders (Lepidoptera: Gelechidae) with its synthetic sex pheromone. *Indian Journal of Entomology*, 50(2): 238, 248.

Siddiqi, J.I. and Khan, M.A., 1983. The secretion and perception of a sex pheromone in the grasshopper *Hieroglyphus nigrorepletus* Bolivar (Orthoptera : Acrididae). *Acrida.*, 1981, recd., 1983, 10(4): 233, 242.

Singh, K. and Majumder, S.K., 1989. Olfactory behaviour response of Angoumois grain moth *Sitotroga cerelella* (Olivier) to (Z, E), 7, 11 hexadecadienyl acetate with an antioxidant. *Indian Journal of Experiemental Biology*, 28: 8783, 784.

Singh, K.N. and Sachan, G.C., 1991. Monitoring of *Spodoptera litura* F. populations with pheromone baited traps at Pantnagar, India. *Journal of Plant Protection in the Tropics*, 8(2): 97, 101.

Tamhankar, A.J., Rajendran, T.P., Bhamburkar, M.W. and Harwalkar, M.R., 1993. Optimum pheromone blend for pink bollworm males: evaluation with two experimental designs. *International Journal of Pest Management*, 39(1): 111, 112.

Taneja, S.L. and Jayaswal, A.P., 1986. Population dynamics of pink bollworm, *Pectinophora gossypiella* (Saunders) on upland cotton. *Insect Science and its Application*, 7(4): 569-573.

Trivedi, T.P., Rajagopal, D. and Tandon, P.L., 1994. Environmental correlates of the potato tuber moth, *Phthorimaea operculella* (Zeller) (Lepidoptera : Gelechiidae). *International Journal of Pest Management*, 40(4): 305-308.

Usui, K., Uchiumi, K., Kurihara, M., Fukami, J. and Tatsuki, S., 1988. Sex pheromone content in female Chilo suppresalis Walker (Lepidoptera : Pyralidae) reared on artificial diets. *Applied Entomology and Zoology*, 23: 97-99.

Chapter 24

Application of Fungicides in Improving Agriculture Production

R.D. Rawal* and C. Chandershekhar
#29, 6th Cross Krishnappa Blocck, Ganganagar Ext.,
Bangalore – 560 032, Karnataka

Introduction

Sustaining agriculture is a key element of sustainable development and essential to the future well being of the planet. Sustainability aims to achieve adequate safe and healthy food production, improved livelihoods of food producers and the preservation of non-renewable resources. Integrated crop management (ICM) and integrated pest management (IPM), are the strategies best suited to preserve our environment and natural resources base in the long term. The demands of a growing world population for food and fibre require world agriculture to produce higher yields from less cultivable land Feeding future populations with to-days crop yield is not viable; it would require drastic expansions of planted acreage. In many parts of the world additional land is unavailable. In others, expansion of the crop area would be environmentally and socially unacceptable. Increasing yields from existing land requires continuous improvement of agricultural technologies including those in crop protection to minimize losses before and after harvesting (CLI). This is possible to achieve through the proper management of insect pests, diseases and weeds. Among the various agencies the losses caused by various agents causing diseases in crop plants are the maximum. Since the earliest have recorded times man has realized that blights, blast and mildew do cause destructive losses to the crops upon which he depended for food, clothing and shelter. Coupled with this has been man's age-old struggle to combat these destructive destroyers of his crops. In earliest historical times agriculturalists were aware of the losses caused by plant diseases, but their knowledge of their nature and their attempted methods of control were based on superstitious beliefs and enormous concepts. Roman agriculture man in his earlier days persisted in his attempts to find "Prevention Of Chemical Nature" to control plant diseases. It is perhaps amusing in the light of modern knowledge of chemical plant protection to recall that Greek agriculturists were advised to sprinkle plants with pure amurca of olive without salt to prevent blight from attacking them (Democritus 470 B.C.). Never the less these early attempts to prevent plant diseases involved the use of fungicides. For many years plant pathology was an avocation of medical practitioners. With DC Barry's proof that fungi were one of the causes of plant diseases, a new science was conceived and after struggling through years of labour pains has achieved the status of modern botanical sciences. The science of plant pathology is thus less than one hundred years young and in the United States dates from the establishment of the division of plant physiology and vegetable pathology by the USDA in 1885. Today all the Agricultural Institutions have established departments of plant physiology.

While the science of applied plant pathology is one of the youngest of the agricultural technologies, the practice of plant disease control is imperfectly understood by most laymen, practical agriculturists and even by highly trained specialists in other phases of agricultural sciences.

Fungicides to most people are complex chemicals with unpronounceable labels that are confusing by virtue of the many different names that are given to the same chemical. Few people fully understand the principles underlying the use of these modern tools for plant disease control. For this reason an attempt has been made to draw aside the confusion that has hidden from many people a workable understanding of the uses of fungicides for plant disease control. Authors have tried to compile the available information through publication and personal contacts.

History of Fungicides

The story of the use of chemicals for disease protection of crop plants upon which man has always depended for food, clothing and for shelter is as old as the sands of time. Plant protection by the use of chemical spray, dusts or seed treatment did not originate in the 20th century but has been practiced in some farm for as long as man has recorded the story of trials, troubles and tribulations.

Ancient Fungicides

Early agriculturists were aware of the seriousness of losses caused by plant diseases. While their knowledge of these maladies was clouded by superstitious beliefs and enormous concepts they nevertheless did attempt to prevent losses to their crops caused by "blight" and "mildew". Most of the earlier attempts to control plant diseases involved ceremonial rituals, but blind attempts were made to utilize chemicals as fungicides.

In the year 1000 B.C., Homes, the Greek epic poet and the traditional author of "lillied and Odyssey" mentioned the "pest averting sulphur with its properties of divine and purifying fumigation".

Modern Fungicides (1930-)

They are called modern fungicides because these are being extensively used in the field till now. In 1931 the USDA established a project to search for non-corrosive substitute for standard. The pioneer work of this project led to the adaptation by U.S. agriculture of the low-soluble coppers as Bordeaux mixture substitute. Horsefall (1932) studied the value of low-soluble coppers for the prevention of damping off the vegetables.

One of the most significant developments in modern plant disease control has been the improvement in application equipment machinery. In 1948, efforts were made to develop high air velocity air blasts, were accepted as a practical method

Systemic Fungicides

Though the concept of use of systemic fungicides started in 1964. Actual use started by Delp and Klopping (1968) who first reported on benomyl in plant disease reporter, dramatic acceptance of Benlate by farmers and widespread development of other systemic fungicides have evolved. Thiobendazole and one or two antibiotics were reported as systemic as early as 1964, but the real escalation in use of systemic fungicides began with the discovery of carboxin and benomyl in late 1960s. A compilation of fungicides currently used in the world lists 44 systemic and 84 non-systemic fungicides.

What 'Systemic' Means?

The word implies movement throughout the plant system, but this is misleading because most systemic fungicides are only partially systemic in plants with the exception of several unregistered experimental fungicides that move systemically into fruit, roots, buds and other plant organs along with the plant photosynthates (sugars), the so called systemic are carried along with the transpiration system. Leaves are the primary transporting organs and movement with plates is from the soil to the expanded leaves. Very young leaves, flowers and fruits do not transpire significant quantities of water and therefore receive only minute quantity of fungicides applied to soil or seed. Actually this is fortuitous, as the chances of residues in fruit are minimal. Certainly no more residues would be expected than with the old conventional "protectants" applied to the surface of plants.

Why is a Fungicide Systemic or Partially Systemic?

Before the discovery of systemic, fungicides were only protectants, remaining on the surface of plants. The protectants are toxic to many processes common in plants and fungi but do not enter the plant and therefore are not phytotoxic.Examples of protectants are maneb (Dithane M-22),Chlorothanonil (Bravo) and Captan (Captan). Systemic fungicides, however, penetrate the plant and are selectively toxic to processes unique to fungi. This selectively is so specific that among the hundreds of different fungi attacking plants, only certain taxonomic groups of fungi are sensitive to each particular systemic fungicide.

Selectivity of Systemics

Fungicides of similar chemical structure are toxic to similar types of fungal organisms. Obviously, as many farmers know, only mildew, rust or any other single disease does not plague a particular crop. Researchers and industry have been busy combining fungicides to control pathogens in a particular crop. For *e.g.*, Ridomil, a new systemic fungicide, looks very effective against late blight of potato caused by *Phytophtora infestans*. However, a farmer must also use manozeb or some other fungicide for early blight caused by *Alternaria solani*, because Ridomil is selectively toxic to *Phytophtora*.

A systemic fungicide appears to be selective because it is toxic to only a single site within the fungus. This can cause problems, since a single change in the fungus can result in resistance to the fungicide. The threat of resistance can be reduced, however, by using different systemic fungicides in combination or alternatively in a pest management programme.

Systemic Fungicides to Use in the World

Common or Chemical Name	*Trade Name*	*Sensitive Fungi*
Chemical Groups		
Acetamides		
N-Cyanoethyl chloroacetamide	Udonkor	Powdery mildew
2-Cyan-N-(ethylaminocarbanyl)-2-(Methoxy imino)-acetamide	Curzate(DPX-3217)	Oomycetes
Acylalamines		
D1-MethylN-(2,6-dimethylphenyl)-N-(2-methoxyacetyl)alaninate	Ridomil/Matco Unilaxyl/ Master etc.	Oomycetes
Acylon	Furalaxye, Fongarid	Oomycetes
Benzimidazole		
Benomyl	Benlate	Botrytis,Fusarium,
Carbendazim	Benlate	Pencillium
Thio-phenate-methyl	Ease out,Topsin-M	Cercosporella
Thiophenate	Carbocin,Topsin	Septoria, Sclerotinia
Thiobendazole	Mertut, Venturia, Verticillium, Fuberidazole/Tecto-90	Powdery mildew Fusarium
Carboximide		
Carboxin	Vitavax	
Furmetamid	BAS 389	Smuts, *Rhizoctonia*
Mephfuroxam	Trivax (H 719)	
Fenfuram	Panoram	
Benodanil	Calirus rust	*Rhizoctonia*
Oxycarhoxin	Plantvax	Rust
Pyracarbolid	Sicarol	Rust
Hydroxy pyrimidones		
Dimethirimol	Milcurb	Powdery mildew of Cucurbits
Ethirimol	Milstem, Milgo E	Powdery mildew of Cereals
Bupirimate	Nimsod	Powdery mildew
Morpholines		
Dodemorph	Meltatox	Powdery mildew of roses
Tridemorph	Calixin	Yellow rust of wheat, Mycosphaerella, Powdery mildew
Organic Phosphates		
1 BP	Kitazin P	
Edilenphos	Hinosan	*Pyricularia*
Pyrazophos	Afugan, Curathil	Powdery mildew
Triamiphos	Wepsyn	
Pyrimidines, Pyridines, Piperidines, Imidazole		

Contd...

Contd...

Common or Chemical Name	Trade Name	Sensitive Fungi
S-n-butyl-S-P-tert butyl-N-3-Pyridyl-dithiocarbonimidate	Denmert	
Fenarimol	Block, Rubigan, Rimidin	Powdery mildew
Imazalil	Imazalil	Fungal flora
Nuarimol	Trimidal,Trimund	Rusts, Venturia
Parimol	Parnond	*Cochliobolus*
Triadimephon	Bayleton	Mycosphaerella
Triadimenol	Baytan	"Fruit rots"
-Butyl-phenyl-IH-midazole-1-propa-nenitrite	Systhane(RH2161) Saprol	Powdery mildews
Triazoles		
n-butyl-1,2,4 triazol	Indar (RH 124)	
Tricyclazole	Beam 75	Brown rust of wheat
Fluotrimazol	Persulon	*Piricularia* Powdery mildew
Others		
Aluminium ethylphosphite	Aliette	Oomycetes
Griseofulvin	Grisiofulvin	Botritis, powdery mildew, *Cladosporium*
Cycloheximide	Actidione	Rusts, Fabrae
Prothiocarb	Pervicur	Oomycetes
Isoprothiolan	Fuziwan	*Pyricularia*
Probenezole	Orizamate	*Pyricularia*

After facing the problems with systemic as target fungicides, the need was felt either to develop a broad spectrum fungicides or a combination of systemic and contact fungicdes.With this in mind very recently another group called strobulirin which primarily has been synthesize from the Pleurotus spp. Of mushroom has come out to the rescue of farmers: In the market these are available as quadris, abound, bankit, priori, ortiva, heritage.Imagine a fungus which secretes a substance that eliminates competing fungi but is otherwise beginning and even considered edible! Nature provides a species of forest-dwelling mushroom with such properties, whose fungicidal activity enables it to compete with other fungal species for nutrients, safeguarding its survival AMISTAR, is a broad-spectrum fungicide, which is safe to the environment and delivers consistent yield and quality benefits in major crops of economic importance around the world. The active ingredient, azoxystrobin, belongs to a new class of fungicides named strobilurins: this new class possesses a new mode of action, which inhibits mitochondrial respiration in fungi, stopping their energy supply.

AMISTAR was first launched in Europe in 1996. It was the fastest development track ever in the crop protection industry.

Broadest spectrum strobilurin with an excellent environmental profile, optimising yield and quality in a wide range of crops. Convenient to use and with a proven track record.A number of other new strobilurin containing fungicides is available now in the market with different trade names. Acanto is available as a straight strobilurin (picoxystrobin). Modem is another straight strobilurin(pyraclostrobin). Straight strobilurins need a non-strobilurin partner in all situations. Opera is a pre-formulated mixture of pyraclostrobin and epoxiconazole and Covershield is a three-way mixture of pyraclostrobin, epoxiconazole and kresoxim-methyl.

They have activity against a wide range of cereal diseases. In particular they give excellent control of net blotch in barley and *Septoria tritici* in wheat. Powdery mildew, Downy mildew and Anthracnose disease in Grapes. They also have been reported to influence physiological effects on crop growth, which can result in yield increases even when disease levels are very low. The new strobilurins, picoxystrobin and pyraclostrobin, have eradicant properties which the earlier strobilurins did not have. While they have this attribute they do not match the eradicant properties of the best triazoles.

The strobilurins (QoI) group of fungicides (which includes the strobilurins and famoxadone) are broad-spectrum fungicides active against a wide range of diseases in many crops. They are excellent inhibitors of spore germination and are consequently first class protectant products. They all have the same mode of action, *i.e.* on entering the fungal cell they affect electron transport in the mitochondria–which are the energy source in cells–depriving the organism of energy and hereby causing cell death.

They largely remain in the waxy layer of plant parts and exert their fungicidal activities there. The various strobilurins do differ in their systemic properties with some of the group being partially systemic and others redistributing themselves around the plant in the wax layer and epidermal cells by vapour action.

They also have other desirable non-fungicidal physiological properties in that they can inhibit ethylene biosynthesis and consequently delay senescence and thus enhance yield in this way. The product remains primarily on the leaf surface bound to the waxy layer but as it has strong translaminar activity it penetrates the leaf where it can eliminate fungal mycelium already present, which is how it exerts its eradicant or kickback action. They can also improve the assimilation of nitrogen into plants.

As the entire QoI group has the same mode of action then if resistance develops in a pathogen to one strobilurin product there will be cross-resistance to all the other members of the group. This has already been reported in the case of wheat powdery mildew in Europe and USA. Consequently, the use of strobilurins in disease control programmes has to be managed to reduce the risk of resistance by other pathogens arising.

Chlorothalonil

Trade Names

Daconil, weather stik, ultrex, Bravo.

This products is registered throughout the world (in more than 65 countries) for the control of 125 diseases. Import of BRAVO-treated crops is allowed in nearly all import nations. BRAVO® is harmless to a wide range of beneficial organisms, including honeybees, earthworms, carabid beetles and parasitic wasps, when used as instructed on the label. Good applicator safety, extremely rainproof, safe for non-target species. Rapid breakdown by the environment, no accumulation in the food chain and well tolerated by an immense range of crops.

Broad spectrum protective fungicide with proven time tested performance. Labeled in over 120 crops in 65 countries in agriculture, horticulture and turf. Bravo has been and continues to be a foundation partner for many fungicide programs globally.

Iprodione

Trade Names

Rovrol 50 per cent WP.

Iprodione is a fungicide, belonging to the cyclic imide family, discovered by Rhome-Poulenc in 1971. it is effective against the spores and mycelium of a number of parasitic fungi and exhibits both protectant and some eradicant activity.

Iprodione affects all stages of the development cycle of fungi *viz.*, spore germination, mycelial growth, spore production etc. Iprodione is used in citrus, pome granate, Grapes, Berries, cereals, oilseeds etc.Iprodione is effective against a range of plant pathogens from the following genera: *Alternaria, Botrytis, Corticium, Fusarium, Heliminthospoium, Monilinia, Phoma, Pleiochaeta, Rhizoctonia, Scleorotinia, Sclereotium, Septoria, Colletotrichum, typhula, Pencillium* and *Rhizopus*. It is effective against Benzimidazole-resistant strains of fungi. Iprodion is safe to the honey bees.

Propineb

Trade Names

Antracol, Fitoraz, Milraz

Propineb is an organic foliar fungicide having Multi-site contact activity, notable for its fast action and its excellent residual activity. It is recommended for Grapes, ornamentals, potatoes, tomatoes, tropical fruits and vegetables. The product has been registered in more than 60 countries. It is especially suitable for protective control of Downy mildew on grapes, tobacco and vegetables as well as for control of *Phytophthora* and *Alternaria* on potatoes and tomatoes. Good results have been obtained with propineb against *Plasmopara* spp., leaf spot diseases of vegetables. Anthracnosis, Scab (*Venturia* spp.), Botrytis. Propineb is outstandingly effective against Sigatoka disease of banana. It has been reported that it is not harmful to the honey bees.

Combi Product of Mancozeb+Carbendezim(Mancozeb 63 per cent + Carbendazim 12 per cent WP)

Trade Names

Companion, Saaf.

It is a mixture of Mancozeb and Carbendazim. Mixture partner Mancozeb remains on the plant surface and acts by contact action (Preventive). It is fungi toxic when exposed to air, converted in to an isothiocyanate, which inactivates the sulphahydral groups of enzymes in fungi, thus, cause disturbance in fungal enzyme functioning. Other partner Carbendazim, which is having systemic activity (absorbed through roots and green tissues), acts as protective and curative. Acts by inhibiting the development of fungal germ tube, formation of appressoria and growth of mycelia.

Triazoles

As mentioned earlier, they are excellent protectant products but have little or no eradicant activity and rely on a partner usually from the triazole group to provide eradicant properties. Triazoles contain a group of chemicals, which have been developed for use as either fungicides or plant growth regulators. The triazoles belonging to a class of compounds known as ergosterol biosynthesis inhibitors, which constitutes the largest and most important group of systemic compounds for the control of fungal diseases of crop plants. The triazoles control a wide range of economically important diseases caused by fungi including powdery mildew, smuts, rusts, Blister blight, Black rot and apple scab.

Triazoles mainly inhibit the C-14 demethylation reactions, also inhibit Cytochrome p-450 component of the mixed function.

Tricyclazole

Trade Names

Beam, Baan.

It is a specialty systemic fungicide. The product is rapidly absorbed by plant and translocated towards leaf tips. It is a protectant fungicide that prevents the fungus from penetrating the plant. Inhibition occurs when the fungus attempts to penetrate and establish an infection site inside the plant.

In rice blast, the melanin pigment is needed for the hardening of the appressorium and inhibition of the pigment formation in appressorium makes it unable to mechanically penetrate the host plant.The best Blasticide available in market. Long duration control of Rice Blast for more than two weeks. Very stable fungicide- not readily destroyed by sunlight and moisture. Upon absorption, it is systemically translocated in plant tissues, protect the entire plant from disease infection. Absorbed/translocated fast in plant–need not spray again if it rains 1 hour after application. Improves grain yield, results in superior grain quality, which are heavier, shiny and upon milling gives highest recovery of full size grains.

Myclobutanil 10 per cent WP

Trade Names

Systhane, Index, Boon.

Highly systemic very effective fungicide for powdery mildew. Its absorption into plant is very fast provides long duration control. When used as preventive does not allow infection to take place at all. It is reported to have a curative effect as even if applied upto 96 hours after infection process has started to kill the infection or germinating spores before infection becomes established. EradicativeMyclobutanil when applied, arrest the infection and sporulation. When applied on developed symptoms after infection it helps in checking the spread of infection. Ensure protection against PM and Black rot.

Propiconazole

Trade Name

Topas.

It is a systemic specialty fungicide. It is rapidly absorbed by the leaves or stems and translocated upward through the xylem. It is a potent ergosterol biosynthesis inhibitor. Ergosterol, a principal sterol in most fungi is an indispensable component in the membrane structures.

Hexaconazole 5 per cent EC Molecular Formula: $C_{14}H_{17}Cl_2N_3O$

Trade Name

Contaf, Sitara, Anvil.

Hexaconazole is an unique highly systemic triazole fungicide, acting as protective, curative and eradicant with strong antisporulant and translaminar action. Hexaconazole is a potent ergosterol biosynthesis inhibitor. Ergosterol, a principal sterol in most fungi is an indispensable component in the membrane structures. Broad-spectrum fungicide controlling wide range of diseases infecting much crops.Gives excellent control of Ascomycetes, Basidiomycetes and Deuteromycetes. It is effective at lower dosages for long period. Broad-spectrum fungicide, which also controls soil borne fungi like *Fusarium, Sclerotium* and *Rhizoctonia*, very effectively. It is a safe fungicide with low toxicity to mammals, fish, birds and natural enemies.

Fosetyl-Al

Trade Name

It was discovered in France in 1973 by Rhone-Poulnc, is a systemic fungicide belonging to the phosphonate family.

It is a foliar fungicide; act as a Plant defense stimulant. It is used for fruits, hops and vegetables. It has been registered over more than 90 countries.A direct effect on the fungus which, when applied as preventive treatment, stops plant infection by inhibiting spore germination and the penetration of the pathogen into the plant or blocks mycelial development and sporulation in the case of a curative treatment. When Fosetyl-Al applied to the foliage is carried to the tips of the root system and to the other parts of the plant, thus protecting the whole growth system.An indirect effect by which the product slows down sufficiently the invasion of the pathogen, by reinforcing the defensive reactions of the plants. This enables the hosts hypersensitive response to seal-off the invading organism.

Carpropamid

Trade Names

Arcado, Win, WinAdmire, Win Admire, Greatam, Win Bariard, Win Gazzet.

It is a Foliar Fungicide, effective against Melanin synthesis in cell wall. Act as a Dehydratase in melanin biosynthesis.It is mainly used in rice it is effective aginst Pyricularia oryzae (leaf, neck and panicle blast) and bacterial diseases.

Bitertanol

Trade Names

Baycor, Baycoral, Proclaim.

It is a Foliar fungicide, Sterol biosynthesis in membranes. C14-demethylase in sterol biosynthesis.Bananas, berries, cut-flowers, ornamentals, pome fruits, stone fruits and vegetables Disease Spectrum: *Blumeriella jaapii, Botryosphaeria* spp., *Cercospora* spp., *Monilinia* spp., *Mycosphaerella* spp., *Puccinia* spp., *Sphaerotheca* spp., *Uromyces* spp., *Venturia* spp.

Triadimenol

Trade Names

Bayfidan®, Baysiston®, Exact®.

Foliar fungicide,Sterol biosynthesis in membranes. C14-demethylase in sterol biosynthesis. Cereals, coffee, cotton, curcubits, grapes, rice, tobacco, top fruits, tomatoes, tree nuts and vegetables. Registered in more than 60 countries worldwide.

Disease Spectrum: *Ceratocystis paradoxa, Erysiphe graminis* spp., *Gymnosporangium haraeanum, Hemileia vastratrix, Mycosphaerella* spp., *Oncobasidium theobromae, Podosphaera* spp. *Puccinia* spp., *Rynchosporium secalis, Sclerotium, Sphaerotheca* spp., *Uncinula necator, Ustilago scitaminea.*

Triadimefon

Trade Name

Bayleton.

Bayleton is a highly effective fungicide with systemic activity. The active ingredient is transported from treated to untreated plant parts in both an acropetal and basipetal direction. Bayleton mainly inhibit the Sterol biosynthesis in membranes. C14-demethylase in sterol biosynthesis. Bayleton is mainly used in Cereals, coffee, fruits, grapes, tea, turf and vegetables.

Bayleton is chiefly effective against powdery mildew, rust fungi, *Rhynchosporium secalis, Diplocorpan rosae* and *Ceratocystis paradoxa*. Bayleton gives protective control and also has a curative and eradicative effect. It has been reported that the active ingredient is not harmful to the honeybees.

Foliar Fungicide

Cyproconazole

Trade Names

Caddy, CaddyArbo, Mohawk.

Cyproconazole is a Foliar fungicide, it is mainly involved in inhibition of sterol biosynthesis in membranes and C14-demethylase in sterol biosynthesis. Cyproconazole is used for Cereals, grapes, stone fruits, sugar beets and vegetables. Cyproconazole is a broad spectrum fungicide with special strength against rust, powdery mildew, *Cercospora* spp. and *Septoria* spp.It is registered in Austria, Belgium, France, Greece, Italy, Netherlands, Spain and United Kingdom.

Tolyfluanid

Trade Names

Elvaron Multi, Euparen Multi, Euparen M.

Tolyfluanid is a Foliar fungicide having multi-site activity. Tolyfluanid is mainly used for Berries, cotton, grapes, ornamentals, pome fruits, soybeans and vegetables.

Tolyfluanid is effective against Botrytis cineria, downy mildew, *Gloeosporium album,* leaf spots, *Helminthosporium* spp., *Nectria galligena, Penicilium* spp., *Phytophthora* spp., scab, *Stemphylium* spp., storage diseases, acaricidal side effect.

Fenamidone

Trade Names

Censor, Consento, Reason, Sereno, Verita.

Fenamidone is a Foliar fungicide, it mainly inhibit Respiration. Complex III: cytochrome bc1 (ubiquinol oxidase) at Qo site. Fenamidone is being Registratered over more than 50 countries worldwide. Fenamidone is chiefly effective against *Alternaria* spp., *Bremia lactuca, Peronospora* spp., *Phytophthora* spp., *Plasmopara* spp., *Pseudoperonospora* spp., *Pythium* spp.

Trifloxystrobin

Trade Names

Flint, Madison, Nativo, Sphere, Stratego, Tega, Twist, Zato.

Trifloxystrobin is a Foliar fungicide, it mainly inhibit Respiration inhibitor. Complex III: cytochrome bc1 (ubiquinol oxidase) at Qo site. Trifloxystrobin is used for the following crops *viz.*,Bananas, cereals, citrus, coffee, corn, cotton, field beans, grapes, hops, nuts, ornamentals, peanuts, pome fruits, potatoes, rice, small fruits, soybeans, stone fruits, sugar beets, sunflowers, tea, tropical fruits, turf, vegetables and various other crops.

Trifloxystrobin has being registered in Argentina, Brazil, Canada, France, Germany, Italy, Paraguay, South Korea, UK, USA and in more than 80 other countries worldwide.

Trifloxystrobin is Broad spectrum fungicide effective against *Alternaria* spp., *Cercospora* spp., *Colletotrichum* spp., *Mycosphaerella* spp., powdery mildews, *Pyricularia* spp., *Rhizoctonia solani*, *Rhynchosporium secalis*, rusts, *Septoria* spp., *Venturia* spp. and various other leaf and fruit diseases.

Tebuconazole

Trade Names

Elite, Falcon, Folicur, Horizon, Matador, Milrod, Nativo, Only One, Prosaro, Pronto Plus, Silvacur.

Tebuconazole is a Foliar fungicide, it mainly inhibit C14-demethylase in sterol biosynthesis. Tebuconazole is used in more than 90 crops such as bananas, cereals, coffee, grapes, canola, peanuts, pome and stone fruits, rice, soybeans and vegetables.

Tebuconazole has been registered more than 100 countries worldwide.

Tebuconazole is highly effective against: *Alternaria* spp., *Blumeriella* spp., *Botryosphaeria* spp., *Botrytis* spp., *Cercospora* spp., *Curvularia* spp., *Erysiphe* spp., *Exobasidium* spp., *Fusarium* spp., *Helminthosporium* spp., *Hemileia* spp., *Leptosphaeria* spp., *Monilinia* spp., *Mycena* spp., Mycosphaerella spp., *Phakospora* spp., *Podosphaera* spp., *Puccinia* spp., *Pyrenopeziza* spp., *Pyrenophora* spp., *Rhizoctonia* spp., *Rhynchosporium* spp., *Sclerotinia* spp., *Sclerotium* spp., *Septoria* spp., *Uncinula* spp., *Venturia* spp. etc.

Difenoconazole

Trade Names

Score : The very potent fungicide belonging to the group of triazoles bogard®, plandom®, armure®, taspa®, sico.

Stops the development of fungi by interfering with the biosynthesis of sterols in cell membranes.Long-lasting preventive activity combined with curative and eradicant properties; fast uptake and high translaminar movement.

A wide range of leaf spot diseases. Key partner for resistance management program can be used on almost any crop.

SCORE is used mainly in Europe, Latin America and Asia. Registered in at least 50 different crops in 57 countries.

Ideally systemic, SCORE gives more to growers: it is reliable and economical, giving a high return.

The ideally systemic triazole for almost any crops against a wide range of diseases: excellent crop tolerance, long lasting disease control and substantial increase in yield.

Propiconazole

Trade Names

Tilt: An ideal tank mix partner with over 30 recommended fungicide partner.Other Names orbit™, banner.

It stops the development of fungi by interfering with the biosynthesis of sterols in cell membranes. Systemic compound with quick uptake and good acropetal translocation in the xylem, resulting in good distribution in the plant tissue and protection from being washed off.

It ensures a curative and preventive control against rusts and leaf spot diseases.

Turf, cereals, peaches, rice, banana.

TILT® is registered for use on more than 40 crops all over the world.

It achieves a high-level control in spraying programs. It contributes significantly to avoiding yield and quality losses. It offers maximum protection at a realistic cost.

Broad spectrum, cost effective fungicide with high level control against rusts and leaf spot diseases and well established as a tank mix partner

Cyprodinil

Trade Names

Unix.

The best and only all-strain eyespot fungicide. The barley strobilurin partner for maximum yields. Inhibits amino acid synthesis and the secretion of hydrolytic enzymes. Eyespot, powdery mildew, net blotch and *Rhynchosporium*

Cereals (Wheat and Barley)

Dimethomoprh

Trade Name

Acrobat.

Dimethomoprh (DMM) is a cinnomic acid derivative which is active against oomycete fungi, and has a unique mode of action. Foliar application of DMM gives good protectant, curative and excellent anti-sporulant activity and provide long residual protection.

Dimethomoprh is systematically translocated in the plant when applied to the roots and is translamianar with local systematicity when applied to the foliage. Dimethomoprh demonstrates activity against peronosporacea and the genus phytophthora. It disrupts the fungal cell wall formation by causing cell wall lysis resulting in death of the fungal cell. Dimethomoprh is active at all stages of the fungal life cycle except during formation of zoospores. Where as sporanigophore and oospore formation stages are particularly sensitive to the effects of dimethomorph.

Tolclofos-Methyl

Trade Name

RIZOLEX 50 WP.

World wide soil born diseases are increasing on most of crops year by year. *Rhizoctonia* and *Sclerotium* diseases especially damage the principle organs of plants and cause serious yield reduction.

Rizolex attacks these invaders from soil and helps to maximum the farmer's income. It should not accumulate in the soil. It can be use by soil incorporation, soil drenching, seed drenching, dipping and seed spray.

Rhizolex used against potato black scurf and stem canker caused by *Rhizoctonia solani*, peanut stem rot caused by *Sclerotium rolfsii*, damping off, seedling rot, root rot, peg rot, and pod rot etc caused by *Rhizoctonia* spp.

Triforine

Trade Names

Saprol, Funginex or triforine.

Triforine affects ergosterol biosynthesis by inhibiting the oxidative C-14 demethylation of 14-methylergostadienol which leads to an accumulation of C-14 methyl sterols and free fatty acids.Triforine affects various stages in the development of fungi. Depending on the genus and the concentration of the fungicide spore germination, mycelial growth, haustoria's development and/or spore production are inhibited. Protective as well as curative control have been shown in many diseases under green house and field conditions.Triforin is a potent fungicide for the control of powdery mildew, scabs, rusts, monilinia and many foliar diseases on top and soft fruits, hops, grapevines, vegetables, cereals, cotton, soybeans and ornamentals as well as post-harvest diseases.

Flusialzole

Trade Name

Nustar.

Flusilazole, a broad spectrum fungicide and a member of the Ergosterol biosynthesis inhibitor class of fungicides and effectively controls many economically-important diseases of fruit and field crops, including cereals, apples, grapes and bananas. Activity is observed against numerous Ascomycetes, Basidiomycetes and Deuteromycetes fungi but not the phycomycete fungi.

Flusialazole is active at low use rates and shows excellent preventive, curative, systemic and residual properties. Its photo stability, vapor action, systemic attributes, rain fastness and compatibility in tank mix combinations allow for excellent disease control and flexibility in spray timing. These attributes high the versality and excellent fit of this new fungicide as an important complement to many disease management strategies.

Flusialazole inhibits the 14α-demethylation step in the ergo sterol biosynthesis pathway of fungi, it is classified as an EBI (ergo biosynthesis inhibiting) and DMI (demethylation imbibing) fungicide.

Flusialazole effective against apple scab, powdery mildew, cedar apple rust, leaf spot, leaf and glum blotch, wheat rust etc.

Penconazole 10 per cent EC

Trade name Tilt affects various stages in the development of fungi. Depending on the genus and the concentration of the fungicide spore germination, mycelial growth, haustoria's development and/or spore production are inhibited. Protective as well as curative control have been shown in many diseases under green house and field conditions,Topas is a systemic fungicide having protective and curative properties for the control of apple scab, powdery mildew and rusts of fruits, vegetables, filed crops and ornamentals. Is is active against a range of ascomycetes, basidiomycetes and deuteromycetes.

It is rapidly absorbed and distributed with in the leaf, acting on the fungus during penetration and haustoria formation phase, by imbibing the biosynthesis of sterols in cell membranes.It is best applied just before or at the first sign of disease development for maximum control and to avoid errevocable loss of crop yield and quality.

Fenarimol

Trade Name

Rubigon.

Rubigon acts at various stages in the development of fungi. Depending on the genus and the concentration of the fungicide spore germination, mycelial growth, haustoria's development and/or spore production are inhibited. Protective as well as curative control have been shown in many diseases under green house and field conditions: Fenarimol is a foliar fungicide which provides excellent control of many economically important diseases of plants, particularly the scab diseases of apple and powdery mildew diseases of apple, grape vines, chillies, peas, cucumber, gram, rose, mango, etc. it is having both protective and curative properties against several phytopathogens. It mainly inhibits ergosterol biosynthesis step by interfering with sterol nucleus, affecting the ring position and demethylation at Carbons 4 and 14. Further, it has been indicated that it interfere with the synthesis of certain fatty acids synthesis. It also disturbs the germ tube by extrusion of the cytoplasmic contents. When applied as droplets it moves systemically both upper and lower surface of the leaves and moves apoplectically towards the tip of the leaf.

Some of the Newly Developed Molecules have been Briefly Tabulated in a Summarized form as the Ready Reckoner for the Readers

Fungicides (Trade)	*Registrant*	*Chemistry*	*Pest Control Spectrum/ Traits*	*Status*
Acibenzolar (ACITIGARD)	Syngenta	Benzothiadiazole (Systemic acquired resistance inducer)	Induces resistance to Blue mold. Bacterial disease, Downy Mildew, *Sclerotinia.*	Reduced risk pesticide, Registration on fruiting vegetables, Brassica leafy vegetables, leafy vegetables and banana. Pending use on cucurbits and wheat.
Azoxystrobin (Heritage) (Quadris) Abound)	Syngenta	Strobilurin	Broad spectrum of pathogens of fungi : *Cladosporium, Venturia, Botryosphaeria, Mycosphaerella, Pyrenospora, Puccinia, Pyricularia, Plasmophora, Guignardis, Psudopeziza, Alternaria, Sphaerotheca, Erysiphe, Levellula, Septoria, Pythium, Uncinula., Didymella, Sclerotium, Colletotrichum, Mycosphaerella, Phytopthora, Rhynchosporium,* Rhizoctonia etc	Reduced risk pesticide. Registered on tree and pistachio, cucurbit vegetables, stone fruits, banana, canola, grapes, peanut, potato, rice, tomato, wheat, barley, citrus, coriander leaves, field and sweet corn, cotton, dry bulb onion, green onion, soybean, leafy vegetables and root and tuber vegetables. Pending use on strawberry and cranberry. Potential use on edible legumes, herbs, Brassica leafy vegetables seed grass, watercress, pepper, blueberry, mint, cranberry, avocado, guava and other tropical fruits.

Contd...

Contd...

Fungicides (Trade)	*Registrant*	*Chemistry*	*Pest Control Spectrum/ Traits*	*Status*
BAS 510 Boscalid	BASF		Broad spectrum activity on *Alternaria*, powdery mildew, and *Botrytis, Sclerotinia Monilinia.*	Candidate reduced risk pesticide: Pending use on beans, peanuts, fruiting vegetables, peas and leaf lettuce and canola.
BAS 516 Boscalid + Strobilurin	BASF		Broad spectrum activity on Anthracnose, *Alternaria,* Downy Mildew, Powdery Mildew, *Botrytis, Sclerotinia* and *Monilinia.*	Candidate reduced risk pesticide: Pending use on grapes, potato, carrot and other root vegetables, bulb vegetables, stone fruit, tree nuts/pistachio, strawberry and berry crop.
Benthia-valicarb	Kumiai		Control downy mildew and oomycetes fungi	Potential use in Brassica, leafy vegetable, fruiting vegetables, cucurbit vegetable, potato and grape.
Copper Octanoate (NEU 1140F)	W. Neudorff	Copper Octanoate	Downy Mildew, Powdery Mildew, blue mold, white rust, anthracnose.	Registered on beans, peas, beets, broccoli, Brussels sprouts, cantaloupes, chive, cucumber, pumpkins, squash, carrot, celery, corn, eggplant, pepper, tomato, lettuce, garlic, leek onion, shallots, ginseng, grape, potato, spinach, chard strawberry, sunflower and turnip.
Cyazofamid (IKF–916)	ISK/BASF	Cyaniomidazole	Effective against oomycetes and Plasmodiophoromycetes fungi	Potential use on potatoes, grapes, tomatoes, cucurbits, onion, lettuce, Chinese cabbage and rice.
Cyflufenamid	Nippon Soda	Phenylacetamid	Excellent activity on various powdery mildew and brown rots	Potential use on stone fruit barley, wheat, grape, black wheat, pepper, strawberry, apple, cucumber, egg plant and melons.
Cymoxanil (Curzate)	DuPont	Acetamide	Downy Mildew, late blight, *Phytophthora, Plasmopara, Pseudoperonospora, Bremia* and *Peronospora.* Should be mixed with other fungicide for resistant management.	Registered on potato and tomato, pending use on hop
Cyproconazole (Alto)	Syngenta	Triazole	Coffee rust	Registered on coffee
Cyproconazole (Vangard)	Syngenta	Anilinopyrimidine	Ascomycetes and Duteromycetes such as : *Botrytis, Alternaria Monilinia, Venturia, Psudocercosporella, Pyrenospora, Septoria, Erisyphe, Rhynchosporium, glomerella, Coccomyces, Colletotrichum.*	Reduced risk pesticide: Registered on almonds, grapes, pome and stone fruit. Pending use on strawberry. Potential use on pistachio, blueberry and watercress.

Contd...

Contd...

Fungicides (Trade)	*Registrant*	*Chemistry*	*Pest Control Spectrum/ Traits*	*Status*
Cyprodinil/ Fludioxonil (Switch)	Syngenta	Anilinopyrimidine and Phenylpyrrole	Controls *Botrytis, Alternaria* and brown rot	Reduced risk pesticide: Pending use on grapes, strawberry, bulb onions, and green onion. Potential use on cane berry, carrot, Brassica leafy vegetable, pear, herbs and lychee.
Difenconazole (Dividend), Score	Syngenta	Triazole	Smut, bunt, *Aspergillus, Fusarium, Penicillium, Septoria, Cochiobolous, Pyrenophora, Pseudocercosporella* and *Gaeumannomyces*.	Registered on banana, wheat (seed treatment) and canola. Pending use on sweet corn.
Dimethimorph Acrobat	BASF	Cinnamic acid derivative	Downy mildew, late blight, *Phytopthora, Plasmopara, Pseudoperonospora, Bremia and Peronospora.* Should be mixed with other fungicide for resistance management.	Registered on potato, hop, grape and tomato. Pending use on lettuce, cucurbits vegetable, tato, onion, cereals and pepper. Potential use on Brassica leafy vegetables.
Dithionon (Delan)	BASF	Nitrite	Scab, downy mildew, rust, leaf spot	Pending on pome fruit and hop.
ELEXA PDB	Safe Science	Carbohydrate-chitin based product, plant defensive booster	Downy mildew, powdery mildew and gray mould	Biopesticide Pending use on cucumber, grapes and strawberry. Potential use on fruiting vegetables, edible legumes and other crops.
Epoxyiconazole (Opus)	BASF	Trizole		Pending use on banana.
Etaboxam (Guardian)	L G chemicals	Thiazole carboxamide	Useful for grape downy mildew, potato and tomato late blight, pepper blight and cucumber downy mildew. Preventive and curative activity	Potential use on grapes, potato fruiting vegetables, cucurbits, vegetables, Brassica leafy vegetable, leafy vegetable, edible legumes and other crops.
DuPont		Oxazolidine-dione	Broad spectrum fungicide, including early blight, downy mildew and other ascomycetes. Can be combined with Cymoxanil to pick up late blight	Candidate reduces risk pesticide. Pending use on potato, fruiting vegetables, grapes, cereals, cucurbits, head lettuce and hop. Potential use on leafy brassicas and onion.
Fenamidone (Reason)	DuPont	Imidazolinone	Foliar protectant and curative activity against oomycetes fungi. Also effective against ascomycetes and *Alternaria*. Inhibits electron transport	Candidate reduces risk pesticide. Potential use on fruiting vegetables, Brassica leafy vegetables cucurbit vegetable, potato grapes, sun flower.

Contd...

Contd...

Fungicides (Trade)	*Registrant*	*Chemistry*	*Pest Control Spectrum/ Traits*	*Status*
Fenbucona-zole (Indar) (Enable)	Rohm and Hass	Trizole	Powdery mildew, rust, apple scab, brown rot, cotton ball, mummy berry (*Monilinia* spp), smut, bunts, Cladosporium, *Mycosphaerella, cercospora, Septoria, Rhizoctonia, Pyrenospora, Helminthosporium* and related genera and a *Colletotrichum* sp.in turf.	Registered on pecans, bananas and stone fruit (except plum). Pending use on grape fruit and blue berry. Potential use on cranberry and pepper.
Fenhexamid (Elevate)	Arvesta	Hydroxyanilide	Non systemic protectant fungicide that is effective against *Botrytis cinerea, Monilinia, Sclerotina, Sclerotiorum* of lettuce	Reduce risk product. Registered on grapes, strawberry, almond, stone fruit group, caneberry sub group,. Pending use on fruiting vegetable group, citrus fruit group, tomatoes, pepper, green house cucumber, head/leaf lettuce. Stone fruit group (Post harvest) pome fruit group, pomegranate and asparagus.
Fenpropimorph	BASF	Morpholine	Controls powdery mildew, rust, *Helminthosporium, Rhyncosporium, Septoria* etc	Potential use in banana, sugar beet and cereals (international registration)
Fluazinam (Omega)	Syngenta/ISK	Pyridinamine	Broad spectrum disease control: *Alternaria, Botrytis, Cladosporium, Colletotrichum, Phytopthora.*	Reduce risk pesticide. Pending use on peanut and potato. Potential use on edible legumes, strawberry, lettuce, grape, onion, citrus and pome fruit.
Fludioxnil (Maxim) (Scholar)	Syngenta	Phenylpyrrole	*Fusarium, Helminthosporium, Rhizoctonia, Aspergillus, Alternaria, Pyrenophora, Telletia. Sclerotinia* and *Septoria.*	Reduce risk pesticide. Registered on most crops as seed treatment. Pending use on pome fruit, stone fruit (post harvest), strawberry and cantaloupe.
Fluoxastrobin	Bayer	Strobilurin	Alternaria Early blight and *Sclerotinia*, can be used as seed treatment or foliar use	Pending use on peanuts, potatoes, wheat and barley.
Fluquinco-nazole (Jockey)	Bayer	Triazole	Controls tale all, rust and a wide range of ascomycetes in cereals other disease	Potential use in cereals (Pending international use)
Flutolanil (Monocut)	Gowan and Nihon Nohyaka	Benzamide	Rusts, sheath blight, damping off and other diseases caused by *Rhizoctonia* and *Verticillium*	Registered on rice. Pending use on peanuts and potato.

Contd...

Contd...

Fungicides (Trade)	Registrant	Chemistry	Pest Control Spectrum/ Traits	Status
Fosetyl–AL Aliette	Bayer	Aluminum phosphate	Controls *Phytopthora, Alternaria* and downy mildew.	Registered on asparagus, avocado, Brassica leafy vegetables, canberry, citrus, cucurbits, leafy vegetable, pineapple,bulb onion, pome fruit, strawberry, tomato, banana, grape and macadamia. Pending on pea and lingon berry. Potential use on leek, green onion and turnip green.
Hexaconazole	Syngenta	Triazole	Controls loose smut and common root rot via seed treatment	Potential use on whet and barley (International registration)
Iprovalicarb (Melody)	Bayer	Amino-acid amide carbamate	Activity on oomycetes fungi, Downy mildew and *Phytopthora*	Potential use on grapes, potato, tomato, cucumber, lettuce, avocado, citrus.
ISK 916	ISK	Cyanoimida-zone	Oomycetes and plasmo-diophorales fungi. Highly effective verus late blight and downy mildew.	Potential use on vegetable crops.
Kresoxim-methyl (Sovran)	BASF	Strobilurin	Mildew, Septoria, rust, scab, phomopsis, black rot. Provides protectant, curative and eradicant control of powdery mildew.	Registered on pome fruit, grapes and pecans. Pending use on cucurbits, cereals, sugar beet and potato.
Mefenoxam	Syngenta	Active isomer of metalaxyl	Same spectrum as Metalaxyl	Reduced risk pesticide. Registered on Alfa Alfa. Almonds, apples asparagus. Avocado, beets, blueberry. Brassica leafy vegetables. Cereals citrus, cotton. Can berry, cucurbits, fruiting vegetables, ginseng, grape, grass, hop, leafy vegetables, edible legumes, bulb onion, green onion, papaya, peanut, pineapple, raspberry, root and tuber vegetables, soybean, stone fruit, strawberry, sunflower, walnuts. Pending use on artichoke, atemony, kiwifruit, sugar apple, mango, sapota, black sapota, papaya and canola.
Mepanipyrim (frupica)	Kumiai chemicals	Anilinopyrimidine	Control Botrytis	Potential use on grape (International registration approved)
Metconazole	BASF and Kurena	Triazole	Broad spectrum	Potential use on cereals and conola (International registration approved)
Metrafenone	BASF			

Contd...

Contd...

Fungicides (Trade)	Registrant	Chemistry	Pest Control Spectrum/ Traits	Status
Myclobutanil	Dow Agro	Triazole	Powdery mildew, rusts, apple scab, brown rot, shot hole, cherry leaf spot, grape black rot.	Registered on apple, asparagus, caneberry, cucurbits, stone fruit grapes, mint, strawberry, tomato, banana, pear and almonds. Pending use on hop, pepper and artichoke. Potential use on lettuce and papaya.
Orysastrobin	BASF	Strobilurin		
Oxolinic acid	Valent	Quinoline	Control gram negative bacteria and fire blight	Potential use on apple, peas, potatoes, beans, fruiting vegetables and rice
Phosphoric acid	Wilbur Ellis	Phosphoric acid	Downy mildew	
Picoxystrobin	Syngenta	Second generation Strobilurin	Wide spectrum of diseases	Potential use on cereals, grape, hop and apple (Pending international registration)
Proquinazid	DuPont			
Prothiconazole	Bayer	Triazolinthone Demethylation inhibitor	Effective on *Septoria* spp, Fusarium spp, *Rhizoctonia.* It has protectent, curative and eradicant activity, can be applied as seed treatment and foliar spray	Pending use on canola, edible legumes, rice and sugar beet, potential use on cereals and oil seed.
Propamo-carb Hypochloride	Bayer	Carbamate	Downy mildew late blight, damping off, *Pythium, Phytopthora* and *Aphano-myces.* Should be mix with other fungicide for resistant management.	Registered on potato. Pending use on tomato, cucurbits, lettuce and pepper. Potential use on other fruiting vegetable, citrus, carrot, sugar beet, onion, spinach and celery.
Pyraclostrobin (Headline) (Cabrio)	BASF	Strobilurin Mitochondrial Electron Transport Inhibitor	Broad spectrum activity on Anthracnose, downy mildew, cercospora leaf spot, rust, powdery mildew, *Septoria, Phytopthora, Pythium, Rhizoctonia.*	Reduced risk pesticide. Pending use on peanut, grape, potato and other tuberous and corm vegetables. Cucurbit vegetable, fruiting vegetable, Brassica leafy vegetable, head and leaf lettuce, wheat, barley, rye, Potential on tropical fruits
Pyrimethanil (Scala)	Bayer	Anilinopyrimidine	Active against *Botrytis* spp. *Venturia* spp, *Alternaria solani, Sphaerotheca Macularis* and *Monilinia* spp.	Potential use on grapes, pome fruit, strawberry, tomatoes, peas, beans, cane berry, onions, pepper, cucumber, citrus, potatoes, banana and stone fruit.
Quinoxyfen (Arius)	Dow Agro Sciences	Quinoline-Disrupts early cell signaling activities	Has shown activity against powdery mildew in a wide range of crops.	Candidate reduced risk pesticide. Pending use on grapes, hop and cherry. Potential use on cereals, fruiting vegetables cucurbit vegetable, apple and other stone fruit.

Contd...

Contd...

Fungicides (Trade)	Registrant	Chemistry	Pest Control Spectrum/ Traits	Status
Silthophane (Latitude)	Monsanto	Carboxamide	Control of take-all via seed treatment	Potential use on wheat and Barley (International used pending)
Simeconazole	Sankyo	Triazole	Effective as seed treatment against Basidiomycetes	Potential use on wheat, barley (International uses approved)
Spiroxamine (Impulse)	Bayer	Morpholine	Powdery mildew, most rusts, leaf blotch, chemical show protective, curative and eradicative effect.	Pending registration on grape and hop
Tebuconazole (Folicure)	Bayer	Triazole	Powdery mildew, rust, smuts, bunts, apple scab, *Pyrenospora, Septoria, cercospora, Ceratocystis, guignardia, Rhizoctonia, Colletotrichum, Botrytis* and *Rhizopus.*	Registration on banana, cherry, grape, peanut. Pending registration on cucurbit vegetable, turnip root and greens, garlic, hop, wheat, tree nuts, cherry mango, plums, barley, sugar beet and sunflower. Potential registration on beans, coffee, garlic, asparagus, lychee, okra, sweet potato and pome fruit.
Tetraconazole	Sipcam Agro Tomen Agro	Triazole	Controls cercospora leaf spot. Powdery mildew. Leaf spots, rust, web blotch and others. New chemistry good for resistance management.	Pending use on sugar beet and peanut. Potential use on blueberry.
Thifluzamid	Rohm and Hass	Thiazole–carbomanilide–inhibit succinic acid metabolism in fungi	Sclerotinia and Rhizoctonia	Potential use on peanut and rice
Tolylfuanid	Bayer		Broad Spectrum contact fungicide with good acaricidal effectiveness. Particularly suitable for control of resistant pathogen population	Potential use on apples, grapes, and hops (International registration pending)
Trifloxystrobin (Flint)	Bayer	Strobilurin	Active against powdery mildew, and leaf spot disease. Also provide significant control of scab, rusts downy mildew and other disease	Reduced risk pesticide. Registered on pome fruit, grape, cucurbits peanut, banana, carrot, celery, citrus, grass seed and stone fruit.
Triflumizole (Procure)	Uniroyal	Triazole	Powdery mildew, rusts, apple scab, *Rhizoctonia, Cylindrocladium. Thielaviopsis, Alternaria, Helminthosporium* and related grasses	Registered on apples, grape and pear. Pending registration on cherry. Potential use on cucurbits, strawberry and filbert.

Contd...

Contd...

Fungicides (Trade)	*Registrant*	*Chemistry*	*Pest Control Spectrum/ Traits*	*Status*
Zozamide (Gavel)	Dow Agro	Amide (Inhibit mitosis by binding to fungal tubulin proteins)	Control of foliar phycomycetes and *Albugo*. Also protectant against oomycetes fungi. Will be mixed with mancozeb for grater broad spectrum activity	Reduced risk pesticide. Pending use on potato grape, tomato and cucurbit, vegetables,(will be reviewed under NAFTA joint data review). Potential use on spinach.

Acknowledgements

Authors are grateful to various pesticides companies and crop life India from supplying the required information during the compilation of this chapter.

References

Dewayne, L.Torgeson, 1967. *Fungicides an Advance Treatise, Vol. 1.* Academic Press, New York and London.

Dunne, B., 1981. *Crop Protection Chemicals: Directions of Future Development,* (Sep. 23, 1981), pp. 83-101.

Hannah Mathers, 2000. *New Weapons in Ornamental Disease Control.* Bulletin of Oihio State University.

Jeffers, S.N., 2003. *Fungicides for Managing Phytophthora Species in Nurseries.* Sudden Oak Death Online Symposium. www.apsnet.org/online/SOD (website of The American Phytopathological Society). doi:10.1094/SOD-2003-SJ.

Moshe, Reuveni, 2004. Activity of the New Fungicide Benthiavalicarb against *Plasmopara viticola* and its Efficacy in Controlling Downy Mildew in Grapevines. *Europ. J. Pl. Pathology,* 109(3): 243-251.

Nene, Y.L. and Thapliyal, P.N., 2002. *Fungicides in Plant Disease Control,* 3rd Edition. Oxford and IBH Publishing Co., New Delhi.

Pathak, A.R., Prajapati, K.S. and Patel, R.C., 2004. Field evaluation of new fungicides against blast of rice. *Pesticide Research Journal,* 16(2): 26-28.

Rawal, R.D., 2007. New molecules of Fungicides a lecture note in the compendium of Summer School on recent advances in Plant Protection held at COA, Raichur.

Reviews of Plant Pathology an IMI Publication.

Ramulu, Sree, 1983. *Chemistry of Insecticides and Fungicides,* 2nd Edition. Oxford and IBH Publishing Co., New Delhi.

Chapter 25

Approaches for Integrated Disease Management of Onion and Garlic

R.K. Mishra[1] and R.P. Gupta[2]
[1]**Assistant Director (Plant Pathology), National Horticultural Research and Development Foundation, Regional Research Station, Salaru Karnal – 132 001, Haryana**
[2]**Director, National Horticultural Research and Development Foundation, Nasik, Maharashtra**

Introduction

Onion (*Allium cepa* L.) and Garlic (*Allium sativum* L) is the most important *Allium* species cultivated in India and used as vegetable, salad and spice in the daily diet by large population. In India, onion and garlic crops are grown almost all over the country, especially in the states of Haryana, Punjab, Madhya Pradesh, Gujarat, Rajasthan, Uttarakhand, Jammu and Kashmir, Bihar, Uttar Pradesh and Maharastra, A.P., Karnataka. M.P. is the leading state in garlic production while in onion, Maharastra is the leading state accounting for more than 22.83 per cent of area and 28.42 per cent of production with an average yield of 12.37 t/ha. In India per hectare yields are highest in Gujarat (22.65 t/ha) followed by Punjab (22.63 t/ha). The area under garlic during 2006-07 is 0.14 million ha and production is 0.75 million tones with average productivity of 5.38 t/ha. In India per hectare yield are higher in Punjab (16.67 t/ha) followed by H.P. (13.14 t/ha) and Haryana (12.38 t/ha). The crop is attacked by many diseases, which vary from region to region, season to season and variety to variety. Various biotic factors like fungi; bacteria, viruses, phytoplasmas and nematodes are associated with garlic at different stages of growth and cause considerable damage/losses in yield as well as quality of garlic. There are many diseases affecting the garlic production through out the country. These diseases can also affect at production, harvesting, processing and marketing stages, which lower the quality, reduce the yield there by increase the cost of production and export potential also. The diseases alter the cropping pattern and affect local and export markets. In the year 1993 in Maharastra state 60-80 per cent losses were reported due to Purple blotch. Consistent use of fungicides and other chemicals to manage the diseases in crop plants not only poses a serious threat to the environment and mankind but also slowly build up resistance in the pathogens. Most of the new generation pesticides are systemic in their mode of action leads to certain level of toxicity in the plant system and thus resulting health hazards. Further, it disturbs complete microbial diversity of whole ecosystem. All these factors have led to new dimension in research for biological control and integrated disease management. The effect of disease can be lessened through various means including cultural, chemical and biological. In recent IDM approach has been found to be one of the most important tools to minimize the incidence of diseases of onion and garlic. Important diseases affecting the onion and garlic and their integrated management are reviewed here.

Integrated Disease Management: Basic Concept

Plant disease control means absolute check of a disease, which is neither possible nor feasible and replaced by more appropriate term management. No pathogen is completely eradicated from natural ecosystem, but its population can be reduced below economic injury level. Disease management does not employ against the only pathogen but also considers the other three components of disease triangle *i.e.* host, environment and time factor which gives the concept of integrated disease management (IDM). The main objective of IDM is to maintain the loss below an economic injury level and minimize the recurrence of disease by interrupting disease cycle, survival period and inoculums source. Pathogen management involves reduction in inoculum, eradication and prevention of inoculum while management of host involves improving plant vigour, induced resistance through nutrition and genetic manipulation and protection by chemical means. The time factor could be managed by adjustment of planting date so that favorable time of disease should not coincide without any adverse effect on yield. The principles of IDM are based on avoidance, exclusion, eradication, protection, host resistance and therapy. Avoiding disease by planting at times when or in areas where inoculum is ineffective due to environmental conditions. It could be also achieved by selecting geographic area for disease free seed production, using disease escaping varieties, selection of disease free seeds and modification cultural practices. Exclusion means preventing the inoculum from entering or establishing in the field or area where it does not exit. Eradication is applied when pathogen has already been entered in area or crop in spite of the above-mentioned precautions. At this conditions, biological control methods, crop rotation, rouging, sanitation, soil treatments, heat and chemical treatments of diseased plants can be applied. In present scenario integrated disease management (IDM) strategies is the only sustainable approach. IDM should be effective against more one disease, long lasting, economically viable and eco-friendly. There is an urgent need to retinalize the use of pesticides and develop IDM packages for onion and garlic diseases that are capable to managing plant pathogens more effectively and have minimum impact on humans, wild life and environment.

Integrated Disease Management: Key Strategies

The success of integrated disease management approach in onion and garlic is only possible when we apply all the following practices and precautions every year on community basis in large area.

- ☆ Avoid indiscriminate use of fungicides and do not mix incompatible pesticides, phytohormones and micro-nutrients at a time.

- ☆ Soil as a reservoir of harmful as well beneficial microorganisms, therefore soil health should be properly managed by timely tillage, summer ploughing, green manuring, optimum C: N ratio, balanced macro and micro-nutrients, aeration etc.
- ☆ Summer ploughing followed by irrigation in summer and then subsequent ploughing is very effective.
- ☆ Seed health must be maintained up to prescribed standard and ensure that seed should be free from internal and external pathogen, contamination of any infected crop debris, sclerotial bodies etc.
- ☆ Field sanitation as essential practices where removal and burning of infected crop debris, alternate and collateral weed hosts should be carried out periodically.
- ☆ Crop rotation is very important for soil borne diseases where non-host crop preferably cereals should be selected for particular pathogen.
- ☆ Avoid off season vegetables and intensive cropping because it prolongs the perpetuation period of a pathogen and do not break the life cycle of the pathogen.
- ☆ Knowledge of correct diagnosis of the diseases and disease cycle of the pathogen is very important to apply different methods of critical phase of life cycle for maximum efficacy.
- ☆ Application of biocontrol agent's particularly resident antagonist accompanied with green manuring, FYM or any organic matter should be maximized.
- ☆ Always prefer to grow tolerant varieties. However, it is difficult to get resistance against diseases with all desirable traits in hybrid
- ☆ Proper post harvest management practices should be followed to avoid any rotting of onion and garlic during storage and transportation.

Knowledge of the following aspects of disease development is essential for effective and economic management:

- ☆ Cause of the diseases.
- ☆ Mode of perennation and dissemination of the infectious causes.
- ☆ Host-parasite relationship and means of secondary spread.
- ☆ Effect of environment on pathogenesis in the plant and spread of the disease in the plant population.

Important Diseases

Pre-harvest Diseases

Damping-off

Pathogens

Pythium sp.

Fusarium sp.

Rhizoctonia sp.

Symptoms

1. Pre emergence
2. Post emergence Damping-off of onion seedlings occurs in two stages:

In pre emergence stage, the younger seedlings are killed before they reach the soil surface. They may, infect, be killed even before the hypocotyls has broken the seed coat. The radicle and plumule, when come out of the seed, undergo complete rotting. Since this happen below the soil surface and the disease often not seen, the failure in emergence of seedlings is attributed to the poor quality of seeds. The post emergence damping off is characterized by the toppling over of infected of infected seedlings anytime after they emerge from the soil. It usually occurs at or before the ground level and infected tissues soft and water soaked. As the disease advances, the stem becomes constricted at the base and plant collapses. It is observed that most of the loss is due to pre emergence damping-off. It is more common during Kharif season when temperature and humidity are very high.

Management

- ✰ Proper drainage is essential.
- ✰ Sowing of clean uninfected seed on raised beds.
- ✰ Crop rotation.

- ✰ Soil solarization of nursery beds with transparent polythene for 30 days before sowing gave good control of damping-off.
- ✰ Seed treatment with Thiram, Captan @ 2.5 g/kg seed or Carbendazim @ 0.1 per cent or *Trichoderma viride* @ 4-5 g/kg is effective.
- ✰ Drenching the nursery by Thiram @2.5 per cent or Carbendazim @0.1 per cent or *Trichoderma viride* @ 4-5g/Litre.

Purple Blotch

Pathogen

Alternaria porri.

Symptoms

This is a most important disease prevalent in all the onion and garlic growing areas in the country. The disease is reported to occur in hot and humid climate. First it appears on the leaves as small, whitish, sunken lesions. These spots later enlarge and eventually encircle the leaf. Later, darkened zones appear on the surface of the leaves, retaining the characteristic purple colour. The leaves and stems fall over gradually. Infection can cause a semi-watery rot of neck of the bulbs, which turn yellow

red in color, bulb tissues eventually become papery. Infected plant debris is the main source of inoculums.

Management

- ☆ Use of healthy seeds.
- ☆ Crop rotation of 2-3 years with non-related crops should be followed.
- ☆ Spraying of Dithane M-45 (2.5gram/liter water), Kavach (2.0g/litre water) should be sprayed at 15 days intervals just after 45 days after transplanting.
- ☆ Spraying of *Trichoderma viride* (5g/litre) and *Baeveria bassiana* (2-3g/litre) at weekly intervals would be effective against this disease.
- ☆ Seed treatment with Captan @ 2.5 g/kg seed, Thiram @2.5 g/kg or *Trichoderma viride* @ 4-5 g/kg seed before nursery sowing.
- ☆ Summer ploughing is reduced the incidence.
- ☆ Grow resistant/tolerant cultivars like, G-1, G-50, and G-323

Stemphylium Blight

Pathogen

Stemphylium vesicarium.

Symptoms

This is a serious problem in Northern parts of the country especially in bulb crops. The disease is also prevalent in some parts of the southern parts of the country. Symptoms appears as small, yellow to orange flecks or streaks on the leaf. These soon develop into elongated, spindle shaped to ovate elongate, diffused spots, often reaching the leaf lips. They usually turn grey at the centre, brown to dark olive brown. The spots frequently coalesce into extended patches blight of the leaves and gradually the entire foliage. The pathogen survives on the dead host tissue and serves as the source of inoculums.

Management

- ☆ Spraying of Mancozeb (2.5g/litre water) along with Dhanuvit is effective to control the disease. Spraying is started just after appearance of disease and repeated fortnightly.

- ☆ Spraying of Sixer (carbendazim + D. M-45) @ 2.5g/litre of water with Dhanuvit (Sticker) at 15 days intervals.
- ☆ Crop rotation with other vegetables and cereals to reduced the disease severity.

Cercospora Leaf Spot

Pathogen

Cercospora duddiae.

Symptoms

The disease appears on the leaves as small ash colored and irregular shaped on leaf lamina. The spots coalesce gradually and result in blighting of the foliage. High temperature and prolonged wet conditions favors the disease development.

Management

- ☆ Spraying of Dithane M-45 (2.4g/litre water) and Copper oxychloride (3.0g/litre water) at 15 days intervals give good results.
- ☆ Spraying of Kavach (Chlorothalonil) @ 2.0g/litre water and sixer (Carbendazim + D. M-45) @ 2.5g/litre water at 15 days intervals.
- ☆ Deep summer ploughing.

Powdery Mildew

Pathogen

Leveillula taurica

Symptoms

Whitish, circular to oblong lesions of variable size occurs on abaxial surface of leaves. Older leaves infected first. The lesions are covered with white powdery mass of fine hypae and conidia.

Management

- ☆ Use of sulpher fungicides @ 0.2 per cent at fortnightly intervals.
- ☆ Proper drainage is essential.

Rust

Pathogen

Puccinia allii or *Puccinia porrii*

Symptoms

This fungus was considered to be minor importance in onion and garlic production. Initial symptoms occur on the foliage and stem as small white flecks that develop into orange spots (spores) or pustules. The bulbs become shrunken and deformed. Heavily infected plants may turn yellow and die. Disease incidence is highest in stressed plants.

Management

- ☆ Use of healthy seeds

- ☆ Spraying of hexaconazole @ 0.1 per cent and propiconazole (Tilt) @ 0.1 per cent at 15 days intervals.
- ☆ Rotate with non-allium crops.

Anthracnose Disease (Twister)

Pathogen

Colletotrichum gloeosporioides

Perfect Stage

Glomerella cingulata.

Symptoms

Curling and twisting of leaves, chlorosis and abnormal elongation of neck portion are major symptoms. Bulb development is already affected. White oval sunken lesions are developed on the leaves. In advance stage roots become sparse and plant may die. Bulbs are slender and may rot before harvest or during storage.

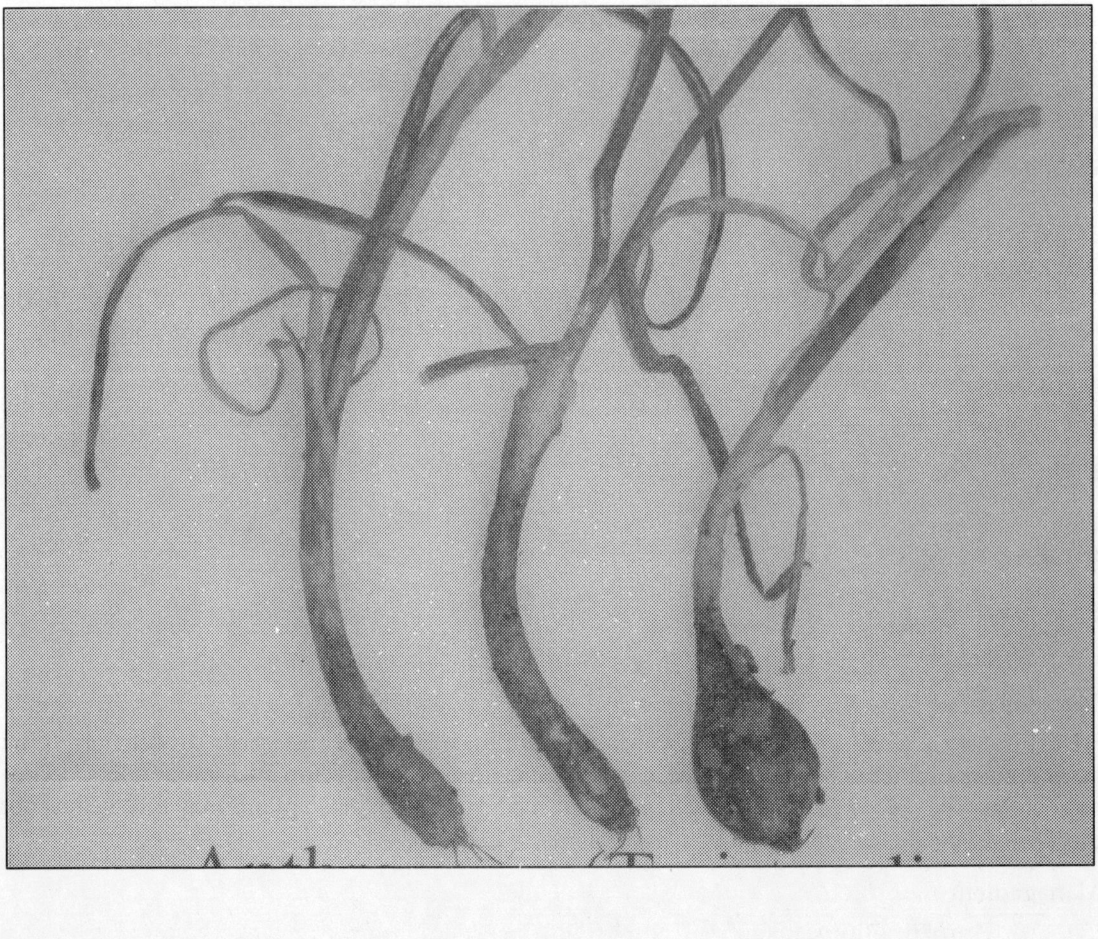

Management

- ☆ Use of good quality of seeds.
- ☆ Seed treatment with Thiram or Capaton (2.0-2.5g/kg seed) before nursery sowing.
- ☆ Seed treatment with *Trichoderma viride* (5.0gram/kg seed).
- ☆ Soil treatment with Benomyl or Benlate (@ 2.0g/m^2) and spraying of D.M-45 (2.5g/lit water) were effective.
- ☆ Sanitation and destruction of infected plant debris helps in reducing the disease incidence.

Garlic Mosaic

Pathogen

Virus

Symptoms

Garlic plants infected with mosaic virus show typical symptoms of chlorotic mottling and strips on the emerging leaf followed by pale yellow broken stripes resulting typical mosaic pattern on matured leaves. Yellowish dots on leaves, whitish leaf margin, or twisting of leaves are also recorded on some cultivars. Generally symptoms are more pronounced in young leaves.

Management

- ☆ Spraying of Monocrotophas @ 0.05 per cent, Endosulphan @0.25 per cent or methyl dematon @ 0.075 per cent at 10-15 days intervals
- ☆ Cloves treatment with Bavistin @ 0.1 per cent before planting
- ☆ Use of HNPV and some neem based botanicals at weekly intervals

Onion Yellow dwarf

Pathogen

Virus (Poty virus)

Symptoms

Onion Yellow Dwarf Virus (OYDV) is an important pathogen of garlic and onion also, causing severe losses in garlic clones. It is an aphid-borne poty virus. It produces symptoms of mild chlorotic strips to bright yellow strips depending on virus isolate and cultivars. Reduction in growth and bulb size also occurs. Infection by other viruses such as Leek yellow stripe, Garlic common latent virus and Shallot latent virus also occurs and may aggravate the symptoms further. However, OYDV is recognized as a major element of the virus disease complex. Enzyme Linked Immunosorbant Assay (ELISA) is the main diagnostic method for large-scale routine detection of OYDV in garlic. The other method of OYDV detection is based on reverse transcriptase polymerase chain reaction (RT-PCR), which is 10-100 times more sensitive than ELISA.

Management

- ☆ Removal of virus infected plants.
- ☆ Alternate spray of systemic insecticides with neem based botanicals.
- ☆ Collect healthy seed from disease free plants.

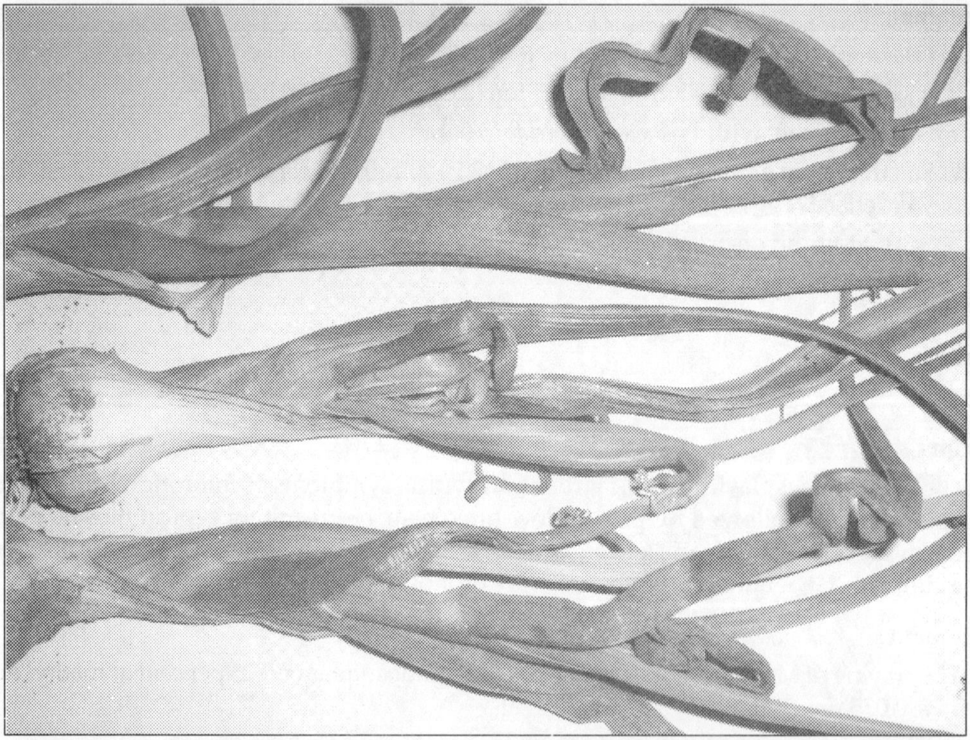

Iris Yellow Spot Virus (IYSV)

Pathogen

Virus (Tospovirus)

Symptoms

The disease symptoms vary among onion bulb and seed crops, but often appear as straw-colored, diamond shaped lesions on leaves and scapes with twisting or banding flower bearing stalks were observed in onion plants. Some lesions have distinct green centre with yellow or tan borders, other lesions appears as concentric rings of alternating green and yellow/tan tissue. Infected plants can be scattered or generalized through out a field. Large necrotic region may develop on scapes and cause a collapse of the escape. Diseased plants may be scattered or wide spread across a field, but the highest incidence of disease is often found on the field edges.

Management

- ☆ Iris yellow spot virus (IYSV) is a top virus, similar to tomato spotted wilt virus, which is currently through to be vectored solely by onion thrips (*Thrips tabaci*), so alternate spray of softer insecticides formulations (spinosad, neem extract) and organic mulches (straw) is effective against IYSV.

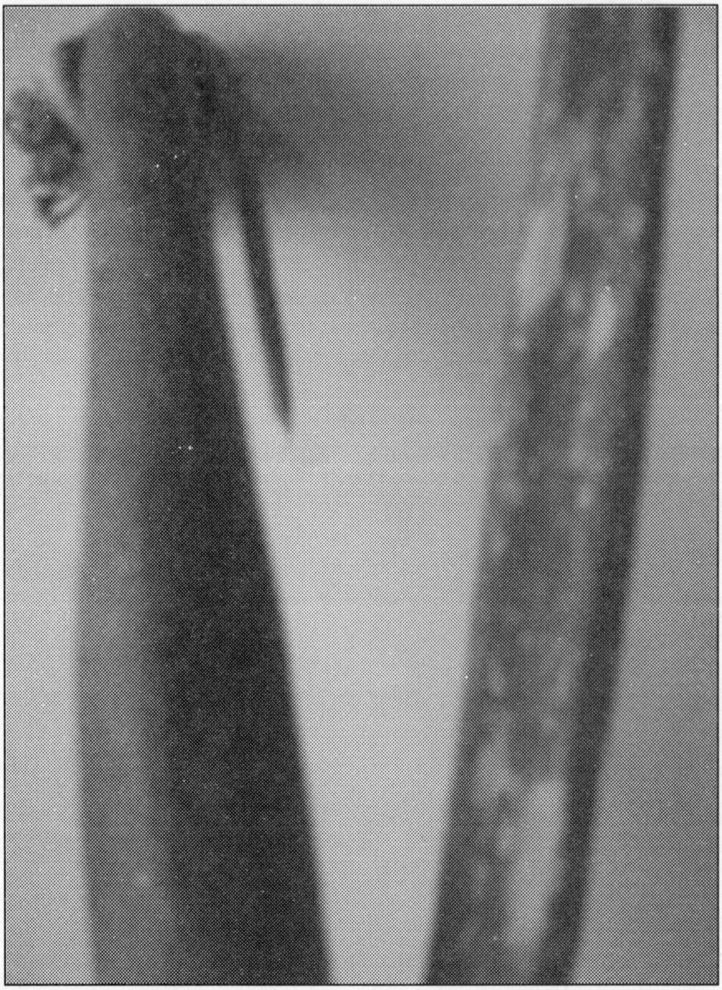

☆ An integrated approach is essential for management of IYSV.

Post-harvest Diseases

White Rot

Pathogen

Selerotium cepivorum.

Symptoms

Pathogen is soil inhabiting and invades the roots and the basal part of the bulb scales. The first symptoms of the disease is yellowing and dying back of leaf tips. The roots are generally destroyed and there is semi watery decay of the scale with abundance of superficial white truly mycelium. Brown to black sclerotia is developed on surface or within tissue. The organism is most active when the temperature is cool. In Northern climates it usually attacks in the spring.

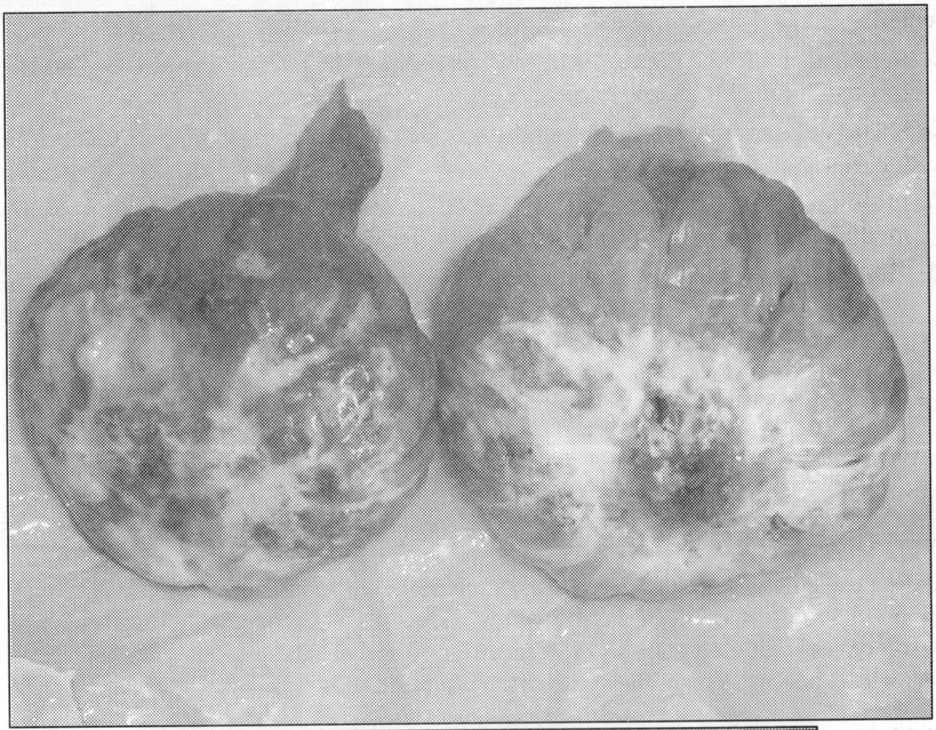

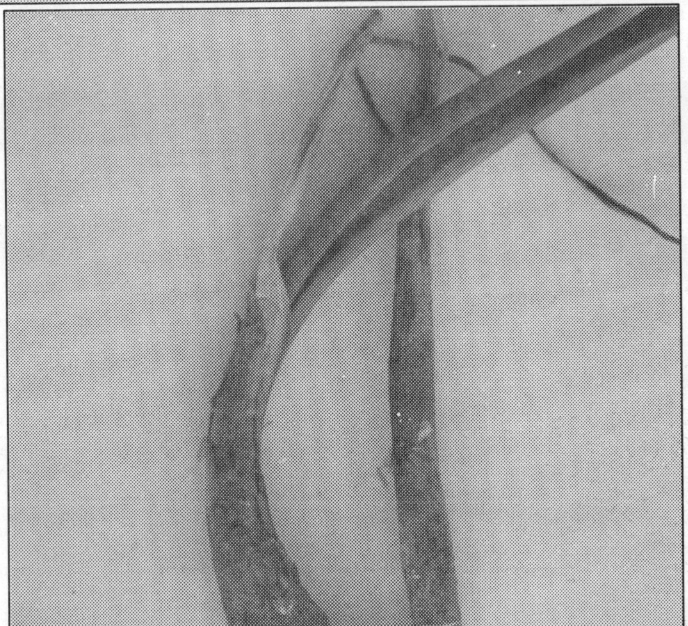

Management

- ☆ Removal of infected plants during season reduces the sclerotial population and also avoids incorporation of the same to the soil.
- ☆ Solarization of soil at high temperature *i.e.* 35ºC for 18hours or 45ºC for 6 hours reduces the incidence.
- ☆ Iprodione (Rovral) @ 0.25 per cent was effective and Benomyl @ 0.1 per cent thiophemate methyl @ 0.1 per cent have reduced disease incidence.
- ☆ The organisms, which have given promising result, are *Trichoderma viride, Gliocladium Zeae, Penicillium nigricans, Bacillus subtilus* and *Trichoderma harzianum* were highly effective against pest harvest diseases.

Basal or Bottom Rot

Pathogen

Fusarium oxysporum f. sp. cepae

Symptoms

In garlic, pre emergence decay of cloves, stem plate and post harvest decay of cloves in stored bulbs are main symptoms. Infected garlic show reduced emergence, yellowing and/or browning (necrosis) of leaves beginning at tips. On stem and bulb early in the season with some discolouration on bulb sheath at harvest. This will eventually wither and die.

Management

- ☆ Use of *Trichoderma viride* @4-5 kg/ha before transplanting is effecting against this disease.

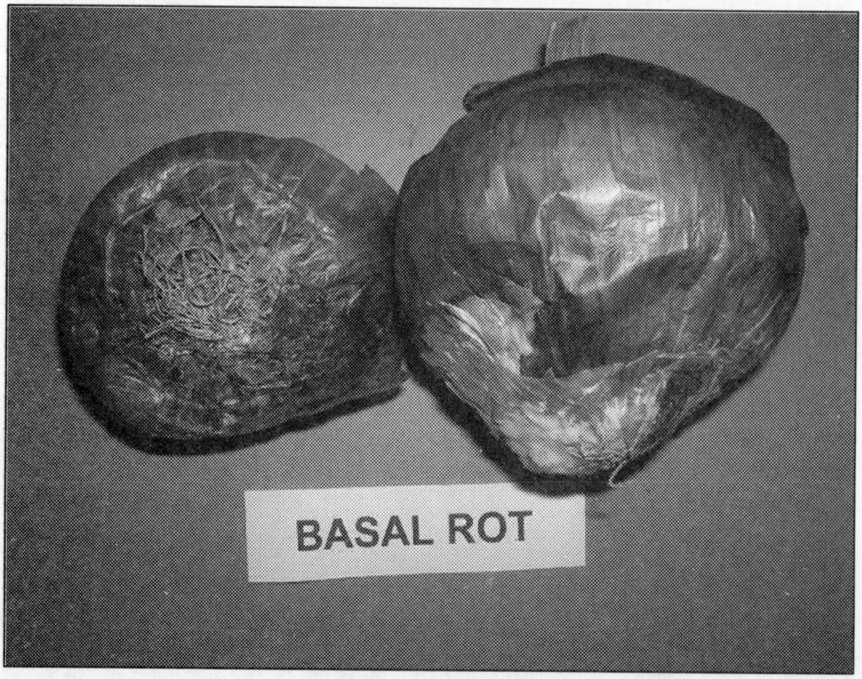

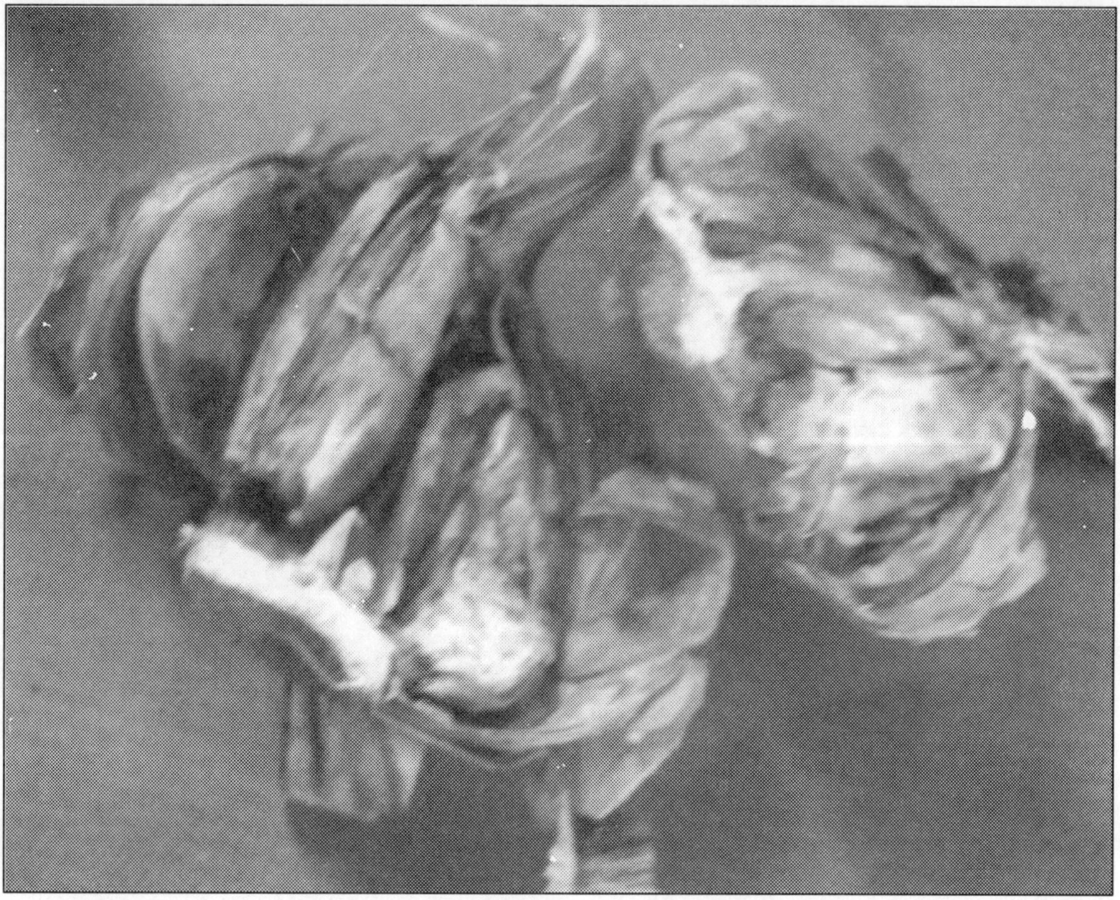

- ☆ Proper drainage is essential during entire cropping period.
- ☆ Proper drying and curing
- ☆ Use of *Trichoderma viride*, *Pseudomonas* spp., *Baeveria bassiana* before transplanting is effecting against this disease.
- ☆ Deep ploughing and avoiding injury during cultural practices.

Brown Rot

Pathogen

Pseudomonas aeruginosa

Symptoms

Dark brown discoloration in bulb scale is the characteristics feature of this disease. Browining of inner scale along with rotting is the main symptoms of disease. The rotting starts from the inner scale and spread to outer scales. Apparently, the bulb seems to be healthy, but when pressed, the white oozing is noticed from the neck. In several cases the whole lot of bulbs gets rotten giving the bad odour in storage.

Management

- ✰ Proper curing is required before storage.
- ✰ Use of maleic hydrazide (20 ml/litre) before one month of harvest.
- ✰ Neck cutting is about 2.5-3.0 cm long above the bulb is reduced the bacterial infection.
- ✰ Light irrigation is required during entire cropping period.

Soft Rot

Pathogen

Erwinia caratovora

Symptoms

Severe discolorations with soft rotting of onion bulbs are the main symptoms of this disease. Severe infection occurs at high temperature. The affected fleshy scale tissues are water soaked and pale yellow to light brown and become soft as the rot progresses. The whole bulb may break down and a watery liquid may ooze from the having foul odour as the disease advanced and, if squeezed.

Management

- ✰ Proper curing is required after harvest.
- ✰ Proper drainage is required during cropping period.
- ✰ Reduced the doses of nitrogenous fertilizers.

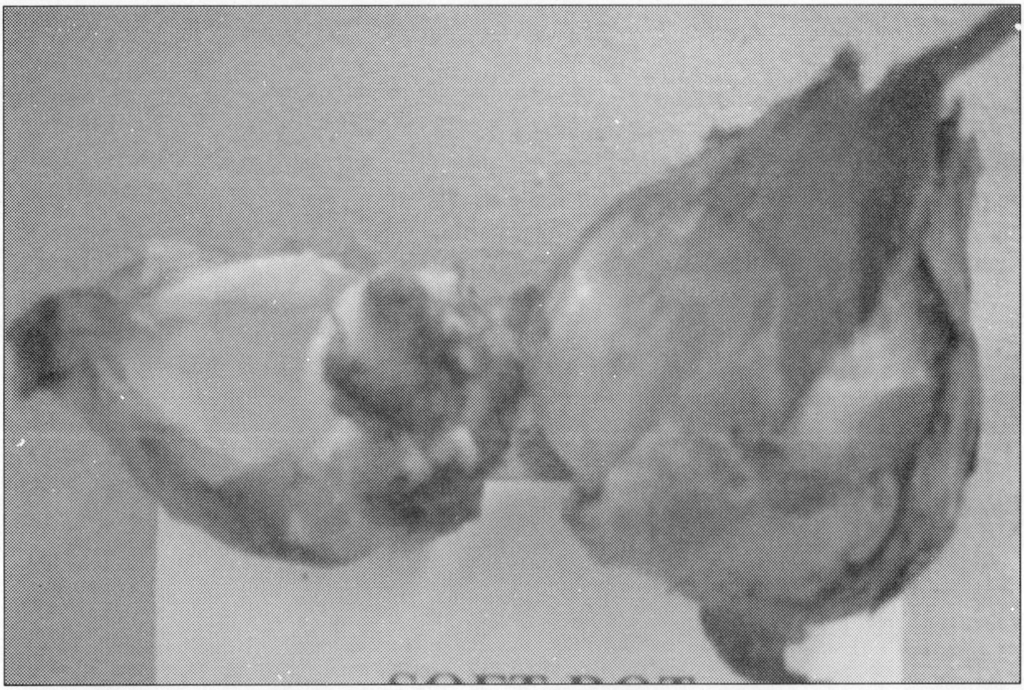

Botrytis Rot

Pathogen

Botrytis allii.

Symptoms

Infection usually takes place through neck tissue and occasionally else where. The first sign is the softening of the effected tissue, which takes a sunken cooked appearance. Later stage, decaying starts, which is separated from healthy tissue by a definite margin. A dense grayish mycelial mat after develops upon the decaying tissue of the scope bearing short conidiophores with conidia. The disease progresses rapidly down the scales of neck tissue. Black sclerotial bodies are developed later. When diseased bulbs are cut open, water soaked brown tissues are seen near the neck region.

Management

- ☆ Seed treatment with benomyl @1 g/kg seed or benomyl + thiram @ 1 g/kg seed reduced the disease incidence.
- ☆ Pre harvest spray of benomyl @0.1 per cent reduced the fungal infection.
- ☆ Proper curing is required before storage
- ☆ Neck cutting is 2.5-3.0 cm above the bulb to reduce the infection.

Pink Root

Pathogen

Pyrenochaeta terrestris

Symptoms

The characteristic symptom is the pink colouration of the roots. The affected roots initially turn yellow but later on become soft and ultimately take distinctly deeper pink color. The new roots, which grow from the infected plants, get immediately infected and become functionless. The affected plants are not killed but development is retarded as leaf number and size are reduced. Bulbing starts earlier, but the size is reduced at maturity. The disease is confined to the roots only.

Management

- ☆ 5-6 years crop rotation gives good control of pink root.
- ☆ Soil solarization also helps in reducing the disease.
- ☆ Use of dichloropropene @ 450 litres/ha increases the yield.

Black Mould

Pathogen

Aspergillus niger

Symptoms

Fungus infects the neck of bulbs and occasionally penetrates from side and balsa end of bulb, where damage to the dry skins has exposed the bulb scales. In advance stages of the symptoms, the entire surface of bulbs turns black and all scales are infected. It reduces the market value of the bulb.

Mamagement

- ☆ Proper drying and curing of the bulbs after harvest.

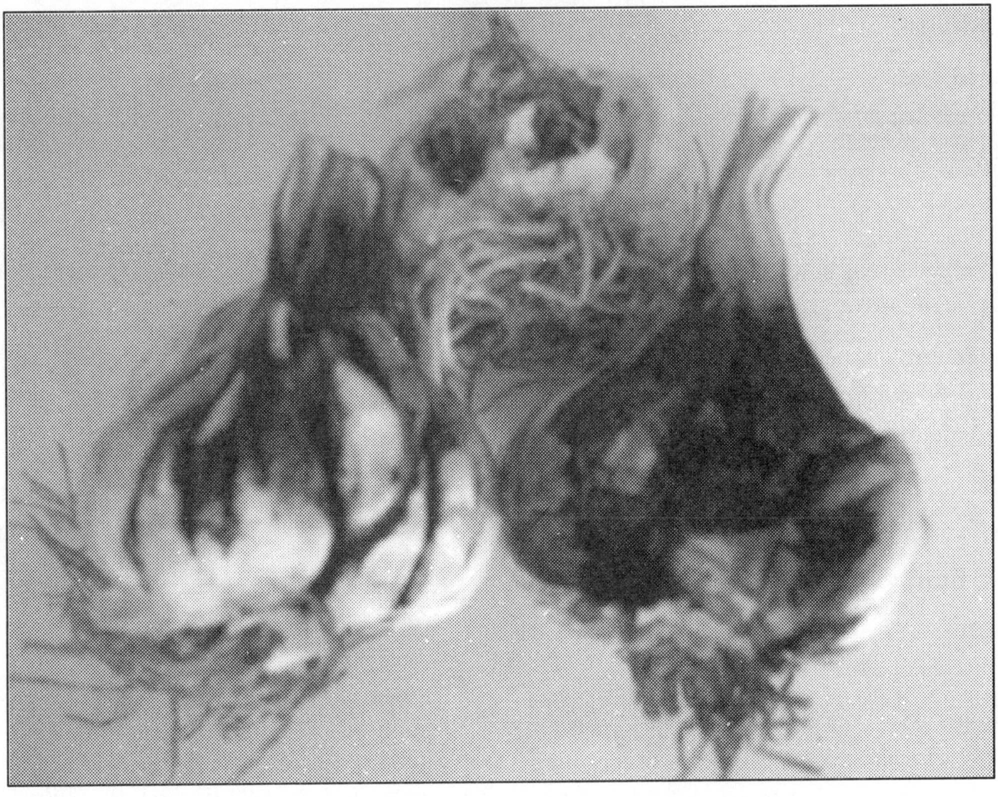

- ✰ Bulbs are dusted with calcium carbonate and godowns will fumigate with nitrogen trichloride @ 430mg/m^2.

Blue Mould

Pathogen

Penicillium corymbiferum.

Symptoms

The first symptoms of the disease are pale yellowish blemishes, watery soft spots, or occasionally a purplish red stain on the scales. A green to blue green mold may develop on the surface of lesions when bulbs are cut longitudinally, one or more of the fleshy scales may appear water soaked and exhibit a light tan or gray colour.

Management

- ✰ Proper drying and curing after harvest and before storage.
- ✰ Injury should be avoided during post harvest handling.
- ✰ Mercurial dip of bulb/cloves was practiced before drying.
- ✰ Bulbs stored at temperature of 5ºC or less with RH as low as possible may present infection.

Garlic Bulb Canker

Pathogen

Embellissia allii.

Symptoms

This is new disease recorded in garlic bulbs (variety: G-313) produced in Himachal Pradesh. The initial symptoms of the disease are small blackish dark on the outer scales of the bulbs, which later

enlarges and cover whole bulbs with black powdery mass. Under advance conditions, decay of the cloves has been observed. It was recorded by NHRDF for the first time in India.

Management

- Spray of Carbendazim (1.0g/litre) should be given before 15-20 days before harvest of the crops.
- Proper curing of bulbs should be taken before storage.
- Seeds (Cloves) will be treated with bavistine (1.0 g/kg seed) before planting.
- The exotic material of garlic not is used for planting.

Future Strategies

- A series of cultural and chemical control methods for garlic pests and diseases with targets varying with crop growth stage.
- Monitoring of new diseases and appropriate management strategies should be developed.
- Exploitation of disease tolerance observed in the screened improved lines.
- Large-scale use of fungal and bacterial bio-agents in a consortium to achieve better disease management.
- Studies on the use of Plant growth promoting Rhizobacteria (PGPR) in improving plant growth and offering protection from diseases.
- Develop techniques for quick diagnosis of viral diseases.
- Conversion of onion and garlic waste in to an antimicrobial agent.
- Develop IPM modules for garlic diseases.
- Since many diseases of garlic reported in other parts of the world and not present in our country, strict embargo on import of seed, planting material and the produce for domestic consumption should be followed. Quarantine and legislation should be mandatory in these crops.

Chapter 26

Biocontrol Strategies for the Management of Vegetable Pests using Entomopathogenic Fungi

T. Jiji[1], A. Naseema[2] and R. Praveena[3]
[1]Professor, Department of Agricultural Entomology,
[2]Professor, Department of Plant Pathology,
[3]Senior Research Fellow, Department of Agricultural Entomology
College of Agriculture, Vellayani

Introduction

Vegetables are infested by an array of pests belonging to different groups like caterpillar, bugs, beetles etc. Use of chemical pesticides has become an indispensable component in modern day intensive vegetable cultivation. It leads to many environmental problems like residue hazards, disruption of natural control etc.Vegetable s are often consumed in non processed form, making the problem of pesticide residues more prominent. Realising the adversities of chemical pesticide application on nature, steps are being taken to give more importance to sustainable agriculture and organic farming using eco-friendly pest management strategies.

Almost all insects have their natural enemies in the form of predators, parasites and pathogens. The identification of suitable natural enemy and their incorporation in the integrated pest management is an important step in any successful pest management programme. Microbial agents *viz.* fungi, bacteraia, virus, protozoa and actinomycets are of considerable importance in the field of biological control. Entomopathogenic fungi due to its specificity in action and non pathogenicity to non target organism play an important role in the biological control of insects.

Fungal Pathogens

Some insect species, comprising many pests, are particularly susceptible to infection by naturally occurring, insect-pathogenic fungi. These fungi are very specific to insects, often to particular species, and do not infect animals or plants. Fungal growth is favoured by moist conditions but fungi also have resistant stages that maintain infection potential under dry conditions. Fungi have considerable epizootic potential and can spread quickly through an insect population and cause its collapse. Because fungi penetrate the insect body, they can infect sucking insects such as aphids and whiteflies that are not susceptible to bacteria and viruses. Several fungal species have potential as microbial insecticides and are commercially available in formulations that can be applied using conventional spray equipment.

Entomopathogenic fungi causing mycosis in insects are identified from most of the insect orders *viz.* Orthoptera, Isoptera, Hemiptera, Diptera, Coleoptera, and Lepidoptera. Important entomopathogenic species belonging to the genera *Beauveria, Entomophthora, Fusarium, Metarhizium, Nomurea, Paecilomyces* and *Verticillium* are known for biological control.

Beauveria spp.

The effectiveness of *B. bassiana* was proven against *Aphis craccivora* (Zaki,1998) *Leptinotarsa decemlineata* (Todorova *et al.*, 2000), *Spodoptera litura* (Gloriana *et al.*, 2001) *Spilosoma obliqua* (Mondal and Bhattacharya, 2003) *and Sylepta derogata* (Ramesh *et al.*, 1999)

Beauveria bassiana produces the toxin beauvericin, bassionolide and beauverolides which causes cellular alterations. Hafez. *et al.* (1994) conducted bioassay using *B. bassiana* against *Pthorimoea operculella* and they obtained LC_{50} value of 4. 7 x 10^8 conidia/ml. In brinjal it was also observed that direct spraying of *B. bassiana* killed 58.1 percent first instar larva and 35.2 per cent pre pupal stage larvae. The adults were not susceptible *to B.bassiana* (Rajendran and Gopalan, 1999).

A potent local isolate of *B. bassiana* was reported by Jiji *et al.* 2005. The same was identified as a new strain and deposited as ITCC 6063. The virulence of the same on pests vegetable crops was reported by Jiji *et al.*, 2006 a. Pathogenicity tests of the new strain of *B.bassiana* was conducted in the laboratory. *B. bassiana* was found to infect pests of *bhindi* (*Sylepta derogata*), cucurbits (*Anadevidia peponis, Diaphania indica, Henosepilachna vigintioctopunctata,Bactrocera cucurbitae*), Amaranthus (*Psara basalis*) and brinjal (*Psara bipunctalis, H. vigintioctopunctata, Urentius hystricellus*).

Larvae of lepidopteran pests showed various symptoms when sprayed with the spore suspension of *B. bassiana.* During the initial stages the infected larvae were sluggish in their movement, became cream in colour and died from the next day onwards. The infected larvae failed to moult and early pupation of late instar was observed. In the case of early instars the food uptake was reduced considerably. Malformed adults were emerged from the pupae, developed from infected larvae. In some cases adults did not emerge from the pupae. These pupae were black in colour, compared to brown colour of normal pupae and white and puffy mycelial growth of *B. bassiana* appeared on pupae. Mean mortality percentages of 93.33, 86.67, 66.67 and 86.67 were recorded for the larvae of *S. derogata, P. bipunctalis, A. peponis* and *D. indica,* respectively (Jiji *et al.*, 2007).

Epilachna beetle (*H. vigintioctopunctata*), a voracious feeder of cucurbits and brinjal was found to be susceptible to the local isolate of *B. bassiana*. The fungus infected the eggs of epilachna and as a result the colour of the eggs changed from yellow to brown, later white mycelial growth appeared on these eggs. Grubs failed to emerge from infected eggs. Epilachna grubs were also susceptible to the infection by the fungus. Infected grubs showed less apetite. Later the colour of the spines changed to brownish black and finally whole grub became black in colour and dead. On dead grubs white mycelial growth of the fungus appeared. Adult beetles when sprayed *with B. bassiana* did not develop any disease symptoms. A mean mortality of 63.33 per cent was recorded on epilachna beetles. Sucking pests *Urentius hystricellus* and *D. cingulatus* were susceptible to the fungus. Affected insects showed less feeding, later became sluggish and white mycelial growth of the fungus appeared on the dead insects(Jiji *et al.*, 2006d) (Plates 26.1–26.4; Figure 26.1 and 26.2)

The safety of the pathogen was tested on important cultivated plants of Kerala as well as on predators in the vegetable ecosystem. The fungus did not develop any disease symptoms on the crop

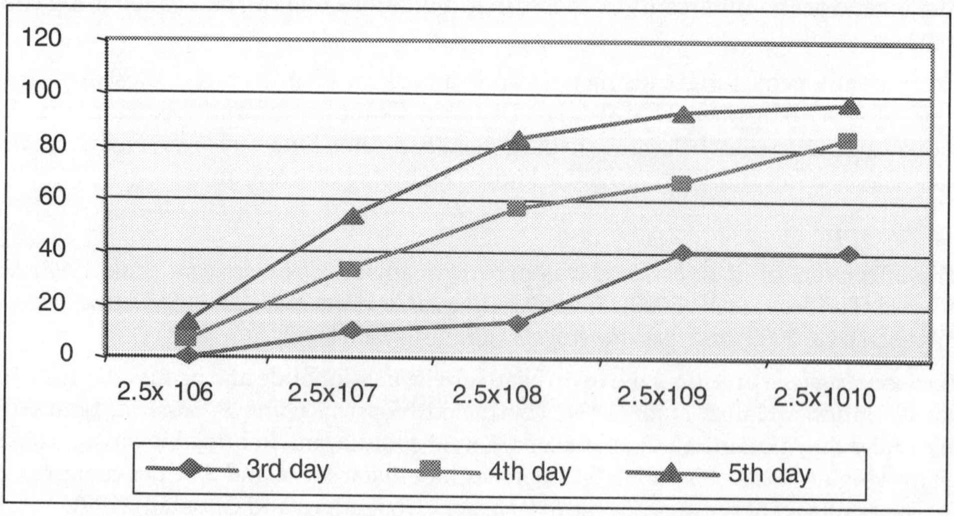

Figure 26.1: Cumulative Per cent Mortality of *Sylepta derogata* Treated with Different Concentrations of *Beauveria bassiana*

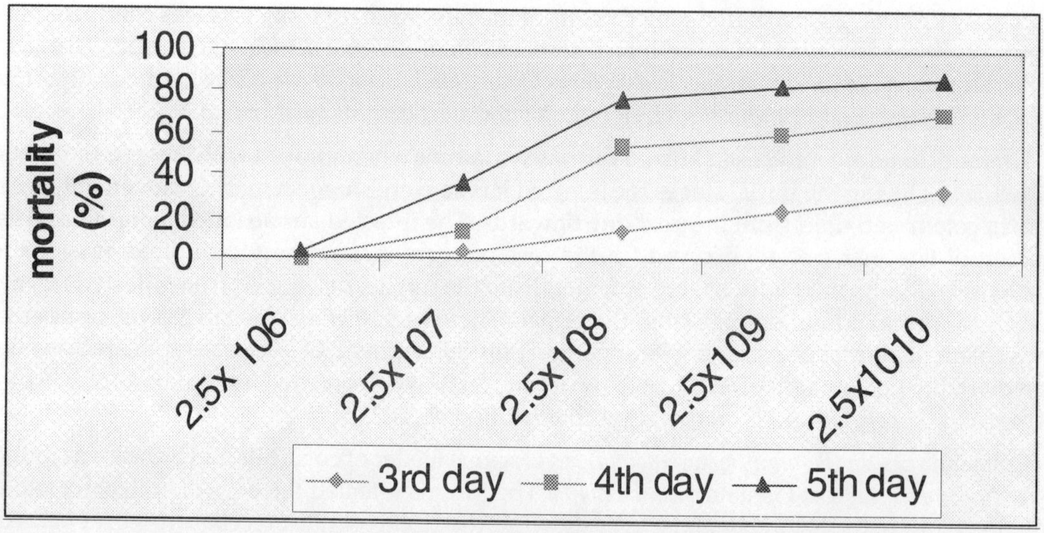

Figure 26.2: Cumulative Per cent Mortality of *Diaphania indica* Treated with Different Concentrations of *Beauveria bassiana*

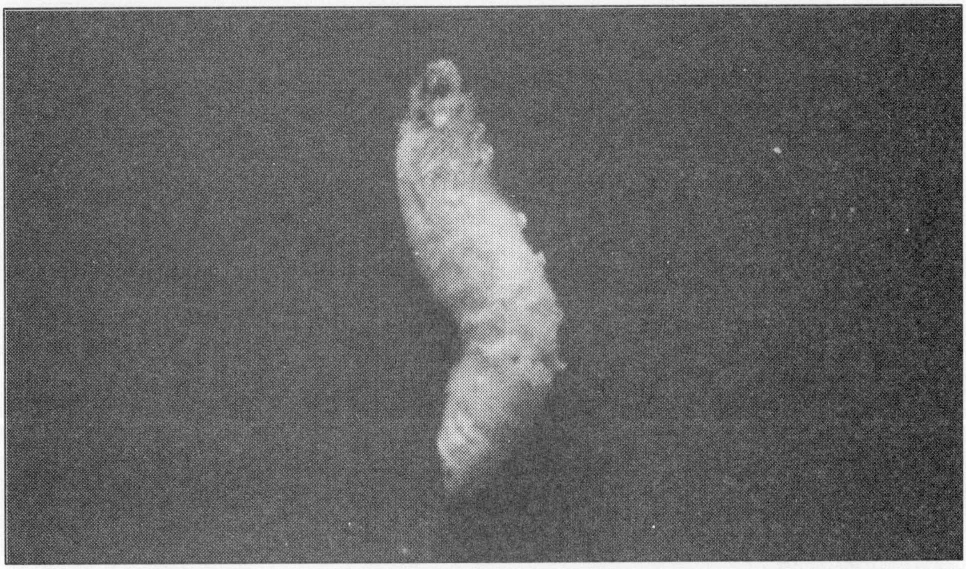

Plate 26.1: *Sylepta derogata* larva Infected with *B. bassiana*

Plate 26.2: *Sylepta derogata* Healthy Adult and Deformed Adult Emerged Out from Infected Larvae

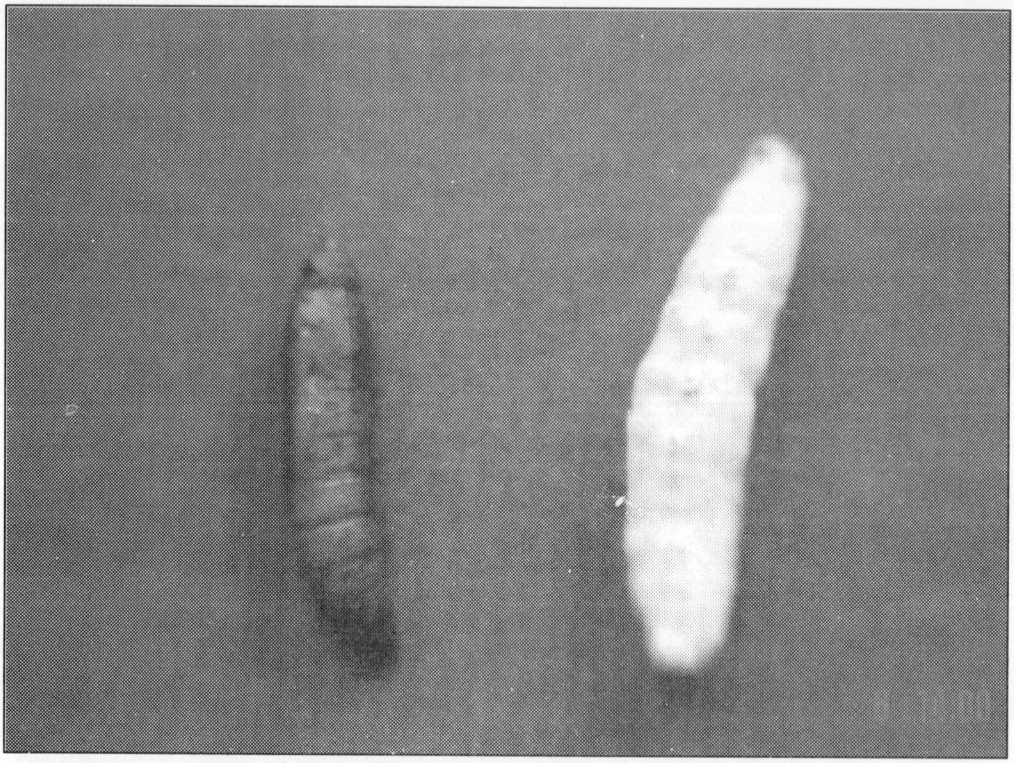

Plate 26.3: *Diaphania indica* Healthy and Infected Pupa

Plate 26.4: *P. lilacinus* Infected *B. cucurbitae*

plants tested. No fungal infection was observed on the eggs and grubs of coccinellids and larvae of syrphids. The fungus was found to be compatible with commonly used insecticides (chlorprifos, imidachlorprid, lamda cyhalothrin, Amritneem, malathion, carbaryl, dimethoate and acetamiprid) and fungicides (mancozeb and carbendazim) for vegetable cultivation (KAU 2007).

Fusarium spp.

Different species of *Fusarium viz. Fusarium equisetti, Fusarium moniliformae* var. *subglutinans, Fusarium semitectum* and *Fusarium pallidoroseum* were known as insect pathogens. The infectivity of *F. equisetti* on *Coccidohystrix insolita* (Gopinathan *et al.*, 1982) and *F. pallidoroseum* on *Aphis craccivora* (Hareendranath and Nair 1987) were recorded. In the bioassay using F. *pallidoroseum* on A. *craccivora* in cowpea revealed that LC_{50} of fungus was 3.408 x 10^6 spores/ml and it was not found pathogenic to *M.sexmaculatus*. The importance of *Fusarium pallidoroseum* as a biocontrol agent of *Aphis craccivora* was elucidated (Faizal, 1996, Sunitha, 1997). The pathogen caused death of *A. craccivora* both by invasion of host tissue by spore germination and multiplication in the haemocoel and also by the production of the toxin, Fusaric acid (Rejirani, 2001). The pathogen @ 7x 10^6 spores/ml. was effective in controlling *A. craccivora*

Metarhizum anisopliae (Metsch.) Sorok.

Metarhizum anisopliae, the green muscardine fungus is pathogenic to grasshoppers, bugs, beetles, fruit flies and lepidopteran caterpillars. Some of the hosts are *Ceratitis capitata* (Ekesi *et al.*, 2002) *Helicoverpa armigera* (Gundannavar and Lingappa, 2003), *Plutella xylostella* (Ma, 2000) *Scotinophora lurida* (Pillai *et al.*, 2003) *Spodoptera litura* (Dayakar and Kanaujia, 2001) *Thrips tabai* (Maniani *et al.*, 2003)

Metarhizium anisopliae infects the insects through cuticle, spiracles and pores of sense organs. The insects are killed mainly by histolytic action of the fungus. The growth of the mycelium within the body creates mechanical blockage of the gut and other internal organs. Toxins *viz.* dextruxin A and B are produced by *M. anisopliae* and it produces pathological changes in haemolymph and other vital sites. Of the two varieties of *Metarhzium, M. anisopliae* var. *anisopilae* (spore size 5-8 µm) and *M. anisopliae* var. *major* (spore size 10-14 µm). Lin and Roberts (1986) found that *M. anisopliae var. anisopliae* is the major source of mycotoxin, dextruxin A compared to *M. anisopliae* var. *major.* According to Roberts (1981) that the importance of toxins to the virulence of an entomopahogenic fungus is difficult to evaluate because toxin production is preceded by several activities of the fungus. The biological effects of dextruxin were investigated by Samuels *et al.* (1988). When a purified mixture of dextruxin A (80 per cent), dextruxin B(10 per cent) desmethyl destruxin B and destruxin A2 was applied to a semi isolated ventro longitudinal muscle from larvae of *Manduca sexta*, it produced a sustained dose dependent muscle contraction and depolarization of muscle fibre membranes, which induced muscle contraction.

The pathogenicity of *M. anisopliae* to many insect species has been reported by many workers. Loss of appetite and sluggishness are the major symptoms of infection. Initially white fungal colonies appeared at the joints, the integument and it later turned greenish following sporulation of the fungus. Finally the cadaver became black. Spore concentration varying from1.5x 10^5 spores/ml. to 1. 5 x10^{10} spores/ml. were reported effective against various pests (Wickramatileke *et al.*, 2000., Rajendran, 2002; Pandey, 2003). Pathogenicity of *M.anisopliae* to *D. cingualtus* was reported by Jij I *et al.*, 2006 d.

Paecilomyces spp.

Among fungal pathogens the genera *Paecilomyces* also have a role in microbial control of insects. *Paecilomyces* species occurs naturally in populations of infected and dead insects and randomly in soils around the world. The active ingredient can be cultured readily on various culture media and exists in several morphological stages, depending upon the culture medium. On solid medium, conidiophores, conidia and mycelia are predominant while submerged culture produces blastospores and mycelia. They are infective to lepidopteran larvae, plant hoppers, white flies and aphids. *P.lilacinus* is a potent pathogen of *Bemisia tabaci* (Osborne and Landa,1992) Economically important pests *viz B. argentifolii* (Vega *et al.*, 1999) *Plutella xylostella* (Gopalakrishnan *et al.*, 2002) and *Trialeurodes vaporarium* (Feng *et al.*, 2004) were reported susceptible to infection by *P.fumosoroseus F. Monilfromae var. subglutinnas* was found pathogenic to spotted beetle *E.vigintioctopunctata* (Jacob *et.al,1978*). The fungal active ingredient, *Paecilomyces fumosoroseus* Apopka strain 97, has been reported to be originally isolated in 1986 from a mealy bug in a greenhouse in Apopka, FL. The pure culture was identified in 1988 and deposited at the American Type Culture Collection (ATCC 20874). In 2005 a local isolate of *Paecilomyces lilacinus* was identified and deposited in the Indian type culture collection (ITCC 6064) (Jiji *et al.*, 2005). The pathogenicity of he same was proven among different insect host infesting vegetable (Jiji, *et al.*, 2006 b; Jiji *et al.*, 2006 c) (Plate 26.4).

Verticillium lecanii (Zimmerman) Vieges

Verticillium lecanii is known entomopthogen against homopteran bugs though they are infective to other groups of insects. Reports on the host range of insects include infection against many insects. In chillies Easwaramoorthy and Jayaraj (1977) found *Verticillium lecanii* to be effective in controlling *M.persicae.* Pathogenicity of *V. lecanii* to other pests was also reported by many; workers (*Bemisia argentifolii*, Gindin *et al.*, 2000; *Planococcus citri*, Mote *et al.*, 2003; *Pieris rapae*, Zhang *et al.*, 2003). Pathogenicity of entomopathogenic fungi, *Verticillium lecanii* to *Scirtothrips dorsalis* was reported by Gangavisalakshy *et al.*, 2006.

Rhizopus spp.

Occurrence of *Rhizopus oryzae* on several hemiptern insects was reported from India. The hosts include *Aphis craccivora, Anchon pilosum Coptosoma cribraria, Leptocorisa acuta, Nezara viridula, Pseudococcus lilacinus* and *Riptortus pedestris*, (Narayanasamy,1998; Hebsy bai *et al.*, 1999; Mathai,1999). The pathogenicity of the *Rhizopus oryzae* on *Riptortus pedestris* on fiel d conditiaon was reported by Jiji *et al;* 2006 (Plate 26.5)

Nomurea spp.

Nomurea rileyi is a promising entomopathogenic fungi used for biological control. *N rileyi* produces metabolites active against insects (Ignoffo *et al.*, 1976). *Nomurea rileyi* attacks cabbage loopers causing them to become chalky white in appearance.

Entomphthora

Strawiinski 1930 first reported frequent epizootics caused by entomogenous fungus *Entomophthoral sphaerosperma* on *Pieris brassicae* In Europe in chillies under favourable conditions the entomophagus fungi were found to be effective against aphids. Dickson and Laid (1962) observed that the population of *M.persicae* was kept down by entomophagus fungi in wet spring. *Entomophthora gammae*, attacks loopers causing them to turn black.

Other Pathogens

The pathogen *Mucor* spp. is infective to *Aphis craccivora.*

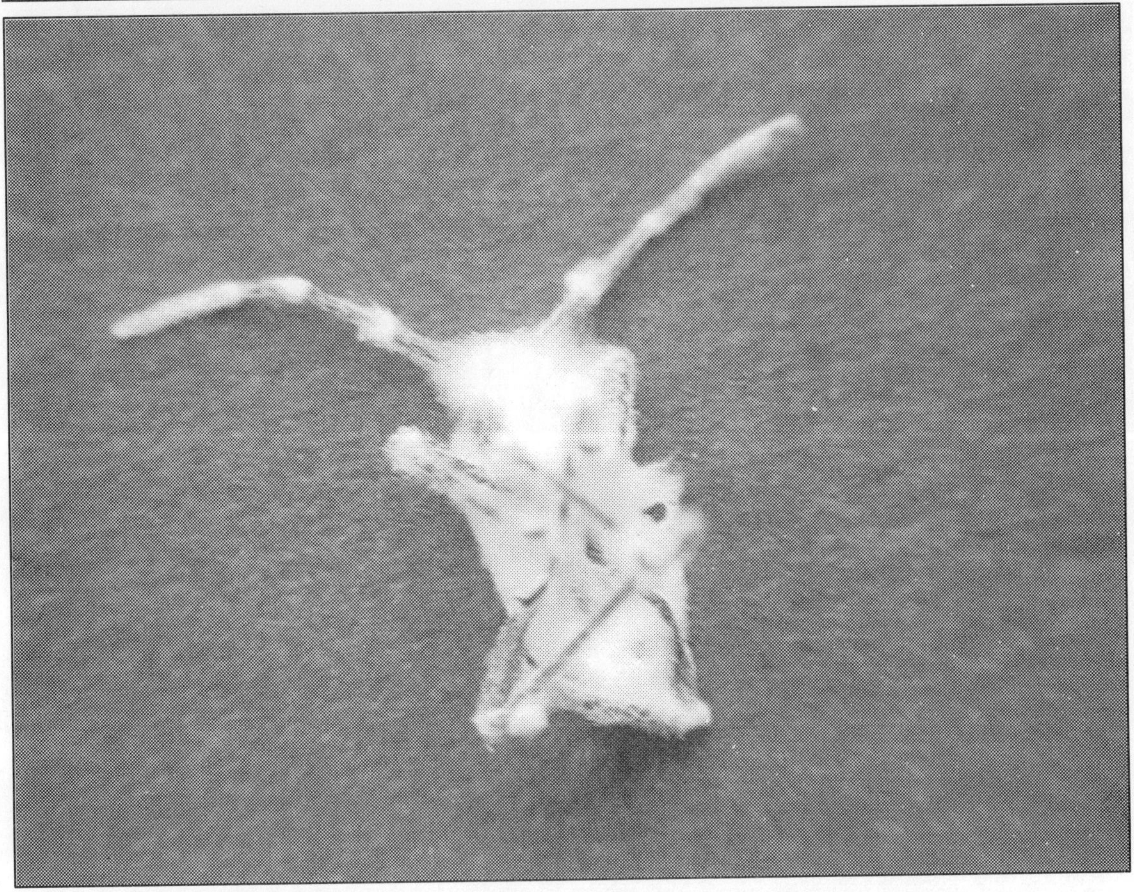

Plate 26.5: *Dysdercus cingulatus* Adult Infected with *Rhizopus oryzae*

In brinjal *Aspergillus flavus* Link parasitized and controlled the population in all stages. The spores of the fungus when sprayed on the insects killed them in 3 days. The crude toxin filtrate and the sore of *Aspergillus flavus* have a quick knock down effect. The crude extract killed the insects and was used in dispersal of the fungus in the fields, where it proved to be so effective that it may be used in integrated control (Krishnamoorthy and Shankar Naidu, 1971).

Aspergillus flavus A. niger, Cladosporium sp. Entomophthora delphacis, Nomuraea rileyi Pandora delphacis, Zoophthora radicans are some of the entomopathogenic fungi which cause epizootics in insects (Ambethgar, 2002).

The pathogenicity of *A.candidus* to *B.cucurbitae* was reported by (Jiji *et al.*, 2006) (Plate 26.6).

Molecular Characterisation of the Fungi

Molecular characterisatioan of the fungal strains help to distinguish the new strains.

Molecular characterisati on of new strains of *B.bassiana and P.lilacinus* was done using RAPD markers.

Plate 26.6: *B. cucurbitae* Infected with *Aspergillus candidus*

Out of the ten primers screened, four were selected for further amplification based on the DNA profile of the products. The primers selected were OPA4, OPA10, OPA14 and OPA18. The primer OPA04 amplified 10 bands from *P. lilacinus* (Strain1), and 4 from *P. lilacinus* (Strain 2), all the bands were polymorphic. OPA10 amplified 4 bands from *P. lilacinus* (Strain 1) and 2 from *P. lilacinus* (Strain 2), out of which two were monomorphic and two bands were present only in the new strain. OPA14 amplified three bands from the strain1, but none from the new strain. OPA18 amplified three bands from the strain1 and only one band from the new strain (2). One band was monomorphic and two bands were present only in the new strain.

OPA04, OPA14 and OPA18 did not amplify the strain of *B. bassiana*(3), but OPA10 amplified two faint bands. OPA 14 failed to amplify the strain (1), but produced three bands from the new strain (strain 4). OPA04, OPA14 and OPA18 amplified one, three and two bands respectively from the new strains, but OPA10 did not amplify the new strain.

All the four primers clearly distinguished the two strains of *P. lilacinus* and *B. bassiana* (Plate 26.7) (Jiji *et al.*, 2006)

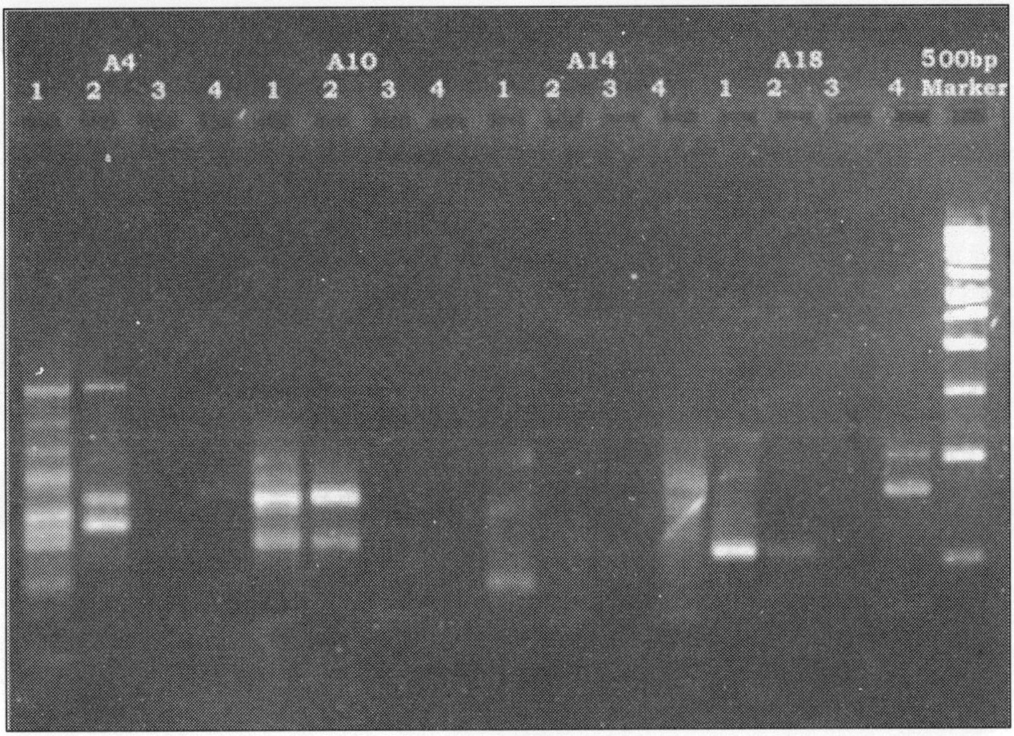

Plate 26.7
Strain 1: *P. lilacinus*; 2: *P. lilacinus*; Strain 3: *B. bassiana*; Strain 4: *B. bassaina* (new)

Mass Production

For implementation of successful biocontrol programem in field, availability of bioagents should be available in sufficient quantities. Hence, cost effective production technology for production of bioagents through simple procedures is essential. Cheaper substrate like coconut water molasses, yeast broth, rice bran, wheat bran etc. were identified as suitable substrates for mass multiplication of fungi (Danger *et al.,* 1991; Faizal and Mathai, 1996; Haraprasad *et al.,* 2001). Technique for mass production of *B. bassiana* was standardized by Ibrahima and low 1993. According to Haraprasad *et al.,* 2001 *B.bassiana* grew well on rice bran, rice bran coarse and wheat bran yielding a maximum sporulatiaon Techniques for the mass multiplicatiaon of *M anisopliae* using coconut water and cassava chips have also been developed (CPCRI, 1999). Technique for large scale production of *F. pallidoroseum* in rice bran in polypropylene bags was also developed. Waste materials obtained after rearing rice moth in biocontrol laborotories was also found suitable for the growth of the fungus (Faizal and Mathai, 1996; Sudharma *et al.,* 2001) The beneficial effects of coconut water for mass production of *F. pallidoroseum* was reported by Rejirani 2001.

The suitability of locally available and cheaper substrates, especially the agricultural byproducts were evaluated for the mass multiplication of *B. bassiana,*. Rice bran, coconut water and rice bran extract were found to be the best substrates for growth and sporulation (Tables 26.1 and 26.2) (KAU, 2007).

Table 26.1 Growth of fungi in different solid substrates

Substrate	*Mycelial Growth of Fungi*				
	Beauveria bassiana	*Metarhizium anisopliae*	*Aspergillus candidus*	*Rhizopus oryzae*	*Paecilomyces lilacinus*
Rice bran	+++	+++	+++	+++	+++
Wheat bran	+++	+++	+++	+++	+++
Guinea grass	++	+	+++	+++	++
Gingelly cake	++	+	+++	+++	++
Neem cake	–	+	–	+	–
Coir pith	+	–	+	+	+

+++: Profuse growth; ++: Moderate growth; +: Slight growth; –: No growth.

Table 26.2: Growth of Fungi in Different Liquid Substrates

Substrate	*Mycelial Growth of Fungi*				
	Beauveria bassiana	*Metarhizium anisopliae*	*Aspergillus candidus*	*Rhizopus oryzae*	*Paecilomyces lilacinus*
Coconut water	+++	+++	+++	+++	+++
Rice bran extract	+++	++	++	++	+++
Raw rice water	++	+	++	++	++
Starch solution	++	+	++	++	+++
Coconut cake extract	+	+	+++	+++	++
Neem cake extract	+	-	-	-	+

+++: Profuse growth; ++: Moderate growth; +: Slight growth; –: No growth.

B. bassiana showed profuse mycelial growth in coconut water and rice bran extract Moderate growth of fungus was observed in raw rice water and starch solution, while it exhibited slight growth in coconut cake extract and it did not grow in water.

Coconut water recorded maximum sporulation (2.78 x 10^6 spores/ml) 5 days after inoculation,which drastically increased to 33.9 x10^7 spores/ml, 10 days after inoculation and thereafter it gradually decreased. Rice bran extract recorded an initial spore count of 5.8 x 10^5 spores/ml which increased drastically to 2.91 x 10^8 spores/ml by 10th day and was on par with coconut water. Neem cake extract recorded a very low sporulation during initial days of observation and there after no sporulation was observed. Among the liquid substrates tested rice bran extract and coconut water were found to be the best substrates.

The fungus *M.anisopliae* exhibited profuse mycelial growth in rice bran and wheat bran Growth in guinea grass, gingelly oil cake and neem cake were slight. The fungal growth was absolutely nil in coir pith.

Mycelial growth and sporulation of fungus differ significantly in different liquid substrates. Among the different liquid substrates tried, the growth was profuse in coconut water. Moderate growth was observed in rice bran extract. In raw rice water,starch solution and coconut cake extract fungus showed slight growth. The fungus did not grow in neem cake extract and water (KAU,2007).

In the case of *A. candidus* rice bran,wheat bran and starch water showed profuse mycelial growth of the fungus. Moderate growth was observed in gingelly cake and rice bran extract. No growth was recorded in neem cake.

R. oryzae showed profuse mycelial growth in rice bran, wheat bran, gingelly cake, coconut cake and starch water. Moderate growth was observed in rice bran extract and absolutely no growth in neem cake (Plates 26.8 and 26.9) (KAU,2007).

N rileyi can also be grown in artificial media produces conidia green in color. But the mycelium growth preceding the formation of conidia is white in color.

Formulations

Formulations have been developed for the large scale use, easiness of trasportation and storage. Formulations are developed by the appropriate combination of ingredients. Dust, granule, bait, oil based formulations *etc.* have been developed.

The shelf life of a biopesticide is of utmost importance in gaining market and further its use in the field. Sandhu *et al.* (1993). found that *B. bassiana* remained viable for 24 months. The spores of *B. bassiana* formulated as wettable powder gave a spore germination rate of more than 85 per cent after 8 months of storage under refrigeration. The pathogenicity was maintained even after storage for one year at room temperature. Fifty fold increase in viability of spores of *Verticillium lecani* was found in the wettable powder formulation of the fungus compared to its blastospore suspension in water (Ravensburg *et al.*, 1990). Andersch (1992) patented the process of formulating *Metarhizium anisopliae* as pellets.

A long shelf for *M. anisopliae* has been identified by Alves *et al.* (1988) who evaluated 64 formulations of *M. anisopliae*. Moore *et al.* (1995) evaluated oil formulations of *M. flavoviridae* and

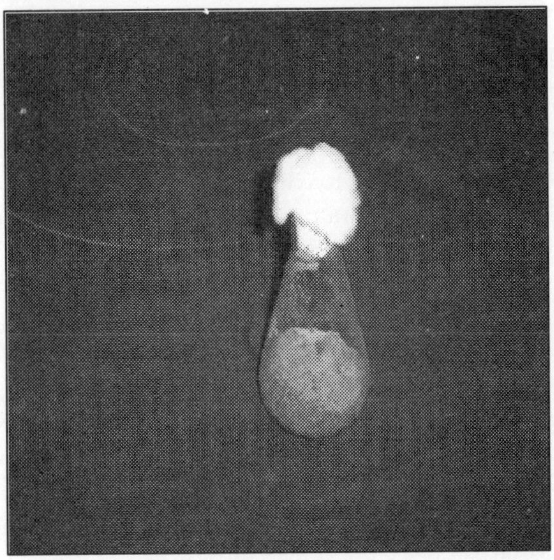

Plate 26.8: ***B. bassiana*** **Solid Substrate Rice Bran**

Solid substrates
Rice bran
Wheat bran

Liquid substrates
Rice bran
Coconut water

Plate 26.9: *Paecilomyces lilacinus*

found that the formulations retained their virulence for 30 months. Addition of sachets of silica gel could improve the conidial survival by 33 per cent (Moore,1996)

Kapdia *et al.*, 2006 reported the efficacy of two myco insecticides, *B.bassiana* and *Verticillum lecanii* (powder and liquid formulations), in the laboratory, effective against brinjal jassids, *Amrasca bigutula biguttula*. *V.lecanii* is compatible with natural enemies of cabbage aphid and *T.chilonis* and is harmless ot silk worm, therefore can be used for integrated pest management of cabbage aphid (Ali Derakhshansh *et al.*, 2006)

B. bassiana and *P.lilacinus* were formulatedand tested as wettable powder and against vegetable pests under *in vitro* conditions and field conditions The spore count of the formulation was estimated as 2.5X 10^8 spores/ml. Wettable powder formulation *of B. bassiana* at 10 per cent, found effective under *in vitro* condition, was used for the field trial. Under field conditions the wettable powder formulation of *B. bassiana* recorded a mean mortality of 65.27, 64.36 and 60 per cent against *A.peponis*, *D. indica* and *H. vigintioctopunctata*, respectively. Under field conditions, the wettable powder formulation of *B. bassiana* showed a mean mortality of 69.58 per cent against *S. derogata*. The fungus recorded a mean mortality of 76 per cent against epilachna on brinjal plants.

Under room temperature (28 ± 4ºC) formulation retained the viability of spores for two months, There was a reduction in the viability of spores from the initial count of 3.83 x 10^8 to 5.27 x 10^6 spores/ ml by eight weeks after storage, then increased drastically. Storage life of formulation could be increased to four months under refrigerated condition. Mean Spore count at the end of the fourth month was 4.15 x 10^6 spores/ml.

The formulation of *P. lilacinus* (5 per cent) was tested under field condition against *S. derogata, A. peponis, D. indica,* and *B.cucurbitae* pupae under field condition and the fungus recorded a mean mortality per cent of 60.3, 58.66 and 62 against, *D. indica A. peponis,* and *E.vigintiopunctata* respectively. Jiji *et al.*, 2007 (Plate 26.10) (KAU, 2007).

Plate 26.10: Wettable Powder Formulation of *B. bassiana*

Conclusions

Biological control is increasingly becoming popular as an ecofriendly method for the management of crop pests and it is important in the context of sustainable agriculture and organic farming. Entomopathogenic fungi due to its specificity in action and non pahogenicity to non target organism are of very much potential and are to be exploited for pest management. More potent local isolates of entomopathogenic fungi are to be identified for practical use. Mass production technology using locally available substrates should be made popular for the use among farmers.Commercial production of good quality formulations, having long shelf life is necessary and should be made available to the farmers. The entomopathogenic fungi can be successfully used for biocontrol of pests in coastal belts, where high humidity and temperature favours its survival.

References

Alves, S. B., Neto, S, Pereira, R. M. and Macedop, N., 1988. Study of formulations of *Metarhizium anisopliae* (Metsch.) Sor. in different storage conditions. *Ecossistema* 12: 78- 87.

Ambethgar, 2002 Record of entomopathogenic fungi from Tamil Nadu *J.Ent. Res.* 26 (2)161-167.

Andersch, W., 1992. Production of fungi as crop protection agents. In.: *Pflanzenschutz.–Nachrichtn* Bayer 45: 129-142.

CPCRI., 1999. *Biological suppression of coconut pests*. Technical Bulletin No. 37. Central Plantation Crops research Institute, Regional Station, Kayangulam. p.13.

Danger, J. K., Geetha, L., Jayapal, S. P. and Pillai, G. B., 1991. Mass production of entomopathogen, *Metarhizium anisopliae* in coconut water wasted from copra making industry *J. Plantn. Crops* 19: 54-69.

Dayakar, S. and Kanaujia, K. R., 2001 Susceptibility of tobacco caterpillar, *Spodoptera litura* F. (Lepidoptera, Noctuidae) to different isolates of entomopathogenic fungi.

Ekesi, S., Maniania N.K.and. Lux. S.A 2002. Mortality in three African fruit fly puparia and adults caused by the entomopathogenic fungi *Metarhizium anisopliae and Beauveria bassiana. Biocontrol Sci. Technol.*, 12(1&2): 7-17.

Faizal, M. H. and Mathai, S., 1996. Evaluation of different substrates for massproduction of *Fusarium pallidoroseum*, an entomogenous fungus. *Indian J. Ent.*58: 99-102.

Feng, M.G., Chen B.and Ying S.H., 2004. Trials of *Beauveria bassiana, Paecilomyces fumosoroseus* and imidacloprid for management of *Trialeurodes vaporarirum* (Homoptera: Aleyrodidae) on greenhouse grown lettuce. *Biocontrol Sci. Technol.*, 14(6): 531-534.

Gloriana, A.S., Raja N, . Seshadri, S. Janarthanan S and Ignacimuthu S., 2001. Pathogenicity of entomopathogen *Bacillus thuringiensis* sub sp. *kurstaki* and *Beauvera bassiana to the larvae of Spodoptera litura (F) and Pericallia ricini* (F.). *Biol. Agric. Hortic.*. 18(3): 235-242.

Gindin, G., Geschtovt N. U, Raccah B.and Barash I., 2000. Pathogenicity of *Verticillium lecanii* to different developmental stages of the silver leaf whitefly *Bemisia argentifolii* Genn. *Phytoparasitica*, 28 (3): 229- 239.

Gopalakrishnan, C. K.Narayanan and Anusuya., 2002. Determination of LD 50 and LT 50 of *Paecilomyces farinosus* (Holmskiold) Brown and Smith on larval instars of *Plutella xylostella* L. *Entomon* 27 (3): 249-254.

Gopinathan P. B., Beevi, S, N., Nair, M. R. G. K., 1982. Occurrence of *Fusarium equsetti* (Corda.) Sacci as a fungal pathogen of brinjal mealy bug, *Coccidohystrix insolita* (Green) *Entomon* 7: 120.

Gundannavar, K. P. and Lingappa, S., 2003. Study of dose mortality response between instars of *Helicoverpa armigera* and *Metarhizium anisopliae* (Metscnikoff) Sorokin. *Natl. Symp. on Biomanagement of insect pests*, 29to 31, March, 200. Annamalai University, Annamalainagar, Abstract: 64.

Hafez, M. Zaki, F. N. Moursy, A. and Sabour, M., 1994. Biological effect of the entomopathogenic fungus *Beauveria bassiana* on the potato tuber moth *Pthorimoea operculella* (Sneller). *J. Islamic Acad. Sci.* 7: 132-136.

Haraprasad, N., Niranjana S.R., and shetty, H. S., 2001. Mass production on solid substrates and enhancement of conidiation *of Beauveria bassiana* for control of coffee berry borer. *Proc. Symp. Biocontrol based Pest Mgmt Quality Crop Prot. Curr. Millenium* July 18-19, 2001 (eds. Singh, D., Mahal, M. S. Sohi, A.S., Dilawari, V. K. Brar, K. S. and Singh, S.P. Punjab Agricultural University pp. 132-133.

Hareendranath, V. Nair, V. K., 1987. *Fusarium pallidoroseum* (Cooke) Sacc. as the fungal pathogen of *Aphis craccivora* Koch. *Entomon* 12:392- 394.

Hebsy bai, Mathai S and Gokulapalan, C., 1999. *Rhizopus oryzae* Went and Gerlings, a new fungal pathogen of *Dysdercus cingulatus* (F). *Insect Environ.* 4: 129-130.

Ibrahim, Y.B.and Low, W., 1993. Potential of mass production and field efficacy of isolates of the entomogenous fungi *Beauveria bassiana* and *Paecilomyces fumosoroseus* against *Plutella Xylostella.Int. J. Pest Mgmt.*39:288-292.

Jacob, A., Kuruvia, S., Philip, B.M. and Asari, P.A.R., 1978. *Fusarium moniliformae* var.subglutinans Wollen and Reink Pathogenic to the spotted beetle Epilachna vigintioctopunctata.*Aric. Res. J. Kerala.16:262-263.*

Jiji, T., Praveena, R., Anitha, N. and Naseema, A 2005. *Beauveria bassiana* (Bals) Vuill.on *Sylepta derogata* Fab. and its cross infectivity on other major pests of vegetable. *Insect Environment* 11(4):144.

Jiji, T., Praveena, R., Kavitha Babu and Naseema, A., 2006a. Evaluation of two fungi for biological control of fruit flies. *Insect Environment*. 12(1):25-26.

Jiji, T., Praveena, R., Kavitha Babu, Naseema, A. and Anitha, N., 2006b. Three promising fungal strains pathogenic to fruit flies. *Presented in Seventh International Symposium on Fruit flies of Economic Importance*, 10th-15th September, Salvador, Bahia, Brazil.

Jiji, T., K. B. Soni, P. Prajeesh, Kavitha Babu and K. Rajmohan., 2006c. Molecular characterisation of promising fungal strains of *Beauveria bassiana* and *Paecilomyces lilacinus* using RAPD markers. *Insect Environment* 12 (1): 33-34.

Jiji, T., Praveena, R., Kavitha Babu, Naseema, A. and Vijayaraghavakumar., 2006d. Some promising fungal strains for the biological control of sucking pests infesting vegetables. *Presented in the National symposium on Biological control* held at Bangalore, 26th-27th May.

Jiji, T. Praveena, R., Babu, K. and Naseema, A., 2006. Occurrence of *Paecilomyces lilacinus* on Melon fly *Bactrocera cucurbitae* Coq. and its cross infectivity on *B. dorsalis* (Hendel) and Bhindi leaf roller *Sylepta derogata* Fb. *Abst. Colloquium on Nanoscale Science and Arthropod Bioresources*, Trivandrum, India p. 18.

Jiji, T. Praveena, R., Naseema, A. Anitha, N and Simna, T., 2007. Entomopathogenic fungus *B.bassiana* (B.) V.for the management fo key insect pests of vegetables.*Green farming* Vol. 1 (10-11): 54-56.

KAU 2007. ICAR *ad hoc* scheme final report, Kerala Agricultural University p75.

Lin, K.J. and Roberts, D.W., 1986. *Metarhizium anisopliae major*: Investigation of production of dextruxin by culture *J. Invertebr. Pathol.* 53: 307-314.

Ma, J., 2000. Laboratory susceptibility of *Plutella xylostella* to *Metarhizium anisopliae* and *Nomuraea rileyi*. *Acta Entomologica Sinica*, 7(1): 53-57.

Maniani, N. K., Sithanantham S., Ekesi, S. Among-Nyarko, K., Baumgartner, J. Lobr, B. Motoka, C. M., 2003. A Field trial of the entomogenos fungus *Metarhizium anisopliae* for control of onion thrips, *Thrips tabaci Crop Prot.*, 22 (3&4) 553- 559.

Mathai, S., 1999. Host range of a potent entomopathogen ic fungus *Rhizopus oryzae* Went and Gerlings. *Insect Envior*. 4: 129-130.

Masuda, J., 2000. Microbial control of the diamond back moth, *Plutela xylostella* by an entomopathogenic fungus *Beauveria bassiana*. Effects of temperature on mycoses and conidial invasion time. *J. appl. Ent*. 44: 177-182.

Moore, D., Batemann, R.P. Carey, M. and Prior, C., 1995. Long term storage of *Metarhizium flavoviride* conidia in oil formulations for the control of locusts and grass hoppers. *Biocontrol Sci. Technol.*, 5: 193-199.

Moore, D. Douro-kpindou, o. k., Jenkins, N. E. and Lomer, C. J., 1996. Effect of moisture content and temperature on storage of *Metarhizium flavoviridae* conidia *Biocontrol Sci. Technol.* 6: 51-61.

Mondal, P. and Bhattacharya. A.K., 2003. Differential sensitivity of *Spilosoma oblique* to two isolates of *Beauveria bassiana. Indian J. Ent.*, *65*(3): 324-329.

Mote, U.N., Kadam, J.R. and Tambe, A.B., 2003. A potent bioinsecticide *Verticillium lecani* for effective control of mealy bugs on fruit crops *Natl. Symp. on Biomanagement of insect pests* 29- 31. March 2003 Annamalai University Annamalainagar Abstract: 103.

Narayanaswamy, P., 1998. Mycoparasites check rice brown plant hopper (BPH) in India. *Insect. Environ* 104-106.

Osborne and Landa Z., 1992. Biological control of whiteflies with entomopathogenic fungi. *Fla Enomol.* 75: 456- 471.

Pandey, A.K., 2003. Susceptibility of egg and pupae of lepidoptera to *Beauveria bassiana* (Balsamo) Vuillemin *and Metarhizium anisopliae* (Metschnikoff) Sorokin *Insect Environ* 9: 123-124.

Pandey, A. K. and Kanaujia, K. R., 2004. Effect of temperature on pathogenicity of *Beauveria bassiana* (Balsamo) Vuillemin and *Metarhizium anisopliae* (Metschnikoff) Sorokin against *Spodoptera litura Fab. larvae Indian J. Plant Prot.*, 32(2): 45-47.

Pawar U. M, Thombre, U. T. and Borikar, P. S., 2001. *Advances in Biological Control*. Punjab Agricultural University, Ludhiana. pp. 75-82.

Pillai, M. A. K., Anandhi, P. and Johnson, N. S. B., 2003. Entomopathogenic fungi (Deuteromycotina) in the control of black bug of rice, *Scotinophora coarctata* (Hemiptera: Pentatomideae). *Natl. Symp. on Biomanagement of Insect pests*, 29- to 31 March., 2003 Annamalai University, Annamalainagar, Abstract, 34.

Rajendran, P., 2002. Preliminary studies on the effect of green muscardine fungus *Metarhizium anisopliae* (Metsch.) Sorokin on eggplant spotted beetle, *Henosepilachna vigintioctopunctata* Fab. (Coleoptera: Coccinellidae) *Pest Mgmt. hort. Ecosyst* 8 11-14.

Ramesh, K., Mohan, H., Lakshmi, K.A., Padmavathi, J. and Devi, K.U., 1999. *Beaveria bassiana* (Bals.) Vuill. (Hyphomycetes: Moniliales) in cotton pest management: A field trial on the cotton leaf roller, *Sylepta derogate* Fabricius (Lepidoptera: Pyraustidae). *J. Ent. Res.*, 23(3): 267-271.

Ravensberg, W. J., Nalais, M. and Van der Schaaf, D. A., 1990. *Verticillium lecanii* as a microbial insecticide against glass house whitefly. In. *Proc. Brighton Crop Potection Conference.–Pests and Diseases* Vol. 3. British Crop Protection Council, Farnham, pp 265-268.

Rejirani, O. P., 2001. Production and evaluation of the fungus, *Fusarium pallidoroseum* (Cooke)Sacc. as a biopesticide against the pea aphid, *Aphis craccivora* Koch. *Ph.D. Thesis,* Kerala Agricultural University, Thrissur pp 136.

Roberts, D. W., 1981. Toxins of entomopathogenic fungi. In: *Microbial control of Pests and Plant Diseases* pp. 441-464 ed. H. D. Burges Academic Press London.

Sandhu, S.S., Rajak, R.C. and Agarwal, G.P., 1993. Studies on prolonged storage of *B. basssiana* conidia. Effects of temperature and relative humidity on conidia viability and virulence against chickpea borer, *Helicoverpa armigra Biocontrol Sci. Technol.* 3: 47-53.

Samuels, R.I., Reynolds, S.E. and Charnley, A.K., 1988. Calcium channel activation of insect muscle by dextruxins, insecticidal compounds produced by entomopathogenic fungus *Metarhizium anisopliae Comparative Biochemistry and Physiology* 90:403- 412.

Sudharma, K. and Hebsy Bai, 2000. Utilisation of biopesticides for the management of pea aphid *Aphis craccivora* Koch. *Entomocongress 2000: Perspectives for the New Millennium.* Association for the Advancement of Entomology, Trivandrum Kerala 14 p.

Sudharma, K., Reji Nair and Peethambaran C, K., 2001. Mass production of *Fusarium pallodoroseum* (Cooke) Sacc. utilizing waste materials from biocontrol laboratories. *Symp. On Biocontrol based Pest Mgmt. for Crop Prot. In the Current Millennium.* July 18-19, PAU, Ludhiana, p. 117. Eds. Darshan Singh, Mahal, M. S, Sohi, A. S, Dilawari, V.K., Brar, K. S and Singh, S. P.

Sudharma, K and Rani, J., 2005. *Halyomorpha* sp. nr. Picus (Fab) (Hemiptera: Pentatomidae) and *Pagria signata* (Motschulsky) (Coleoptera: chrysomelidae) two new pests of vegetable cowpea. *Entomon* 30 (3): 235- 236.

Sunitha, 1997 *Fusrium pallidoroseum* (Cooke) Sacc. as a biocontrol agent for the pea aphid *Aphis craccivora* Koch. Msc. Ag. Thesis Kerala Agricultural University.

Todorova, S.I., Coderre, D. and Cote, J.C., 2000. Pathogenicity of *Beauveria bassiana* isolates toward *Leptinotarsa decemlineata* (Coleoptera: Chrysomelidae), *Myzus persicae* (Homoptera: Aphididae) and their predator *Coleomegilla maculate lengi* (Coleoptera: Coccinellidae). *Phytoprotection,* 81(1): 15-22.

Vega, F.E., Jackson, M.A. and Guire, M.R., 1999. Germination of conidia and blastospores of *Paecilomyces fumosoroseus* on cuticle of the silverleaf whitefly, *Bemisia argentifolii. J.Invertebr. Pathol.,* 47(1): 33-45.

Visalakshy, P.N.G., Kumar, A.M.and Krishnamoorthy, A., 2004. Epizootics of a fungal pathogen, *Verticillium leacanii* Zimmermann on *Thrips palmi* Karny. *Insect Environ.,* 10(3): 134-135.

Wickramatileke, W.S.N., Ahangama, D. and Banda, D.M.A., 2000 Evaluation of the effectiveness of entomopathogenic fungus *Metarhizium anisopliae* var. major on cabbage semilooper (*Chrysodeixis erisoma* Doubt.) *Trop. Agric. Res.* 12: 177-185.

Zaki, F.N., 1998. Efficiency of the entomopathogenic fungus, *Beauveria bassiana* (Bals) against *Aphis craccivora* Koch and *Bemisia tabaci* Gennandius. *J. Appl. Ent.,* 122: 397-399.

Chapter 27

Disease and Pest Management of Pomegranate in the Semi Arid Region of the Western India

B.G. Bagle[1], H.K. Joshi[1], and T.A. More[2]
[1]Central Horticultural Experiment Station, Vejalpur (Godhra) – 389 340, Gujarat
[2]Central Institute for Arid Horticulture, Bikaner (Rajasthan)

Pomegranate, *Punica granatum* Linnaeus, is a fruit plant of Iranian origin and is grown extensively in Spain, Morocco and other Mediterranean countries as also in Egypt, Iran, Afghanistan, Arabia and West Pakistan (Singh *et al.*, 1963). It is also grown to some extent in India, Bangladesh, Burma, China, Japan and USA. In India, it is confined mostly to Gujarat and Maharashtra states and a small area in Uttar Pradesh, Andhra Pradesh, Karnataka and Tamil Nadu. It thrives best in areas having hot dry summer with irrigation facilities. It can also withstand frost to some extent but not humid weather. A wild type is also grown along the foot of Himalayas; the seed of this variety is dried and used as spices (anardana). The juice of pomegranate is cool and refreshing and has many medicinal properties, specially for patients suffering from leprosy. The rind is also used as dyeing material for cloth.

Besides being a delicious table fruit, it is also liked for sweet-sour juice, and for making sharbat and squash, and paste (chatney), etc. Its sour varieties commonly grown in hilly areas are widely used for making anardana. The fruit rind contains tannin and used in the tanneries (leather industry) The rind of the fruit, as also the bark of the trees are well known for their use as medicine for leprosy, diarrhoea, cold, cough and several other disorders (Chundavat, 1990).

For its growth, it requires moderately cool winters and dry and warm summers. It requires warm and sunny weather for fruit maturity whereas in humid-climate its quality is affected. That is why the semi-arid western Indian region suits to this tree very much. Though the tree survives very harsh vagaries of weather, for its commercial production, some irrigation during fruiting is suggested to be necessary. Owing to a variable climatic requirement, this tree is affected at various stages by a number of diseases, insect-pest, mites, nematodes, vertebrate pests and even non-parasitic disorders. These factors thwart the success of cultivation to such an extent that some times very harsh or unpalatable measures are suggested *e.g.* application of pesticides, abandoning more affected areas, avoiding a particular cropping season, imposing laws for banning sale of unhealthy planting materials, etc.

These pests/diseases affect the plant at all stages starting from nurseries, to during growth in field, during fruit maturity, and during post-harvest stage. Over 50 fungal diseases organisms are recorded on pomegranate, the major fungal diseases include *Cercospora* and *Colletotrichum* leaf and fruit spots, bacterial blight, fungal fruit rots due to *Aspergillus* spp. and *Phytophthora nicotianae* etc. Among major insect-pests, anar butterfly, girdling insects, leaf eating caters pillars, thrips, mites and aphids etc. are included. Besides, these important diseases and pests like mites, nematodes (*Meloidogyne* sp.), rodents, birds, and non-parasitic disorder like fruit cracking are also found damaging to the crop.

Looking at the severity of some pests, it has become obligatory on the part of plant protectionists to suggest harsh measures stated above, which may sound unpleasant to environmentalists or orchardists. Another aspect of the pest and disease management is that sometimes, even minor problems also compound in favorable condition to such an extent that it becomes unmanageable *e.g.* occurrence of fruit rot due to *Phytophthora nicotianae* which occurs during long spells of rainy days when spraying is not effective. Compared to pests, diseases multiply fast during favorable weather and turn severe before any adequate control strategies are applied, and bring about appreciable losses. Further, information on minor pests/diseases is not available or it is very meager, sparse, and discrete, hence, control strategies do not work efficiently. Some measures like banning the sale of bacterial blight affected planting material is often unpalatable to nursery owners.

In view of these facts, in the present article, major diseases and pests are described with specials reference to their relative importance, causal organism, epidemiology, and control measures. The account on major diseases/pests is based on the studies conducted in CHES,Vejalpur, besides other relevant information is also compiled for the benefits of the readers.

Diseases

Colletotricum Leaf and Fruit Spots

The disease is also known as anthracnose and is a common problem in most of the areas where the crop is grown. The disease occurs in high humidity conditions of August and September and is one of the reasons for recommendation for this area to avoid *Mrug bahar* crop and to go for *Hast bahar*.depressed. The lesions often cover only a part of the surface of young fruits but some times; entire fruit surface may be affected. Minute, black dot made of acervulus of the fungus, can be seen on the affected area. On mature fruits, several small, roundish, brown to black, depressed spots are seen evenly spread on entire surface, which are often surrounded by a yellow halo. The disease is caused by a melonconiale fungus *Colletotrichum gloeosporioides* (Chandra and Tandon, 1965), which produce acervulus (a mat of hypha on saucer shaped fruiting body, on the tip of hyphae hyaline, single celled, cylindrical conidia are borne). The disease spreads through reinfection by wind-blown conidia. In Gujarat, the disease starts in first half of August month and reaches to peak by the end of September to middle of October, depending on climatic conditions. The disease thrives well when the minimum temperature ranges between 14-28°C, maximum temperature varying between 30-38 °C and average morning humidity ranging between 55-72 per cent (mean 75 per cent). The pathogen persists in plant debris, in the infected fruits and living and dead branches for several, months and causes fresh infection by conidia whenever conducive condition occurs.

The disease can be managed by fungicidal and other means as follows:

1. Sprays with appropriate fungicide like mancozeb 75 wp (0.2 per cent), copper oxichloride 50 wp (0.2 per cent),captan 50 wp (0.2 per cent),benomyl 50 wp or carbendazim 50 wp (0.1 per cent),Bordeaux mixture (0.8 per cent) and haxagonazole 10 EC (0.1 per cent) twice or thrice during August-September months at 15-20 days interval. Cuman-L (ziram) (0.3 per cent), aureofungin (100ppm), or thiophenate methyl 50 wp (0.1 per cent) are also very effective in disease management (Anonymous, 2004).
2. Removal of infected debris (dead twigs, diseased fruits) and its destruction, and deep ploughing during the April month helps in reducing the inoculums. If these practices are followed by one spray with copper oxychloride, the disease is brought into control.
3. Incorporation of *Trichoderma viridae* spore suspension sprays @$2x10^6$ cfu in the soil helps in reducing the disease severity. Spray with neem oil and 0.1 per cent KH_2PO_4 as foliar spray, also decreases the spread of disease.

At Godhra; during 1994-95 another kind of anthracnose caused by another common species of *Colletotrichum*,(*C.capsici*) was also found occurring nursery seedlings. This fungus can be distinguished from *C.gloeosporioides* in having sickle shaped conidia devoid of droplets (as compared to drop containingcylindrical conidia of the later)., more frequent dark setae formation, and its specific symptoms. *C.capsici* produced large brownish lesions, which were semi-circular to irregular in shape, and initiated from leaf margin and tips and later covered large area of leaf lamina. Acervulus dots were frequently seen in the later. The disease was found to be easily manageable by spray with copper oxychloride 50 wp (0.3 per cent) or carbendazim 50 wp (0.1 per cent).

Cercospora Leaf and Fruit Spots

bring about defoliation. In some varieties the defoliation is very conspicuous and is responsible for their sickly look. On fruits, round grey spots are seen on rind, which are not depressed.

The casual organism (*P.punicae* P.Henn.) (Vasudeva, 1963) is a dematiaceous fungus, which produces cylindrical to sub-fusoid, sub-clavate, 2-5 septate, hyaline or slightly pale-olivacious conodia of the size 40-50x3-3.5 micron. Conidiophores are amphigenous with epiphyllous fruiting and are short, olivaceous brown fasciculate, sparingly septate and measure 20-40x3.0 micron.

The pathogen survives on diseased leaves. The disease generally appears after rainy season and is favoured by high humidity and frequent rainfall.

For the disease management prophylactic fungicidal sprays, cultural practices and selection of genetic resistance are suggested (Anonymous, 2004):

1. Sprays with thiophanate methyl 50 wp (0.1 per cent), copper oxichloride 50 wp (0.3 per cent),carbendazim 50 wp (0.1 per cent) or chlorothalonil 25 wp (0.2 per cent) commencing from first fortnight of August and followed by a repeat spray at fortnightly to triweekly interval effectively manages the disease.
2. Sometimes, because of prolonged rainy spells, spraying becomes ineffective. In such conditions it is advisable to go for *Hasta bahar* (winter season) crop and avoid *Mrug bahar* (rainy season crops).

3. Cultural practices like deep ploughing in summer, removal of debris and pruning and destruction of dead branches, spray with neem oil, incorporation of *Trichoderma viridae* spores in soil and use of *Azotobacter* (bio-fertilizer) helps in reducing the disease intensity.
4. Varieties Jyoti, Ramnagaram, P-16, P-23, P-26, etc. were found moderately resistant to disease. Commercial cultivar Ganesh was found susceptible to the disease.

Pomegranate Wilt

This disease has become an emerging problem in Karnataka and Maharastra where substantial loss due to mortality of thousands of plants was reported (Somasekra, 1999). However, its etiology (causal organism) is found to differ at both the places hence it requires confirmation of the casual organizer for a given locality.

The disease manifests itself by general pale-yellow appearance of foliage sick and wilted look of the plant and rapid defoliation, which initially starts from one or two branches. The vascular region of roots shows brown discoloration, and often disintegration of bark is seen associated with the diseases.

The disease at Maharastra was found to be caused by *Fusarium oxysporum* (Kore and Metkar, 1993) whereas at Karnataka, it was found due to a *Cerotocystis fibriata* (fungus)-*Euwallacea fornicatus* (scolytid beetle) complex (Somasekara, 2002).

For want of precise information on its etiology, its management is a difficult proposition, however, proper sanitation, maintenance, of plant vigour and good soil health, use of soil antagonists, (*Bacillus subtilis* and *Pseudonomonas fluorescens*) and pesticidal spray (Somasekara, 2002). etc. have been recommended.

Bacterial Leaf and Fruit Spots

This important disease of bacterial etiology was first reported in 1952 from Delhi. (Hingorani and Mehta, 1952) Subsequently, it was reported that its causal organism is *Xanthomonas campestris pv.punicae*. During the recent years, it has evoked a renewed interest of plant pathologists as it has become a severe constraint in successful cultivation of pomegranate in Karnataka, Maharashtra, Andhra Pradesh and Tamilnadu (Anonymous, 1996). In this states, it is known as oily spots or nodal blight, the names conforming to their common symptoms. The disease is suspected to be spread through infected nursery material.The disease is characterized by appearance of minute (2-5 mm) dark, translucent leaf spots which are initially roundish but letter turn irregular and have characteristic water soaked margins. Small spots enlarge in size, coalesce, and from irregular patches which are seen surrounded by chlorotic halo. The pathogen migrates from leaf tip or margins towards mid-rib and forms bands or long vein bound patches, which are irregular in shape. From mid-rib, the pathogen migrates to petioles and enters main stem and becomes systemic (Chand *et al.*, 1993). On main stem, it impact mostly to nodal region where dark necrotic lesions are seen and cracking of the bark is resulted. Infection streaks from petioles to nodes can be seen on tender shoots. The disease also causes breaking of shoots due to weakening of the stem at nodal region. The disease also affects fruits on which roundish to irregular lesions surrounded by an oily margin are seen which later cause creaking of the rind exposing the inside tissue of pericarp in a typical x or y shape.

The causal organism is a typical bacterial pathogen, which invades the leaf through stomata; during rainy season (when stomata are generally open) and colonises the intercellular spaces and forms translucent (oily) spots, which after migration of pathogen to stem leads to nodal blight stage.

The pathogen survives in infected debris for 3-4 months and on availability of conducive situation attacks new host. Its spread is believed to be by rain splash or insects. Pathogen invades new areas through infected planting materials.

The disease management of the bacterial blight is difficult problem since the bacterium is a soil inhabitant and survives in debris for several months. Field sanitation cultural modifications, and use of disease-resistant varieties for improvement are the strategies, which can be incorporated with chemical control.

1. The infected debris should be collected and destroyed from time to time.
2. Field sanitation measures like avoiding irrigation water from bacteria-infested areas to disease-free area, use of disease-free planting material, use of insecticides to ward-off sucking pests which may carry pathogen to otherwise disease free leaves, etc. help in disease management.
3. Deep ploughing in summer season.
4. Spray with 1 per cent Bordeaux mixiture.
5. Possible use of soil antagonist bacteria like *Pseudonomous fluorescens*.
6. Spray with streptocycline (500 ppm) +copper oxychloride (0.2 per cent) three times during rainy season at fortnightly interval (Suriachandraselvan *et al.*, 1993).
7. Evolving of resistant varieties should be given importance.

Minor Leaf Spots and Fruit Spots

Besides cercosporae, colletotrichums and bacterial pathogens, several other minor diseases contribute to the factors responsible for dwindling the farmers margin, specially when they assume more severe form. Important leaf and fruit infecting fungi are *Fusarium* leafspots, *Phomopsis* leafspots,leafspots caused by *Sphaceloma punicae, Phyllosticta* sp., *Pestolotia* sp, *Alterneria alternata, Curvularia, Drechslera rostrata, etc.* Since these are minor pathogens with limited economic importance, not much work has been reported on their epidemiology and control. Most salient features of these diseases are described here.

Phomopsis Leafspots

The disease was reported from Jodhpur (Rajasthan) by Shreemali(1972). The causal organism is *Phomopsis aucbicola.* Buff brown colored lesions can be seen near margin or may be found anywhere on lamina. Black pycnidia can be seen on the spots on upper surface of the leaf. Variety Jeolicola was found relatively free of disease incidence. This disease is also noticed in Gujarat.

Alterneria Leaf and Fruit Diseases

The disease is of common occurrence in various states, but its importance is limited to some pockets.

The disease is characterized by appearance of minute, brown spots, which are scattered on entire leaf surface specially near margin. The spots enlarge irregularly and are surrounded by a light colored halo and cover the large area to render a blighted look to the leaf. On fruits, several small, greyish spots are seen on the rind and give ugly look to the fruits. Fruits rarely may develop concentric rings in the spots, or become rusty and corky (Nallathambi *et al.*, 2007).

The causal fungus is *Alterneria alternata,* a hyphomycetous fungus. The disease can be managed by proper sanitation avoidance of growing susceptible crops in the near vicinity of pomegranate, since they act as colletoral hosts of the fungus, and sprays with copper oxychloride 50 wp (0.3 per cent),captafol (0.2 per cent) and mancozeb(0.2 per cent).

Curvularia Leafspots

This is a minor disease of the pomegranate, which is characterized by small, roundish, and ashey grey leaf spots lamina. Causal fungus is *Curvularia pallescens;* a hyphomycetous fungus. Prophylactic spray with Mancozeb (0.2 per cent) or Chlorothalonil 50 wp (0.2 per cent) may be able to contain the disease.

Pestalotia Leafspots

The disease appears during and after rainy season and is characterized by circular or semi-circular, light brown spots generally starting from margins. Causal fungus is *Pestalotia* sp., a member of melanconiales. Spray with copper fungicides, Mancozeb or carbendazim help manging this disease.

Sphaceloma Leafspots

The disease causes light brown lesions with dark margins on leaves and also on fruits. It is of minor importance, not much is known about its epidemiology and control.

Phytophthora Leaf, Fruit and Flower Lesions

(Metalaxyl 8 per cent + mancozeb 72 wp) @ 2g/L may be given for its management. Affected fruits and branches should be pruned away to remove source of fresh infection.

Die Back

It is known to occur in Gujarat. It is characterized by death of young, tender shoots from tip downwards. It is caused by a fungus *Ceuthospora phyllosticta.* Besides this, *Phoma, Botryodiplodia theobromae* and *Colletotrichum* sp. are also suspected to bring about dieback symptoms. Pruning of dead branches, spraying with copper fingicides or mancozeb or carbendazim, and drenching near collar region with carbendazim help in its management.

Fruit Spots and Fruit Rots

Several fungi are responsible for rotting of fruits; some them fungi cause complete decay of the fruits whereas others cause superficial spots on rinds.

Aspergillus Fruit Rots

Many species of *Aspergillus* are responsible for causing fruit rot in pomegranate.

A. versicolor

It produces small brownish patches on rind which increase in size, turn blackish in center and brings about disintegration of the central tissue resulting in irregular depression in the rind surface and exudation of shiny mass can be seen from spots.

A. nivens

It causes circular, light, yellow patches on the rind beneath which the tissue turns soft and dark colored and produce yellow fluid and emits foul smell (Rao, 1966).

A. flavus

Causes brownish discoloration of arils, which turn grey and blakish, become slimy, develop soft rot and emit fermented odor (Srivastava and Tandon, 1971).

A. foetidus

A. foetidus has been reported from Orissa causing fruit rot. (Ghosh and Dutta, 1960).

Phoma Fruit Rot

This is caused by *Phoma versoniana* fungus and is air borne. It is found in all tropical regions where pomegranate is grown. The pathogen hibernates in infected crop debris.

Beltaraniella humicola Fruit Spots

The disease is reported from Maharastra (Sherkar and Ulikar 1982B). Black circular spots with reddish brown margin appear on entire surface of the fruit rind, spots gradually enlarge and coalesce to form bigger lesions.

The fruit rots may be managed by practices like fungicidal sprays followed by covering the fruit with proper bags, sprays with insecticides to protect from insects which may be responsible for making the fruit prone to infection, protecting the fruit from abrasion, removal of low hanging fruits, etc.

Major Insect Pests of Pomegranate

The most obnoxious enemy is pomegranate butterfly, next in order being bark eating caterpillars. Besides, termites attack the roots; aphid, whiteflies, scale insects, thrips and leaf defoliators feed on foliage; a few bugs borers, moths and fruit flies damage the fruit and some scavengers attack even the rotten and over-ripe fruits.

Pomegranate Butterfly

Virachola isocrates (Fabricius) it is a polyphagous pest having a very wide range of host plants, including, aonla, apple *ber*, citrus, guava, litchi, loquat, mulberry, peach, pear, plum, pomegranate, and tamarind. The female lays eggs singly on calyx of flowers or small fruits. On hatching, the caterpillars bore inside the developing fruits and are usually found feeding on pulp and seeds just below the rind. As many as eight caterpillars may be found in a single fruit. Subsequently, the infested fruits are also attacked by bacteria and fungi causing the fruits to rot. The conspicuous symptoms of damage are offensive smell and excreta of the caterpillars coming out of entry holes, may also be seen plugged with the anal end of a caterpillar. The affected fruits ultimately fall down and are of no use; even if the fruits be picked before falling down; these fruits have no market value. Adult butterflies are medium sized, glossy-bluish-violet(males) to brownish-violet(females) in colour;females have conspicuous orange patch on the forewings, Wing expanse is 40 to 50mm. The pest breeds throughout the year with four overlapping generations in a year.

Strategy of Control

To reduce the incidence of this pest, remove and destroy all the affected fruits. The only other effective though expensive method is bagging of fruits (Pruthi, 1969) and may be practiced, if the number of fruit trees is limited. Alam (1962) from Bangladesh, suggested spraying with 0.03 per cent phosphamidon just when the fruit formation starts.

Pomegranate fruits are also damaged by pomegranate borer. *Deudorix epijarbas* (Moore), castor capsule borer, *Dichocrocis punctiferalis* Guenee, pyralid borer, *Euzophera punicaella* Moore.

Table 27.1: Insect Pests of Pomegranate

Sl.No.	Common Name	Scientific Name	Locations						
			Anant-pur	Arrup-kotai	Godhra	Hissar	Jobner	Rahuri	Udai-pur
1.	Aphid	*Aphis punicae*	–	–	√	–	√	√	–
2.	Thrips	*Scirtothrips Dorsalis*	–	–	√	–	–	√	–
3.	White fly	*Siphoninus phillyreae*	–	–	–	–	√	√	√
4.	Mite	*Brevipalpus phoeniis*	–	–	–	–	√	–	–
5.	Leaf curling mite	*Aceria granati*	–	–	–	–	√	–	–
6.	Mealy bug	*Ferrisia conssobrina* *Nipaecoccus viridis*	–	–	–	–	–	√	√
7.	Fruit bug	*Bathycoelia indica*	–	–	–	–	–	√	√
8.	Scale insects	*Parasaissetia nigra*	–	–	–	–	–	√	–
9.	Hairy Cater pillar	*Euproctis frterna*	–	–	√	–	–	√	√
10.	Semilooper	*Achaea janata*	–	–	√	–	–	–	√
11.	Bark eating caterpillar	*Indarberla quadrinotale*	–	–	√	–	√	√	√
12.	Fruit borer	*Conogethes punctifertis*	√	–	–	–	–	√	√
13.	Anar butterfly	*Virachola isocratis*	–	√	√	√	–	√	–
14.	Stem borer	*Ceosterna scabrator*	–	–	–	–	–	√	–
15.	Beetle	*Carpophilus multilalus*	–	–	–	–	–	√	–
16.	Fruit sucking moths	*Othreis* spp. *O. trihaca* *O. fullonica* *O.materna*	√	–	√	–	–	√	–
17.	Shot hole borer	*Xyleborus fornicatus*	–	–	–	–	–	–	–
18.	Termites	*Odontotermes obsesus*	–	–	–	–	–	–	√
19	Bag worm	*Claria cromori*	–	–	–	–	–	–	√
20.	Leaf weevil	*Myllocerus maculosus*	–	–	–	–	–	–	√

Bark Eating Caterpillars

Indarbela tetraonis (Moore) and *I.quadrinotata* (Walker) have been recorded boaring the bark of pomegranate trees and feeding inside. Older trees and the trees in orchards that are not well maintained are more prone to the attack of these pests. Usually there is only one caterpillar in each hole but there may be 10 to 12 holes in a badly infested tree. Such trees will bear no fruits.

A female lays 500 to 1000 eggs in 1 to 2 weeks. Eggs are laid in strings on the bark, branches or stems. On hatching, the larvae bore, usually at the joints between leaf-stalk or twig and main stem, and tunnel straight downwards. The larvae also cut circular holes at various places through which they eject the frass etc. Pupation takes place inside the tunnels. Moths are white with pairs of small black dots on thorax; numerous small black spots and streaks on forewings and a few black spots on posterior edges of hind wings.

Table 27.2 : Distribution of Common Diseases of Pomegranate in Various Centers of Arid and Semi-arid Region

Sl.No.	Location	Disease	Causal Organism
1.	Anantpur(A.P.)	Fruit spot	*Alternaria alternata*
2.	Aruppukottai(T.N.)	Bacterial fruit and leaf spot	*Xanthomonas campestris pv pnicae*
3.	Bikaner (Rajasthan)	–	–
4.	Godhra (Gujarat)	Leaf and fruit spot (cercospora leaf spot)	*Pseudocercospora punicae* *Colletotrichum gloeosporiodes*
		Leaf and fruit rot (anthranose)	*C. capsici* *X. campestris* *Pestalotia* sp.
		Bacterial leaf spots	*Fusarium* sp. *Sphaceloma* sp. *Phomopsis* sp. *Alternaria* sp. *Phyllosticta* sp. *Curvularia* sp.
		Fungal leaf spots	*Aspergillus niger* *A. flavus* *A. versicolor*
		Fungal fruit spots and fruit rots	*Aspergillus* sp.
5.	Hisar (Haryana)	Bacterial leaf and fruit spots	*Xanthomonas campestris pv punicae*
6.	Jobner (Rajasthan)	Bacterial leaf and fruit spots	*X. campestris pv punicae*
		Fungal leaf and fruit spot	*Pseudocercosporella* sp.
		Leaf spot/blight	*Alternaria* sp.
		Nematode	*Meloidogyne* sp.
7.	Jodhpur (Rajasthan)	Fruit rot	*Mucor* sp. *Colletotrichum* sp. *Fusarium* sp. *Alternaria* sp.
8.	Rahuri (Maharastra)	Fruit spot	*Alternaria alternate*
		Leaf and fruit spotting fungi	*Colletotrichum gloeosporioides* *Cercospora* sp. *Drechslera rostrata*
9.	Sardar Krushi Nagar (Gujarat)	Fruit spot	*Alternaria alternata*
10.	Udaipur (Rajasthan)	Leaf spot	*Glomerella cingulata* *Pseudocercosporella punicae* *P. granati* *Phytophthora nicotianae* *Phomopsis* sp.

Table 27.3: Consolidation of Pest Management Strategies in Panchmahals

Pest	*Nature of the Damage*	*Management Strategies*
Thrips *Scirtothrips dorsalis*	Nymphs and adults lacerate the leaves, flower stalks petals and sepals and rasp the sap that oozes out resulting into leaf tip curl and dry away while flowers are shed.	1. Removal of wild creepers of Menospermaceae reduces the incidence of fruit sucking moths. 2. Avoid fruiting in rainy season and destroy the fallen fruits.
Aphids *Aphis punicae*	Both nymphs and adults suck the cell sap from tender leaves and twigs. The affected parts get discolored and disfigure. They secrete honeydew on which sooty mould develops.	3. Bagging of fruits before maturity prevents the damage by moths and borers.
Mealybugs *Planococcus lilacinus*	Both nymphs and adults suck the cell sap and devitalize the plant resulting into shedding of buds and flowers and smaller size fruits.	4. Injection of petrol, CS_2, Diclorvos in holes and plugging the holes with mud controls the damage by bark eating caterpillar.
Leaf eating cater pillar *Achaea janata*	Freshly hatched larvae congregate on leaves later they segregate and feed voraciously devouring the entire leaf lamina. Adults are destructive, suck the juice of fruits.	5. Inudndative release of *Trichgramma chilonis* is quite effective against anar butterfly.
Fruit sucking moths *Othreis* spp.	The moths are nocturnal in habit and are found during rainy season. The adults pierce the ripening fruits and the damaged fruits start rotting as the punctured region are easily infected with bacteria and fungi, ultimately the fruits are dropped.	6. Sprays with Monocrotophos, Dimethoate, (0.05 per cent) followed by Endosulfan (0.07 per cent) or Fenvalerate (0.05 per cent) effective against major pests or pomegranate.
Anar Butterfly *Virachola isocrates* Fabr. *Deodorix epikarbas*	On hatching the caterpillars bore inside the developing fruits subsequently the infested fruits are attacked by bacterial causing fruit to rot. The conspicuous symptoms of damage are offensive smell and excreta coming out of entry hole	
Bark eating caterpillar *Inderbela quadrinotata*	On hatching the larvae usually bore at joints between leaf stalk and twig and main stem and tunnel straight down ward. The larvae also cut circular holes at various places through which they eject frass.	
Shoot hole bore *Xyleborus fornicatus*	A stem boring ambrosia beetle causes damage to main trunk and benches.	
Termites *Odonototerms obesus* Rambur	Maximum damage to the stem upto 30 cm height from ground level.	

Strategy of Control

Keeping the orchards clean and avoiding over-crowding of trees helps in minimizing the attack by these borers. In case of infestation, clean the affected portions by removing all webs, etc. and insert in each hole, swab of cotton-wool soaked in carbon bisulphide, petrol or even kerosene and seal the holes with mud.

Stem Boring Beetles

Celosterna spinator Fabricius has been reported boring the stems and trunk of pomegranate trees. The grubs bore inside the trunk and feed on sapwood. Adult beetles are active by day and feed by gnawing the green bark of shoots. The life-cycle is annual with an extended emergence period from May to November. Generally no control measures are required. Nevertheless, the measures suggested against bark eating caterpillars will prove effective in controlling these beetles as well.

Sap Sucking Insects

Drosicha *mangiferae* (Green), *Hemiaspidoproctus cinera* (Green), *Ferrisia virgata Icerya purchasi* Maskell, *Planococcus Iilacinus* (Cockerell), *Hemiberlesia punicae* (Signoret) are some of the mealy bugs and scale insects reported from various parts of India on pomegranate leaves. Most of these are polyphagous, having a wide range of host plants. Due to damage by these insects, the trees are devitalized, resulting in shedding of buds and flowers and smaller sized fruits.

Strategy of Control

To prevent the attack from spreading, prune and destroy the affected parts in the initial stage of attack. If the infestation becomes severe, spray with 0.05 per cent dimethoate or monocrotophos.

Pomegranate aphid, *Aphis punicae* and pomegranate whiteflies *Siphonimus finitimus* Silvestri and S. *phillyreae* (Haliday) are also minor pests recorded occasionally. They feed by sucking the cell sap from leaves and tender twigs. The affected parts get discoloured and disfigured. These insects also secret copulas amount of honey dew on which sooty mould derelops hindering photosynthetic activity of plants. Thrips Scirtothrips, dorsalis Hood infest the flowers. Nymphs and adults lacerate the leaves, flower stalks, petals and sepals and rasp the sap that ooze out of these wounds. As a result, the leaf tips curl and dry away while flowers are shed and ultimately the fruiting capacity of the tree is adversely affected.

Strategy of Control

To control the aphid, whiteflies and thrips spray with 0.03 per cent dimethoate, oxydemeton mehyl or phosphamidon. Two spraying at an interval of 10 to 15 days will be sufficient to keep completed check of these pests.

Leaf Eating Caterpillars

Castor semilooper, *Achaea janata* (Linnaeus) is an important pest of castor. It has also been reported damaging ber, citrus-grapevine, guava, pomegranate etc. Freshly hatched caterpillars congregate on the leaves of various weeds and a few economic crops, including pomegranate, ber and castor, and feed on the chlorophyll. Later, they segregate and feed voraciously devouring the entire leaf lamina. In case of severe infestation, the young trees may be completely defoliated. The moths are also destructive. They suck the juice of various fruits, like citrus, grapes, guava, mango etc. The caterpillars show conspicuous color variations-some are grey with red or brown lateral strips while other are bluish-grey specked with blue-black and having yellowish lateral strips.

Strategy of Control

To check the damage caused by these semiloopers, collect and destroy the caterpillars mechanically. In case of severe infestation spray with 0.05 per cent dichlorvos or endosulfan. This will also kill other leaf eating caterpillars, if any. Hairy caterpillar, Euproctis flava (Bremer), plum hairy caterpillar, *E. fraterna* (Moore) and castor hairy caterpillar, *E. lunata* Walker are the species reported damaging pomegranate trees. Thes are highly polyphagopus pests, found all over Indian sub-continent. On hatching, the caterpillars feed gregariously on epidermis of leaves; later they segregate and feed voraciously, defoliating the entre trees.

Bagworms are widly distributed all over India. The species recoreded feeding on pomegranate leaves, include, *Clania crameri* Westwood and *Acanthopsyche* species. The caterpillars construct a case over their body and live within, nibbling leaf lamina. Females are apterous and devoid of antennae, mouth parts and legs. Males are winged and have bipectinate and short mouth parts.

Beetles and Weevils

Anomala dimidiata Hope, *Myllocerus laetivirens* Mashall and M. *undecimpustulatus maculosus* Desbrocher have been occasionally reported feeding on pomegranate foliage. These are polyphagous pests, destructive mostly to temperate fruit trees and of minor importance in case of pomegranate. The grubs live in the soil and feed on the roots and other organic matter. It is only the adults that come out of the soil at night and feed on foliage. Generally no control measures are required against theses pests.

Termites

Odontotermes obesus (Rambur) has been reported attacking pomegranate trees in Rajasthan (Kushwaha, 1960) and Andhra Pradesh (Krishnamoorthy and Ramasubhiah, 1962). This sub-terranean enemy is highly polyphagous and feeds on a large number of economic and wild crops. Its attacking is more pronounced in sandy and sandy-loam soils than in clayey or black soils. The termites cannot thrive under conditions of bad aeration and poor drainage. Drenching with chlorpryriphus @ 5 ml/ litre is quite effective in warding off the attack of this pest should be regularly practiced atleast once in a year in termite infested orchards.

Scavengers

Scavengers beetles (calvicorn) feed on fermenting or decaying vegetable matter, particularly those with exuding sap, souring fruits and withering flowers, decomposing bark and sapwood. The species recorded on pomegranate are, *Amphicrosus* species and *Carpophilus dimidiatus* Linnaeus. These polyphagous beetles feed on rotten fruits that have been already damaged by the caterpillars of pomegranate butterfly or fruit flies.

Fruit Sucking Moth (*Othreis fullonica, othreis maternal*)

The adults pierce the ripening fruits and the damaged fruits start rotting as the punctured regions are easily infected with bacteria and fungi, ultimately fruits are dropped.

Management Strategy

1. Preventing the moths from approaching the fruits by smokes or by spraying crude oil emulsion or fish oil rosin soap mixture.
2. Bagging the fruits with 300 guage alkathene bags with punchivers on the bottom.
3. Collecting adult moths by light traps or keeping pieces of fruits near the tree basin for attracting sluggish moths in the evening.
4. Barking with malathion 0.05 per cent + 1 per cent crude sugar alongwith fruit juice or vinegar.
5. Destruction or eradiction of alternate lost.

Besides these pathogens, a nematode *Meloidogyne incognita* is also infecting pomegranate in Gujarat and causes stunting in growth (Anonymous, 1990). It was first reported from Jobner (Rajasthan).

Use of neemcake, granular soil pesticides, trap crops like merigold and deep ploughing in summer may help in its management.

Many birds, especially parrots, are known to damage the crop at maturity stage as a result of which fruits are rendered unmarketable. Bird scaring ribbon is effective in their managements.

The problem of fruit cracking is also severe at certain places, which is managed by providing irrigation at regular interval and applying boron (Borex, 0.3 per cent) as spray application on bushes. Jalore Seedless variety is less susceptible to cracking.

Acknowledgements

Thanks are due to Sh.B.J.Patel, (Photograther), Sh.R.K.Solanki (PA), and Sh.D.C.Joshi for their assistance in preparing manuscript

References

Anonymous, 1990. A decade of Research on diseases of Horticultural crops under AICRP (1980-1989). *Group discussion of Plant pathologists working in Co-ordinated projects of Horticultural crops.* Held on 14-15 June, IIHR, Bangalore.

Anonymous, 1996. Co-ordinated Fruit Research in Indian arid zone: A two decades profile (1976-1995). Eds O.P.Pareek and Vishal Nath. Published by Director, NRC for Arid Horticulture, Bikaner Rajasthan.

Anonymous, 2004. Two decades of research and development 1979-80 to TB/RC/08/20041999-2000. Central Horticultural Experiment Station, Vejalpur (Godhra) 389 340 Panchamahals. Gujarat, Published by D.G. Dhandher, Director, Central Institute for Arid Horticulture, Bikaner.

Bagle, B.G., 1993. Seasonal incidence and control of thrips Scirtothips dorsalis Hood in pomegranate. *Indian J. Ent.*, 55(1): 148-153.

Bindra, D.S., 1969. Controlling citrus pests. *Indian Hort.*, 13(2): 28-31.

Butani, D.K., 1979. *Insect and Fruits.* International Book Distributors, Dehradun and Periodical Experiment Book Agency, Delhi pp415.

Chand, R., Ram, K., and Kishan, R., 1993. Systemic movement of *Xanthomonas campestris pv.punicae* (Hingorani and Singh) Dye from leaf to node in pomegranate. *International Journal of Tropical Plant Diseases*, 11: 85-90.

Chandra, S. and Tondon, R.N., 1965. Control of leaf spot of pomegranate with fungicides. *Science and Culture*, 31: 536.

Chundawat, B.S., 1990. *Arid Fruit Culture.* Oxford and IBH Publ. Co. Pvt. Ltd., New Delhi, pp. 1-208.

Ghosh, G.R. and Dutta, B.G., 1960. Soil fungus from Orissa (India)-I. *Mycologia*, 52: 915-918.

Hellenppanarvar, N.L., 1955.The pomegranate fruit-borer. *Farmer Bombay*, 6(12): 56-60.

Hingorani, M.K. and Mehta, P.P., 1952. Bacterial leaf spots of Pomegranate. *Indian Phytopathology*, 5: 55-56.

Kakar, K.L., Dogra, G.S. and Amitnath, 1967. Incidence and control of pomegranate fruit borer, *Virachola isocrates* (Fabr.) and *Deudorix epijarbas* Moore. *Ind. J. Agric. Sci.*, 57(10): 749-752.

Kore, S.S. and Mitkar, P.L., 1993. Dry root-rot disease of Pomegranate incited by *Fusarium solani. Journal of Maharastra Agricultural Universities*, 18: 256-258.

Kushwaha, K.S., Sharma, J.C. and Sharma L.S., 1963. Common lepidopterean fauna of Udaipur (Rajasthan). *Indian Journal of Entomology*, 25: 142.

Mote, U.N., Tambe, A.B. and Patil, C.S., 1992. Management of fruit sucking moths in pomegranate garden. *Maharashtra J. Hort.*, 6(1): 20-24.

Nallthambi, P., Umamaheshwari, C., Nagraja, A. and Dhandar, D.G., 2007. Pomegranate diseases and management. *Technical Bull.* Central Institute for Arid Horticulture, ICAR, Beechwal, Bikaner, Rajasthan, p. 1-30.

Pruthi, H.S. and Batra, H.N., 1960. *Some Important Fruit Pests of North-West India* pp. 1-3, Bulletin 80, Indian Council of Agricultural Research, New Delhi.

Rao V.G., 1966 An account of market and storage disease of fruits and vegetables in Bombay, Maharashtra. *Mycopath. et Mycol.Appl.*, 28: 165-176.

Ruscoe, C.N.E., 1977. The new NRDC pyrethroids as agricultural insecticides. *Pesticides Science*, 8: 236-242.

Sharma, R.B., Sinha, B.P. and Roy, A.N., 1981. Post-harvest fruit rot of *Punica granatum:* Two new records. *Indian Phytopath.*, 34: 69-70.

Sherker, B.V. and Utikar, P.G., 1982a. *Fusarium fusarioides*: A new leaf spots disease of Pomegranate *Indian J. Mycol. Plant. Path.*, 12:51.

Sherker, B.V. and Utikar, P.G., 1982b. *Betraniella humicola:* A new fruit spotting fungus on Pomegranate. *Indian J. Mycol Plant Path.*, 12: 50.

Shreemali, J.N., 1972. Some new members of sphaeropsidales from India. *Indian Phytopath.*, 25: 58-60.

Shukla, R.P. and Prasad V.G., 1983. Comparative efficacy of various treatments for controlling pomegranate fruit-borer, *Virachola isocrates* (F.). *Entomon.*, 8(4): 381-383.

Snedecor, G.W. and Cochran, W.G., 1961. *Statistical Methods.* The Iowa State College Press Ames, Iowa, USA, pp. 285-289.

Somasekara, Y.M., 1999. New record of *Ceratocystis fimbriata* causing wilt of pomegranate in India. *Plant Disease*, 83: 400.

Somasekaras, Y.M., 2002, Application of *Bacillus subtitis* in the management of Pomegranate (*Punica granatum* Linn.) wilt (*Ceratocystis fimbriata* Fll and Halot). *Disease Research on Crops*, 3: 202-203.

Srivastava, M.P.and Tondon, R.N., 1971. Aspergillus rot of Pomegranate. *Indian Phytopath.*, 24:172.

Suriachandraselvan, M., Jaysekhar, M. and Anbu, S., 1993. Chemical control of bacterial leaf and fruit spot of Pomegranate. *South Indian Horticulture*, 41: 228-229.

Thanakamma, L., 1983, *Phytophthora* species on eight indigenous host species in south India and their pathogenicity on rubber. *Indian Phytopath.*, 36: 17-23.

Vasudeva, R.S., 1963. *Indian Cercosporae*. ICAR, New Delhi, pp. 1-245.

Yadava, C.P.S., 1969. Combating fruit sucking moths. *Indian Hort.*, 13(3): 32-34.

Zaka-ur-Rab, M., 1980. The corenlian, *Deudorix epijarbas* (Moore) (Lepidoptera: Lycaenidae) as a serious pest pomegranate fruits in Kashmir. *Journal of Entomological Research*, 4 (2): 233-235.

Chapter 28

Fungal Biocontrol of Plant-Parasitic Nematodes

D. Prasad
Division of Nematology, IARI, New Delhi – 110 012

Introduction

Plant-parasitic nematodes often cause plant diseases resulting in significant losses in crop yield and quality. However, uneven growth, stunting, chlorosis, poor quality and quantity of yield are general symptoms of severe infection and damage caused by plantparasitic nematodes. Among the plant-parasitic nematodes, root-knot nematodes belonging to the genus *Meloidogyne* are regarded as serious pests constituting an important major limiting factor in the production of agricultural and horticultural crops. Though they attack almost every type of crop, causing considerable losses of yield and affecting the quality of produce, vegetable crops mostly suffer greatest damage. Bhatti and Jain (1977) found an estimated loss due to *Mincognita* throughout India in tomato is 46.2 per cent. Thus, various measures have been in use for control of phytonematodes in the past are chemical, physical, regulatory, use of host resistance and biological methods. Fungi, bacteria and plant-parasitic nematodes constitute the significant component of soil ecosystem.

A number of chemicals both soil fumigants and systemic were successfully used for the management of plant-parasitic nematodes. However, most of the chemicals are banned and out of market due to their toxic effect on beneficial flora and fauna including man. They cause pollution hazards and their influence on non-target organisms have caused ecological imbalance which are sometimes difficult to be restored. Search for some alternative methods of nematode control thus received greater attention. In addition to the safety problems of pesticides, new crop protection technique, the IPM or Integrated Pest Management methods including organic amendments, fungal bioagents, botanical antagonists and reduced dose of nematicides has been proposed in order to reduce the effects of agrochemicals on the ecosystem. This is intended to maintain pest density at economic threshold injury level using a combination of various pest control methods.

Attempts are in progress in India to try fungal bioagents with very low dose of nematicides or organic amendments etc. as a part of integrated approach for the management of plant parasitic nematodes. The first report on fungal biocontrol agents was suggested by Lohde (1874) with observations on the endoparasite, *Harposporium anguillulae* in Germany. Kuhn (1877) reported *Catenaria auxillaris* in the females of beet cyst nematode, *Heterodera schachtii.* Contributions of Goswami and Rumpenhorst (1978), Kerry (1980), Khan (1990) and Leiz *et al.* (1992) clearly showed that the. biocontrol agents, particularly the nematode destroying fungi are common and abundant in both natural and agricultural soil as also, in all kinds of decaying organic materials.

Fungal bioagents of phytonematodes are grouped into three categories:

1. Predaceous or trapping fungi
2. Endozoic or endoparasitic fungi
3. Egg parasitic or opportunisitic fungi

Predaceous or Trapping Fungi

Predaceous or nematode trapping ones mostly belong to the order Zoopagles (Zygomycetes) and from order Moniliales (Deuteromycetes) which capture nematodes by various devices as (*a*) Sticky and (*b*) mechanical traps. Sticky traps are of 3 types (1) sticky branches *e.g., Dactylella lobata* (2) sticky network *e.g., Arthrobotrys oligospora* and (3) sticky knobs *e.g., Dactylella ellipspora* while the mechanical traps are of 2 types (*a*) nonconstricting rings *e.g., Dactylaria candida* and (*b*) constricting rings *e.g., Dactylella bembicodes.* Sucrose decomposition in the soil stimulates and increase in both the population of free living nematodes and the activity of indigenous nematodes trapping fungi and soil conditions viz., pH, temperature, moisture and availability of nutrients (Mankau, 1968). The use of nematophagous fungi as bioagents of nematodes was shown by Upadhyay and Dwivedi (1988)

Endozoic or Endoparasitic Fungi

Endozoic or endoparasitic fungi which are natural enemies of plant-parasitic nematodes produce either simple spores or flagellate spores as infectious agents and instead of producing hyphal development outside the body of the host, germination of spores are within the nematode body. Till they make contact with the host the spores remain viable but dormant. Simple spores are ingested by the nematodes, reach the oesophagus/buccal cavity where they germinate, penetrate the oesophagus and colonise the body cavity *e.g., Harposporium anguillulae.* The sticky spores stick to the nematode at any point but mostly cause infection in the head region. They germinate, penetrate directly through the nematode cuticle and produce infective hyphae within the body cavity. *E.g., Meria coniospora, Nematoctonus bisporus.* The flagellate spores stick to the nematode and encyst before germination, penetration and colonisation of the host *Catenaria unguillulae, C. vermicola, Harposporium anguillulae* among the fungi was to be found effective in checking nematode population (Lodhe, 1874). *Nematophthora gynophyla* in addition to *Globodera rostochiensis* and *Heterodera avenae* also parasitised females of *H. trifolii, H. schachtii,* etc. (Kerry and Crump, 1977). *In vitro* observations on the infection of *Meloidogyne incognita* eggs by *C. anguillulae* was reported by Wyss *et al.* (1992) whose embryos were observed to be killed within a few minutes within mass aggregation and encystment of flagellate spores of fungus. However, the obligate nature of both predaceous and endozoic fungi difficulty in mass production and inability to establishment in soil are the characteristic that put a serious question mark *for* their candidacy as biocontrol agents.

Fungal Parasities or Opportunistic Fungi of Eggs/Cysts

Several soil fungi found to be parasitic on eggs of plant parasitic nematodes colonise the reproductive structures of the nematodes particularly *Meloidogyne, Heterodera* and *Globodera* species whose sedentary stages are most vulnerable to fungal attack.. These fungi occur in genera as *Paecilomyces, Verticillium, Fusarium, Phoma, Gliocladium* etc. These opportunistic fungi attack nematode eggs and/or cysts of the group Heteroderidae and those deposited in gelatinous matrix. Plant parasitic nematodes that become sedentary upon their maturity and have oviposition nature are considered to be the most vulnerable to attack to these fungi. As soon as these fungi contact egg masses or cysts, they rapidly grow and colonise these eggs that are in early embryonic developmental stages. Arai *et al.* (1973) isolated peptide antibiotics lencinostatin, lilacinin from *Penicillium lilacinum. P. lilacinus* exhibits chitinase (Okafor, 1967; Gintis *et al.*, 1983) and proteolytic activity (Endreeva *et al.*, 1972). It also produces a peptidal antibiotic P168 which exhibited wide range of toxicity on fungi, yeast and gram-positive bacteria (Isogai *et al.*, 1980a, 1980b, 1981).

The embryonated eggs treated with culture filtrates showed morphological changes in eggs and within 2–4 days, development of eggs stopped, embryo disintegrated and vacuole formed (Fitters *et al.*, 1993). The death of mature eggs, colonised internally was due to fungal metabolites (Carneiro and Gomes, 1993). *P. lilacinus* can grow well in temperature ranges between 151°C and 30°C which is similar to its hosts. It can also survive well in wide range of soil pH, thus showing its competitive nature in most agricultural soils (Jatala, 1986). This fungus is reported to be compatible with many fungicides and nematicides. (Davide and Batino, 1985). *P. lilacinus* showed growth at 3.0 to 11.0 while optimum mycelial growth occurred at pH 6.0 to 9.0 and sporulated at pH 4.0 to 6.0. The good sporulation was observed at 30°C and mycelial growth inhibited at 10°C and 40°C. The alternate light and darkness provided better fungal growth (Villanueva and Davide, 1984).

Cabanillas *et al.* (1989b) recorded maximum fungal growth of *P. lilacinus* between the temperature ranges of 24°C and 30°C and least was at 12°C and 36°C. The effect of soil temperature on fungal parasitism has not been studied so far. However, it has significant impact on the survival and biocontrol activity of *P. lilacinus* isolates (Cabanillas *et al.*, 1989b). Fioretto and Villacorta (1989) found that optimum temperature for *P. lilacinus* growth was 24°C to 25°C but biostatic at per cent C and 35°C. The lower threshold temperature was –10.9°C for development of the fungus. This attribute could be useful for storage of *P. lilacinus.* Liu *et al.* (1995) observed that among the 20 isolates of *P. lilacinus,* optimum temperature (25°C to 30°C) requirement for growth and sporulation was not different among the isolates. Certain chemicals *viz.*, ethoprophos, fenamiphos, aldicarb, oxamyl and carbofuran were reported to toxic to *P. lilacinus* in PDA-culture (Equiguren-Carrion, 1995). Amoncho and Sasser (1995) found that benomyl stopped the *P. lilacinus* growth and chlorothalonil, cpatan and PCNB suppressed the radial growth but fenamiphos slightly enhanced the fungal growth.

The biocontrol potential of the fungus, *V. chlamydosporium* for root-knot nematodes has been reviewed by Leij and Kerry (1991). The fungus required some external nutrient for its establishment in soil. The colonisation of nematode egg masses depended on fungal inoculum and on galling caused by nematodes (Leij *et al.*, 1992). The efficiency of *P. lilacinus* in controlling *M. incognita* and *G. pallida* was demonstrated by Franco *et al.* (1981). The eggs were deformed by the fungus with the help of toxic metabolites. Mature females of root-knot nematodes were usually penetrated by the fungus through the anus or vulva while cysts of *G. pallida* through the vulva and exposed or broken neck region (Jatala, 1986). They reported that hyphal penetration of egg shells were brought about by mechanical pressure and/or enzymatic activity. Rate of fungal penetration into egg shell of *Meloidogyne* eggs is faster than *Globodera* where the egg shells show more complexity (Jatala, 1985). Work has been carried out for managing root-knot nematode infecting vegetables in India (Goswami and Mittal, 2000). Khan and Hussain (1986) reported that females, eggs and juveniles of *Mincognita* were infected by *Fusarium solani.*

There are various ways for soil-borne fungi to suppress nematode multiplication. A detailed review of fungi as biocontrol agents against plant-parasitic nematodes has been published (Kerry and Jaffee, 1997). In summary, there are four mechanisms that fungi use to suppress nematodes. Some of these interactions are direct whereas others are indirect. The direct mechanism is performed by (1) fungi that feed on nematodes directly, known as nematophagous fungi: fungi interact with nematodes in an indirect manner by several mechanisms including; (2) fungi that kill nematodes by mycotoxin or (3) through the destruction of the feeding sites of sedentary nematodes in roots; (4) fungi that are nonpathogenic to plants, but compete with nematodes in roots and siginificantly reduce nematode multiplication. Many of these fungi are used as potential nematode biocontrol agents; (5) Mycorrhizal fungi improve the growth of nematode infected plants and may also affect nematode development.

This chapter also summarizes groups of nematophagous fungi according to their feeding habits:

Nematode-Trapping Fungi

Facultative fungi that form trapping structure to trap nematodes. There are 6 types of traps reported by Barron (1977).

Adhesive hyphae

e.g. *Zygomycotina*

Stylopage

Cystopage

Adhesive traps

e.g. *Deuteromycota*

Monacrosporium cionopagum (branches)

Mellipsosporum (knobs)

Arthrobotrys oligospora (networks)

Non-adhesive Traps

e.g. *Deuteromycota*

Arthrobotrys dactyloides (constricting ring)

(non-constricting ring)

Facultative Parasitic Fungi Attacking Sedentary Stages of Nematodes (Kerry and Jaffee, 1997)

These are facultative fungi that are commonly soil saprophytes and are opportunistic fungi isolated from the sedentary stages (female and egg stages) of sedentary nematodes such as *Heterodera, Globodera* and *Meloidogyne.* They do not form specialized infection structures except appressoria. They can survive and proliferate in soil in the absence of nematodes.

e.g. *Hyphomycotina*

Acremonium

Cylindrocarpon
Fusarium
Paecilomyces
Verticillium

Endoparasitic Fungi (Kerry and Jaffee, 1997)

These are obligate parasitic fungi that have limited growth in soil outside the colonized nematode cadaver. They can infect vermiform nematodes by producing adhesive spores attached to cuticle of passing nematodes.

e.g. *Hyphomycotina*
Hirsutella rhossiliensis
Drechmeria coniospora
Verticillium spp.

Some can infect vermiform nematodes by producing conidia spores that can be ingested by nematodes.

Harposporium anguillulae

Some can infect vermiform nematodes by producing motile zoospores that encyst on the nematode's surface.

e.g. *Oomcota*
Myzocytium spp.
Lagenidium spp.
Chytridiomycota
Catenaria anguillulae

Some can infect sedentary nematodes when the nematodes were exposed on the root surface.

e.g. *Oomycota*
Nematophora gynophila

Methods for the Isolation of Fungal Bioagents

Stirling and White (1982) used baiting technique to isolate obligate parasites in which usually a large number of susceptible nematodes from laboratory culture are added to soil and then extracted after a period of incubation and assessed for infection. The egg-parasitic fungi which are consistently associated with the egg masses of root-knot nematodes are isolated, sub-cultured and attempted for mass culture after testing their potentiality as biocontrol agent through *in vitro* studies and pot expts. But, for the isolation of facultative parasites sprinkling techniques is suitable or in some cases selective media *i.e.* carrot extract, Potato Dextrose agar, etc. can be used.

Maintenance of Stock Culture

The culture of toxic and egg-parasitic fungi are maintained on Potato Dextrose Agar medium slants in test tubes and Potato dextrose broth in 250ml conical flasks after incubating in BOD incubator at 25±2°C for 10–15 days. Spores of source inoculum containing same medium serves as a seed culture for further mass multiplication.

Mass Production of *P. lilacinus*

Biocontrol programmes include enhancement of naturally occurring antagonists or augmentation of their activity by introducing other antagonist for suppression of target pathogen or pest of crops. Such augmentation could be achieved through release of mass

produced parasites or predator. Sharma and Trivedi (1987) observed maximum growth of *P. lilacinus* on sesame cake followed by cotton, mustard and groundnut cake. Among the waste materials, mung bean husk exhibited maximum growth followed by cotton seed, guar powder, gram powder and rice husks. The oilcakes had additive effect because of their nematotoxic properties. Bansal *et al.* (1988) evaluated agro-industrial wastes for mass production of *P. lilacinus.* The rice husk produced of *P. lilacinus.* The rice husk produced least spores while the wheat bran provided maximum spores which was comparable with rice grain. The sporulation was found enhanced in rice husk and rice bran with addition of molasses amended with nitrogen and phosphorus.

Zaki and Bhatti (1988) recorded higher spore counts (55.5 × 10/g) in neem leaves and concluded that incorporation of neem leaves infested with fungus may have synergistic or additive effect on nematode population. Mani and Anandam (1989) also evaluated plant leaves, oil cakes and agro-industrial wastes as substrate material for mass production of *P. lilacinus.* The leaves of *Leucaena leucocephala* and neem were found suitable and supported higher spore load. Mani *et al.* (1989) observed that wheat, bajra, jowar and rice as substrate media were suitable for mass production of *P. lilacinus. T. semipenetrans* population was markedly decreased with increasing the level of fungal inoculum and the spore suspension was found more effective than fungus infested wheat grams.

Vicente *et al.* (1989) compared among the available rice grains for mass production of *P. lilacinus* and found that pounded commercial rice and unpounded ground rice were most suitable on the basis of spore count while unpounded commercial rice had least spore load. Zaki and Bhatti (1991) found that *P. lilacinus* sporulated profusely on maize, gram, oats, rice. and wheat grains. The maximum number of spores/g was observed on rice (52.8 × 10) followed by gram (24.5 × 10). The fungus grown on gram seed was the most effective against *M. javanica* infecting tomato plants. The reduction of gall index (78.6 per cent), population of *M. javanica* (94.2 per cent) and higher percentage of egg destruction were observed.

Patel *et al.* (1991) tested various organic amendments for their effects on growth and sporulation of *P. lilacinus* and found that karanj, neem and mahua cakes were most effective while pressmud and FYM were least effective. Bansal *et al.* (1992) observed that wood charcoal was a good carrier of *P. lilacinus* for applying in the field. The carrier material in low-density polyethylene pouch could support up to 10 spores/g for at least six months. The spore viability in charcoal nackets was not affected by constant temperature 28°C and ambient temperature 14°C to 19°C or even under aerated or unaerated condition.

Mode of Action

The combined effect of both fungal bioagents *T. viride* and *P. lilacinus* for the control of root-knot nematodes is mainly based on the toxic effect of former followed by the egg parasitic ability of the latter. Thus, a reasonable number of the infective second stage juveniles occurring in the soil around the root zone of affected plants would be reduced. The remaining juveniles expected to be less in number as also possessing lower rate of infectivity due to the toxic effect of *T. viride* would invade the roots. Out of these reduced number of juveniles which in turn would invade the fresh egg masses, very few would develop into adult which are attacked by *P. lilacinus* resulting in the production of mostly unviable eggs. Many of the eggs due to this fungus are found empty.

Method of Application (with dose)

For transplantable crops *viz.*, tomato, brinjal, chilli, etc. after deep ploughing treat the nursery bed with organic amendment like oilseed cakes etc. 10 days before sowing. At the same time, treat the bed with both the packets along with 400g/packet of the carrier material *i.e.* saw dust. Before transplanting the seedlings, a root-dip treatment may be given for 5 minutes in dense conidial suspension of fungal bioagents. Spot treatment may also be given after 20-25 days of transplantation. The dose is 10 packets (5 each) for 3 × 3 square meter nursery beds. But, for directly seeded crops like cucurbits, bhindi and pulses apply at the time of sowing with a sticky substance like gum arabica etc. along with appropriate amount of carrier material (saw dust) @ 400g/packet. In these spot treatment with both after 20–25 days of sowing is given at @ 2 packets of each for 1 kg seed.

Zaki and Maqbool (1991) obtained most effective control of *M. javanica* in chickpea with application of *P. lilacinus* (2g /pot) one week before nematode inoculation. Zaki (1994b) suggested that 4g of gram infested *P. lilacinus* per kg soil was optimum for significant reduction of gall index, and *M. javanica* population in tomato plants. The egg infection and destruction were 58 per cent and 66 per cent respectively. Jonathan *et al.* (1995) found greater reduction of *M. incognita* population on Piper beetle with application of *P. lilacinus* infested rice substrate @ 8g/kg soil. Patel *et al.* (1995) observed that soil application of *P. lilacinus* at 3 per cent to 5 per cent (w/w) on neem cake was effective against *M. javanica* infecting groundnut whereas seed treatment was ineffective. Davide and Zorilla (1995a) found that tuber-dip and soil-mix method of application of *P. lilacinus* were equally effective against *G. rostochiensis* on potato and their combined application enhanced nematode control efficacy.

Work done by several researchers have proved that some microorganisms, especially fungi produced metabolites that are toxic to root-knot nematodes (Goswami and Mittal, 2000). Thus, Mankau in 1968 reported nematicidal activity of culture filtrates of *Aspergillus niger*. Singh and Sitaramaiah in the same year showed parasitisation of *M. javanica* by *Curvularia* species. Later, Singh *et al.* (1983) reported that culture filtrates of *Aspergillus niger*, *Curvularia lunata*, *Trichoderma viride* and *T. lignorum* obtained from rhizosphere of tomato plants proved to be nematoxic and inhibited hatching of *M. incognita* larvae. Tabreiz *et al.* (1884) studied the effect of culture filtrates of 8 species of *Aspergillus* on hatching and mortality of *M. incognita* out of which *A. niger*, *A. terreus* and *A. fumigatus* were more toxic than the other species. Sharma and Saxena (1992) observed that culture filtrates of *Rhizoctonia solani* and *Trichoderma viride* adversely influenced hatching of *M. incognita* larvae. Both fungus showed relatively more toxicity. Khan *et al.* (1988) observed inhibition of hatching of *M. incognita* juveniles at different concentrations of fungal culture filtrates and complete inhibition was recorded at 50 per cent and 100 per cent concentrations. The filtrates suppressed the root galling of tomato plants caused by *M. incognita*.

Chawla and Goswami (1998) in an vitro evaluation of undiluted *A. niger* culture filtrate, demonstrated the complete suppression of larval emergence for the first three days and the total death of juveniles within 24 hrs in case of *M. incognita*. Siddiqui and Mahmood (1995) while working on the management of root-rot disease of chickpea, demonstrated than filtrate of *A. niger* markedly reduced number of nematode in the soil which in turn was attributed to the concentration of oxailc acid produced by the fungus. Moreover, autoclaved culture filtrate also immobilized the nematodes, the heat stability of toxin principle.

Cayrol *et al.* (1989) reported nematicidal properties of culture filtrates of *P. lilacinus*. The toxins production were different on different media and their effect differed on various nematode species. The mechanism of toxic activity was considered to be neurotropic. Acetic acid was identified from culture filtrates of *P. lilacinus* and it affected the movement of nematode juveniles (Djian *et al.*, 1991). Caroppo *et al.* (1990) found ovicidal activity of *P. lilacinus, P. maraquandi, P. variotii* and *P. coleus* against *M. incognita*. The culture broths of these fungi reduced 78 to 93 per cent egg hatching under exercised root culture of tomato. Khan and Khan (1992) observed that concentration of culture filtrates of *P. lilacinus* had direct relation to the inhibition of hatching and percentage mortality of *M. incognita.*

Siddiqui and Hussain (1991) tested that culture filtrate of *P. lilacinus* was quite effective against *M. incognita* and *Macrophomina phaseolina* while Zaki (1994a) reported that culture filtrate of *P. lilacinus* had immobilising effect on *M. javanica* and inhibited hatching of eggs. Oduor and Waudo (1995) also reported that *P. lilacinus, P. herbarum* and *Fusarium oxysporum* had suppressive effect on hatching of *Meloidogyne* spp. but hatched juveniles were not parasitized. A reduced hatching ranging between 21 per cent and 56 per cent was observed in *Heterodera glycines* by *P. lilacinus* (Chen-Sen Yu *et al.*, 1996).

Experimental results from the infested fields in different countries have indicated that *P. lilacinus* effectively controls *M. incognita* and even better than number of commonly used nematicides (Candanedo *et al.*, 1983; Jatala, 1983). The effect of single and multiple application of *P. lilacinus* against *M. incognita* under field trials revealed that only one inoculation was sufficient to establish it and to get substantial degree of nematode control (Jatala *et al.*, 1981). However, Cabanillas and Barker (1989) indicated that effective biocontrol for *M. incognita* required more than one application and at proper time.

Davide and Zorilla (1985) while investigating the biocontrol potential of *P. lilacinus* against *M. incognita* on okra found the fungus to be quite effective and economically better than nematicides. According to Shahzad and Ghaffar (1987) carbofuran @ 1kg a.i./ha was less effective than *P. lilacinus* against *M. incognita.* This fungus also effectively controlled nematodes on okra and mung in subsequent seasons but nematicide was no longer effective. Similarly, in the fungus treated plots, higher yields were recorded than carbofuran @ 2.5 kg a.i./ha. The efficacy of *P. lilacinus* had been comparable to that of most commonly used nematicides (Jatala, 1985). The fungus, *P. lilacinus* reduced the soil population of *M. incognita* by 66 per cent to 77 per cent where as isazophos reduced it by 89 per cent (Davide and Zorilla, 1986).

Dube and Smart (1987) obtained higher crop yields and reduced *M. incognita* population when *P. lilacinus* and *Pasteuria penetrans* were applied together in field microplots. Maheshwari and Mani (1988) also found that simultaneous inoculation of *P. lilacinus* and *Pasteuria penetrans* were more effective and capable of significant reduction of soil population of *M. javanica.* The introduction of *P. lilacinus* 10 days before at planting of tomato resulted in great protection against *M. incognita* and midseason inoculation also provided higher percentage of egg mass infection (Cabanillas and Braker, 1989).

Khan and Esfahani (1990) found that *P. lilacinus* was more effective for controlling *M. javanica* infecting tomato when fungus was added prior to nematode or simultaneous inoculation. A great number of nematode eggs were infested, juvenile development inhibited, eggs devoid of juveniles and only filled with fungal mycelium. The developed juveniles were also found attacked and mycelial growth over their bodies were observed.

Walia *et al.* (1991) reported that application of *P. lilacinus* through wheat bran was better than carbofuran @ 1 kg a.i./ha in gall reduction on okra plants infected with *M. javanica.* Oduor *et al.* (1996) observed plant performance and low gall index with the application of *P. lilacinus* in combination with aldicarb or nematicidal plants.

The effect of *Calotropis procera* and *G. fasciculatum* for the management of *M. incognita* in tomato was studied by Rao *et al.* (1996). Significant reduction of galls, eggs/egg mass observed in interactive effect of these components. Higher colonisation of mycorrhiza in root of tomato inoculated with *G. fasciculatum* and *Calotropis* leaf was observed. Gaonkar and Sreenivasan (1994) observed a positive influence of locally available organic amendments on proliferation of *G. fasciculatum.* Devi and Goswami (1992) and later Bhagawati *et al.* (2000) demonstrated that both *G. fasciculatum* in case of cowpea while *G. etunicatum* in tomato respectively together with mustard cake helped in reducing the disease severity caused by *M. incognita* fungus complex in both the above hosts. They observed that the pre-establishment of VAM fungus checked the entry of *M. incognita* larvae as also colonisation of pathogenic fungi *Macrophomina phaseolina* and *F. oxysporum* f.sp. *lycopersici* respectively.

Goswami and Mittal (2002) studied the effect of some fungal bioagents on root knot nematode, *Meloidogyne incognita* infecting brinjal *in vitro* studies showed that *Aspergillus niger* was highly toxic followed by *Aspergillus fumigatus* exhibiting 85 per cent mortality at 1 : 10 dilution while *Paecilomyces lilacinus* and *Geotrichum candidum* showed least mortality. *Aspergillus niger* and *A. fumigatus* showed maximum inhibition of hatching and *G. candidum* the least. However, remarkably high egg parasitization was observed by *Paecilomyces lilacinus* (75 per cent) and *Geotrichum candidum* (72 per cent). *In vivo* studies showed that *Aspergillus niger* and *Paecilomyces lilacinus* along with mustard cake gave maximum reduction in root-knot nematode population both in soil and root.

Hemlata *et al.* (2001) studied the efficacy of biocontrol agents (*Trichoderma harzianum, Paecilomyces lilacinus* and *Aspergillus niger*) for the management of rootknot nematode on chickpea. Crop growth was best with *P. lilacinus* followed by *T. harzianum, A. niger* and a combination of all three biocontrol agents. Maximum reduction of root-knot nematode was recorded with *P. lilacinus* followed by a combination of all biocontrol agents, *T. harzianum* and *A. niger.*

Ruqiya *et al.* (2000) studied the female parasitism of *Aspergillus* species and their effect on root penetration by root-knot nematode, *Meloidogyne javanica.* Maximum reduction in the root penetration of okra by *Meloidogyne javanica* was recorded with *A. niger.*

Rao *et al.* (1997) studied the management of root-knot nematode, *Meloidogyne incognita* on tomato by integration of *Trichoderma harzianum* with neem cake. Significant increase in plant growth and reduction in root galling and final population of *M. incognita* were observed in tomato seedlings transplanted in neem cake amended soil incorporated with *T. harzianum.*

Among the fifteen different fungi tested *in vitro* by Khan *et al.* (2000) *Paecilomyces lilacinus* and *Trichoderma harzianum* exhibited the best performance followed by *Verticillium chlamydosporium, Aspergillus niger, Aspergillus flavus, Fusarium solani, Penicillium* spp, *Cephalosporium* spp, *Fusarium oxysporum, Harposporium helicoids, Rhizoctonia solani, Arachniotus* species and *Pythium ultimum, P. lilacinus* gave the maximum larval mortality of juveniles followed by *Trichoderma harzianum.*

Combination of Fungal Bioagents

In recent years with the development of a number of fungal bioagents isolated from egg masses of *Mincognita* infecting vegetables and categorization of two types *i.e.* (*a*) toxic and (*b*) egg parasitic ones. Attempts have been to test the performance of two types alone as well as together. The combined effect of a toxic bioagent *Aspergillus terreus* with an egg parasitic one, *Paecilomyces lilacinus* showed better performance in reducing nematode population with improved plant health of tomato than when either of the above fungi was used alone. (Goswami and Sharma, 1999).

Commercialization of Nematophagous Fungi

In France two commercial agents, Royal 300 and Royal 350 were sold for control of *Ditylenchus myceliophagous* on mushrooms and *Meloidogyne* spp. on vegetables which being at a very high cost, have not been widely used. More recently, the facultative parasite, *P. lilacinus* has been produced in the Philippines as Bicon which can be applied for the control of several nematode species including root-knot nematodes. The American firm Mosanto is going to launch two commercial products one each of

P. lilacinus and *V. chlamydosporium* respectively in the near future. *Trichoderma viride* has also been mass cultured and tried in farmer's (mainly in nursery) in India. Attempts for commercialisation of *T. viride* and *P. lilacinus* and other potential fungi are in progress.

Future Prospects

1. Integrating ecofriendly methods like organic amendments, fungal bioagents and botanical antagonists to phytonematodes would prove a very promising management package and also profitable to our farmers.
2. Since many of the biopesticide *viz., T. viride, P. lilacinus* and *A. niger* has already been commercialized both in India and abroad, it is advisable to search for an isolate which would prove effective against both important fungal diseases as well as nematode diseases in under field conditions.

References

Amoncho, A. and Sasser, J.N., 1995. Biological control of *Meloidogyne incognita* with *Paecilomyces lilacinus. Biocontrol*, 4: 51-61.

Arai, M., Mikami, Y., Fukushima, K., Utsumi, T. and Yazawa, K., 1973. A new antibiotic, leucinostain, derived from *Penicillium lilacinum. J. Antibiot.*, 26: 157-161.

Barron, G.L., 1977. *The Nematode-Destroying Fungi*. Canadian Biological Publication Ltd.

Bansal, R.K., Walia, R.K and Bhatti, D.S., 1998. Evaluation of some agro-industrial wastes for mass production of nematode parasitic fungus, *Paecilomyces lilacinus. Nematol. Medit.*, 16: 135-136.

Bansal, R.K., Walia, R.K. and Bhatti, D.S., 1992. Wood charcoal powder, a carrier of *Paecilomyces lilacinus* spores. *Nematol. Medit.*, 20: 5-7.

Bhatti, D.S and Jain, R.K., 1997. Estimation of loss in okra, tomato and brinjal yield infested with *Meloidogyne incognita. Indian J. Nematol.*, 1: 37-41.

Bhagwati, B., Goswami, B.K. and Singh, C.S., 2000. Management of disease complex caused by root-knot nematode and wilt fungus through bioagents. *Indian J. Nematol.* (In press).

Cabanillas, E. and Barker, K.R., 1989. Impact of *Paecilomyces lilacinus* inoculum level and application time on control of *Meloidogyne incognita* on tomato. *J. Nematol.*, 21: 115-120.

Cabanillas, E., Barker, K.R. and Nelson, L.A., 1989b. Growth of isolates of *Paecilomyces lilacinus* and their efficacy in biocontrol of *Meloidogyne incognita* on tomato. *J. Nematol.*, 21: 164-172.

Candanedo, E., Lara, J., Jatala, P. and Gonzales, F., 1983. Control biologico del nematodo *Meloidogyne incognita* con el hongo *Paecilomyces lilacinus*. Abstr. XXIII Meeting, APS Caribbean Division.

Carneiro, R.M.D.G. and Gomes, C.B., 1993. Methodology and pathogenicity tests effect of *Paecilomyces lilacinus* and *P. fumosoroseus* isolates to eggs of *Meloidogyne javanica. Nematologia Brasileira*, 17: 66-75.

Caroppo, S., Perito, B. and Pelagatti, O., 1990. *In vitro* evaluation of nematicide activity by several fungi against *Meloidogyne incognita* eggs. *Redia.*, 73: 451-462.

Cayrol, J.C., Djian, C. and Pijarowski, L., 1989. Study of the nematicidal properties of the culture filtrate of the nematophagous fungus, *Paecilomyces lilacinus. Revue. Nematol.*, 12: 331-336.

Chawla, G. and Goswami, B.K., 1998. *In vitro* evaluation of culture filtrate of soil fungi on hatching of egg and mortality of *Meloidogyne incognita* and their *in vivo* effect on root-knot nematode infecting mungbean. In: *Proceedings of the National Symposium on Rational Approaches in Nematode Management for Sustainable Agriculture*, November, 23-25, B.A. College of Agriculture, Anand Campus, Anand, India, pp. 17.

Chen-Sen Yu, Dickson, D.W., Mitchell, D.J and Chen, S.Y., 1996. Pathogenicity of fungi to eggs of *Heterodera glycines. J. Nematol.*, 28: 148-158.

Davide, R.G. and Batino, E., 1985. Biological control of root-knot nematodes on cotton through the use of fungi *Paecilomyces lilacinus* (Thorn) Samson and *Gliocladium roseum* Bainler as seed treatment. *Phil. Agric.*, 68: 159-167.

Davide, R.G. and Zorina, R.A., 1985. Evaluation of a fungus *Paecilomyces lilacinus* for the biological control of root-knot nematodes *Meloidogyne incognita* on okra as compared with nematicide isazofos. *Phil. Agric.*, 68: 493-500.

Davide, R.G. and Zorina, R.A., 1986. Evaluation of a fungus *Paecilomyces lilacinus* for the biological control of root-knot nematodes *Meloidogyne incognita* on okra compared with a nematicide isazofos. *Int. Nematol. Network Newsl.*, 3: 32-33.

Davide, R.G. and Zorina, R.A., 1995a. Evaluation of *Paecilomyces lilacinus* and some nematicides for the control of potato cyst nematode, *Globodera rostochiesis. Biocontrol*, 1(3): 69-76.

Devi, T.P. and Goswami, B.K., 1992. Effect of VA-mycorhiza on the disease incidence due to *Macrophomina phaseolina* and *Meloidogyne incognita* on cowpea. *Annals of Agric. Res* B. K. 13: 253-265.

Djian, C., Pijarowski, L., Ponchet, M., Arpin, N. and Favre-Bonvin, J., 1991. Acetic acid: a selective nematicidal metabolite trom culture filtrates of *Paecilomyces lilacinus* (Thorn) Samson and *Trichoderma longibrachiatum* Rifai. *Nematologica*, 37: 101-112.

Dube, B. and Smart Jr., G.C., 1987. Biological control of *Meloidogyne incognita* by *Paecilomyces lilacinus* and *Pasteuria penetrans. J. Nematol.*, 19: 222-227.

Eguiguren-Carrion, R., 1995. Control of *Meloidogyneincognita* and *Nacobbus* sp. with *Paecilomyces lilacinus* in the greenhouse and effect of nematicides on the fungus. *Biocontrol*, 4: 41-49.

Endreeva, N.A., Ushakova, V.I. and Egorov, N.S., 1972. Study of proteolytic enzmes of different strains of *Penicillium lilacinum* (fhom) in connection with their fibrinolytic activity. *Microbiologiya*, 41: 364-368.

Fioretto, A.M.C. and Vinacorta, A., 1989. Thermal requirements for the development of the nematogenic fungi *Paecilomyces lilacinus. Pesquisa Agropecria Brasileira*, 24: 975978.

Fitter, P.F.L., Belder, E-den and Den, Belder E., 1993. A time-lapse technique to study the effect of fungal products on embryogenesis of nematode eggs. Mededelingen Van de Faculteii Landbouww-etenschappen, Universiteit-Gent., 58: 751-756.

Franco, J., Jatala, P. and Bocangel, M., 1981. Efficiency of *Paecilomyces lilacinus* as a biocontrol agent of *Globodera pallida. Journal of Nematol.*, 13(4): 338-339.

Gintis, B.O., Morgan-Jones, G. and Rodriguez-Kabana, R., 1983. Fungi associated with several developmental stages of *Heterodera glycines* from an Alabama soybean field soil. *Nematropica*, 13: 201-213.

Go swami, B.K. and Mittal, A., 2002. Effect of some fungal bioagents on root-knot nematode, *Meloidogyne incognita* infecting brinjal. *Pakistan J. Nematol.*, 20: 155-159.

Goswarni, B.K. and Rumpenhorst, H.J., 1978. Association of an unknown fungus with potato cyst nematode, *Globodera rostochiensis* and *G. pallida. Nematologica*, 24: 251–256.

Goswarni, B.K. and Sharma, S.B., 1999. Management of root-knot disease of tomato with integration of fungal bioagents (*Aspergillus terreus* and *Paecilomyces lilacinus*). In: *Abstract, Nat. Symp. on Pl. Dis. Scenario Under Changing Agroecosystems*, October 29-30, HPKV, Palampur, H.P., 42 pp.

Goswarni, B.K. and Mittal, A., 2000. Effect of *Aspergillus fumigatus* and *Fusarium solani* alone and in combination along with neem oilseed cake against *Meloidogyne incognita* infecting tomato. *Annals of Agric. Res.* (in press).

Gaonkar, S.B.N. and Sreenivasan, M.N., 1994. Effect of inoculation with *Glomus fasciculatum* in conjunction with different organic amendments on growth and yield of wheat. *Microbiol. Res.*, 4: 419-423.

Hernlata, P. and Pandey, G., 2001. Efficacy of biocontrol agents for the management of root-knot nematode on chickpea. *Annals of Plant Protection Sciences*, 9(1): 157-159.

Isogai, A., Suzuki, A., Higashikawa, S., Kuyama, S. and Tamura, S., 1980a. Constituents of a peptidal antibiotic P168 produced by *Paecilomyces lilacinus* (Thorn) Samson. *Agr. Biol. Chem.*, 44: 3029-3031.

Isogai, A., Suzuki, A., Higashikawa, S. and Tamura, S., 1980b. Structure of peptidal antibiotic P168 produced by *Paecilomyces lilacinus. Agr. Biol. Chem.*, 44: 3033-3036.

Isogai, A., Suzuki, A., Higashikawa, S., Kuyama, S. and Tamura, S., 1981. Isolation and biological activity of a peptidal antibiotic P168. *Agr. Biol. Chem.*, 45: 1023-1024.

Jatala, P., 1981. Biological control of *Meloidogyne* species: Methodology for preparation and establishment of *Paecilomyces lilacinus* for field inoculation. *Int. Meloidogyne Proj. Proc. 3rd Res. Plann. Conf. on Root-knot Nematodes, Meloidogyne* spp. Jakarata Region VI, p. 228-231.

Jatala, P., 1983. Potential and prospects of biological control of nematodes. Abstr. XXIII meeting APS Caribbean Division.

Jatala, P., 1985. Biological control of nematodes. In: *An Advanced Treatise on Meloidogyne, Vol. 1: Biology and Control*, (Eds.) J.N. Sasser and C.C. Carter. N.C.S University Graphics, Raleigh, p. 303-308.

Jatala, P., 1986. Biological control of plant parasitic nematodes. *Ann. Rev. Phytopathol.*, 24: 453-489.

Jonathan, E.I., Padmanabham, D. and Ayyamperumal, A., 1995. Biological control of root-knot nematode on betelvine, Piper beetle by *Paecilomyces lilacinus. Nematol. Medit.*, 23: 191-193.

Kerry, B.R., 1980. Biocontrol: fungal parasites of female cyst nematodes. *J. Nematol.*, 12: 253-259.

Kerry, B.R. and Crump, D.H., 1977. Observations on fungal parasites of females and eggs of the cereal cyst nematodes, *Heterodera avenae* and other cyst nematodes. *Nematologica*, 23: 193-201.

Kerry, B.R. and Jaffee, B.A., 1997. Fungi as biological control agents for plant-parasitic nematodes. In: *The Mycota,* (Eds.) K. Esser and P.A. Lemke, Vol. IV, pp. 204-218. Springer Verlag, Berlin, Heidelberg.

Khan, M.W., Asharf, S.S and Haider, S.R., 1988. An analysis of biocontrol efficacy of *Paecilomyces lilacinus* in management of root-knot nematodes. *Int. Conf Res. Plant Sciences and its Relevance to the Future,* Delhi, India, p. 67.

Khan, M.W. and Esfahani, M.N., 1990. Efficacy of *Paecilomyces lilacinus* for controlling *Meloidogyne incognita* on tomato in greenhouse in India. *Pakistan J. Nematol.,* 8: 95-100.

Khan, S.T. and Khan, T.A., 1992. Effect of culture filtrates of soil fungi on the hatching and mortality of root-knot nematode, *Meloidogyne incognita. Curr. Nematol.,* 3: 53-60.

Khan, M.W. and Hussain, S.I., 1986. Parasitism of *Meloidogyne incognita* by *Fusarium solani. Intern. Nematol. Netw. Newsl.,* 3(2): 11-13.

Khan, H.U., Waqar, A., Ahmed, R. and Khan, M.A., 2000. Evaluation of culture filtrates of different fungi on the larval mortality of *Meloidogyne incognita. Pakistan J. Phytopathology,* 12(1): 46-49.

Kuhn, J., 1877. Vorlaufiger Bericht uber die bisherigen ergebnisse der selt Jahre 1875 in auftrage des Vereins fur rubezucker Industrie Z. ver. Rubenzucker Indus. Disch. Reiches. 452-457.

Leiz, De. F.A.A.M. and Kerry, B.R., 1991. The nematophagous fungus, *Verticillum chlamydosporum* as a potential biocontrol agent for *Meloidogyne arenaria. Revue de Nematologie,* 14(1): 157-164.

Leiz, De. F.A.A.M., Kerry, B.R. and Deunchy, J.A., 1992. The effect of fungal application rate and nematode density on the effectiveness of *Verticillum chlamydosporum* as the biocontrol agent for *Meloidogyne incognita. Nematologica,* 38(1): 112-122.

Liu, X.Z., Liu, W.M and Zhang, D.S., 1995. Biological characteristics of *Paecilomyces lilacinus* associated with soybean cyst nematode. *Chinese J. Biological Control,* 11: 70–74.

Lodhe, G., 1874. Einge heven parasitishehen pilze. Tageblatt der 47 Versamunhung deutscher haturforseher und Aerzste Berslan, p. 203-206.

Maheshwari, T.U. and Mani, A., 1988. Combined efficacy of *Pasteuria penetrans* and *Paecilomyces lilacinus* on the biocontrol of *Meloidogyne javanica* on tomato. *Int. Nematol. Network Newl.,* 5: 10-11.

Mani, A., Murthy, I.R., Rao, P.K. and Prasad, V.D., 1989. Growth of *Paecilomyces lilacinus* on natural substrates and its efficacy against citrus nematode, *Tylenchulus semipentrans. J. Biological Control,* 3: 59-61.

Mankau, R., 1968. Reduction of root-knot disease with organic amendments under semifield conditions. *Plant Dis. Reptr.,* 52: 313-319.

Oduor, O.P., Sikora, R.A., Waudo, S.W. and Schuster, R.P., 1996. Effects of aldicarb and mixed cropping with *Datura stramonium, Ricinus communis* and *Tagetes minuta* on the control and integrated management of *Meloidogyne javanica. Nematologia,* 42: 127130.

Oduor, O.P. and Waudo, S.W., 1995. Effects of fungi on hatching and parasitism of eggs, juveniles and females of root-knot nematode. *Pakistan. J. Nematol.,* 13: 119-127.

Okafor, N., 1967. Decomposition of chitin by microorganisms isolated from a temperate and tropical soil. *Nova Hedwigia,* 13: 209-226.

Patel, R.R., Patel, D.J. and Patel, B.A., 1991. Influence of organic amendments on growth and sporulation of nematophagous fungus, *Paecilomyces lilacinus* (Thorn) Samson. *Curr. Nematol.,* 2: 39-40.

Patel, D.J., Vyas, R.V., Patel, B.A. and Patel, R.S., 1995. Bioefficacy of *Paecilomyces lilacinus* in controlling *Meloidogyne javanica* (Pathotype 2) on groundnut. *Int. Archais Newsl.*, 15: 46-47.

Rao, M.S., Reddy, P.P. and Das, S.M., 1996. Effect of irrigation of *Calotropis procera* leaf and *Glomus fasciculatum* on the management of *Meloidogyne incognita* infesting tomato. *Nematol. Medit.*, 74(1): 59-61.

Ruqiya, A., Zareen, A., Zaki, M.J. and Khan, S., 2000. Female parasitism of *Aspergillus* species and their effect on root penetration by root-knot nematode *Meloidogyne javanica. Pakistan J. Phytopathology*, 12(2): 115-117.

Rao, M.S., Reddy, P.P. and Nagesh, M., 1997. Management of root-knot nematode, *Meloidogyne incognita* on tomato by integration of *Trichoderma harzianum* with neem cake. *Zeitschrift-fur-Pflanzenkrankheiten-und-Pflanzenschutz*, 104(4): 423-425.

Shahzad, S. and Ghaffar, A., 1987. Field application of *Paecilomyces lilacinus* and furadan for the control of root-knot disease of okra and mung. *Int. Nematol. Ntework Newsl.*, 4: 33-34.

Sharma, A. and Trivedi, P.C., 1987. Screening substrates suitable for the growth of *Paecilomyces lilacinus. Int. Nematol. Network Newsl.*, 4: 24-26.

Sharma, M. and Saxena, S.K., 1992. Effect of culture filtrates of *Rhizoctonia solani* and *Trichoderma viride* on hatching of larvae of root-knot nematode, *Meloidogyne incognita. Curr. Nematol.*, 3: 61-64.

Siddiqui, Z.A. and Hussain, S.I., 1991. Control of *Meloidogyne incognita* and *Macrophomina phaseolina* on chickpea by fungal filtrates. *Pakistan J. Nematol.*, 9: 131137.

Siddiqui, Z.A. and Mahmood, I., 1995. Management of *Meloidogyne incognita* race 3 and *Macrophomina phaseolina* by fungus culture filtrates and *Bacillus subtilis* on chickpea. *Fundm. Appl. Nematol.*, 18: 71-76.

Singh, S.P., Veena, P., Khan, A.M. and Saxena, S.K., 1983. Inhibitory effect of culture filtrates of some rhizosphere fungi of tomato as influenced by oil-cakes on the mortality and larval hatch of *Meloidogyne incognita. Nematologia Medit.*, 11(2): 119-123.

Stirling, G.R. and White, A.M., 1982. Distribution of a parasite of root-knot nematodes in South Australian vineyards. *Pl. Dis.*, 66: 52-53.

Tabreiz, A., Khan, T.A., Hussain, S.I. and Azam, M.F., 1884. Effect of culture filtrates of eight species of *Aspergillus* on the hatching and mortality of *Meloidogyne incognita. Indian J. Nematol.*, 14: 50-74.

Upadhaya, K.D. and Dwivedi, K., 1988. Role of fungi in biological control of plant parasitic nematodes. *Brit. Crop. Prod Conn.*, 3: 1205-1208.

Vicente, N.E., Acosta, N. and Sanchez, L.A., 1989. Rice (*Oryza sativa* L.) substrates and growth of the fungus *Paecilomyces lilacinus*, a nematode control agent. *J. Agric. Univ. Puerto Rico*, 73: 79-82.

Villaneva, L.M. and Davide, R.G., 1984. Influence of pH, temperature, light and agar media on the growth and sporulation of a nematophagous fungus, *Paecilomyces lilacinus* (Thorn) Samson. *Phil. Agric.*, 67: 223-231.

Walia, R.K., Bansal, R.K. and Bhatti, D.S., 1991. Effect of *Paecilomyces lilacinus* application time and method in controlling *Meloidogyne javanica* on okra. *Nematol. Medit.*, 19: 247-249.

Wyss, U., Vossand, B. and Janson, H.B., 1992. *In vitro* observation on the infection of *Meloidogyne incognita* eggs by the zoospore fungus, *Catenaria anguillulae. Fundm. Appl. Nematol.*, 15(2): 113-139.

Zaki, F.A., 1994a. Effect of culture filtrates of *Paecilomyces lilacinus* on *Meloidogyne javanica. Nematol. Medit.*, 22: 41-43.

Zaki, F.A., 1994b. Dose optimisation of *Paecilomyces lilacinus* for the control of *Meloidogyne javanica* on tomato. *Nematol. Medit.*, 22: 45-47.

Zaki, F.A. and Bhatti, D.S., 1988. Economical method for mass culturing of *Paecilomyces lilacinus* (Thorn) Samson. *Curr. Sci.*, 57: 135.

Zaki, F.A. and Bhatti, D.S., 1991. Effect of culture media on sporulation of *Paecilomyces lilacinus* and its efficacy against *Meloidogyne javanica* in tomato. *Nematol. Medit.*, 19: 211-212.

Zaki, M.J. and Maqbool, M.A., 1991. *Paecilomyces lilacinus* controls *Meloidogyne javanica* on chickpea. *Int. Chickpea Newsl.*, 25: 22-23.

Chapter 29

Indigenous Knowledge to Manage Insect Pests and Diseases of Cotton Based Cropping Systems in India

A.K. Kanojia* and Sumitra Arora
National Centre for Integrated Pest Management,
Lal Bahadur Shastri Bhawan, IARI Campus, PUSA, New Delhi - 110 012

Introduction

Agriculture occupies a dominant position in India's economic structure. In view of the fact that nearly 30 per cent of the potential food production is lost in India due to insect pests, plant pathogens, weeds, rodents, birds and in storage, the use of pesticide is vital. It is estimated that one-third of world's food crop is destroyed by these pests annually. Farmers use pesticides/fungicides to prevent their crops, foods grains, vegetables or fruits from attacks by pests during growth in field or in post harvest storage conditions. Today, pesticides have become an important part of our everyday lives. The indiscriminate use of pesticides for controlling pest population has led to serious environmental concerns. Some of these substances are now however playing havoc with the lives of humans and environment because of their widespread environmental contamination, residues detected far from their site of application. In addition to their persistent nature, they are also toxic to non-target insect species and can be potential polluters of water, soil and air. Pesticides are usually non-biodegradable and so, minuscule quantities are left either in the soil or they enter the food chain. Moreover, application of high inputs in agriculture, in terms of chemical fertilizers and pesticides has endangered the sustainability of production system. Today organic compounds and their formulations have become the leading pesticides world over. They are less hazardous, environmentally sound; nature friendly and economically feasible and are required in much smaller quantities than inorganic chemicals to obtain the same effect. So there is a need to divert from synthetics to indigenously available technologies adopted by the farmers.

Indigenous knowledge, which entirely depends on natural resources and having depth of experience and knowledge, is an important means of crop husbandry through which insect pests and diseases are managed knowingly or inadvertently. Indigenous Technical Knowledge (ITK) is an important bank-source of knowledge conserved by different cultural groups followed through the centuries in different agro-climatic conditions in the country. The cotton based cropping system is one of the systems where such indigenous people are engaged in practicing ITKs controlling and managing pests and diseases of crop plants. Since cotton crop consumes huge amount of pesticides, it attracts the attention of everybody concerned with the development of agriculture towards the consequences of vicious cycles of chemical inputs, which are not only polluting our environment and posing economical burden on farmers but also killing natural enemies of the pests. This very reason has prompted the authors to focus on the indigenous knowledge in this cropping system. Indigenous knowledge of agriculture, is the result of farmers' thousands years of experience with nature. Indigenous practices are known to the farmers and are helpful in maintaining and enhancing the quality of the environment. With the dissemination of modern practices the indigenous practices have started to loose their ground and have been eroded to a large extent(Joshi and Singh 2006). It is the information base for a society, which facilitates communication and decision making. Indigenous information systems are dynamic and are continually influenced by internal creativity and experimentation as well as by contact with external systems (Flavier *et al.* 1995).

Biological control of plant pests and pathogens continues to inspire research and development in many fields. Plant pathogens are just one class of targets of biological control, which also is designed to limit other pests such as insects, parasitic nematodes, and weeds (Gardener 2002). Targeting the major plant pathogen through plant based botanicals-ITKs as biological control may help in a big way.

The information given in this chapter is based on the farmers' knowledge and their experience with the components and application methodology of ITKs and these are in practice by groups of farmers on a large scale at some places and in restricted areas by limited number of farmers at other places. However, it needs not to be adopted as such and proper scientific validation for field application in different agro-ecological regions is sought for fullest utilization of the given useful tactics in the area of plant protection in cotton based cropping system.

What Indigenous Technical knowledge (ITK) is?

Indigenous Technical knowledge (ITK) is also called as Indigenous Traditional Knowledge or Indigenous Know how. It is the local knowledge–knowledge that is unique to a given culture or society (Warner 1991). It contrasts with the international knowledge system generated by universities, research institutions and private firms. It is the basis for local-level decision making in agriculture, health care, food preparation, education, natural-resource management, and a host of other activities in rural communities (Warren 1991). Such knowledge is the consequence of practical engagement in everyday life, and is constantly reinforced by experience and trial and error. It is, as Hunn (1993) neatly puts it, 'tested in the rigorous laboratory of survival'. It is really the product of many generations of intelligent reasoning and time tested, having the confidence of the indigenous people. This knowledge is empirical rather than theoretical knowledge. It is shared to a much greater degree than other forms of knowledge, including global science. This is why; it is sometimes called 'people's science', an appellation which also arises from its generation in contexts of everyday production. However, its distribution is still segmentary, that is socially clustered (Hobart 1993). The knowledge usually goes silently side by side without any external intrusion and users evolve their own way of using the available resources by adopting appropriate methodology for preparation and application. Confidence and for doing better than other farming fellows in the same settings of ecological environment energizes the innovators in acquiring and selecting the specific knowledge based technical skill. Such propounders are reluctant to share their knowledge as it is believed that science based knowledge is advanced and valid and what these fellows are practicing is outdated and not relevant to modern know-how. This kind of attitude hinders the natural development for transfer of the available knowledge and even farmers to farmers technology transfer does not take place in many of the cases. Sincere efforts are to be put forward for natural blending of traditional with modern in general and farmers technical knowledge in crop protection with the latest synthesized packages of pest management. "Indigenous knowledge is the knowledge held by people, no matter how recent their arrival, who happen to live in a certain place. Generally the expression "indigenous knowledge" is used to make reference to knowledge held by indigenous people, where "indigenous" stands for aboriginal, native or autochthonous (Heyd 1995).

Cotton, the white gold is a highly valued commercial crop and is prone to pest attack at all the stages of growth. Though it has been reported that about 162 insect pests attack on cotton in India, only a dozen are major and half of them are key production constraints which cause losses to the extent of 30-80 per cent. Jassids, *Amarasca biguttula biguttula Ishida*; Thrips, *Thrips tabaci* (L.); Aphids, *Aphis gossypii* (Glover) and Whiteflies, *Bemisia tabaci* (Genn.) are the important sucking pests start to de-sap the cotton crop at seedling stage and cause heavy losses (Udikeri *et al.*, 2007). Crop loss estimated at Rs 406 cr; 6 per cent of farmland hit. About six per cent of the cotton crop has been affected. The production losses are about 13,67,130 tonnes.

This chapter throws light on some useful Indigenous Knowledge (IKs) which are in practice by the farmers in India to manage insect pests and diseases of cotton crop. These tactics are not only cheaper and evolved by incorporating local and easily available material but environment friendly certainly. Such tactics are of important value and meaningful to provide alternative means in plant protection in the country and elsewhere as scientific community is in search of alternative ways to manage and control pest problem.

Useful and Promising ITKs Adopted by Farmers

Control of pests in Cotton (i) Mechanical control measures include cotton seed treatment with cow dung resulted in delineating of the seed (fibre free seed), followed by identification and removal of pink boll worm infested seeds and hand collection, destruction of larvae and infested plant parts leads to reduction in insect pest population. (ii) Cultural control measures includes field sanitations–clean cultivation during the entire cropping seasons which results in the reduction of insect fauna and weed flora in the filed and intercropping or mixed cropping of cotton with sorghum/maize/greengram/blackgram for augumentation and conservation of natural enemies of major insect pests. (iii) Application of botanical pesticides include aqueous decoction of *neem* leaf (*Azadirachta indica*) or neem seed kernel or tobacco leaf (*Tobaccum solanum*) or alcoholic extracts of *neem* seed kernel against pest complex of cotton. (iv) Alternatively for pest management, application of ash water in which leather is soaked by cobbler against sucking pest or fermented butter milk or mixture of fermented farm animal urine + distillates of *neem* seed kernel or application of kerosene oil is practiced against insect pest complex of cotton in West Nimar, Madhya Pradesh (Paliwal *et al.*, 2002).

Non-pesticidal management technologies include deep ploughing in summer, use of bonfires/ light traps, nipping of the terminal buds to destroy eggs and larvae of spotted bollworm, selecting tolerant varieties for sucking pest complex, spraying of tobacco decoction, spraying of 5 per cent neem seed kernel suspension, chilli-garlic spray and spraying of cattle dung and urine, bird perches, collection and destruction of affected bolls. The farmers who practices this traditional knowledge gained more income as compared to the farmers who treated with pesticides (non NPM) in Warangal district of Andhra Pradesh. The farmers cultivating cotton in Warangal district with high pesticide load are prone to risk of negative returns as compared to the traditional practices. Farmers in Jalgaon/ Jalna spent more money for pesticides which enhanced cost of production while traditional NPM technologies reduced the cost of production by 25 per cent (Chary *et al.*, 2000).

Sufali (*Mundulea suberosa*) to Check Aphids

Sufali (*Mundulea suberosa*) leaves and neem leaves are crushed to obtain extract. Cow urine and water are added to the obtained extract in an earthen pot. This pot is placed underground in a dark place by digging a pit to accommodate the pot for 7-8 days. Then it is taken out and the fermented formulation so obtained is applied as spray @ 1ltr/10ltr water on cotton, chickpea, pigeon pea, and vegetables crops (Figure 29.1). It is believed that is practice is able to check the attack of sucking pests. Materials required for this tactis are: Sufali (*Mundulea suberosa*) leaves 1 kg; Cow urine 5 ltr; Neem leaves extract 2.5 ltr; Water 10 ltr. (Kanojia *et al.* 2005)

Figure 29.1

Calotropis and Castor Oil Against Aphids

Leaves of *Calotropis* plant are immersed in water channel during irrigation to minimize aphid infestation practiced in Banaskantha, Gujarat (Thakor 1991), while a little quantity of castor oil is added during irrigation in the water channel for reducing aphid population in Ahemdabad, Gujarat (Patel 1991a).

Control of Pest Complex in Cotton

Helicoverpa, spotted bollworm and aphid are generally found in cotton. Aphids are also common in brinjal, chilli and Lucerne in Amreli district of Saurashtra. A mixture was prepared from dried tobacco leaves (250 g) *hirakasi* (300 g) and citric acid by boiling in 1-2 litres of water and filtered. Approximately 250 ml of the decoction was mixed in 15 litres (one pump full) water and sprayed over crop. Depending upon the growth and density of the crop, total quantity required varied; however three to four pumps were enough for cotton crop. After a week decoction was sprayed again. It could control pest complex in cotton and the growth of the plants also boosted in Amreli, Gujarat (Kalyanbhai 1995).

Control of Whitefly

Whitefly (Bemisia tabaci) is usually found on the dorsal surface of the leaves. The nymphs are small and flat with visible beaks that suck the plant sap. To get rid of these flies, papers pasted with castor oil/grease are hung at five or six places in the cotton field. Then air is blown using a sprayer over the crop. The disturbed adult flies (white fly) come out of the plant and stick to the oily paper. By this method 90 per cent of white fly can be successfully reduced in Pondicherry (Balasubramanian 2000-2001).

Calotropis for Control of Caterpillars

Calotropis for control of caterpillars in cotton Caterpillar infestation can severely damage a cotton crop. Reportedly the latex of *Akda* (*Calotropis gigantia*, when diluted with 15 parts water and sprayed on the crop, effectively controls the pest within three days. The new growth after treatment is also free from infestation in Bhavnagar, Gujarat (Jani 1992).

Control of Greenworm by Dhatura

Of all the pests occurring in the cotton crop, greenworm (*Heliothis* spp.) is one of the most harmful pests. Leaves and stem of Dhatura (*Datura metel*) are used to control green worm. About 250 to 300 g of Dhatura' leaves along with stem are dipped in 1 litre of lukewarm water. After cooling down, 250 to 300g of the solution is mixed with 15 litres of water and sprayed on the crop. Pests perish within a period of six to seven hours. Spraying the mixture when the crop is of one-month-old yields better results in Amreli district, Gujarat (Kaachhadiya 1998).

Okra (*Ablemoschu s esculentus*) as a Sacrifice Crop

Use of okra (*Ablemoschu s esculentus*) as a sacrifice crop for pest control in cotton. Farmers grow 2-3 lines of okra plant surrounding the cotton field. Cotton is more susceptible to insect pests like spotted bollworms (*Earias vitella*), worm (*Platyedra gossypiella*), Jassid Amarasca biguttula) etc. Farmers believe that pests prefer okra plant as compared to cotton plant and attack them first. Pest is controlled by simply eradicating/destroying the okra plants attacked by insect and pests in Surendranagar district, Gujarat (Patel 1991g).

Neem leaves 2.5 kg. are put in 10 ltr cow urine and kept in an earthen pot underground for 21 days. Fermented solution is taken out after 21 days and it is filtered. The filtrate is then boiled in a copper utensil vigorously so as to reduce to one fourth. It is applied as spray on cotton, chickpea, pigeon pea, tomato, brinjal and soybean crops @ 100 ml/10 ltrs water for insect control (Figure 29.2). Its results are claimed to be very promising on insect pests and diseases of the crops, besides it provides many nutrients through leaves and acts as a tonic to the plants (Kanojia *et al.* 2005).

Salt with Manure to Increase Cotton Yield

Increasing yield of Cotton by application of salt mixed with manure is claimed by taking 20 to 25 kg of common salt and mixing with about 10 cart loads of natural manure. It is then applied in farm

Figure 29.2

during *Jyeshtha-Vaishakh* (May). It has bee observed to be more effective if the mixture is applied uniformly in all the furrows. 10 maunds of cotton/*bigha* is received (1/3 of acre). It is believed that the manure increases the crop's resistance to pests. This knowledge is from Surendranagar in Gujarat (Sanghani 1998).

Nerium Arali for White Fly

White fly control in cotton by seed extract of *Nerium Arali* (*Nerium oleander*) *is practiced by* pulverizing and soaking its seeds in water overnight and filtering. This filtrate is diluted in water and sprayed on cotton fields. This practice provides 70 per cent control in Thiruvadagam (Murugesan 1994).

Dry Sowing of Cotton and Pigeonpea

For getting advantage of early seeding in *Kharif* season, farmers resort to dry seeding of cotton (local variety), pigeonpea and also rice. Seeding is generally done 10-12 days before the anticipated date of onset of monsoon. In this practice, dry seeded crops give the advantage of 8-10 days earlier sowing than normal. When sufficient rains are received, the dry seeded grains start germinating. In dry seeding, however, there is a risk of re-sowing in the event of receipt of inadequate rainfall. Otherwise it gives 10 per cent more yield. Dry seeding of cotton and pigeonpea is in practice in Dhule, Nandurbar and Jalgaon districts of Maharashtra. Dry seeded rice is taken in Taloda and Akkalkuwa tehsils of Nandurbar district in Maharashtra. This knowledge as been opined by Directorate of Extension Education, Mahatma Phule Krishi Vidyapeeth, Rahuri, Ahmednagar, Maharashtra.

Sufali (*Mundulea suberosa*) to Control Borers

To control the borers of tomato and chilli there is a practice to make a paste of *Mundulea suberosa* (*Sufali*) leaves of 3 kg which is mixed with 10 ltr water in an earthen pot and kept overnight. The formulation so prepared is applied next evening @ 200 ml/15 ltrs water on standing crop(Figure 29.3).

Materials required for this ITK are: leaves of *Sufali* 3 kg; water 10 ltrs. The *sufali* (*Mundulea suberosa*) plants are locally available at the elevated lands in and around Ahmadnagar during rainy season

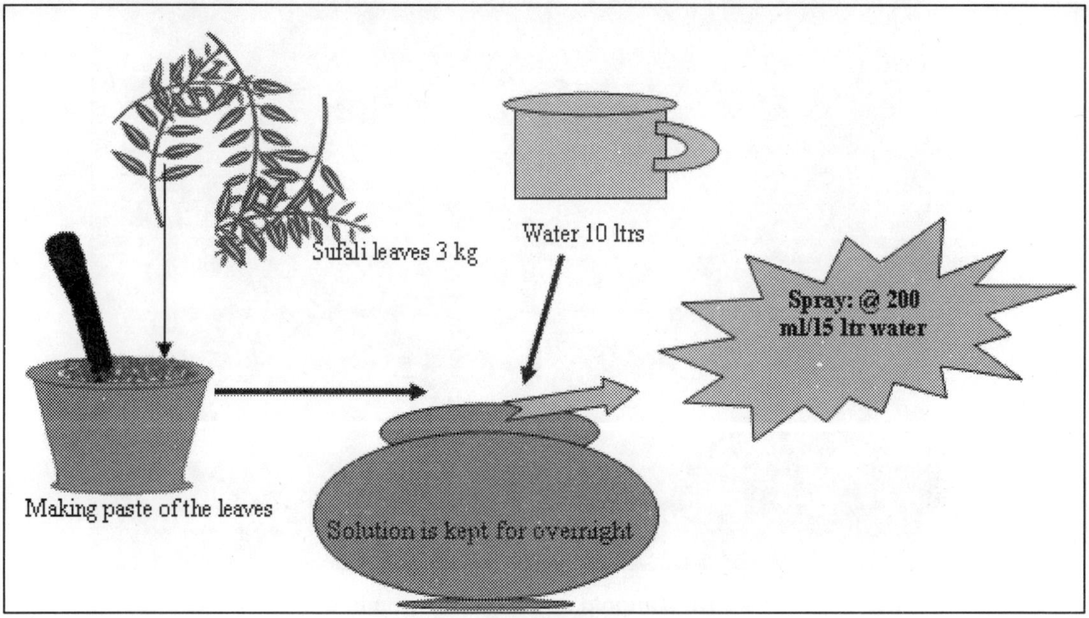

Figure 29.3

and it is said by the farmers that the tribal community of the region practiced the same leaves to kill fish by taking the extract of the leaves and mixing with the balls of edible flour. This knowledge inspired the villagers to use the leaves in plant protection (Kanojia *et al*. 2005).

Raising of Sorghum as Mixed Crop

Raising of sorghum as mixed crop in cotton as bird percher Iis adopted in Jalgaon, Nadurbar, Dhule and Ahmednagar districts of Maharashtra. Farmers raise sorghum as a mixed crop scattered in cotton fields. The grain of sorghum attracts the birds and served as a perch for the birds to reach the insects of cotton plants. Director of Extension Education, Mahatma Phule Krishi Vidyapeeth, Rahuri, Ahmednagar (Maharashtra) 413 722.

Spraying of Mixture of Garlic Extract and Monocrotophos to Control Bollworm in Cotton

Farmers of Sindhkheda and Shirpur tehsils of Dhule district and Kopargaon tehsils in Ahmednagar district are using this practice. Garlic bulbs are crushed and soaked in water to draw extract. This extract is added to monocrotophos and it is sprayed on cotton to control cotton bollworm, explored by Mahatma Pule Krishi Vidyapeeth, Rahuri, Ahmednagar, Maharastra.

Marigold to Manage *Helicoverpa* in Cotton

Marigold plants are sown as trap crop around cotton plants which attract *Helicoverpa* (Figure 29.4). It prefers these plants and gets attracted to the plants. It starts laying eggs on the top of the marigold flowers. Farmers pluck such flower and sell those in the market. Thus, practicing this knowledge reduces pest load.

Figure 29.4: Marigold as a Boarder Crop in Cotton

Garlic and *Chilli* Extract to Control Pests

Use of garlic and *chilli* extract to control pests in cotton is in practiced by using 500g garlic and red *chillies.* The extract is added to 100 litres water, mixed with 200g soap and sprayed on cotton to control sucking as well as other pests. This practice is in vogue in Dhule and Ahmednagar districts of Maharastra (Inventory of Indigenous Knowledge 2003).

Mixture of Ash, Cow Urine, Kerosene and Soap Against Aphids and Jassids

Dusting of ash, spraying of cow urine and kerosene + soap mixture for control of aphids, jassids and whitefly on vegetable crops and cotton The tribal farmers of Nandurbar and Ahmednagar districts of Maharashtra treat the seeds of vegetable crops by dusting ash and spraying of cow urine. It is very rarely applied on *chilli,* brinjal and cucumber to control aphids and jassids. Some farmers spray cow urine in cotton for control of aphids and jassids. In tribal areas some farmers also spray kerosene + soap on cotton crop to control whitefly attack. This is an economic practice and the cost involved is much less as compared to chemical insecticides, explored by Mahatma Pule Krishi Vidyapeeth, Rahuri, Ahmednagar, Maharastra.

White Fly Control in Cotton by Jaggery Solution

Some farmers use gur (jaggery) solution to control white fly and for this they take 1 kg gur and dissolve it in 10-12 ltrs of water and filter through a thick cotton cloth. Approximately 5-6 ltrs solution is sprinkled in one acre with the help of sprayer pump gently on the foliage. The tiny insects of white fly stuck on the leaf surface due to stickness of the solution. All the stuck eggs of white fly are kileed on the leaf surface which increase the possibility of occurrence of Madhiya disease. Hence they prefer to use this practice once only. This practice can achieve 40-50 per cent control of white fly. This technique is used in the months of November-December in Bharuch, Gujarat (Ahmedabadi 1992).

Control of Bollworm in Cotton by Extract of *Mahua* and Tamarind

Bollworms like spotted bollworm and pink bollworm of cotton crop attack cotton plants frequently. These pests can be controlled by making a stock solution of 500ml taking equal amount of *mahua* (*madhuca indica*) and tamarind (*Tamarindus indica*) bark. This solution is filled into a pump of 15 ltr capacity and sprayed on the infested crop. If the attack is severe then the same solution is sprayed once again after a week. If it is used more than twice then there are chances of root burning (Kodarvi 1991). This practice has been used for over 30 years in Banaskantha in Gujarat, however, this needs to be used judiciously.

Control of *Spodoptera litura* and *H. armigera*

To combat the menace of *Spodoptera litura* and *H. armigera,* farmers use 4 kg of *Kuvarpathu* (*Aloe vera*), 50 ml of neem oil and 500 g of tobacco powder by boiling together these ingredients in 20 lts water for 3-4 hours to make a 5-ltr solution.Then 50 g of aritha (*Sapindus emarginatus*) powder is added to the solution and mixed thoroughly. It is applied @ 100-150 ml/15 lts water on the crop and it is practiced in Northern Gujarat region (Patel 2001).

Ratooning of *Kharif* Cotton

Some farmers in pimpri village of Dhule Districts in Maharashtra have started taking ratoon crop of Kharif cotton. The farmers having irrigation facilities take the ratoon. It is taken after 3-4 pickings are done. The crop is provided with chemical fertilizers 2-3 irrigations are given till February–March. This practice is followed to obtain additional returns from the same crop. An additional 10-15 q/ha is obtained from the same crop if it is properly managed. The practice has been in use for 8-10 years. About 10-15 farmers are practicing this technology. This has been opined by Prof. P. S. Patil, Training Organiser, Krishi Vigyan Kendra, Dhule, Maharashtra.

Shaking of Pigeon Pea Plants for Managing *H. armigera*

Shaking of pigeon pea plants and mass collection of *Helicoverpa* for their destruction is practiced to control population of *Helicoverpa* (Figure 29.5).

Fungal Disease Control in Cotton and Chillies

Fungal disease control in chillies, cotton and also lemon fruits is controlled by spraying a mixture, locally prepared called Panchagavya. The mixture is prepared by mixing 5 parts milk, 2 parts curd, ½ part ghee, 2 parts cow urine, ½ part cowdung and 10 parts water. The mixture is kept for 7 days before it is sprayed in field where fungal disease is predominant. Spray of the mixture also controls leaf shedding and enhances flower setting. Ten farmers of Periakovilankulam village follow this practice for the last three years, opined and evidenced by Shri. S. Murugesan Periakovilankulam Sankarankovil, Tirunelveli, Tamil Nadu.

Jaggary to Check Pod Borers and Pod Sucking Bugs in Pigeonpea

To check pod borers and pod sucking bugs in pigeon pea, many farmers in Nanded district of M.S. dissolve 1kg jaggary in 40 litres of water and use the formulation as spray on pigeon pea plants when they observe eggs of pod borer and pod sucking bugs. The sweet taste of the formulation remains on plants for around one week, which is quite able to attract big sized ants, which prey on these eggs. Thus, the visiting ants check the population of pod borers and pod sucking bugs naturally (Figure 29.6). The formulation is sufficient for an area of 1 acre and according to the density of eggs more jaggary may be added in water and vice versa. (Kanojia *et al.* 2002 and Kanojia *et al.* 2005).

Control of Cutworm in Cotton

Citrullus grandiflora (periya kumuttikai), *Cissus quadragularis* (perandi kodi), Tecoma stans (yellow arali seeds) and *Azadirachta indica* (neem leaves/seeds) are ground and pasties prepared. The paste is allowed to ferment as such for 10 days. Paste from *Sesbania grandiflora* is made separately and fermented. After 10 days, the fermented pastes are filtered though a thin cloth, diluted with required quantity of water and sprayed over the foliage. This controls the cutworms in cotton very efficiently with no side effect evidenced by Shri S. Pidhai Muniandi Koil Street, Jayanangalam Periyakulam, Theni, Tamil Nadu.

Suggested Methodology for Validation of ITKs

There are some ways to be followed while dealing with indigenous knowledge and the broad lines of such ways have been given as suggested methodologies, these steps are very handy and beneficial for scientifically validation of the associated and claimed potentials of ITKs. The given steps are self explanatory and need not be explained in detail:

- Exploration of the concerned ITKs
- Documentation
- Selection of crop
- Lab validation of crude formulation against major insect pests and diseases of the concerned crop
- Dose calculation
- Field validation at experimental plots of crude formulation including pest monitoring using pheromone traps and agronomical practices
- Lab validation of organic extracts
- Economics of the formulation
- Validation at farmer's field
- Large scale validation
- Recommendations for incorporation in national programmes for pest management
- Commercialization of the formulation
- Benefits to farming communities, nations, environment etc.

Validation of the Explored ITKs

This has also been suggested and emphasized in previous pages of this chapter that validation of ITKs in laboratory before taking them to fields for the claimed potential regarding their efficacies is important. It involves selection of target insect pests and disease causing damages to the concerned crop. In the following example laboratory testing of one ITK has been explained and such testing opens new avenues for further quest. In this example validation of ITK in laboratory against *Rhizoctonia solani* and *Rhizoctonia bataticola*, is presented.

An insect control decoction, which is prepared from cow urine and Neem leaves by putting neem leaves 25kg in cow urine 10ltrs and kept in an earthen pot underground for 21 days. Fermented solution is then boiled in a copper utensil vigorously to make one fourth. This formulation has been taken for the mentioned test organisms. Its efficacy shows that it can be used for seed treatment of cotton and other crops to protect against seed rots and seedling mortality (Kanojia *et al.* 2005).

Figure 29.5

Figure 29.6

Test organism were: *Rhizoctina solani* and *Rhizoctonia bataticola*

Treatment: Control 0 ml/ltr PDA(Potato dextrose agar)

10 ml/ltr PDA

25 ml/ltr PDA

50 ml/ltr PDA

100 ml/ltr PDA

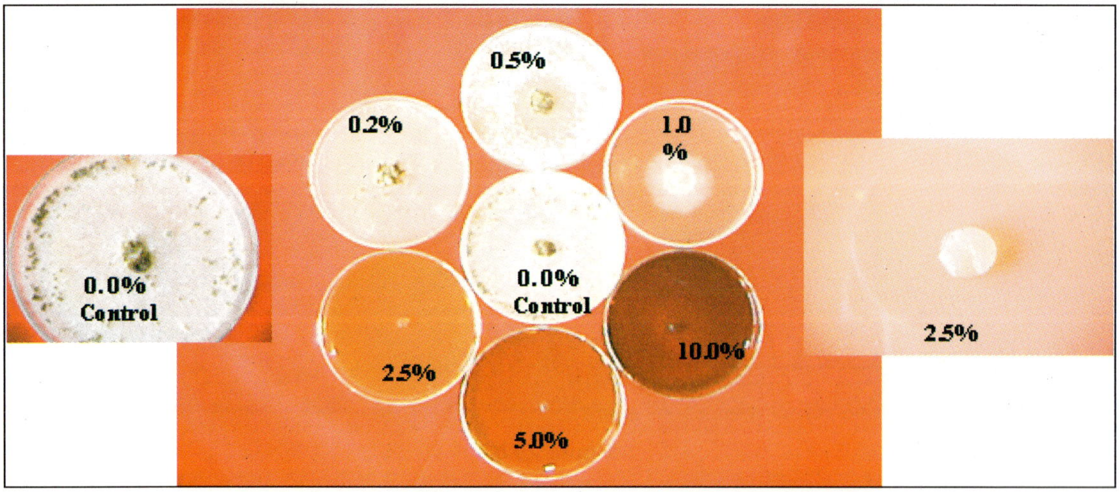

Figure 29.7: Formulation at 0.0 , 0.2, 0.5, 1.0, 2.5, 5.0 and 10.0 per cent Concentration in PDA Against *Rhizoctonia solani*

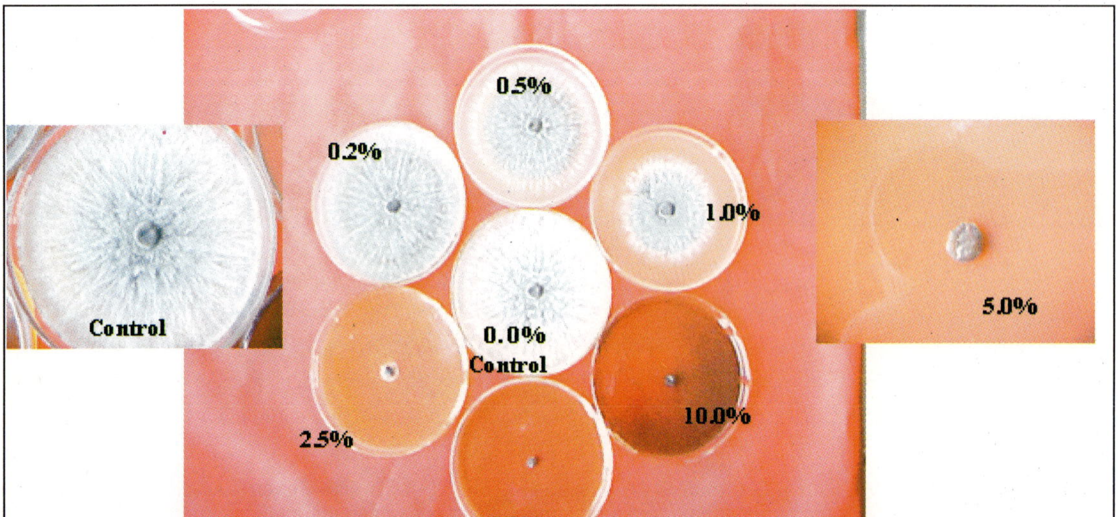

Figure 29.8: Formulation at 0.0 , 0.2, 0.5, 1.0, 2.5, 5.0 and 10.0 per cent Concentration in PDA Against *Rhizoctonia bataticola*

Through this testing it was found that at 1 per cent concentration the formulation started checking the growth of *Rhizoctina solani* and controlled it cent percent at 2.5 per cent (Figure 29.7), while in the case of *Rhizoctonia bataticola*, cent percent inhibition was observed at 5 per cent (Figure 29.8).

Conclusion

This chapter on the discussed indigenous technical knowledge is an effort to provide such information to facilitate scientists, extension workers, planners, farmers and administrators for harnessing the potential by utilizing suitable ITKs as valuable components in the national IPM programme for the sustainability of Indian agriculture and alleviation of rural poverty. It is expected to enhance knowledge base and stir up a light of interest in simple and small tips of indigenous origin for managing in pests and provide new dimension in the generation of IPM technologies at national level abridging the knowledge and the communication gaps between researchers and community involved in pest management in cotton based cropping system.

An escape from rich indigenous knowledge systems may not be spared and wise by treating them outdated and unscientific just because being practiced by the resource poor and under privileged farmers belonging to the underdeveloped regions. Since the ITKs are a Basket of sustainable technological options rather than fixed packages and therefore these may be promoted to strengthen the ongoing pest management programmes in the country and else where creating an environment of respect for local people engaged in practicing indigenous knowledge using sound and thorough knowledge of local inhabitants with locally and easily available resources and home made preparation and reducing the social distance by interacting frequently ofr greater involvement and further improvement protecting the agricultural environment. The available information on indigenous technical knowledge in managing and controlling pests of crops must be explored and documented.

References

Ahemdabadi, D.K., 1992. White fly control. *Honey Bee*, 3(1): 7.

Balasubramanian, K., 2000-2001. Beyond clinical pesticides in cotton. *Honey Bee*, 11(4) and 12(1): 21.

Chary, M.S., 2000. Traditional innovations for sustainable pest management. *Proceeding of National Symposium on Sustainable Agriculture Production: Lesson from Traditional Technologies*. Jan-30-31, Vigyan Bhawan, New Delhi, pp. 24-29.

Flavier, J.M., 1995. The regional programme for the promotion of indigenous knowledge in Asia, pp. 479-487. In: Warren, D.M., slikkerveer, L.J. and Brokensha, D. (eds.). *The Cultural Dimension of Development: Indigenous Knowledge Systems*. Intermediate Technology Publication, London

Hobart, M. (ed), 1993. *An Anthropological Critique of Development: The Growth of Ignorance*. Routledge, London.

Inventory of Indigenous Knowledge, 2003. Use of *chilli* and garlic extract for *helicoverpa* management in cotton. Pest and Disease-code1804. *Inventory of Indigenous Knowledge*, pp. 156, ICAR, New Delhi.

Jani, B., 1992. *Calotropis gigantia:* Caterpillars in cotton. *Honey Bee*, 3(3 and 4):17.

Joshi, C. P. and Singh, B.B., 2006. Indigenous Agricultural Knowledge in Kumaon hills of Uttaranchal. *Indian Journal of Traditional Knowledge*, 5(1): 19-24.

Kaachhadiya, D.D., 1998. To control green worm (*Heliothis* spp.) in cotton crop. *Honey Bee*, 9(3): 15-16.

Kalyanbhai, K.R., 1995. Control of cotton and chili pest complex. *Honey Bee*, 6(1):13.

Kanojia, A.K., Singh, R.V., Kumar, Surender, Bambawale, O.M. and Trivedi, T.P., 2001. Exploration, documentation and mapping of ITKs in pest management. *Paper Presented in the Seminar on Vedic Agriculture,* held at Rajasthan Agriculture University, ARS, Durgapura, Jaipur on August 10, organized by Asian Agri-History Foundation, Udaipur (Raj.) and published in the Abstr., pp 28-29.

Kanojia, A.K., Singh, R.V., Bambawale, O.M. and Trivedi, T.P., 2002. Exploration, documentation and mapping of indigenous technical knowledge in pest management in cotton based cropping system in Central India. *Annual Report 2001-02,* NCIPM, New Delhi, pp. 14-15.

Kanojia, A.K., Singh, Amerika, Singh, R.V. and Kumar, Surender, 2001. Indigenous Technical Knowledge in Integrated Pest management. *Paper Presented in the Seminar on Vedic Agriculture,* held at Rajasthan Agriculture University, ARS, Durgapura, Jaipur on August 10, organised by Asian Agri-History Foundation, Udaipur (Raj.) and published in the Abstr., pp. 30-32

Kanojia, A.K., Singh, R.V., Bambawale, O.M. and Trivedi, T.P., 2005. Explored Indigenous Technical Knowledge in Pest Management. *Tech. Bulletin, National Centre for Integrated Pest Management,* New Delhi, pp. 3-5, 17-18 and 28.

Knowledge and Policy, 1995. *The International Journal of Knowledge Transfer and Utilization,* Spring, 8(1): 63-73.

Kodarvi, A.D., 1991. Tamarind and *Mahua* scare pests of cotton. *Honey Bee,* 10(3): 14.

McSpadden Gardener, B.B. and Fravel, D.R., 2002. Biological control of plant pathogens: Research, commercialization, and application in the USA. Online. *Plant Health Progress* doi:10.1094/PHP-2002-0510-01-RV.

Murugesan, P., 1994. White fly control in cotton. *Honey Bee,* 5(1):15.

Paliwal, M., Das, S.B. and Veda, O.P., 2002. Indigenous technology for effective pest management of cotton-adopted by tribal farmers of West Nimar, Madhya Pradesh. *International Seminar on Traditional Knowledge, Health and Environment,* 23-24 February at Bhubaneshwar, Abs., p. 42.

Patel, M.H., 1991. Plant Protection measures aphids castor oil. *Honey Bee,* 2(1):16.

Patel, R.K., 2001. Controlling worms in crop. *Honey Bee,* 12(2): 9

Patel, U.T., 1991a. Cotton pest complex–spraying of botanical pesticide. *Honey Bee,* 2(1):15.

Sanghani, A.T., 1998. Increasing crop yield. *Honey Bee,* 9(1):15.

Thakor, D.D., 1991. Aphids- leaves and twigs of *Calotropis* in irrigation channel. *Honey Bee,* 2(1):16.

Thomas Heyd, 1995. *Indigenous Knowledge, Emancipation and Alienation.*

Udikeri, S.S., Patil, S.B., Krishnanaik, L., Rachappa, V., Farook Nimbal and Guruprasad, G.S., 2007. Poncho 600 FS: A New Seed Dressing Formulation for Sucking Pest Management in Cotton. *Karnataka J. Agric. Sci.,* 20(1): 51–53.

Warner, K., 1991. *Shifting Cultivators: Local Technical Knowledge and natural Resource management in the Humid Tropics.* FTP. Community Forestry Note 8. FAO. Rome.

Warren D.M., 1991. *Using Indigenous Knowledge in Agricultural Development.* The World Bank, Washington DC.

Chapter 30

Predators and Parasitoids in Pest Management

S.K. Jalali
Project Directorate of Biological Control, Post Bag No. 2491, H.A. Farm Post, Bellary Road, Bangalore – 560 024, Karnataka

Introduction

Predators and parasitoids are one of the important components of pest management and have been proved to be an excellent alternate to chemical pesticides in many ecosystems. In the field, both natural enemies in combination of abiotic factors are responsible in reducing the pest population by up to 80 per cent. However, predators and parasitoids are active in the field under natural conditions only after the damage to crop has already been done. The history of biological control dates back to the seventeenth century and since then a great deal of information has been generated on biological methods of pest suppression. Biological control has been accepted as an important component of Bio-intensive Integrated Pest Management (BIPM). Much of the information developed on BIPM remains at the level of researchers and information flow from researchers to farmers has been neglected. Unfortunately, biological control has not been practised to the extent it should be due to various reasons and the voluminous research findings have not percolated down to the farmers' level. The folly of sole dependence on pesticides has been realised the world over and the concern for preserving a cleaner world for posterity has increased. Still, it has been a challenge to make biological control work in the field for the benefit of the small and marginal farmers, especially in the developing world.

Available Natural Enemy Diversity

India is one of the rich biodiversity countries with rich wealth of natural enemies of crop pests. It is not possible to list each of the natural enemy but attempt has been made to list important ones with their reported status. The information is provided for important predators and parasitoids of crop pests on few selected crops like sugarcane, rice, maize, cotton, tobacco, fruits, vegetables and plantation crops in Tables 30.1–30.8.

Predators and Parasitoids in Pest Management

Sugarcane

Biological control has gained maximum acceptance among sugarcane farmers of India. Sugarcane crop is infested by number of borers and sucking pests from sowing to harvest. Use of *Trichogramma* species for the management of sugarcane borers has been followed in a major way. Sugar mills have their own co-operative parasitoid production units and have contributed in a big way in adoption of biological control. Doomra *et al.* (1994) recorded least infestation of *Scirpophaga excerptalis* and highest yield in areas where *T. japonicum* was released @ 50,000/ha/10 days during April–June with mechanical control of top borer infested shoots compared to furadan treated areas in Punjab. Some of the earlier work for two decades on the releases of *T. chilonis* evolved around *Emmalocera depressella* and highly satisfactory control was achieved (Narayanan and Mukherji, 1953). Gupta (1951) reported that releases of *T. chilonis* controlled *E. depressella* in UP. Similarly for *Chilo sacchariphagus indicus* releases of *T. chilonis* @ 40,000/ha at fortnightly interval six times resulted in lower borer infestation in treated area (Kalyanasundram *et al.*, 1993). Varma *et al.* (1991) released *T. chilonis* in an area of 500 hectares and reported highly significant results in reducing incidence of *C. auricilius* in Punjab. Release of *Sturmiopsis inferens* has decreased the population of *C. infuscatellus* in Tamil Nadu, Orissa and on *Chilo auricilius* in Haryana (David, 1980). Similarly inoculative releases of *Isotima javensis* have given good results in the control of top borer in many parts of the country.

Table 30.1: Natural Enemies of Sugarcane Pests*

Pest	Natural Enemies	
	Number Recorded	Status
Chilo infuscatellus	50	*Trichogramma* spp. parasitised 4–5 per cent eggs, *Cotesia flavipes* 4–14 per cent and *Sturmiopsis inferns* 6–26 per cent larvae.
C. sacchariphagus indicus	39	*Trichogramma* spp. parasitised 2–78 per cent eggs, *C. flavipes* 4–14 per cent.
C. tumidicostalis	09	*Trichogramma* spp. parasitised 30–71 per cent eggs, *C. flavipes* 16–35 per cent.
C. auricilius	33	*Cotesia flavipes* parasitised 10–14 per cent and *Campyloneurus mutator* 60–78 per cent larvae
Scirpophaga excerptalis	61	*Trichogramma* spp. parasitised 4–5 per cent and *Telenomus* spp. 5–90 per cent eggs, *C. flavipes* 4–14 per cent larvae and *Isotima javensis* 5–40 per cent pre-pupae.
Acigona steniellus	18	*Cotesia flavipes* 2–4 per cent and *Stenobracon nicevillei* 2–33 per cent larvae, *Tetrastichus* spp. 16 per cent pupae.
Pyrilla perpusilla	31	*Ooencyrtus papilionisi* parasitised 71–79 per cent and *Tetrastichus pyrillae* 3–41 per cent eggs and *Epiricania melanoleuca* 15–100 per cent nymphs and adults
Melanaspis glomerata	23	*Adelencyrtus mayurai*, *Botryoideclava bharatiya*, *Chilocorus nigrita*, *Pharoscyumns horni* are key parasitoids and predators
Holotrichia serrata	23	*Scolia* sp., *Campsomeriella collaris* are key natural enemies
Ceratovacuna lanigera	25	*Dipha aphidivora* and *Micromus igrotus* are key predators

*: *Source*: David *et al.* (1983); Pawar (1979); Singh (1993).

Amongst sucking pests, notable success has been achieved in the bio-suppression of the hopper, *Pyrilla perpusilla* in some states by the colonisation/redistribution of the lepidopteran parasitoid, *Epiricania melanoleuca*. Misra and Pawar (1983) released 8000–12000 cocoons or 800000–1000000 eggs/ha when pest density was 5–7 individuals/leaf and obtained significant control all over India. The releases of *Chilocorus nigrita*, *Pharoscymnus horni* and *Sticholotis madagassa* @ 800 adults/acre reduced infestation by 48 per cent (Ansari *et al.*, 1989).

Maize

Chilo partellus is most destructive pest of maize across the country. The release of *Trichogramma chilonis* @ 100,000/ha resulted in parasitism of *C. partellus* eggs to the tune of 75.2 and 62.6 per cent in 1st and 90.4 and 78.4 per cent in 2nd generation egg laying in three and five days' interval parasitoids releases plots. Stem tunnelling by larvae was significantly less as compared to untreated control plots. Considering cost factor, releases of *T. chilonis* at 5 days interval 3 times in 1st and 2 times in 2nd generation of the pest is the best treatment for effective control (Jalali, 2000).

Table 30.2: Natural Enemies of Maize Pests*

Pest	*Natural Enemies*	
	Number Recorded	*Status*
Chilo partellus	32	*Trichogramma chilonis* parasitised 29–38 per cent eggs. *Cotesia flavipes* parasitised 5–35 per cent larvae during June–July in TN, up to 100 per cent hibernating larvae during November to February in UP. *Stenobracon deesae* parasitised 20 per cent larvae and *S. nicevillei* parasitised 41 per cent larvae. An unidentified tachinid fly is reported to parasitised 50 per cent larvae.

*: *Source*: Katiyar (1962); Chandy (1955); Subba Rao *et al.* (1969); Issac (1941); Butani (1957); Saxena (1972); Jalali and Singh (2003)

Rice

In the rice growing areas in India, yellow stem borer (*Scirpophaga incertulas*) and leaf folder (*Cnaphalocrocis medinalis*) and leafhoppers have been recorded as the major pest constraints.

Among the borers *Scirpophaga incertulas* (Walker), *Chilo* spp. and *Sesamia inferens* Walker are the important pests. For the biological suppression of borers, conservation of indigenous natural enemies and inundative release of egg parasitoid *T. japonicum* at 50,000/ha/week during the egg laying periods is advocated. The release start when I egg mass or adult/m^2 or first moth on light trap is recorded. With the release of *Trichogramma* the overall egg parasitism increases considerably.

For leaf and plant hoppers, colonization of mirid predator *Cyrtorhinus lividipennis* for which now rearing technique is available (over 1,500 predators could be reared on 1cc *Corcyra* egg), may prove effective if releases are carried out @ 100 mirid bugs or 50-75 egs/m^2 at 10 day interval. Weeds like *Cyperus* sp. help in off-season survival of mirid bug through harbouring plant hoppers. Predation by mirid bug was more on BPH resistant rice variety PTB 33. The presence of any combination of 3 nos./ hill of spider *Lycosa preudoannulata*, *Oxyopus javanus* and *Tetragnatha* sp. checked the population of BPH and WBPH.

Cnaphalocrosis medinalis (Guenee) and *Pelopidas mathias* (Fabricius) are the important leaf feeders recorded on paddy. As far as the augmentation is concerned only inundation of *Trichogramma chilonis* has been tried and it has not only increased the natural parasitism of eggs but also the overall population of parasitoids and predators in the rice ecosystem.

For the suppression of rice gall midge *Orseolia oryzae* (Wood Manson) conservation of parasitoid *Platygaster oryzae* Cameron and use of gall midge resistant varieties are recommended. Unfortunately techniques for mass multiplication of *P. oryzae* are not available otherwise augmentative release of the parasitoid during the early part of the crop growth, could be attempted Gunathilagaraj and Ganesh Kumar, (1997).

Table 30.3: Natural Enemies of Rice Pests*

Pest	*Natural Enemies*	
	Number Recorded	*Status*
Scirpophaga incertulas	72	*Trichogramma* spp. and *Telenomus* spp. are common egg parasitoids. *Ipobracon* sp. parasitised 7–20 per cent, *Isotima javensis* 7–20 per cent.
Cnaphalocrocis medinalis	40	*Trichogramma* spp. and *Macrocentrus* spp are common parasitoids.
Pelopidas mathias	41	*Apantelea* sp. parasitised 23 per cent, *A.* baoris 33 per cent, *Rhysipolis hesperidis* 45 per cent, *Charops bicolor* 30 per cent and *Halidaya* luteicornis 52 per cent larvae.
Leaf and plant hoppers	73	*Anagrus* spp. and *Oligosita* spp. are most common parasitoids of plant hoppers. In AP, *Gonatocerus* sp. and *Paracentrobia andoi* are common on leafhoppers. *Haplogonatopus* sp., *Echthrodelphax fairchildii*, *Elenhus* sp. and *Hexamermis* sp. are the most common nymphal agents. *Cyrtorhinus lividipennis*, *Coccinella arcuata*, *Micraspis discolor*, *Microvelia astrolineata* and atleast 10 species of spiders are also common in paddy ecosystem.

*: *Source*: Israel and Padmanabhan (1976); Rao *et al.* (1970); Manjunath (1990); Bentur and Kalode (1985a); Manjunath *et al.* (1978b); Samal and Misra (1975)

Cotton

Cotton is another crop, which suffers from number of insect pests, major ones being bollworms, sucking pests and leaf eating caterpillar. Therefore management becomes very difficult because these pests are associated with the crop till harvest. The farmers spray approximately 10 rounds in rainfed cotton and 20 rounds of insecticide in irrigated cotton in India. Therefore attempts were made to develop BIPM schedules. BIPM schedule comprising the use of trap crop, releases of *Trichogramma chilonis*, *Chrysoperla carnea*, spray of *Ha* nuclear polyhedrosis viruses and neem based formulations integrated with insecticides when tried in Karnataka, Maharashtra and Gujarat has given better results against pest complex compared to insecticidal sprays alone. The results indicated that insecticidal sprays were reduced by half and yield was increased by up to 10 q/ha in BIPM area (Lingappa *et al.*, 2001). These findings could convince the farmers that they could obtain higher returns by adopting BIPM than purely insecticidal control measures.

Table 30.4: Natural Enemies of Cotton Pests*

Pest	*Natural Enemies*	
	Number Recorded	*Status*
Earias spp. (*E. vittella* and *E. insulana*)	27	*Trichogramma* spp parasitised up to 56 per cent eggs. *T. achaeae* parasitised 4–61 per cent, *T. chilonis* 5–15 per cent and *Trichogrammatoidea guamensis* 5–42 per cent eggs. *Rogas aligarhensis* parasitised up to 80.0 per cent larvae, *Bracon greeni*, *Brachymeria nephanitidis* and *Agathis* sp. has often been recorded.
Pectinophphora gossypiella	43	*Trichogramma achaeae* parasitised 6–27 per cent and *Trichogrammatoidea guamensis* 21–25 per cent eggs. *Apanteles angaleti* was the most common larval parasitoid recorded. Ectoparasitic mite, *Pyemotes ventricosus* destroyed upto 30 per cent diapausing larvae in Punjab and 48 per cent in Haryana. *Goniozus* sp., *Apanteles pectinophorae*, *Microbracon gelechidiphaga* and *Chelonus* sp. have been commonly recorded in Tamil Nadu.
Helicoverpa armigera	53	*Trichogramma* spp parasitised up to 56 per cent eggs. *T. achaeae* parasitised 4–61 per cent, *T. chilonis* 5–15 per cent and *trichogrammatoidea guamensis* 5–42 per cent eggs. *Rogas aligarhensis* parasitised up to 80.0 per cent larvae, *Bracon greeni*, *Brachymeria nephanitidis* and *Qagathis* sp. has often been recorded on *Earias* spp. in different parts of the country.
Spodoptera litura	42	*Telenomus rowani*, *Chelonus formasanus*, *Apanteles africanus* and *Peribaea orbata* have been found promising.
Aphis gossypii	34	*Cheilomenes sexmaculata*, *Coccinella septempunctata*, *Scymnus* sp. and *Chrysoperla carnea* are important predators. *Aphelinus* sp., *Aphidencyrtus* sp. and *Aphidius* sp. are important parasitoids.
Bemisia tabaci	17	Among the parasitoids *Encarsia transvena*, and *Eretmocerus mundus* are important, providing 85 per cent parasitism in Tamil Nadu and 54–63 per cent in Maharashtra. *Brumoides suturalis*, *Serangium parcesetosum*, *Scymnus* sp., *Chrysoperla carnea*, *Mallada boninensis*, *Geocoris bicolor and Orius* sp. are important predators.

*: *Source*: Thontadarya and Jai Rao (1978); Sekhaon and Verma (1983); Deshpande and Nadkarny (1936); Sundaramurthy and Basu (1985); Patel (1980)

Tobacco

In tobacco combinations of several natural enemies are tried to control pest complex. An IPM strategy comprising castor as a trap crop, NPV of *S. litura*, neem seed kernel extract, *Bt*, release of *C. carnea*, *Telenomus remus* and *Cotesia africanus* for the management of *S. litura* and the aphid, *Myzus persicae*, has been formulated and found effective in Andhra Pradesh. The results of IPM trial resulted in cost-benefit ratio of 1:1.92–1:2.74 compared to 1:1.52–1:2.25 in insecticidal treatment both in nursery and open field (Anonymous, 1995–2001). In Karnataka, BIPM was compared to a recommended insecticidal approach. In BIPM castor was grown as trap crop, 0.05 per cent quinalphos was sprayed on 30th day after transplanting, *Sl* NPV was sprayed on 45th day and NSKE 1.0 per cent on 60th day after transplanting. In the chemical control, 3 sprays of quinalphos were given. The yield obtained was 1311 kg/ha in BIPM compared to 980 kg/ha in insecticidal treated area (Shivayogeshwara *et al.*, 1991). Tobacco IPM has been successfully field demonstrated in farmers' fields in Andhra Pradesh.

Table 30.5: Natural Enemies of Tobacco Pests*

Pest	*Natural Enemies*	
	Number Recorded	*Status*
Spodoptera litura	42	*Telenomus rowani*, *Chelonus formasanus*, *Apanteles africanus* and *Peribaea orbata* gave 24 per cent larval parasitism.
Helicoverpa armigera	06	*Chelonus heliopae*, *Campoletis chlorideae* and *Carcelia illota* parasitised 64 per cent larvae.
Scrobipalpa heliopa	03	*Chelonus heliopae* parasitism reached 100.0 per cent in early August in Gujarat. *Cyrtopeltis tenuis* was a key natural enemy during winter.
Myzus persicae	25	None important under nursery conditions.
Bemisia tabaci	34	Among the parasitoids *Encarsia* sp., *Eretmocerus mundus* and *Amitus aleurolobi* and *Aphelinus flavipes* are common. *Serangium parcesetosum*, *Apertochrysa* sp. and *Geocoris ochropterus* are most commonly recorded.

* Source: Chari and Rao (1990); Rao *et al.*, (1981); Patel *et al.* (1959); Chari *et al.* (1982).

Table 30.6: Natural Enemies of Coconut Pests*

Pest	*Natural Enemies*	
	Number Recorded	*Status*
Opisina arenosella	41	*Apanteles taragamae*, *Bracon hebetor*, *Goniozus nephantidis*, *Elasmus nephanitidis*, *Brachymeria nosatoi*, *B. nephantidis*, *Xanthopimpla punctata* and *X. nana* were the common parasitoids. *Parena nigrolineata* and *Calleida splendidula* and several spiders were predatory on larvae. *Goniozus nephantidis*, *E. nephanitidis*, *B. nosatoi* were particularly effective in the summer.
Oryctes rhinoceros	45	Predators such as *Santalus parallelus*, *Scarites* sp., *Harpalus* sp., *Pheropsophus* spp. and *Agryphus* sp. were frequently associated in the breeding sites.

*: *Source*: Pillai (1985); Singh (1993).

Fruits

Apple

Quadraspidiotus perniciosus is one of the most serious pests of apple and many other deciduous fruit trees in Kashmir, Himachal Pradesh and Uttar Pradesh. A number of coccinellid beetles such as *Chilocorus infernalis* (*C. bijugus*), *C. rubidus*, *Pharoscymnus flexibilis* and *Sticholotis marginalis* have been recorded feeding on this pest, but these beetles are unable to give the desired natural control. These beetles have been introduced from Jammu and Kashmir to Thanedar area of Himachal Pradesh. Both the predators have established and provided reasonable control of San Jose scale (Rao *et al.*, 1971; Singh, 1989).

In 1958 and 1960, three strains of *Encarsia perniciosi viz.*, Californian, Russian and Chinese, were introduced. In addition, *Aphytis diaspidis* (origin: Japan) was introduced from California. All the strains were established; the Russian strain of the parasitoid gave 89 per cent parasitism in Himachal Pradesh. *A. diaspidis* in combination with *E. perniciosi* gave 86.5 per cent parasitism. In the beginning, low parasitism by Chinese strain was attributed to small number of parasitoids released. In Kashmir, the Russian and Chinese strains appeared to be superior to Californian and Ilinois strains. American and Chinese strains of *E. perniciosi* were also released in the Kumaon hills of Uttarakhand; the population of the pest was reduced by about 95 per cent. In Kashmir, releases of *E. perniciosi* and *A. proclia* resulted in an increase of parasitism from 8.9 to 64.3 per cent. Studies on the biology of *E. perniciosi* revealed that the multiplication rate of the parasitoid was over 10 times. In apple, release of *Encarsia perniciosi* or *Aphytis proclia* @ 2000/infested tree and *Chilocorus infernalis* @ 20/infested tree gives effective control of San Jose scale (Rao *et al.*, 1971; Singh, 1989).

In India, the first consignment of the exotic parasitoid *Aphelinus mali* to control woolly aphid was imported in 1930 from England. The parasitoid was not effective in Kumaon hills due to the activity of *Coccinella septempunctata,* which fed on parasitised and unparasitised aphids alike and provided nearly complete control. But the parasitoid was quite effective in Kullu valley of Himachal Pradesh. The parasitoid has since then spread to Kashmir valley. *A. mali* was also released around Shimla and it has established well. However, the current status of this parasitoid indicates that it gives good control in valleys, but is less effective on slopes. It has established in most of the woolly aphid infested orchards (Rao *et al.*, 1971; Singh, 1989).

Cydia pomonella is a serious pest of apple and other fruits in Ladakh district of Jammu and Kashmir. Two exotic egg parasitoids *Trichogramma embryophagum* and *T. cacoaciae pallidum* were evaluated. Based on the results, weekly releases of the egg parasitoids are recommended. A survey for two consecutive years brought out the need to initiate control measures by monitoring the pest by large scale use of pheromones, conservation of natural enemies, synchronization of the first release of *T. embryophagum* with the first appearance of the moth, and subsequent releases at weekly intervals at 2000 per tree. Fortunately, the codling moth has not crossed the Zojila Pass and it may be possible to check the pest in Ladakh region itself if proper measures are adopted (Rao *et al.*, 1971; Singh, 1989).

Indian Gypsy moth, *Lymantria obfuscata* is a common pest of apple, poplar, willow, walnut, peach, plum, cherry, etc. in Kashmir. It has been found to cause severe damage to many tree and fruit crops in temperate regicns of Himachal Pradesh and Uttar Pradesh. Egg parasitoid *Anastatus kashmirensis;* larval parasitoids *Glyptapanteles indiensis, G. flavicoxis, G. liparidis, Cotesia melanoscela, Hyposoter lymantriae, Rogas indiscretus, Exorista rossica, Palexorista conspicua;* pupal parasitoids *Brachymeria intermedia* and *B. lasus* are the dominant parasitoids of Indian gypsy moth. In addition to these pupal parasitoids, *Pimpla laothoe, Theronia atalantae* and *Calosoma himalayanum* have been recorded as population regulatory factors of the Indian gypsy moth (Rao *et al.*, 1971).

Amblyseius finlandicus was the most common predator on mite *Tetranychus urticae* attacking apple in Kashmir. The life cycle of the host synchronised with that of the predator (Rao *et al.*, 1971).

Citrus

On citrus butterfly *Papilio demoleus*, egg parasitoid *Trichogramma chilonis* parasitised up to 75.86 percent and *Telenomus* sp. nr. *incommodus* 78 per cent in February (Krishnamoorthy and Singh, 1988; Jalali and Singh, 1990). *Distatrix papilionis* is the dominant parasitoid of caterpillars and *T. chilonis*, *T. incommodus* and *D. papilionis* caused a cumulative parasitism of 88 per cent (Krishnamoorthy and Singh, 1988). Most of the parasitoids of *P. demoleus* also attack another common species of butterfly *P.*

polytes L. in other places. *T. chilonis, Melalophacharops* sp. and *D. papilionis* could be utilised for the biological suppression of butterflies attacking citrus.

Table 30.7a: Natural Enemies of Fruit Crop Pests–Citrus*

Pest	Natural Enemies	
	Number Recorded	Status
Papilio demoleus	16	*Melalophacharops* sp. and *Apanteles papilionis* were the most important larval and *Holocogopa coelopyga* and *Pteromalus puparum* pupal parasitoid. *Trichogramma chilonis* and *Telenomus* sp. nr. *Incommodus* parasitised 75.0 and 78.0 per cent eggs in February.
Diaphorina citri	15	*Tamarixia radiate* was the most important nymphal parasitoid, parasitising 95 per cent.
Coccus viridis	23	*Aneristus ceroplastae* and *encyrtus lecaniorum* parasitised 43.0 per cent nymphs in July.
Aonidiella citrina	04	*Pharoscyumns horni* is most important predator.
Aleurocanthus woglumi	16	*Encarsia divergens* and *E. merceti* provided 90.0 per cent parasitism in Gauahati. *E. clypealis* 80.0–90.0 per cent in Pune, *E. clypealis* and *Amitus hesperidium* 80.0 per cent in Bangalore.
Dialeurodes citri	11	*Encarsia lahorensis* has been recorded as the key parasitoid.
Eutetranychus orinetalis	13	*Amblyseius aerialis, A. finlandicus* were most common predatory mites.
Planococcus citri	11	*Cryptolaemus montrouzieri, Leptomastix dactylopii* and *Coccidoxenoides perigina* are most important natural enemies.

*: *Source*: Singh (1985); Krishnamoorthy and Singh (1988).

The eggs of fruit sucking moth, *Othreis fullonia* (Cl.) are successfully parasitised by *Trichogramma chilonis*, which suggests the possibility of utilising *T. chilonis* for the control of this pest (Dodia *et al.*, 1986).

Coccus viridis can be controlled by releases of *Cryptolaemus montrouzieri* @ 10 beetles/tree. *Aneristus ceroplastae* and *Encyrtus lecaniorum* caused 43 per cent parasitism in July (Tandon, 1985). These can be utilised to achieve control of this pest.

Mealybug *Planococcus citri* can be controlled in the fields by releasing *C. montrouzieri* and preventing patrolling of ants on the plants by applying 5 per cent diazinon granules about one foot around the trunk. Ten beetles (starved overnight) per plant were released 24 hours after the application of diazinon. The impact was visible within a week and by the 5th week the population of mealybugs was reduced to a negligible level (Singh, 1989). Releases of the predator may have to be repeated annually whenever the mealybugs appear.

An exotic parasitoid, *Leptomastix dactylopii* was introduced from the West Indies in 1983. It is a fairly specific parasitoid of *P. citri*, possessing excellent host searching ability. Field release of *Leptomastix* resulted in its establishment in mixed plantations of citrus and coffee, and also in citrus orchards in several parts of Karnataka, resulting in control of *P. citri* within 3-4 months. No insecticidal sprays were required subsequently for the control of *P. citri* in the following season (Manjunath, 1985; Krishnamoorthy and Singh, 1987; Nagarkatti *et al.*, 1992).

Over a dozen parasitoids have been recorded on the mealybug *Nipaecoccus viridis*. However, *Anagyrus dactylopii* was dominant, parasitising up to 90 per cent in the field (Ali, 1957; Subba Rao *et al.*, 1965). A severe infestation of *N. viridis* on citrus was completely controlled by releasing *C. montrouzieri* @ 10 beetles per tree in Andhra Pradesh (Tirumala Rao and David, 1958).

Icerya purchasi was reported to be under control by the introduced ladybird predator *Rodolia cardinalis* (Rao and Kamath, 1966). Release of 8-10 beetles on each moderately infested plant has been suggested whenever required.

Guava

The striped mealybug *Ferrisia virgata* is a very serious pest of guava in south India. *Cryptolaemus montrouzieri* has controlled the pest effectively within 30 days of its release (Mani *et al.*, 1989). The efficacy of *Cryptolaemus* when released @ 20 adults per tree in controlling green shield scale, *Chloropulvinaria psidii* was demonstrated in farmer's field (Mani and Krishnamoorthy, 1990). In another trial, the parasitoid, *Aenasius advena* and the coccinellid *Scymnus coccivora* complimented the performance of *C. montrouzieri* in the complete control of *F. virgata* (Mani *et al.*, 1989). *Planococcus citri* can be brought under effective control by releasing 10 *C. montrouzieri* per plant (Mani and Krishnamoorthy, 1990e).

Table 30.7b: Natural Enemies of Fruit Crop Pests–Grapevine and Guava*

Pest	*Natural Enemies*	
	Number Recorded	*Status*
Grapevine		
Maconellicoccus hirsutus	14	*Anagyrus dactylopii* provided 62.5 per cent parasitism and *Scymnus coccivora* was most common predator.
Sylepta lunalis	02	Larval parasitoid *Apanteles clita* and the pre-pupal parasitoid *Cardiochiles fulvus* are most common.
Guava		
Aphis gossypii	05	Coccinellid and syrphid predators give complete suppression of the pest in undisturbed orchards.
Chloropulvinaria psidii	09	*Apanteles ceroplastae* is most promising parasitoid. *Cryptolaemus montrouzieri* is most effective predator giving complete control.
Ferrisia virgata	10	*Blepyrus insularis* and *Spalgis epius* are most promising and give complete control.
Planococcus citri	10	*Plesiochrysa lacciperda*, *Spalgis epius* and *Cryptolaemus montrouzieri* controls the pest almost completely.

Source: Mani *et al.* (1987); Singh (1978).

Grape

Maconellicoccus hirsutus has become a major pest in south India in recent years. Release of 10 beetles of *C. montrouzieri* per vine could effectively suppress the grape mealybugs in about 75 days of release (Mani and Thontadarya, 1988a). Dichlorvos, chlorpyriphos and all the commonly used fungicides at recommended concentrations were safe to all the stages *of C. montrouzieri*. Application of these pesticides can be combined with *Cryptolaemus* in the pest management programme (Mani and Thontadarya, 1988b; Babu, 1986).

Cashew

Egg parasitoids, *Telenomus* sp., *Chaetostricha* sp. and *Erythmelus helopeltidis* have been found promising against tea mosquito bug (Singh, 1996).

Arecanut

On arecanut scales (*Chrysomphalus* spp. and *Lepidosaphes* spp.), extensive releases of *Chilococrus nigrita* have shown promise (Singh, 1996).

Cardamom

On shoot and capsule borer of cardamom (*Conogethes punctiferalis*), *Friona* sp., *Eriborus trochanteratus*, *Agrypon* sp. and *Xanthopimpla australis* are important parasitoids (Singh, 1996).

Vegetables

Cabbage

Cotesia plutellae is the most common parasitoid of *Plutella xylostella* on cabbage and cauliflower in plains. *Diadegma fenestrale* and *D. semiclausum* are the dominant larval parasitoids at high altitudes. On cabbage butterfly, *Pieris brassicae*, *Cotesia glomerata* parasitises up to 65 per cent larvae. Coccinellids *Coccinella septempunctata*, *Hippodamia variegata* and *Harmonia dimidiata*, parasitoids *Diaeretiella rapae* and the syrphid, *Eupeodes confrater* are important bioagents of cabbage aphids. The pest management strategy adopted for cabbage and cauliflower pest *P. xylostella*, includes raising Indian mustard as trap crop. Neem seed kernel extract (4 per cent) and pongamia oil can be sprayed twice to control the pest infestation in cabbage. Larval parasitoid *C. plutellae* and egg parasitoid *Trichogrammatoidea bactrae* can be conserved in the plains and *Diadegma semiclausum* in the higher altitudes by avoiding pesticidal sprays (Singh, 1993; Jalali *et al.*, 2001).

Table 30.8a: Natural Enemies of Vegetable Crop Pests–Cabbage*

Pest	*Natural Enemies*	
	Number Recorded	*Status*
Plutella xylostella	29	*Cotesia plutellae* is most common parasitoid throughout country. *Diadegma fenestralis* and *D. semiclausum* are most effective in hills. *Oomyzus sokolowskii* parasitised 32.0–70.0 per cent pupae in Bangalore and 18.0–68.0 per cent at Coimbatore and *Brachymeria excarinata* parasitised 60.0 per cent pupae. Ants and spiders also cuase considerable mortality during early larval instar stages.
Crocidolomia binotalis	10	*Palexorista solennis* and *Apanteles crocidolomia* were the key parasitoid.
Hellula undalis	04	*Bracon* sp. is most promising parasitoid.
Lipapis erysimi *Myzus persicae* Brevicoryne brassicae	51	Several parasitoids and predators are recorded but not very effective.

*: *Source*: Jayarathnam (1977); Nagarkatti and Jayanth (1982)

Tomato

Trichogramma spp. and *Campoletis chlorideae* are important biotic agents of *Helicoverpa armigera*. *H. armigera* can be suppressed through inundative releases of egg parasitoid, *T. chilonis*/*T. brasiliense*/*T. pretiosum* @ 50,000/ha/week. A nucleopolyhedrovirus of *H. armigera* at 10^9 POBs/ha/week and *Bacillus thuringiensis* var. *kurstaki* at 0.5 kg/ha can also be used. Trap crop of African marigold (yellow-tall) can be utilised in the management of *H. armigera* on tomato in 1:14 row combination. *H. armigera* females prefer to lay eggs on marigold buds which also attract parasitoids or *T. brasiliense/T. pretiosum/ T. chilonis* Bio H_1 could be released (Krishnamoorthy and Mani 1990; Singh, 1993).

Table 30.8b: Natural Enemies of Vegetable Crop Pests–tomato*

Pest	*Natural Enemies*	
	Number Recorded	*Status*
Helicoverpa armigera	120	*Trichogramma* spp., *Campoletis chlorideae* and *Carcellia illota* are important parasitoids. Some *Trichogramma* spp. parasitised upto 80.0 per cent eggs.

*: Source: Krishnamoorthy and Mani (1990); Manjunath *et al.* (1989).

Beans

Pod borers *Adisura atkinsoni* and *H. armigera* on field beans can be suppressed by applying NPV *of A. atkinsoni* 10^9 at weekly interval. *Tetranychus* spp. can be controlled by releasing the predator; *Phytoseiulus persimilis* @ ten adults per plant, 30 days after germination or as and when infestation is noticed (Singh, 1993).

Tea

On tea, *Micromus timidus, Calleida nilgiriensis, Opistoplatys* sp., *Pseudopachybrachius guttus, Jauravia pubescens* and *Cheilomenes sexmaculata* are the predators, which regulate the population of tea flushworm, *Cydia leucostoma.* On leaf roller of tea, *Caloptilia theivora, Sympiesis dolichogaster* parasitises up to 67 per cent larvae. *Telenomus* spp. is a very important parasitoid of tea mosquito bug, *Helopeltis theivora.* Combined action of reduviid bug and parasitoids (*Aphytis* sp.) keeps the diaspine scale (*Fiorinia theae*) population very low. *Comperiella bifasciata* and *Aphytis chrysomphali* are important parasitoids and *Chilocorus circumdatus* is the most efficient predator of purple scale *Chrysomphalus aonidum* on tea. Syrphids and coccinellids play a major role in the natural regulation of *Toxoptera aurantii.* Parasitoids–*Aphidius, Lipolexis, Trioxys, Aphelinus,* etc.–regulate the aphid population in different seasons. Coccinellid *Stethorus gilvifrons* and predatory mites *Agistemus* sp. are very efficient predators of phytophagous mites on tea (Singh, 1996).

Coffee

On coffee, parasitoids *Coccophagus bogoriensis, C. cowperi,* and predator *Chilocorus nigrita* are effective on green scale, *Coccus viridis. Coccidoxenoides peregrinus* on *P. citri, Tetracnemoidea indica* on *P. lilacinus, Blepyrus insularis* on *F. virgata* and *Anagyrus dactylopii* on *Nipaecoccus viridis* are the potential parasitoids. Exotic coccinellid predator *C. montrouzieri* and exotic parasitoid *L. dactylopii* have been found to be effective against *P. citri* on coffee (Singh, 1996).

Pepper

On black pepper scales, *Aspidiotus destructor* and *Lepidosaphes* sp., predatory beetles, *Cybocephalus* sp. and *Chilocorus circumdatus* and the parasitoid *Encarsia lounsburyi* can be released for effective management (Singh, 1996).

Coconut

A preliminary attempt was made for biological control of *Opisina arenosella* Walker, a leaf-eating caterpillar of coconut by releasing two *Trichogramma* species *viz. T. chilonis* Ishii and *T. embryophagum*

Hartig @ 1000 to 4000 parasitised eggs per palm in Bangalore, India. The reduction in larval population by 77.8 and 88.9 per cent, 100.0 per cent, 66.7 per cent and 75.0 per cent was recorded in *T. chilonis* and *T. embryophagum* treated palms during 1st, 2nd, 3rd and 4th generation of the pest compared to untreated control. There was non- significant difference between various release rates, indicating that lower release rate of 1000 parasitised eggs/palm as adequate rate. On sentinel egg cards, egg parasitism was low 2.0–5.0 per cent, 0.0–2.0 per cent and 0.0 per cent in *T. chilonis, T. embryophagum* released palms and untreated control palms, respectively. The field trial gave initial indication of usefulness of releases of *Trichogramma* and its possible use on a large scale in endemic areas to authenticate efficacy (Jalali, 2005).

References

Ali, M. S., 1957. Some biological studies on *Pseudococcus vastator* (Maskell). *Indian J. Ent.*, 19: 54-58.

Anonymous, 1995–2001. *Annual Report*. All India Co-ordinated Research Project on Biological Control of Crop Pests and Weeds. Project Directorate of Biological Control (PDBC), Bangalore.

Ansari, M. A., Pawar, A. D., Murthy, K. R. K. and Ahmed, S. N., 1989. Sugarcane scale, *Melanaspis glomerata* Green and its biocontrol prospects in Karnataka. *Plant Prot. Bull.*, Faridabad, 41: 21-23.

Babu, T.R., 1986. Population density of grape mealybug *Maconellicoccus hirsutus* (Green) and its control by a predator *Cryptolaemus montrouzieri* Muls. Ph.D. Thesis, APAU, Hyderabad.

Bentur, J. S. and Kalode, M. B., 1985. Rearing techniques for the predatory mired bug *Cyrtorhinus lividipennis* on eggs of *Corcyra. Curr. Sci.* 54: 513-514.

Butani, D. K., 1957. A tachnid fly parasite of *Chilo zonellus* Swinhoe. *Indian J. Entomol.*, 19: 62-63.

Chandy, K. C., 1955. A note on *Apanteles flavipes* Cam., a braconid parasitie of the cholam stem borer *Chilo zonellus* Swinhoe. *J. Bombay Nat. Hist. Soc.* 53: 69.

Chari, M. S. and Rao, R. S. N., 1990. Biological suppression of tobacco pests. In: Proceedings of Indo-USSR workshop on Biological Control of Crop Pests and Diseases. (Eds: Singh, S.P. and Narasimham, U.), PDBC, Bangalore, pp. 77-87.

Chari, M. S., Patel, S. N. and Bharpoda, t. M., 1982. The predatory habits of *Nesidiocrocis tenuis* Reuter. *Tobacco Res.* 8: 192-196.

David, H., 1980. Tachnid parasites to control moth borers in India. In: *Biological Control of Sugarcane Pests in India.* (S. Sithanantham and A. R. Solayappan–Eds.). Tamil Nadu State Federation of Cooperative Sugar Factories Ltd., Chennai, pp. 61-66.

Deshpande, B. P. and Nadkarny, N. T., 1936. The spotted bollworm of cotton, *Earias fabia* Stoll and *Earias insulana* Boisd. In South Gujarat, Bombay Presidency. *ICAR Sci. Monograph*, 10: 1–208.

Dodia, J. F., Yadav, D. N. and Patel, R. C., 1986. Management of fruit sucking moth, *Othreis fullonica* L. (Lepidoptera : Noctuidae) in citrus orchards at Anand (Gujarat) *GAU. Res. J.*, 11: 72-73.

Doomra, S., Maninder, S. and Varma, G. C., 1994. Studies on the effect of *Trichogramma japonicum* Ashnead (Hymenoptera: Trichogrammatidae) on the control of *Scirpophaga excerptalis* (Walker) (Lepidoptera: Pyralidae). In: *Biological control of Insect pests*, 143-144, 1994, PAU, Ludhiana

Gunathilagaraj, K. and Ganesh Kumar, M., 1997. Rice pest complex: An appraisal. *Madras Agril. J.*, 84: 249-262.

Gupta, B. D., 1951. A note on the scope of biological control of sugarcane pests. In: Proceedings of the 2nd Biennial Conference of Sugarcane Research and Development Workers, India, Part 2, pp. 229–235.

Israel, P. and Padmanabhan, S. Y., 1976. Biological control of stem borers of rice in India. Final technical bulletin (US PL 480 Project), Cuttack, India, pp. 1–155.

Issac, P. V., 1941. Report: 1939-40 of second entomologist (Dipterist) in charge, scheme for research on insect pests of sugarcane. *Sci. Rep. Agric. Res. Inst.*, New delhi, pp. 115-119.

Jalali, S. K. and Singh, S. P., 1990. A new record of *Ooencyrtus papilionis* (Hymenoptera: Encyrtidae) on the eggs of *Papilio demoleus* (Linn.) from India. *J. Biol. Control.* 4: 1, 59-60.

Jalali, S. K., Singh, S. P. and Venkatesan, T., 2001. Choice of suitable trichogrammatid species for suppression of diamondback moth *Plutella xylostella* (Linnaeus) (Lepidoptera: Yponomeutidae) on cabbage. *Shashpa* 8: 161–166.

Jalali, S. K. and Singh, S. P., 2003

Jayarathnam, K., 1977. Studies on the population dynamics of the diamondback moth, *Plutella xylostella* (Linnaeus) (Lepidoptera: Yponomeutidae) and crop loss due to the pest in cabbage. Ph.D. thesis, University of Agricultural Sciences, Bangalore, 215pp.

Kalyanasundram, M., Justin, C. G. L., Swamiappan, M., Sundarababu, P. C. and Jayaraj, S., 1993. Efficacy of *Trichogramma chilonis* against sugarcane internode borer *Chilo sacchariphagus indicus* in Tamil Nadu. *Indian J. Plant Prot.* 21:

Krishnamoorthy, A. and Mani, M., 1990. Feasibility of managing *Heliothis armigera* through parasitoids, pp., 195-205. In Jayaraj, S., Uttamasamy, S., Gopalan, M. and Rabindra, R. J. (eds). *Heliothis Management–Proceedings of National Workshop.* Centre for Plant Protection Studies, Tamil Nadu Agricultural University, Coimbatore, 328 pp.

Krishnamoorthy, A. and Singh, S. P., 1987. Biological control of citrus mealubug with an introduced parasite, *Leptomastix dactylopii* in India. *Entomophaga*, 32: 143-148.

Krishnamoorthy, A. and Singh, S. P., 1988. Observational studies on the occurrence of parasitoids of *Papilio* spp. in citrus. *India J. Plant. Prot.* 16: 79-81.

Katiyar, R. N., 1962. Preliminary trial of the field experiment in the biological control of maize and jowar stem borer *Chilo zonellus* (Swinhoe). *Agra Univ. Res Sci.* 11: 17-21.

Lingappa, S., Yadav, D. N. and Brar, K. S., 2001. Augmentative biocontrol within cotton IPM–Indian scenario. In: *Augmentative Biocontrol–Proceedings of the ICAR–CABI Workshop.* (S. P. Singh, S. T. Murphy and C. R. Ballal–Eds.), pp. 57–92, PDBC, Bangalore.

Mani, M. and Krishnamoorthy, A., 1990a. Evaluation of the exotic predator *Cryptolaemus montrouzieri* Muls. (Coccinellidae, Coleoptera) in the suppression of green shield scale *Chloropulvinaria psidii* (Maskell) (Coccidae, Homoptera) on guava. *Entomon,* 15: 45-48.

Mani, M. and Krishnamoorthy, A., 1990b. Natural suppression of mealybugs in guava orchards. *Entomon,* 15: 245-247.

Mani, M. and Thontadarya, T. S., 1988a. Field evaluation of *Cryptolaemus montrouzieri* Muls. in the suppression of *Maconellicoccus hirsutus* (Green) on grapevine. *J. Biol. Control,* 2: 14-16.

Mani, M., Thontadarya, T. S. and Singh, S. P., 1987. Record of natural enemies on the grape mealybug, *Maconellicoccous hirsutus* (Green0. *Curr. Sci.* 56: 624-625.

Mani, M. and Thontadarya, T.S., 1988b. Response of *Cryptolaemus montrouzieri* Muls. (Coccinellidae, Coleoptera) to commonly used pesticides in vineyards. *J. Biol. Control*, 2: 17-20.

Mani, M., Krishnamoorthy, A. and Singh, S.P., 1989. The impact of the predator, *Cryptolaemus montrouzieri* Muls. on the pesticide resistant populations of the striped mealybug *Ferrisia virgata* (Ckll.) on guava in India. *Insect Sci. applic.* 11: 167-170.

Manjunath, T. M., 1990. Biological Conrol of rice pests: An overview. In: Proceedings of Indo-USSR workshop on Biological Control of Crop Pests and Diseases. (Eds: Singh, S.P. and Narasimham, U.), PDBC, Bangalore, 1-294.

Manjunath, T. M., Rai, P. S. and Gowda G., 1978. Parasites and predators of *Nilaparvata lugens* in India. *PANS*, 22: 265-269.

Manjunath, T. M., Bhatnagar, V. S., Pawar, C. S. and Sithanantham, S., 1989. Economic importance of *Heliothis* spp. in India and an assessment of their natural enemies and host plants. In: Proceedings of the Workshop on Biological Control of *Heliothis*: Increasing the effectiveness of natural enemies, (Eds. E. G. King and A. D. Jackson), FERRO, USDA, New Delhi, pp., 197-198.

Manjunath, T. M., 1985. *Leptomastix dactylopii* in India. *Biocontrol News Inform.*, 6: 297.

Misra, M. P. and Pawar, A. D., 1983. Biocontrol trials against sugarcane pyrilla, *Pyrilla perpusilla* Walker in eastern Uttar Pradesh and elsewhere in India. In: *Biological Control of Sugarcane Pests in India.* (S. Sithanantham and A. R. Solayappan–Eds.). Tamil Nadu State Federation of Cooperative Sugar Factories Ltd., Chennai, pp. 75–78.

Nagarkatti, S. and Jayanth, K. P., 1982. Population dynamics of major insect pests of cabbage and their natural enemies in Bangalore district (India). In: Proceedings of International Conference of Plant Protection in Tropics, Malaysian Plant Protection Society, Kaula Lampur, Malaysia, pp. 325-347.

Nagarkatti, S., Singh, S.P., Jayanth, K.P. and Bhumannavar, B.S., 1992. Introduction and establishment of *Leptomastix dactylopii* How. against *Planococcus* spp. in India. *Indian J. Plant Prot.*, 19: 102-104.

Narayanan, E. S. and Mukherji, P. B., 1953. Two decades of field trials in India with *Trichogramma* against the sugarcane stem borer *Argyria* sp. and the root borer, *Emmalocera depresella*. In: Proceedings of the 40th Indian Science Congress. 10: 148-149.

Patel, R. C., 1980. Role and feasibility of natural enemies in integrated pest management of cotton. *Andhra Agric. J.* 27: 35-40.

Patel, H. K., Patel, R. C. and Patel, R. M., 1959. Life history, epidemiology and seasonal history of tobacco stem borer (*Gnorimoschema heliopa* how.). *Indian Tob.* (July–September): 175-184.

Pawar, A. D., 1979. Efforts made for the biological control of sugarcane insect pests in India. *Plant Prot. Bull.* 31: 138-147.

Pillai, G. B., 1985. Biological control of pests of plantation crops. *Proc. Natl. Se. Ins.*, Calicut, pp. 232-238.

Rao, R. S. N., Joshi, B. G., Satyanarayana, S. V. V. and Soundarajan, V., 1981. Notes on new additions to the natural enemies of *Spodoptera litura* F. and *Myzus persicae* Sulz. On FCV tobacco in Andhra Pradesh. *Sci. Cult.* 47: 98-99.

Rao, V. P., Chacko, M. J., Phalak, V. R. and Rao, H. D., 1970. Leaf eating caterpillar of paddy and their natural enemies in India. *J. Bombay Nat. Hist. Soc.* 66: 455-477.

Rao, V.P., Ghani, M.A., Sankaran, T. and Mathur, K.C., 1971. A review of the biological control of insects and other pests in South East Asia and the pacific region. *CIBC Tech. communication*. NO. 6, 149 pp.

Rao, V. P. and Kamath, S. M., 1966. Record of cottony cushion scale *Icerya purchasi* Maskell in the Cannore district of Kerala and its control. *Indian J. Ent.*, 27: 552.

Samal, P. and Misra, B. C., 1975. Spiders, the most effective natural enemies of the BPH in rice. *Rice Entomol. Newl.*, 3: 31.

Saxena, A. P., 1972.Studies on *Stenobracon desae* Cam. and *S. nicevillei* Bing. (Hym.: Braconidae). *Zeit. Ange. Entomol.* 71: 145-151.

Sekhon, B. S. and Verma, G. C., 1983. Parasitoids of *Pectinophora gossypiella* (Lep.: Gelechiidae) and *Earias* spp. (Lep.: Noctuidae) in Punjab. *Entomophaga*, 28: 45-53.

Shivayogeshwara, B., Mallikarjunaiah, H., Krishnaprasad, N. K. and Shetty, M. V. N., 1991. Integrated pest management (IPM) of *Spodoptera litura* (Fab.) (Noctuidae: Lepidoptera) in FCV tobacco crop. *Tob. Res.* 17: 59–61.

Singh, S. P., 1978. Propagation of coccinellid beetle for the biological control of citrus and coffee mealybugs. *Scientific Seminar*, CPA, 2p.

Singh, S. P., 1989. Achievements of AICRP on Biological Control. *Tech.Bull.* No. 2. Biological Control Centre, NCIPM, Faridabad, 20 pp.

Singh, S. P., 1985. Biological control of insect pests of horticulture crops. In: National Seminar on Entomophagous insects and other Arthropods and their potential in biological control, Calicut, pp. 221-231.

Singh, S. P., 1993. Biological control of vegetable pests. In-*Advances in Horticulture, Vol.6-Vegetable Crops*: Part 2 (1993), K.L. Chadha and G. Kalloo (Eds.). Malhotra Publishing House, New Delhi, pp. 887-920.

Singh, S. P., 1993. Biological control of vegetable pests. In: Advances in Horticulture–Vegetable Crops, Vol. 3. (Eds. K. L. Chadda and D. P. Pareek), Malhotra Publishing House, New Delhi, pp. 1591-1615.

Singh, S. P., 1993.Integrated pest management in horticulture crops. *Indian Hort.* 38: 25–28, 37,39-40.

Singh, S. P., 1996. Biological control in India, pp.216-328. In-*IPM System in Agriculture Volume 2 Biocontrol in Emerging Biotechnology* Rajiv K. Upadhyay, K.G. Mukherjee and R.L. Rajak (Eds.) Aditya Books Pvt Ltd., New Delhi, India, 620 pp.

Singh, S. P., 1996. Biological control, pp. 89-116. In-*50 years of Crop Science Research in India–Golden Jubilee Year of India's Independence 1947-1996*, R.S.Paroda and K.L.Chadha (Eds.), Indian Council of Agricultural Research, New Delhi, 796 pp.

Subba Rao, B. R., Sangwar, H. S. Abbasi, O. A., Singh, Y. and Ksheer Sagar, A. M., 1965. New records of hymenopterous parasites of *Nipaecoccus vastator* (Maskell) (Homoptera : Coccidae), a serious pest of *citrus* spp from India. *Indian J. Ent.*, 27: 109-110.

Subba Rao, B. R., Singh, R. H., Saxena, J. D. and Sharma, A. K., 1969. Bionomics of *Apanteles flavipes* (Cameron), a parasite of *Chilo zonellus* at Delhi with special reference to the mode of overwintering of the parasite. *Indian J. Entomol.* 31: 7-12.

Sundaramurthy, V. T. and Basu, A. K., 1985. Management of cotton insect pests in polycrop systems in India. *Outlook Agriculture* 14: 79-82.

Tandon, P. L., 1985. Spatial distribution sampling technique and population dynamics of citrus green scale *Coccus viridis* (Green) on mandarin. Ph.D. thesis, UAS, Bangalore.

Tirumala Rao, V. and David, L.A., 1958. The biological control of a coccid pest in South India by the use of beetle *Cryptolaemus montrouzieri* Muls. *Indian J. Agric. Sci.* 28: 545-552.

Thontadarya, T. S. and Jai Rao, K., 1978. Preliminary observations on the integrated control of cotton bollworms. *Proc. Seminar on Integratesd Pest Management System for Cotton*. CICR, Nagpur *Tech. Bull.* 4: 30-24.

Varma, G. C., Rataul, H. S., Shenmar, M., Singh, S. P. and Jalali, S. K., 1991. Role of inundative releases of *Trichogramma chilonis* Ishii in the control of *Chilo auricilius* Dudgeon on sugarcane. *J. Insect Sci.* 4: 165–166.

Chapter 31

Recent Advances in Management of Insect Pests of Vegetables Crops

T.M. Shivalingaswamy, S. Satpathy and A.B. Rai
Indian Institute of Vegetable Research,
Varanasi - 221 305

Introduction

During the last two decades India has made enormous progress in terms of vegetable production. Introduction and adoption of new technologies particularly hybrids played a vital role in boosting the production. The existing area under vegetable cultivation in India is around 6.09 million ha with a production of 93.92 million tons. There is an additional need of 56 million tons of vegetables to feed 1.3 billion India's population by 2020 AD. The biotic stresses however, cause considerably high amount of crop loss in vegetables, which may be due to pest infestations directly or due to viral diseases vectored by insects. The extent of crop losses in vegetables due to pest infestation is viable depending on the plant type, agro climatic conditions, damage potential of the pest involved and cropping season followed in different parts of the country. On an average 40 per cent losses are noted in vegetable crops in India.

Abrupt changes in crop husbandry practices, selection of varieties and improper pest management practices have accentuated the problem as increase in pest status of vegetable pests both in terms of damage and number. Recently high input based cultivation of vegetable hybrids has further aggravated the pest problem because of their susceptibility to pests and diseases. Most of the plant protection recommendations in vegetable so far indicated the calendar-based, ill conceived and haphazard application of insecticides. Consequently, the many fold problems like resurgence of sucking pests; threat of new pest insects, and occurrence of pests with high severity, biodiversity reduction residue problems emerged. Through the rate of pesticide consumption in India is low (500 g/ha) as compared to over 10-12 kg/ha in Japan, several pesticide induced pest outbreaks have been reported in the past in India. Vegetables consumed 13-14 per cent of the total pesticide grown in 2.6 per cent of total cropped area in the country, it is imperative to stress upon Integrated Pest Management in vegetables.

Growing concern among the public for environmental issues, particularly pesticide uses, the concept of integrated pest management (IPM) emerged. It stresses the need for adoption of compatible multiple techniques of pest management based as much as possible on biological and non-chemical methods to avoid the selection pressure of surviving pest population towards pesticide use.

IPM indeed is a component of the agro-ecosystem management technology for sustainable crop production. It is a knowledge intensive system and background information regarding the pest, abiotic and biotic factors, agro-ecosystem and management tactics is required for execution of IPM programmes. First and formost thing in IPM programme is to determine the level of pest population, which is going to cause economic damage. In this context, economic threshold level (ETL) and economic injury level (EIL) have been accepted as index for initiating economically viable control measures. Need based control strategy will not only help in reducing the pesticide consumption and environmental abuse, but also helps in maintaining minimum pest population for survival of natural enemies.

Different components of IPM include cultural, mechanical, biological and chemical means of management. In the present context of environmental issues, emphasis should be given on cultural control and biocontrol agents followed by judicious use of safer insecticides. Insecticides with least harmful effect on natural enemies, less persistency and residues can be judiciously used in pest management of vegetables.

A brief biology, nature of damage, spectrum of natural enemies identified and pest management methods adopted for each of the important pest in different vegetable crops is discussed in this chapter.

Pests of Brinjal

Shoot and Fruit Borer

Leucinodes orbonalis Guenee: (Lepidoptera: Pyralidae)

This pest is reported from regions of brinjal cultivation in Africa, south of the Sahara, and South-East Asia, including India, Bangladesh, Sri Lanka, China and Philippines.

Young caterpillars are creamy white, but full-grown are light pinkish and measure about 18-23 mm in length. The moth is medium sized with white wings having triangular brown and red markings on the fore wings. The moth measures bout 20-22 mm across the wings.

Life Cycle

A female moth lays about 8-120 creamy white eggs, singly or in batches of 2-4 on the underside of leaves, on green stems, flower buds or the calyces of fruits during its life span of 2-5 days. The incubation period is 3-6 days. The young caterpillars bore into the tender shoots near the growing points, flower buds or the fruits. There are five larval instars and mature in 9-28 days depending on the prevailing temperature and humidity. A single caterpillar may destroy as many as 4-6 fruits. The larva comes out of the damaged shoot or fruit and pupates in a tough gray cocoon in soil or plant debris. The pupal period is 6-17 days. During the active season, the total life cycle is completed in 20-43 days. There are five overlapping generations in a year.

Damage

The larvae bore into the petiole and midribs of large leaves and tender shoots and cause wilting.

In the later stage, they also bore into the flower buds and fruits. The flower buds shed due to attack. The fruit is damaged considerably and the entry hole is plugged with excreta. Attacked fruits become unfit for consumption and marketing. Infestation up to 70 percent has been recorded.

Natural Enemies

Trathala flavoorbitalis, Goryphus nursei, Eriborus argenteopilosus, and *Bracon* sp. are some of the larval parasitoids of this pest.

Management

- ✫ Selection of less susceptible cultivars like Arka Kesav, Arka Kusmakar, Pusa Purple Cluster, Pusa Purple Round, Punjab Barsati, Kalyanpur-2, Punjab Chamkila, Gote-2 reduce the extent of infestation
- ✫ Removal and destruction of early-infested shoots and fruits regularly helps to reduce the larval population.
- ✫ Installation of sex pheromone traps @100/ha (10 x10m) and lure replacement after 45 days
- ✫ 4-5 Releases of *Trichogramma chilonis* @ 50,000 parasitized eggs/ha coinciding with infestation
- ✫ Application of NSKE 4 per cent or Bt formulation @500g/ha
- ✫ Need based application of cartap hydrochloride @ 500g a.i./ha

Stem Borer

Euzophera perticella Rag. : (Lepidoptera: Pyralidae)

Occasionally this pest attains serious status. Infestation is noticed usually in the later stages of the crop growth.

Life Cycle

Grown up caterpillars are creamy white with a few bristly hairs and measure about 20-22 mm in length. The moth is small having pale straw yellow fore wings and whitish hind wings. The insect is active from March to October and hibernates as larva in the stem of old plant from November to the beginning of March. The moths emerge sometime in second half of March and soon after mating, the females start laying eggs. The eggs are cream in color and are laid singly or in batches on tender leaves, petiole, and branches. The incubation period is 3-10 days. The young larvae feed for a few minutes on exposed parts of plants and then bore into the stem making longitudinal tunnels. They pupate inside the feeding galleries after making silken cocoons or in cracks and crevices in the soil. Moths emerge in 6-8 days. The insect completes its life cycle in 35- 75 days.

Damage

The caterpillars cause the damage and feed exclusively in the main stem. It enters the main stem and makes tunnel, which result either in stunting of growth or wilting of whole plant, prior to yellowing of leaves.

Natural Enemies

Pristomerus testaceus and *P. euzopherae* are parasitic on the larvae of this pest.

Management

- ☆ Uproot and destroy the infested plants.
- ☆ If the pest is of regular occurrence, apply neem cake @ 500 kg/ha or carbofuran @1 kg a.i./ ha at the time of transplanting around the base of the plant.

Hadda Beetles

Epilachna vigintioctopunctata (F.): (Coleoptera: Coccinellidae)

Major Host Plants

Brinjal, tomato, beans and occasionally on cucurbits.

The grubs are about 6mm long, yellowish in colour and have six rows of long branched spines. Beetles measure 8-9 mm in length and 5.5 mm in width and are deep red and usually have 7-14 black spots on each forewing whose tip is somewhat pointed.

Life Cycle

The activity of beetle starts during March-April. A single female can lay up to 400 eggs in her life time. The eggs are yellow, cigar-shaped, laid mostly on the underside of the leaves, in batches of 5-40 each. The incubation lasts for 3-5days; a larval period of 7-18days and pupal period of 5-12 days have been recorded.

Damage

Both beetles and the grubs cause the damage. They feed on the upper surface of leaves. The leaves are eaten between the veins, sometimes being completely stripped to the mid-rib. The leaves present a lace-like appearance; turn brown, dries up and falloff. Thus, the plant is completely skeletonized.

Natural Enemies

Pediobius foveolatus and *Uga menoni* are the two important parasitoids on the grubs.

Management

- ☆ Collection and destruction of various stages (eggs, grubs and pupae are visible on the plant surface) of the pest.
- ☆ Spray with malathion or carbaryl @1 kg a.i./ha or dusting with malathion 5 per cent @ 25 kg/ha is effective.

Minor Pests

Pest	Damage
Leaf roller: *Eublemma olivacea*	Leaves are rolled longitudinally with silk by the larvae.
Leafhopper: *Amrasca biguttula biguttula*	Both nymphs and adults suck the sap from the leaves.
Aphids: *Aphis gossypii*	Infests the younger foliage.
Midge: *Asphondylia* sp.	The pest attacks the flower buds or young fruits.
Mites: *Tetranychus urticae*	Small red mites found feeding on the lower side of the leaves.
Ash weevil: *Myllocerus subfasciatus*	Ash gray weevils feed on the leaves.
Caterpillar: *Spodoptera litura*	Defoliator.
Lace bug: *Urentius hystricellus*	Skeletonize the leaves.

Pests of Tomato

Fruit Borer
Helicoverpa armigera (Hub.): (Lepidoptera: Noctuidae)

Major Host Plants

Okra, onion, brinjal, potato, chillies, cowpea, beans, cereals and pulses.

The caterpillars are greenish to variable colours with dark broken gray lines along the sides of the body. They measure about 35-45 cm in length and the moth is brown with a V-shaped marking and dull black border on the hind wings.

Life Cycle

Female moth lays eggs on the leaves, flowers and developing fruits. Eggs are yellowish and spherical. A female lays an average of 750 eggs over a week's time. The younger larvae feed initially on the foliage for a while and later bore into the green fruits. The larva while feeding, thrusts its head into the fruit leaving the rest of its body outside. The fully matured larva pupates in an earthen cell. *H.armigera* has a facultative pupal diapause which is induced by short day lengths (11-14 hours/day) and low temperatures (15-23°C) experienced as a larva. The total developmental period varies from 25-40 days depending upon the temperature and the host.

Damage

Early instars feed on leaves and fruiting bodies. Later they bore into fruits. The bored fruits invite secondary infection by other organisms lead to rotting.

Natural Enemies

Bracon brevicornis, Campoletis chlorideae, Trichogramma chilonis, and Trichogrammatoidea sp. are some of the important parasitoids recorded from India.

Management

- ✰ Use of tolerant varieties like Punjab Kesri, Punjab Chuhara, Pant Bahar, Azad, Pusa Hybrid-4, BT-1, BT-32 and T-27.
- ✰ Deep summer ploughing to expose the larvae and pupae to sunlight and predation by birds
- ✰ Planting of marigold (40 days old) as trap crop with every 16 rows of tomato (25 days old) attracts the pest. Collection and destruction of larvae from marigold flowers
- ✰ Use of pheromone traps for early pest detection
- ✰ *Trichogramma brasiliense* @ 2,50,000-parasitised eggs/ha (Inundative release) during peak flowering stage
- ·Spray *Ha*NPV @ 250 LE with jaggery (10g/litre), soap powder (5g/litre) and a tinopal (1 ml/litre) during evening hours or need based application of endosulfan 35 EC@ 2ml/litre of water

Defoliator
Spodoptera litura (F.): (Lepidoptera: Noctuidae)

Major Host Plants

Brinjal, Cabbage, cauliflower, Capsicum, cucurbits, okra, Phaseolus, potato, sweet potato and species of *Vigna* and other wild hosts.

Life Cycle

Commonly known as the tobacco caterpillar, *S. litura*, is one of the most important insect pests of agricultural crops in the Asian tropics. The fully-grown larva is stout, cylindrical, pale-greenish brown with dark markings, which measures about 35-40 mm in length. The moth is stout, dark with wavy white markings on the forewings and white hind wings. The pest is active throughout the year. Fecundity varies from 2000-2600 eggs, and oviposition period vary from 6 to 8 days. The female moth after mating at night, lays about 300 eggs in clusters. These clusters are covered with brown hairs. The incubation period is about 3-5 days. After hatching, the larvae feed gregariously together and later they spread out and feed individually. The larvae go through six instars. The last instar larvae enter the soil for pupation. The pupal period and the longevity of moths are 7-15 days and 7-10 days, respectively. The total life cycle covers 32-60 days. There are 8 generations in a year.

Damage

The caterpillars feed on leaves in the nursery and in the main field and cause serious damage. In crops like tomato and chillies, they bore into fruits and spoil them. Larvae are very active at night. Heavy infestation in the form of outbreaks can severely defoliate the crop.

Natural Enemies

Bracon brevicornis, *Campoletis chlorideae*, *Eriborus argenteopilosus*, *Microplitis prodiniae* and *Peribaea orbata* are the larval parasitoids, *Trichogramma chilonis* an egg parasitod, *Bacillus thuringiensis* (bacteria), *Beauveria bassiana*, *Metarhizium anisopliae*, *Nomuraea rileyi* (fungi) and Nuclear polyhedrosis virus are some of the entomopathogens recorded.

Management

- ✩ Clean cultivation to expose the pupae to natural enemies.
- ✩ Hand picking and mechanical destruction of caterpillars during early stage of attack helps in reducing the infestation.
- ✩ Application of neem kernel extract during the early stages of crop growth
- ✩ Pheromone traps to detect egg laying by the moths for initiating timely control measures
- ✩ Spray *Sl*NPV @ 250LE with jaggery (10g/lit), soap powder (5g/lit) and tinopal (1 ml/lit) during evening hours.
- ✩ Spray malathion 1000g a.i./ha or endosulfan @ 700 g a.i./ha and repeat the spraying at 10 days interval, if necessary.

Serpentine Leaf Miner
Liriomyza trifolii (Burgess) : (Diptera: Agromyzidae)

Major Host Plants

Okra, onion, cucurbits, beans, potato, brinjal, crucifers and other crops.

Life Cycle

Liriomyza trifolii is one of the polyphagous agromyzids causing serious damage on several crops including ornamentals in the recent years. Eggs are inserted just below the leaf surface and they hatch in 2-5 days depending on the prevalent temperature. Many eggs may be laid on a single leaf. Neonate larva is transparent turn to yellow-orange in later instars. After hatching, the larva start mining into the leaves. The typical mine resembles a serpent shape, hence the name. Pupation occurs in the soil beneath the plant. Pupal development will vary according to season. Adult emergence takes place 7-15 days after pupation. Life cycle completes in 12-15 days.

Damage

Damage is due to mining into leaves and petiole by the larvae. The photosynthetic ability of the plants is often greatly reduced as the chlorophyll-containing cells are destroyed. Severely infested leaves may dry and fall.

Parasitoids

Neochrysocharis formosa, Chrysonotomyia rexia and Oomyzus liriomyzae are some of the important eulophid parasitoids recorded on *L. trifilii* from India

Management

- ✰ Judicious application of nitrogenous fertilizers reduces pest build up in endemic areas
- ✰ Infested leaves should be removed and destroyed
- ✰ Application of NSKE 4 per cent with a sticker found to be effective.
- ✰ Spray of imidacloprid 17.8SL @ 0.3ml/lit. of water during early stages of crop growth before flowering.

Whiteflies

Bemisia tabaci Genn. : (Homoptera: Aleyrodidae)

Major Host Plants

Okra, crucifers, cucurbits and solanaceous crops

Life Cycle

Eggs are laid in circular groups, on the undersides of leaves by a pedicel inserted into a fine slit made by the female. Eggs are whitish in colour when laid later turn brown. Each female lays up to 160 eggs. Hatching occurs after 5-9 days at 30°C depending on host species, temperature and humidity. On hatching, the first instar larva (only mobile stage) or crawler moves to a suitable feeding site on the lower leaf surface where it moults and becomes stationary throughout the remaining stages. The first three larval stages last 2-4 days each (depending on temperature). The fourth larval stage is called as puparium. Pupation lasts for about 6 days. The adult emerges through a 'T'-shaped rupture in the puparium. A female may live up to 60 days while the male lives shorter (9 to 17 days). There are 11 to 15 generations in a year.

Damage

Apart from the direct damage to the crops by way of sucking the sap, *B. tabaci* is an important vector of a wide array of plant virus diseases. *e.g.* Tomato yellow leaf curl virus (TYLCV), Bean Golden Mosaic (BGMV), African cassava mosaic virus (ACMV) etc. Infected plants exhibit any one or a combination of the symptoms like vein yellowing, inter-vein yellowing, leaf yellowing, yellow blotching of leaves, yellow mosaic of leaves, leaf curling, vein thickening, leaf enations, leaf cupping and plant stunting.

Natural Enemies

Encarsia brevivena, Eretmocerus corni, Eretmocerus mundus are some of the parasitoids, *Chrysoperla carnea, Mallada boninensis, Coccinella septempunctata* are the predators and *Beauveria bassiana* and *Paecilomyces farinosus* are the pathogens recorded from *B. tabaci.*

Management

- ✰ Many weed plants harbor whiteflies, removal of weed hosts found to reduce both the incidence of whiteflies and associated viral diseases.
- ✰ Seed treatment with imidacloprid 70WS@2.5g per kg of seed provides protection for 25-30 days.
- ✰ Protection of seedlings in nursery by nylon net (200 mesh) covering for 25-30 days followed by need based application of systemic insecticides after transplanting.

Minor Pests

Leafhopper: *Amrasca biguttula biguttula*	Adult and nymphs suck the sap from leaves.
Hadda beetles: Epilachna vigintioctopunctata	Beetles and grubs feed on the foliage.
Thrips: *Thrips tabaci*	Both adult and nymphs suck the sap.
Crucifers	
Cutworms: *Agrotis segetum, A. ipsilon*	Larvae cut the young seedlings at the base.
Semilooper: Trichoplusia ni	Defoliator.
Thrips: *Thrips tabaci*	Both adult and nymphs suck the sap.
Painted bug: *Bagrada cruciferarum*	Both adult and nymphs suck the sap.
Bihar hairy caterpillar: *Spilarctia obliqua*	Defoliator.
Mustard saw fly: *Athalia lugens proxima*	Defoliator.

Pests of Chilli

Thrips
Scirtothrips dorsalis Hood: (Thysanoptera: Thripidae)

Major Host Plants

This is a polyphagous pest. Besides chillies, it feeds on tomato, cotton, castor, sunflower, mango and citrus.

Life Cycle

The pest is active throughout the year except the monsoon season. The nymphs are like adults in shape and colour but are wingless and smaller in size. The adults are slender, yellowish brown, having apically pointed wings and measure about 1 mm in length. The female having long narrow wings with the fore margin fringed with long hairs. A female lay about 45 to 50 eggs over a period of 20-30 days inside the tissues of the leaves and shoots. Entire life cycle completes in 15-20 days. The pest has several overlapping generations in a year.

Damage

The nymphs and adults lacerate and suck the cell sap from leaves and tender parts of the plants and cause the leaves to shrivel. In case of severe infestation, there is a malformation of leaves, buds and fruits. The attacked shoots hardly develop and the leaves falloff. These thrips are also responsible for transmitting the *leaf curl disease* in chillies.

Natural Enemies

Franklinothrips megalops, Mymarothrips garuda, Scolothrips indicus (Predatory thrips), *Orius maxidentex, Geocoris ochropterus* (Predatory bugs) are some of the effective predators.

Management

- ☆ Protecting the nursery from thrips and whiteflies using nylon nets (200 mesh) reduces the leaf curl disease
- ☆ Seed treatment with imidacloprid 70WS @ 2.5g per kg of seed provides protection for 25-30 days.
- ☆ Foliar spray with thiamethoxam @ 50g a.i./ha during the pre-flowering stage and endosulfan 750g a.i./ha during fruit stage.

Broad Mite
Polyphagotarsonemus latus Banks: (Acari: Tarsonemidae)

Major Host Plants

Polyphagous pest with a wide array of host plants belonging to solanaceae and cucurbitaceae.

Life Cycle

Tiny yellow coloured mites are found in large numbers in the ventral side of leaves. They suck the cell sap and cause deformity of leaves creating confusion with leaf curl. On chilli, the developmental period from egg to adult averages 3-4 days. Adult longevity is 10-15 days. Each female lays about 25 eggs. Eggs are laid on the underside of leaves, tender stems, flowers and fruits. Attack occurs during a short period of time.

Damage

Edges of damaged young leaves usually curl. The foliage often becomes rigid and appears bronzed or scorched. Discoloration of tissues due to mite feeding, fruits become deformed or fail to develop. Severely infected fruits fall. When leaves are attacked, the leaf tissues disintegrate and the epidermal layer of the infested leaves thickens. Downward curling of leaves like inverted boat and petiole elongation in older leaves are characteristic symptom of mite damage. Yield is significantly reduced.

Natural enemies

Amblyseius sp. is an efficient predatory mite, which regulates the pest under field conditions.

Management

- ☆ Release and or conservation of native predatory mites if available.
- ☆ Spraying with dicofol @ 3 ml/lit of water should be done at 10 days interval at the initiation of symptoms.
- ☆ Spraying with 0.3 per cent wettable sulphur. Sulphur spraying should be done alternatively with propargite 57EC @ 3.5 ml/lit of water.

Pests of Cole Crops (Cabbage and Cauliflower)

Diamondback Moth
Plutella xylostella (L.): (Lepidoptera: Plutellidae)

P. xylostella is considered as the main insect pest of crucifers, particularly cabbage, broccoli and cauliflower.

Life Cycle

The moths are brown or gray, with distinct white spots on the fore wings. They measure about 8-12 mm in length. When the wings are folded on its body at rest, three white diamond shaped median dorsal patches are seen and hence the name "diamondback moth". The female lays minute yellowish-white eggs singly or in batches of 2-40 on the underside of leaves. A single female lay about 20-300 eggs. The eggs hatch in 3-10 days. The freshly hatched larvae move to the under surface of the leaf, mine into the leaf tissue. The three later instars feed on the underside of the leaf making holes. The full-grown larvae are pale yellow-green with fine black hairs all over the body and measure about 8 mm in length. The larval development is completed in 9-17 days. Pupation takes place inside a barrel-shaped, loose silken cocoon constructed by the larva. Normally, the cocoon is attached to the leaf surface. The pupal period is 4-5 days. The moths may live up to 20 days. The total life cycle is completed in 2-3 weeks. There are 8 to 12 generations in a year.

Damage

The newly hatched larvae cause serious damage to crops like cabbage, cauliflower by scraping the leaf tissues. As they grow older, bite holes in them and render the vegetables unfit for human consumption. A severe infestation results in the formation of under-sized curds in cauliflower and in cabbage, head formation does not take place when the infestation is severe during primordial stage.

Natural Enemies

The larvae are parasitized by *Cotesia plutellae, Diadegma semiclausum, Oomyzus sokolowskii* and the pupae are parasitized by *Brachymeria excarinata.*

Management

- ☆ In endemic areas, early or late cropping should be avoided
- ☆ Clean cultivation, summer polughing reduces the infestatin of DBM.
- ☆ Intercropping of Cole crops with tomato or carrot reduces the infestation
- ☆ Growing of paired mustard rows in both sides of 16 cabbage rows, first row is sown 15 days prior to cabbage planting while second is sown 25 days after planting helps in trapping the DBM adults in mustard. Spray mustard crop with 0.1 per cent dichlorvos to control the pest
- ☆ Spraying neem seed kernel extract 5 per cent or Bt formulation @ 500g ai/ha alternatively also proved to be effective during primordial stage.

Note: Avoid pesticide use during peak activity of the parasitoid (*Cotesia plutellae*)

Leaf Webber
Crocidolomia pavonana (F.): (Lepidoptera: Pyralidae)

Major Host Plants

Cabbage and cauliflower.

Life Cycle

C. pavonana is of great economic importance in Central and Southern India. The larva is green with red head and longitudinal red stripes on the body. It is about 2 cm in length. The moths are small with light-brownish forewings. The moth lays eggs in masses of 40-100 each, on the lower surface of the leaves. The eggs hatch in 5-15 days. Young larvae feed gregariously on the leaf tissues. As they grow older, they start webbing the leaves and feeding on them. The larval stage is completed in 24 to 27 days in summer and in about 50 days in winter. The pupation takes place in the soil in an earthen cocoon. In central and northern India, the pest undergoes diapause in summer (38-45°C). The adult emerges in 14-40 days. The total life cycle is completed in 43-82 days. There may be more than one generation in a year.

Damage

The larvae web the leaves with silken strands and feed on the lower surface of the leaves and completely skeletonize them. In cauliflower, larvae nibble the growing tip of seedlings and bore into the curd resulting in discoloration of curd. Even a single mature larva per plant is capable of causing economic loss to cabbage at pre- and post-heading stages.

Natural Enemies

Microbracon mellus, Apanteles crocidolomiae, Palexorista solennis are the parasitoids on the larvae of this pest.

Management

- ✰ Removal and destruction of webbed leaves with larvae inside.
- ✰ Use of safer insecticides for conservation of these beneficial insects
- ✰ Use of Indian mustard as trap crop for DBM management also attracts the leaf webber moths for egg laying and reduce the infestation in the main crop
- ✰ Foliar spray of neem seed kernal extract (4 per cent) along with sticker or *Bacillus thuringiensis* formulations @ 500g a.i./ha
- ✰ Need based application of cypermethrin 10 EC @ 100g a.i./ha or malathion 50 EC @ 1000g a.i./ha is effective. For increased efficacy, spray the insecticides prior to web formation.

Cabbage Caterpillar
Pieris brassicae (L.): (Lepidoptera: Pieridae)

Major Host Plants

A specific pest on cruciferous crops.

Life cycle

A sporadic pest causes serious damage in northern and eastern parts of India. Newly hatched larvae are pale yellow and turn green yellow later on. Mature larvae measures 40-50mm in length. Adults are medium sized butterflies pale white with black markings. Wings in female have two black circular dots on the dorsal side of each fore wing whereas in male the black spots are on the underside of each fore wing. After mating, the butterflies lay about 100-150 yellowish eggs, which are conical in shape in clusters on the upper or lower side of the leaf. The eggs hatch in 11-17 days in winter and 3-7 days in summer months. The caterpillars feed gregariously in early stages. There are five instars and are full-grown in 2-3 weeks in March-April and in 4-6 weeks during November-February.

Damage

Caterpillars cause the damage. Freshly hatched caterpillars scrape the leaf surface. The older larvae eat up the leaves from margins, leaving intact the main vein. Sometimes all the leaves of entire plant are eaten away.

Natural Enemies

Cotesia glomeratus, Phryxe vulgaris and *Pteromalus puparum* are the hymenopteran parasitoids on the larvae of this pest.

Management

- ✰ Hand picking and mechanical destruction of caterpillars during early stage of attack helps in reducing the infestation.
- ✰ Foliar spray of *B. thuringiensis* formulation @ 500 g ai/ha along with sticker control the larvae effectively.
- ✰ Need based application of cypermethrin @ 50 g ai/ha or malathion @ 100 g ai/ha provides good control of this pest.

Cabbage Head Borer
Hellula undalis F.: (Lepidoptera: Pyralidae)

A major pest of cole crops in tropical and subtropical regions.

Life Cycle

The mature larva is creamy-yellow with a pinkish tinge and has seven purplish brown longitudinal stripes. It measures 12-15 mm long. The adult moth is slender, pale yellowish brown, having gray wavy lines on the fore wings and hind wings pale dusky. The female lays pinkish oval eggs either singly or in groups on the under surface of the leaves. The eggs hatch after 2-3 days and caterpillars feed inside the cabbage and mature in 7-15 days after passing through five larval instars. The full-grown caterpillar forms a cocoon among the leaves or even inside the larval tunnel. The adult emerges in about six days. Total Life cycle is completed in 15 to 25 days.

Damage

Younger seedlings are most susceptible to the attack of this pest. The caterpillar in initial first two instars mine the leaves. Later on, they feed on the leaf surface and bore into the heads of the cauliflower and cabbage. The plants lose their vigour resulting in deformed heads which are unfit for consumption and marketing.

Natural Enemies

Bracon gelechiae and *B. hebetor* are the larval parasitoids.

Management

- ✰ Monitoring of infestation in seedling and early growth stage
- ✰ Collection and mechanical destruction of caterpillars in early stage of attack helps to check the infestation.
- ✰ Spraying of synthetic pyrethroids like fenvalerate, cypermethrin or deltamethrin @ 50 g ai/ha at 10 days interval up to 50-60 days after transplanting controls this pest effectively.

Semilooper
Thysanoplusia orichalcea (F.): (Lepidoptera: Noctuidae)

Major Host Plants

T. orichalcea is a polyphagous pest and defoliates a wide range of crop plants such as cabbage, turnip, lettuce, cotton, chicory, groundnut, sun hemp, tobacco, sunflower, safflower, tomato, potato, chickpea and soybean.

Life Cycle

The caterpillars are pale-greenish, plumpy and move with a characteristic half-loop movement hence the name '*semi looper*'. Females lay eggs on leaves of host plants and the larvae feed individually making holes of varying sizes. Pupation takes place in debris on the ground. The total generation time is 24-35 days. The moth has mottled brown forewings with pale straw-coloured hindwings. The forewing has a conspicuous, roughly triangular, golden patch, which makes the species readily recognizable. The adults are active during evening and are seen visiting in large numbers on flowers.

Damage

The younger larvae inflict damage by biting holes into the leaves. The later stage larvae completely defoliate the plant during severe infestation.

Natural Enemies

Voria ruralis is a larval parasitoid.

Management

Refer to cabbage caterpillar, *Pieris brassicae.*

Flea Beetles
Phyllotreta cruciferae Goeze. : (Coleoptera: Chrysomelidae)

P. cruciferae feeds primarily on crucifers.

Life Cycle

The adults are beetles vary in colour from shiny black to black, have very stout femora, with which they jump like a flea. The female lays 50 to 80 creamy white eggs, in the soil around the host plant. The eggs hatch in 5-7 days and larva is dirty white in color with pale-white head. It measures about 5 mm in length. There are three larval instars. Total larval period lasts for 9 to 15 days and later it pupates in the soil in an earthen cell near the infested plants. The pre pupal and pupal periods lasts for 2-4 days and 8-14 days respectively. Adults of *P. cruciferae* are known to over winter in plant debris on the soil surface or in cracks in the soil. The pest completes 7-8 generations in a year.

Damage

The larvae live in the soil and feed upon the roots of the host plants. The adults feed on the leaves of young plants leading to a characteristic 'shot hole' appearance. Severe infestation results in loss of plant vigour, wilting and death. The stem and flowers may also be attacked. The old eaten away leaves dry up and the young leaves rendered unfit for consumption.

Management

- ☆ Deep ploughing during summer helps to expose and to kill the over wintered population of the pest.

- ☆ In endemic areas, off-season cultivation of cabbage and cauliflower particularly late crop should be avoided.
- ☆ In young seedlings and early planted crop endosulfan (4 per cent) or malathion (5 per cent) dust should be applied @ 10-15 kg/ha.
- ☆ Foliar spray of dichlorvos @ 1000 g ai/ha or endosulfan @ 750 g ai/ha may be sprayed if necessary.

Aphids

Myzus persicae Sulzer and *Brevicoryne brassicae* (L.): (Homoptera: Aphididae)

Life Cycle

Aphids appear in November and remain active till April. Cloudy and moist weather favours rapid multiplication of this pest. Both nymphs and adults are similar in appearance. Nymphs are 1.5 to 2.0 mm long and adults are 2.0 to 3.0 mm. A single female produces 5 to 10 young ones per day during February to May and 2 to 5 per day during July to January. Their activity goes down in extreme hot conditions of May-June. The life cycle is completed in 11 to 18 days during April to March. Under favourable conditions, 11 to 20 generations are completed in one year. During October to March, pre-reproduction, reproduction and post reproduction period last for 8 to 18, 13 to 30 and 5 to 18 days, respectively whereas during April to September these stages occupy only 6 to 8, 4 to 9 and 4 to 5 days, respectively. A female produces on an average 7, 13 and 26 young ones during October to December, January to March and April to September, respectively.

Damage

Direct feeding on young growth leads to wilting of plants. Early attack may lead to stunted growth. Symptoms of viruses transmitted by *B. brassicae* include mosaic, chlorotic and necrotic lesions on leaves. *M. persicae* is the most important aphid virus vector. It has been shown to transmit well over 100 plant virus diseases

Natural Enemies

Chrysoperla carnea (Chrysopidae), *Coccinella septempunctata* (Coccinellidae), *Episyrphus balteatus* (Syrphidae) are some of the important predators helps in pest suppression in nature.

Management

- ☆ Cut and destroy the infested shoot mechanically as soon as aphid infestation appears.
- ☆ Foliar spray with methyl demeton 25 EC @ 300g ai/ha or endosulfan @ 750g ai/ha when 5 per cent plants are infested. Repeat foliar spraying fortnightly if 5 per cent infestation persists.

Cut Worms

Spodoptera litura (F.) (Lepidoptera: Noctuidae)

See under pests of tomato.

Pests of okra

Shoot and Fruit Borer

Earias vittella (F.): (Lepidoptera: Noctuidae)

A major pest on Malvaceous crops like okra and cotton.

Life Cycle

Eggs are spherical, about 0.5 mm in diameter, light bluish green in colour, with ridges. The larva is grayish brown with dark brown to black spots on the body. Full-grown caterpillars are 18-24 mm long, stout, spindle shaped. Pupa is 13-16 mm long and chocolate brown in colour. Adults are medium sized moths 13-15 mm long, forewings are pale white with a broad wedge shaped horizontal green patch in the middle, hind wings are silver creamy white in colour. Wing expansion is 30-35 mm. A female lays on an average 400 eggs, incubation, larval and pupal period lasts for 3-9, 9-20 and 8-12 days respectively. Single life cycle takes 22-25 days extending up to 74 days during winter.

Damage

Caterpillars bore into tender shoots and fruits results in wilting of the shoot and affected fruits become deformed in shape and remain stunted in growth. The damaged fruits become unfit for consumption and marketing.

Natural Enemies

Trichogramma chilonis (Egg parasitoid), *Apanteles angaleti, Bracon lefroyi* and *Goryphus nursei* are some of the larval paraitoids.

Management

- ✰ Deep ploughing during summer to expose the puape to sunlight and predation by birds
- ✰ Avoid overuse of nitrogenous fertilizers
- ✰ Soil application with carbofuran @ 1 kg ai/ha gives protection against shoot damage.
- ✰ Application of endosulfan (0.035 per cent) + NSKE (4 per cent) along with sticker from 40 days after sowing at 10 days interval gives effective fruit borer control.

Leaf Hoppers

Amrasca biguttula biguttula Ishida: (Hemiptera: Cicadellidae)

This pest is widespread in the Indian subcontinent, covering Bangladesh, Nepal and Pakistan. Major host plants include okra, potato, brinjal, beans and cotton,

Life Cycle

The eggs are yellowish white. The female lays 17-38 eggs in the leaf lamina. The incubation period is 8-10 days. There are five nymphal instars, with each instar completed in 3-5 days. The longevity of adults is 11 days. Preoviposition and oviposition periods range from 2-4 and 4-9 days, respectively. Rainfall is a major mortality factor for the nymphs and adults. In Northern India and parts of South India peak populations of these leafhoppers are attained between June and August. Cotton, okra and brinjal, sown in May, suffer severe damage in these regions.

Damage

Both nymphs and adults of leafhoppers are seen on the underside of the leaf sucking the sap. The initial symptoms of leafhopper damage are yellowing of leaves, followed by crinkling around the margins and upward curling of leaves. Severely affected plants have stunted growth.

Natural Enemies

Anagrus empoascae, A. flaveolus, Gonatocerus ater, Oligosita sp. and *Stethynium triclavatum* are important egg parasitoids, *Cheilomenes sexmaculata, Chrysoperla carnea, Mallada boninensis* and *Paederus fuscipes* are the predators recorded on this pest.

Management

- ☆ The varieties/lines Sel. 6-2, AE 30, White Velvet, Clemson Spineless, Early Long Green, AE 27, IC 75, HB 45, IC 7194, Punjab Padmini and New Selection are found to be tolerant.
- ☆ In endemic areas, seed treatment with imidacloprid @ 3g/kg which will protect the crop for 40-50 days after sowing. Or Soil application of carbofuran @ 1 kg ai/ha at the time of sowing.
- ☆ During the vegetative stage, foliar spray of monocrotophos @ 500g ai/ha at 15 days interval is advocated.

Red Spider Mites

Tetranychus urticae Koch: (Acari: Tetranychidae)

Host Plants

The pest is polyphagous and is known to feed on more than 150 species of plants including cucurbits, brinjal and okra.

Life Cycle

The red spider mite is also called the two-spotted spider mite. The mite is active from March to October and passes the winter as gravid female. As the season warms up in March, it spins webs on the undersurface of leaves of various host plants and lays 60-80 eggs. The eggs are spherical and hatch in 2-6 days. The emerging larvae are light brown and have three pairs of leg. Larvae feed on lower surface underneath of the webs and changes into nymph within 3-4 days, which have four pairs of legs. The nymphs grow to maturity within 4-9 days and the adults live for 9-11 days. Red spider mite completes their life cycle within 9-19 days during active period. It is believed that in Punjab red spider mite complete 32 generation in a year.

Damage

The damage is caused both by nymphs and adults. A large number of webs are formed on the leaves under which the pest sucks the cell sap. The affected leaves dry up and finally wither away giving an unhealthy appearance. Webbing interferes with the plant growth and reduction in yield.

Management

- ☆ Destruction of severely attacked plant part mechanically during initial stage of infestation reduces mite population.
- ☆ Spraying with dicofol @ 500g a.i./ha and wettable sulphur 80WP @ 0.3 per cent alternatively at 10 days interval at the initiation of symptoms.
- ·In endemic areas the crop should not be sown adjacent to dusty roads and the field should be kept wet regularly.

Minor Pests

Semilooper: *Anomis flava*	Defoliator.
Aphids: *Aphis gossypii*	Suck the sap from the plants.
Whitefly: *Bemisia tabaci*	Suck the sap from the plants.
Red cotton bug: *Dysdercus cingulatus*	Suck the sap from the plants.

Contd...

Contd...

Fruit borer: *Helicoverpa armigera*	Defoliator.
Leaf miner: *Liriomyza trifolii*	Mines the leaves.
Stem fly: *Melanagromyza hibisci*	Larvae bore into the stem.
Pink boll worm: *Pectinophora gossypiella*	Borer
Caterpillar: *Spodoptera litura*	Defoliator

Pests of Onion

Thrips

Thrips tabaci L. : (Thysanoptera: Thripidae)

These are cosmopolitan, polyphagous pests, often abundant in warm, dry regions.

Major Host Plants

Onion, garlic, cabbage, cucurbits, tobacco and cotton.

Life Cycle

The nymphs are slender, yellowish-brown and look similar to adults but are wingless and slightly smaller in size. The adults are yellowish-brown and measure about 1 mm in length. The males are wingless while the females have long, narrow fringed wings. Female live for 12-30 days and lay 50-60 kidney shaped eggs singly inside leaf tissue with its sharp ovipositor. The incubation period is 4-10 days and the nymphs feed on the plant sap by lacerating the leaf tissues. Usually they congregate at the leaf sheath or in the flowers. Nymphs mature in 4-5 days after passing through four instars and pupate in the soil. There is a pre-pupal stage, which lasts for 1-2 days, and pupal period is 2-4 days. Several generations are completed in a year.

Damage

Both nymphs and adults cause the damage. As a result of its attack, the leaves show a spotted appearance initially and later develop into pale-white patches (Silvery shoot) due to drainage of sap. Subsequently the leaf tips fade and gradually dry up. The infested plants do not form bulbs and there is no seed setting in seed crop.

Natural Enemies

Orius albidipennis and *Scymnus roepkei* are some of the predators and *Aspergillus ochraceus, Beauveria bassiana* and *Metarhizium anisopliae* are the entomopathogenic fungi recorded on thrips.

Management

- ☆ Seed treatment with imidacloprid 70 WS @ 3g/kg
- ☆ Spray malathion 50 EC @ 1000g ai/ha as soon as the pest appears.

Onion Maggot

Delia antiqua (Meigen): (Diptera: Anthomyiidae)

This is a specific pest on onion.

Life Cycle

The flies are slender grayish and large winged and about 6 mm in length. The females lay white and elongate eggs near the base of the plant. The incubation period is 2-8 days. Maggots are small, white and measure about 8 mm in length. The maggots after hatching crawl up and enter the leaf sheath and reach the bulb and feed inside for about 15-20 days and then crawl out of the bulb and pupate in the soil. The adult flies emerge and start the new generation. The last generation of the pest shortly before the harvest leads to rotting of produce in storage.

Damage

The attacked plants become flabby and yellowish. The maggots bore into the bulbs leaving only outer sheath. The infestation varies from 8-10 per cent. The infestation is not completely destructive to the bulbs but it leads to subsequent rotting in the storage.

Management

- ✰ Crop roattion helps in reducing the infestation
- ✰ Infested plants should be removed together with larvae and burned. The soil should be turned to destroy puparia.
- ✰ Application of phorate 10G to the soil followed by light irrigation found to be effective.

Pests of Leguminaceous Vegetables (Beans, Peas and Cowpea)

Whiteflies

Bemisia tabaci Genn. : (Homoptera: Aleyrodidae)

Refer under pests of tomato.

Bean Aphids

Aphis craccivora Koch: (Homoptera: Aphididae)

These aphids are distributed worldwide particularly in the tropics. Leguminaceous crops are the major host plants of this pest.

Life Cycle

Young colonies of *A. craccivora* concentrate on growing points of plants and are often tended by ants (mutualism). This symbiotic association with ants helps in dispersal of the aphids from plant to plant. Parthenogenetic reproduction is common and noticed throughout the year. The aphid is ovo-viviparous, with females retaining eggs inside their bodies and giving birth to nymphs. Males are winged and females are both winged and wingless. The total life cycle varies from 10-12 days depending on the weather conditions.

Damage

Aphids feed on young leaves, shoots, flowers and immature pods by sucking the sap. Symptoms of damage include stunting of the plants and seed shriveling in developing pods. Yield losses are severe when aphid colonies concentrate on the growing tips of plants. Indirect damage caused by aphids is due to the production of honeydew, which hampers plant respiration and photosynthetic efficiency. *A. craccivora* are known to transmit several plant viruses (Bean common mosaic virus, Bean yellow mosaic virus, cucumber mosaic virus etc.) leading to complete crop losses.

Natural Enemies

Trioxys indicus is an important parasitoid, *Cheilomenes sexmaculata, C. septempuncta* (lady bird beetles), *Episyrphus balteatus, Ischiodon scutellaris,* (syrphid larvae) are the predators and *Fusarium pallidoroseum* is a fungal pathogen.

Management

- ☆ Cut and destroy the infested shoot mechanically as soon as aphid infestation appears.
- ☆ Foliar spray with methyl demeton @ 250 g ai/ha or endosulfan @ 750 g ai/ha (safer to natural enemies) when 5 per cent plants are infested. Repeat foliar spraying fortnightly 5 per cent infestation persists.

Pea Stem Fly
Ophiomyia phaseoli (Tryon.): (Diptera: Agromyzidae)

Major Host Plants

Peas, cowpea, beans.

Life Cycle

The larvae are leaf and stem miners. The metallic flies are active in summer and mate 2-6 days after emergence. The female inserts her ovipositor into the leaf and sometimes in the stem and petiole and makes an elliptical cavity under the epidermis. A female lays 14-64 elongate, oval and white eggs. The incubation period is 2-4 days. The maggot mines through the leaf towards the petiole and later on reaches the stem. They pass through three instars. The larval development is completed in 6-7 days in March-April and in 9-12 days in November-December. The larva pupates within its gallery. The pupal period lasts 5-9 days and 18-19days in March-April and November-December respectively. The longevity of adults varies from 8-12 days. There are 8-9 generations in a year.

The maggots bore into the stem causing drying of affected shoots. The adults also cause damage by puncturing the leaves and feed on the liquid that oozes. Owing to the feeding, the injured portion turns yellow. Seedlings are more affected than grown up plants.

Management

- ☆ Removal and destruction of all the affected branches during the initial stage of attack.
- ☆ Seed dressing with thiamethoxam 70 WS @ 3g ai/kg provides effective protection in the initial state of the crop.
- ☆ First application of methyl oxydemeton @ 300g ai/ha should be given just after germination on the crop and repeat twice at an interval of two weeks each.

Leaf Miner
Chromatomyia horticola Gour. : (Diptera: Agromyzidae)

Major Host Plants

Leguminaceous crops, cucurbits, crucifers, tomato, lettuce,

The larvae are minute and slender. The adults are two winged flies having greyish black mesonotum and yellowish frons.

Life Cycle

The pest is generally active from December to May. The rest of the period passes in the soil as a pupa. The adults emerge at the beginning of December and after mating start-laying eggs singly in leaf tissue. The incubation period is 2-3 days. After hatching the larvae feed between the lower and upper epidermis by making zigzag tunnels. They are full-fed in about 5 days and pupate within the galleries. Pupal period lasts in 6 days. The insect completes one generation in about 13-14 days and pass through several generations during the period of its activity.

Damage

The larvae are the only destructive form. They make prominent whitish tunnels in the leaves. The tunnels made by the larvae interfere with photosynthesis, and also normal development of the plants. Mining causes wilting in seedlings.

Natural Enemies

Chrysocharis horticola, *Closterocerus agromyzae*, *Diglyphus isaea*, *Opius exiguus* and *Tetrastichus* sp. are the important parasitoids of this leaf miner.

Management

- ☆ Removal of old miner infested leaves.
- ☆ Foliar spray of NSKE (4 per cent) along with sticker during initial stage of infestation.

Pea Pod Borer
Etiella zinckenella Treits. : (Lepidoptera: Pyralidae)

Major Host Plants

Leguminous crops like peas, beans and pulses

Life Cycle

E. zinckenella is a cosmopolitan pest of worldwide distribution. The larvae are pinkish in colour. The moths are brownish with a white line on the anterior margin and a yellow transverse band of the forewing. Hind wings are pale. The moths are active during night and after mating the females lay eggs on different plant parts including the immature pods. The eggs are whitish and elliptical laid singly. The incubation and larval and pupal periods last for 5-6 and 10-25 and days respectively. The fully matured larvae drop down and pupate in the soil, the pupal period last for 10-15days.

Damage

The larvae feed on the floral parts and seeds of the younger pods resulting in economic loss. The damaged pod has a large emergence hole made by the pupating larva.

Natural Enemies

Bracon pectoralis and *Phanerotoma planifrons* are the parasitoids

Management

- ☆ Spray the crop at flower initiation with carbaryl @ 100g a.i./ha.

Pod Borer
Maruca vitrata F.: (Lepidoptera: Pyralidae)

Major Host Plants

A serious pest of leguminous crops.

Life Cycle

The eggs of are laid individually or in small batches on flowers or flower buds. The eggs hatch in 2- 3 days. Neonate larvae may be found in-group on flowers, later they disperse moving from one flower to another so that each larva damages number of flowers. Young larvae may feed on any part of the flowers or foliage, but later-instar larvae are more common in the pods. The larval stage lasts for 12-15 days. There is a brief pre-pupal for 1-2 days and the pupal stage takes about 6-7 days. Moth in medium sized with brownish-black wings. Fore wings have a conspicuous transverse white elongate mark.

Damage

Larvae make round holes on in the petals of the flowers. The pods are bored and covered with frass.

Natural Enemies

Cedria paradoxa and *Phanerotoma hendecasiella* are the important parasitoids recorded from this pest.

Management

- ☆ Spray the crop at flower initiation with carbaryl 50 WP @ 1000g ai/ha.

Red Spider Mites
Tetranychus urticae Koch :(Acari: Tetranychidae)

Please refer under pests of Okra.

Minor Pests

Hadda beetle: *Epilachna vigintioctopunctata*	Defoliator
Fruit borer: *Helicoverpa armigera*	Fruit borer
Blue butterfly: *Lampides boeticus*	Pod borer
Leaf miner: *Liriomyza trifolii*	Miner
Flea beetle: *Monolepta signata*	Defoliator
Thrips: *Thrips tabaci*	Suck the sap.
Green stink bug: *Nezara viridula*	Suck the sap.
Gundhy bug: *Leptocorisa acuta*	Suck the sap.

Pests of Cucurbits

Melon Fruit Fly
Bactrocera cucurbitae (Coq.): (Diptera: Tephritidae)

Life cycle

B. cucurbitae is a serious pest of cucurbitaceous crops particularly bitter gourd and musk melon and have dark patch on the outer margins of their wings. The female fly makes a cavity with the help of her sharp ovipositor and thrusts the white elongate eggs (cigar shaped) singly or in groups of 4 to 10 into the flowers and tender fruits. After laying the eggs, the female secretes a gummy secretion on the ovipopsition puncture and makes the entrance waterproof. During the life span of 15-60 days, a female on an average lays 58-95 eggs. The incubation period is 1-9 days. The freshly hatched maggots bore into the pulp, forming galleries. Depending on weather, the larval period extends from 3 days in summer to 21 days in winter. The full-grown maggots come out of the fruits and drop to the ground and pupate in the soil at a depth of 1.5 to 15 cm. The pupal period is 5 to 9 days in summer and may extend to 30 days in cold weather. The total Life cycle occupies 12- 34 days depending on weather conditions. There are several generations in a year.

Damage

The larvae tunnel in the fruit, contaminating them with frass and providing entry points for fungi and bacteria, which cause the fruit to rot. Owing to feeding on pulp, there is a premature dropping of fruits and make them unfit for consumption. The young infested fruits can be destroyed in a few days, but older fruits show less obvious symptoms but on splitting open, a mass of maggots in pulp is found. The infestation varies from 50-100 per cent in different cucurbitaceous crops.

Management

- ☆ Earthing up of the soil around the vine helps in exposing the pupae for desiccation and predation by birds.
- ☆ Ploughing the infested field after the crop is harvested can help in killing the pupae.
- ☆ Apply the bait spray containing 20 ml Malathion 50 EC + 10g protein hydrolysate+ 500g of molasses/jaggery in 20 litres of water per acre. When the attack is serious, it should be repeated at weekly intervals.

Red Pumpkin Beetle
Aulacophora foveicollis (Lucas): (Coleoptera: Chrysomelidae)

Host Plants

Pumpkin, cucumber, ash guard, bitter gourd and melons

Life Cycle

The grubs feed on the roots and sometimes on the stems of the host plants. They are creamy white, with a slightly darker oval shield at the back. When full grown, they measure about 12mm in length. The beetles are oblong and 5.8 mm long. Their dorsal body surface is brilliant orange red and the ventral surface is black. The female beetle lay about 250-300 oval, yellow eggs singly or in batches of 8-9 in moist soil, near the base of the plant during their life span of 60-85 days. The incubation period lasts 6-15 days. The larval and pupal periods are 13-25 and 7-17 days, respectively. The pupation takes place in earthen chambers in the soil at a depth of about 20-25 cm. On emergence, the beetles began to feed on leaves and flowers. The Life cycle is completed in 26- 3 7 days. There are five generations from March to October.

Damage

The grubs live in soil feeding on roots and underground stems of vines and sometimes on fruits in contact with the soil. The beetles feed on aerial parts of the plants like cotyledons, flowers and foliage by biting holes into them. The damage by beetles is detrimental to the cotyledon stage of the seedlings.

Management

- ☆ Deep ploughing after harvest to kill the grubs and pupae in the soil.
- ☆ Apply 2.75 kg of carbofuran 3G per acre 3-4 cm deep in the soil near the base of the plant after germination and irrigate or spray 75-150g carbaryl 50 WP in 50-100 litres of water per acre.
- ☆ Spray the crop with carbaryl @ 1000g ai/ha during the primordial stage of the crop.

Leaf Eating Caterpillar
Diaphania indica (Saunders): (Lepidoptera: Pyralidae)

It is an irregular and sporadic pest of cucurbits becoming serious currently on cucumber, gherkins and bitter gourd.

Life Cycle

The adults are medium sized moths; forewings are white and transparent with big brown patches on the margin. Females have a tuft of orange coloured hair at the caudal end. The caterpillars are elongated bright green and bear a pair of narrow longitudinal white streaks along the mid-dorsal line. The female lays up to 360 eggs, singly or in groups on the lower surface of leaves. The egg, larval and pupal periods last for 3-6, 9-14 and 5-13 days respectively. The pupation takes place in coccons formed in the leaf folds.

Damage

Initially after hatching, caterpillars feed on the leaves by binding them together. Some time the flowers are also infested. The vines are cut from one side and the terminal parts become yellowish and dry up. The fruits show clear holes made by the larvae. The bored fruits become unfit for consumption.

Natural Enemies

The gregarious endoparasitoid, *Apanteles taragamae, A.machaeralis, Elasmus brevicornis, Trathala flavoorbitalis* are the larval parasitoids and the entomopathogenic fungus, *Paecilomyces farinosus* cause excellent natural mortality in the field.

Management

- ☆ The pest should be carefully monitored in the field. In the early stage infestation the caterpillars and infested leaves along with the larvae inside can be handpicked and destroyed.
- ☆ Spraying with *Bt.* formulations @ 500 g a.i./ha.
- ☆ Foliar spray of cypermethrin @ 100 g ai/ha or endosulfan @ 500 g a.i./ha or carbaryl @ 1000 g a.i./ha once or twice based on pest pressure.

Minor Pests

Weevil: *Baris granulipennis*	Damage the fruits.
Leaf miner: *Liriomyza trifolii*	Mines the leaves.
Thrips: *Thrips tabaci, T. palmi*	Suck the sap.
Whitefly: *Bemisia tabaci*	Suck the sap.
Aphids: *Aphis gossypii*	Suck the sap.

References

Solanaceous Crops

Brinjal shoot and fruit borer (Leucinodes orbonalis *Guen.*)

Alam S.N., Rashid M.A., Rouf F.M.A., Jhala R.C., Patel J.R., Satpathy S., Shivalingaswamy T.M., Rai, S., Wahundeniya I., Cork A., Ammaranan C. and Talekar N.S., 2003. Development of an integrated pest management strategy for egg plant fruit and shoot borer in South Asia. Shanhua, Taiwan: AVRDC–the World Vegetable Center. Tech. Bull. No. 28, AVRDC Publication No. 03-548. pp. 56.

Behera T.K., Singh N., Kalda T.S. and Gupta S.S., 1999. Screening for shoot and fruit borer incidence in eggplant genotypes under Delhi condition. Indian J. Entomol. 61(4): 372-375.

Chakraborti S., 2001. A biorational approach for the management of brinjal fruit and shoot borer, *Leucinodes orbonalis* Guen. J. Entomol. Res. 25(1): 73-76.

Chaudhary O.P. and Kashyap R.K., 1992. Effect of some management practices on the incidence of insect pests and yield of eggplant (*Solanum melongena* L.) in India. Trop. Pest Managt. 38(4): 416-419

Gupta S.C., Singh P.P. and Kumar A., 1999. Effect of intercropping with spices on the incidence of *Leucinodes orbonalis* Guen. in brinjal and its economics. Shashpa. 6(1): 45-48.

Kumar P. and Johnsen S., 2000. Life cycle studies on fruit and shoot borer (*Leucinodes orbonalis*) and natural enemies of insect-pests of eggplant (*Solanum melongena*. J. Appl. Biol. 10(2): 178-184.

Patil P.D., 1990. Technique for mass rearing of the brinjal shoot and fruit borer, *Leucinodes orbonalis* Guen. J. Entomol. Res. 14(2): 164-172

Singh S.V., Singh K.S. and Malik Y.P., 2000. Seasonal abundance and economic losses of shoot and fruit borer, *Leucinodes orbonalis* on brinjal. Indian J. Entomol. 62(3): 247-252.

Srinivasan G. and Babu P.C.S., 2000. Sex pheromones for brinjal shoot and fruit borer, *Leucinodes orbonalis*. Indian J. Entomol. 62(2): 225-226.

Sudhakar K., Punnaiah K.C. and Krishnayya P.V., 1998. Influence of organic and inorganic fertilizers and certain insecticides on the incidence of shoot and fruit borer, *Leucinodes orbonalis* Guen. infesting brinjal. J. Entomol. Res. 22(3): 283-286.

Brinjal Stem Borer (*Euzophera perticella* Rag.)

Gill C.K., Dhooria M.S. and Chadha M.L., 1984. Relative susceptibility of some varieties of brinjal to *Euzophera perticella* Ragonot. Indian J. Entomol. 46(2): 239-240.

Rizvi S.M.A. and Sanyal P.K., 1977. *Euzophera perticella* Rag. (Lepidoptera: Pyralidae) as a root borer of brinjal. Indian J. Pl. Prot. 5(2): 213.

Epilachna Beetle (*Epilachna vigintioctopunctata* F.)

Dhamdhere S.V., Koshta V.K. and Rawat R.R., 1990. Effect of food plants on the biology of *Epilachna vigintioctopunctata* Fabr. (Coleoptera: Coccinellidae. J. Entomol. Res. 14(2): 142-145.

Grewal J.S., 1988. Seasonal fluctuations in population of *Epilachna vigintioctopunctata* on brinjal (*Solanum melongena*) crop. Bull. Entomol. New Delhi. 29(1): 73-75.

Mehta P.K., Vaidya D.N. and Kashyap N.P., 1995. Antifeedant properties of some plant extracts against brinjal hadda beetle, *Henosepilachna vigintioctopunctata* F. J. Entomol. Res. 19(2): 147-150.

Rajagopal D. and Trivedi T.P., 1989. Status, bioecology and management of epilachna beetle, *Epilachna vigintioctopunctata* (Fab.) (Coleoptera: Coccinellidae) on potato in India: a review. Trop. Pest Managt. 35(4): 410-413.

Rajendran B. and Gopalan M., 1999. Effect of biopathogens (*Bacillus thuringiensis* Berliner and *Beauveria bassiana* Vuillemin) on different stages of egg plant spotted beetle, *Henosepilachna vigintioctopunctata* (Fab.) (Coleoptera: Coccinellidae. J. Entomol. Res. 23(1): 9-14.

Tomato

Tomato Fruit Borer [*Helicoverpa armigera* (Hub.)]

Brar K.S., Raman H., Dhaliwal H.S. and Cheema D.S., 1995. Field screening of different species of *Lycopersicon* against tomato fruit borer, *Helicoverpa armigera* (Hubner. J. Insect Sci. 8(2): 196-197.

Gopal S. and Senguttuvan T., 1997. Integrated management of tomato fruit borer with insecticides, neem products and virus. Madras Agril. J. 84(2): 82-84.

Kashyap R.K., Banerjee M.K., Kalloo G. and Verma A.N., 1990. Survival and development of fruit borer, *Heliothis armigera* (Hubner. (Lepidoptera: Noctuidae) on *Lycopersicon* spp. Insect Sci. Appl. 11(6): 877-881.

Katole S.R., 1992. Effect of various larval food plants on the development and reproduction of *Heliothis armigera* Hb. Comparative Physiology and Ecol. 17(3): 97-101.

Patil S., Kotikal Y.K., Revanappa and Patil D.R., 1997. Effects of intercropping tomatoes (*Lycopersicon esculentum* Mill) on the infestation of tomato fruit borer, *Helicoverpa armigera* Hubner. Adv. Agril. Res. India. 8: 141-146.

Rai S., Satpathy S. and Shivalingaswamy T.M., 2005. *Helicoverpa armigera* management in vegetables– present scenario and future strategies.In: Saxena, H., Rai, A.B., Ahmad, R.and Gupta, S. (Eds.) Recent advances in *Helicoverpa* management. pp 250-264.

Satpathy S., Rai S. and Chattopadhyay M., 1999. Field evaluation of NPV and insecticides against tomato fruit borer, *Helicoverpa armigera* (Hubner. Insect Environ. 5(3): 117-118.

Sekhar P.R., Venkataiah M. and Rao N.V., 1995. Life-table studies on *Helicoverpa armigera* (Hubner) in the Krishna-Godavari zone of Andhra Pradesh. J. Insect Sci. 8(1): 27-30.

Singh D., Singh S., Cheema D.S. and Brar K.S., 1994. Resistance to *Helicoverpa armigera* (Hubner) in tomato genotypes. J. Insect Sci. 7(1): 28-31.

Sivaprakasam N., 1996. Influence of weather factors on the attraction of *Helicoverpa armigera* (Hubner) on tomato. Pest Mangt. Hort. Ecosyst. 2(2): 87-88.

Sivaprakasam N., 1998. Management of fruit borer, *Helicoverpa armigera* on tomato. Madras Agril. J. 85(1): 68-69.

Srinivasan K. and Murthy P.N.K (1991. Mustard plants trap major cabbage pests. Indian Farming. 40(11): 11-12.

Srivastava C.P., Pimbert M.P. and Reed W., 1992. Monitoring of *Helicoverpa* (= *Heliothis*) *armigera* (Hubner) moths with light and pheromone traps in India. Insect Sci. Applic. 13(2): 205-210.

Tobacco Caterpillar [*Spodoptera litura* (F.)]

Ahuja D.B. and Noor A., 1991. Effect of different host plants on the development of *Spodoptera litura* (Fab.. J. Insect Sci. 4(2): 176-177.

Chaudhari S. and Ramakrishnan N., 1980. Field efficacy of baculovirus and its combination with sub-lethal dose of DDT and endosulfan on cauliflower against tobacco caterpillar *Spodoptera litura* (Fabricius. Indian J. Entomol. 42(4): 592-596.

Dhandapani N. and Jayaraj S., 1989. Efficacy of nuclear polyhedrosis virus formulation for the control of *Spodoptera litura* (Fab.) on chillies. J. Biol. Contrl. 3(1): 47-49

Dhandapani N., Kareem A.A. and Jayaraj S., 1986. Life-fecundity tables of *Spodoptera litura* (Fabricius) on chilli. Indian J. Agril. Sci. 56(4): 290-293

Dwivedi S.C. and Mathur B., 1997. Evaluation of food preference of fourth instar larvae of *Spodoptera litura*. Insect Environ. 2(4): 138-139

Narayanan K., 1985. Susceptibility of *Spodoptera litura* (F) to a granulosis virus. Curr. Sci. India, 54(24): 1288-1289.

Ramakrishnan N., Saxena V.S. and Dhingra S., 1984. Insecticide-resistance in the population of *Spodoptera litura* (F.) in Andhra Pradesh. Pesticides, 18(9): 23, 27.

Rao S.V., Azam K.M. and Ahmed K., 1989. Mating disruption through behavioural manipulation of *Spodoptera litura* Fabricius. Curr. Sci. 58(16): 933-935.

Zaz G.M., 1990. Relative effectiveness of *Bacillus cereus* Frankland and Frankland, *B. thuringiensis* Berliner and endosulfan against *Spodoptera litura* (Fabricius) on cauliflower. Indian J. Pl. Prot. 18(1): 85-88.

Serpentine Leafminer [*Liriomyza trifolii* (Burg.)]

Dash P.C., 1994. Studies on the host range of American serpentine leafminer in Orissa. Orissa J. Agril. Res. 7: 92-94.

Jagannatha R. and Viraktamath C.A., 1997. Host preference by the female serpentine leafminer, *Liriomyza trifolii* (Diptera: Agromyzidae. Insect Environ. 2(4): 137.

Patanik H.P., 1997. Studies on neem formulations against serpentine leafminer, and tobacco caterpillar, on tomato. Insect Environ. 3(1): 10.

Srinivasan K., Viraktamath C.A., Gupta M. and Tewari G.C., 1995. Geographical distribution, host range and parasitoids of serpentine leafminer, *Liriomyza trifolii* (Burgess) in South India. Pest Managt. Hort. Ecosys. 1(2): 93-100.

Shivalingaswamy, T.M., Satpathy, S., Rai, A.B., Gracy, R.G., Kumar, A and Rai, M., 2008. *Neochrysocharis formosa* Westwood (Hymenoptera:Eulophidae), a potential parasitoid of tomato leafminer, *Liriomyza trifolii. In: Proceedings of the 2nd Congress on Insect Science: Contributed Papers, Joia, B.S., Sharma, D.R., V.K.Dilawary and P.C.Pathania (Eds.)Feb.21-22, 2008, INAIS, PAU, Ludhiana, India*

Thulsiram K., Nadagouda S., Patil B.V. and Somasekhar (1997. Efficacy of different designs of yellow sticky traps in attracting the serpentine leafminer, *Liriomyza trifolii* adults. Indian J. Entomol. 59(4): 437.

Whitefly (*Bemisia tabaci* Genn.)

Banerjee M.K. and Kalloo G., 1991. Evaluation of tomato breeding material against tomato leaf curl virus: technique of inoculation. Veg. Sci. 18(1): 109-112.

Dhanju K.C., Cheema S.S., Kapur S.P., Arora S.K. and Ghai T.R., 1995. Occurrence of enation leaf curl disease of okra in Punjab. Pl. Dis. Res. 10(2): 157-159.

Mallick S.C. and Chowdhury A.K., 1996. Virus diseases of some vegetable crops in Nadia. J. Mycopathol. Res. 34(1): 29-33.

Singh D. Jaglan R.S. and Baswana K.S., 2002. Screening of brinjal genotypes against whitefly (*Bemisia tabaci* Genn.. Haryana J. Hort. Sci. 31(3-4): 286-289.

Chilli

Chilli Thrips (*Scirtothrips dorsalis* Hood)

Chandrasekaran M. and Veeravel R., 1998. Field evaluation of plant products against chilli thrips, *Scirtothrips dorsalis*. Madras Agril. J. 85(2): 120-122.

Dhandapani N. and Jayaraj S., 1982. Effect of chilli seedling root dip in insecticides for the control of sucking pests. Pestology. 6(3): 5-10.

Kumar N.K.K., Aradya M., Deshpande A.A., Anand N. and Ramachandar P.R., 1996. Initial screening of chili and sweet pepper germplasm for resistance to chili thrips, *Scirtothrips dorsalis* Hood. Euphytica. 89(3): 319-324.

Kumar N.K.K., 1995. Yield loss in chilli and sweet pepper due to *Scirtothrips dorsalis* Hood (Thysanoptera: Thripidae. Pest Mangt. Hort. Ecosys. 1(2): 61-69.

Rao R.D.V.J.P., Iizuka N., Ragunathan V. and Joshi N.C., 1980. Occurrence of tomato spotted wilt virus on tomato in Andhra Pradesh. Indian Phytopathol. 33(3): 436-439.

Chilli Mite [*Polyphagotarsonemus latus* (Banks)]

David P.M.M., 1991. Resurgence of yellow mite *Polyphagotarsonemus latus* (Acarina: Tarsonemidae) on chilli following application of insecticides. Madras Agril. J. 78(1-4): 88-91.

Karmakar K., 1997. Effect of micronutrients on the biology of *Polyphagotarsonemus latus* (Banks) (Acari: Tarsonemidae. Environ. Ecol. 15(3): 699-701.

Mallapur C.P., 2000. Screening of chilli genotypes against thrips and mites. Insect Environ. 5(4): 154-155.

Rai A.B., Solanki V.Y., 2003. The rate of natural increase of tarsonemid mite *Polyphagotarsonemus latus* Banks (Acari: Tarsonemidae) when reared on chilli leaf (*Capsicum annum* L.. *Shashpa.* 10(1): 13-18.

Ram S., Patnaik N.C., Mohapatra A.K.B., Sahoo S. and Samal K.C., 1997. Reaction of some chilli varieties to yellow mites *Polyphagotarsonemus latus* banks under Eastern Ghat high land zone of Orissa. Environ. Ecol. 15(1): 199-201.

Solanki V.Y., Rai A.B. and Desai H.R., 2003. Biology of chilly mite, *Polyphagotarsonemus latus* Banks (Acari: Tarsonemidae) under laboratory condition. *Gujarat J. Appl. Hort.* 3 (1&2): 91-98.

Sudharma K. and Nair G.M., 1999. Assessment of loss caused by *Polyphagotarsonemus latus* Banks on chilli. Entomon. 24(1): 97-99.

Ukey S.P., Naitam N.R. and Patil M.J., 1999. Determination of economic threshold level of mites on chilli crop. J. Soils Crops. 9(2): 268-270.

Cole Crops

Diamondback moth [*Plutella xylostella* (L.)]

Chauhan U., Bhalla O.P. and Sharma K.C., 1997. Biology and seasonality of the diamondback moth, *Plutella xylostella* (L.) (Lepidoptera: Yponomeutidae) and its parasitoids on cabbage and cauliflower. Pest Mangt. Hort. Ecosys. 3(1): 7-12.

Chelliah S. and Srinivasan K., 1986. Bioecology and management of diamondback moth in India. Diamondback Moth Management. Proceedings of the First International Workshop, Tainan, Taiwan, 11-15 March, 1985. pp. 63-76.

Dilawari V.K., Khanna V., Gupta V.K., Dhaliwal H.S. and Dhaliwal G.S., 1996. Toxicity of *Bacillus thuringiensis* var. *kurstaki* delta-endotoxin proteins from HD-1 and HD-73 strains against *Plutella xylostella* (L. Allelopathy J. 3(2): 267-272.

Justin C.G.L., Rabindra R.J. and Jayaraj S., 1990. *Bacillus thuringiensis* Berliner and some insecticides against the diamond-back moth, *Plutella xylostella* (L.) on cauliflower. J. Biol. Contrl. 4(1): 40-43.

Kandoria J.L., Singh G. and Singh L., 1999. Effect of intercropping cauliflower with tomato on the incidence of diamondback moth. Insect Environ. 5(3): 137-138.

Moorthy P.N.K., Kumar, N.K.K., Selvaraj C. and Daniel J.S., 1998. Neem seed kernel extract applications for diamondback moth management: transfer of technology for mechanised farming. Pest Mangt. Hort. Ecosys. 4(2): 128-130.

Rabindra R.J., Justin C.G.L. and Jayaraj S., 1995. Relative susceptibility of two larval populations of diamondback moth (*Plutella xylostella*) to insecticides and *Bacillus thuringiensis*. Indian J. Agril. Sci. 65(2): 152-153.

Raju S.V.S. and Singh H.N., 1995. Resistance levels in the field population of *Plutella xylostella* (L.) to certain commonly used insecticides. Indian J. Entomol. 57(2): 164-166.

Reddy G.V.P. and Urs K.C.D., 1997. Mass trapping of diamondback moth *Plutella xylostella* in cabbage fields using synthetic sex pheromones. Internat. Pest Contrl. 39(4): 125-126.

Sannaveerappanavar V.T. and Viraktamath C.A., 1997. Management of insecticide resistant diamondback moth, *Plutella xylostella* L. (Lepidoptera: Yponomeutidae) on cabbage using some novel insecticides. Mysore J. Agril. Sci. 31(3): 230-235.

Srinivasan K., Krishnakumar N.K. and Krishnamoorthy P.N., 1988. Action thresholds for management of diamondback moth, *Plutella xylostella* and leaf webber, *Crocidolomia binotalis* on cabbage. Indian J. Pl. Prot. 16(1): 21-25.

Srinivasan K. and Veeresh G.K., 1986. The development and comparison of visual damage thresholds for the chemical control of *Plutella xylostella* and *Crocidolomia binotalis* on cabbage in India. Insect Sci. Applic. 7(4): 547-557.

Srinivasan K.and Murthy P.N.K (1991. Mustard plants trap major cabbage pests. Indian Farming. 40(11): 11-12.

Cabbage Head Borer (*Crocidolomia pavonana* Zell.)

Chavan A.P., Pawar D.B., Warade S.D. and Kalhapure S.P., 2004. Effect of dates of transplanting on the incidence of aphid and leaf webber on cabbage. J. Maharashtra Agril. Univer. 29(1): 93-94.

Jayanth K.P. and Nagarkatti S., 1981. An artificial diet for rearing *Crocidolomia binotalis* Zeller and *Hellula undalis* Fb. (Lepidoptera: Pyralidae), two major pests of cole crops in India. Entomon. 6(2): 95-97.

Krishnaiah K., 1980. Methodology for assessing crop losses due to pests of vegetables. In: Govindu H.C., Veeresh G.K., Walker P.T., Jenkyn J.F. (Eds.. Assessment of crop losses due to pests and diseases. pp. 259-267.

Peter C., Singh I., ChannaBasavanna G.P., Suman C.L. and Krishnaiah K., 1988. Loss estimation in cabbage due to leaf webber *Crocidolomia binotalis* (Lepidoptera: Pyralidae. J. Bombay Natural History Soc. 85(3): 642-644.

Rai P.S. and Chandra B.K.N., 1976. Bionomics of the cabbage leaf-webbing caterpillar, *Crocidolomia binotalis* Zeller. Indian J. Entomol. 38(3): 233-235.

Srinivasan K. and Moorthy P.N.K., 1991. Indian mustard as a trap crop for management of major lepidopterous pests on cabbage. Tropical Pest Mangt. 37(1): 26-32.

Suman C.L. and Wahi S.D., 1983. Sequential sampling plan for cabbage leaf webber, *Crocidolomia binotalis* (Zell. Entomon. 8(1): 61-65.

Cabbage Caterpillar (*Pieris brassicae* L.)

Kakar K.L. and Sharma J.P., 1991. Role of natural enemies on the mortality of cabbage butterfly, *Pieris brassicae* (L.) in cauliflower seed crop. Indian J. Pl. Prot. 19(2): 210-212.

Lal O.P., 1991. Studies on field resistance of different cultivars and crosses of cabbage against cabbage caterpillar, *Pieris brassicae* L. (Lepidoptera: Pieridae. Zeitschrift-fur-Pflanzenkrankheiten-und-Pflanzenschutz. 98(3): 293-300.

Sood A.K., Bhalla O.P., Verma A.K. and Sharma K.C., 1995. Development of life tables for cabbage white butterfly, *Pieris brassicae* (L.) (Lepidoptera: Pieridae) in cabbage and cauliflower crop ecosystem. J. Entomol. Res. 19(2): 127-133.

Thakur N.S.A. and Deka T.C., 1997. Ecological studies on the cabbage butterfly, *Pieris brassicae* (Linnaeus) in north eastern India. Pest Mangt. Hort. Ecosys. 3(1): 13-15.

Cauliflower Head Borer (*Hellula undalis* F.)

Bhalani P.A., 1984. Biology and seasonal incidence of cauliflower head borer, *Hellula undalis* Fabricius under Junagadh (Gujarat State) condition. Gujarat Agril. Univ. Res. J. 9(2): 63-66.

Brar K.S., Sharma B.R., Thakur J.C. and Khattra A.S., 1993. Note on the resistance in cauliflower to stem borer, *Hellula undalis* Fabricius. Indian J. Entomol. 55(3): 333-335.

Lal O.P., Gill H.S., Sharma S.R. and Singh R., 1991. Field resistance in different cultivars of cauliflower against head borer, *Hellula undalis* Fabricus (Lepidoptera: Pyralidae. J. Entomol. Res. 15(4): 277-281.

Peter C., Balasubramanian R. and Prasad V.G., 1987. Studies on the biology of the cabbage borer *Hellula undalis* Fabricius (Lepidoptera: Pyralidae) in Karnataka. Mysore J. Agril. Sci. 21(3): 309-312.

Aphids *Myzus persicae* (Sulz.)

Haseena Bhaskar and Viraktamath C.A., 2002. Diversity and abundance of aphidophagous coccinellids in cabbage field. Insect Environ. 8(1): 31.

Rajavel D.S. and Veeraraghavathatham D., 1989. Efficacy of certain plant extracts on the major pests of cabbage. South Indian Hort. 37(3): 186-188.

Sood A. and Singh G., 1997. The intrinsic rate of *Myzus persicae* (Sulzer) (Homoptera: Aphididae) on cabbage *Brassica oleracea* var. *capitata*. Annal. Entomol. 15(2): 21-26.

Sood A. and Singh G., 1999. Biology of *Myzus persicae* (Sulzer) on winter and summer cabbage in Pantnagar. Annal. Agril. Res. 20(2): 243-244.

Brevicoryne brassicae (L.)

Debaraj Y. and Singh T.K., 1996. Aerial population fluctuation of cabbage aphid, *Brevicoryne brassicae* (Linnaeus. Ann. Agril. Res. 17(3): 308-310.

Duhra M.S. and Hameed S.F., 1990. Evaluation of toxicity and persistence of residues of some insecticides on cabbage [cauliflower] against *Brevicoryne brassicae* L. Indian J. Entomol. 52(1): 146-154.

Kotwal D.R., Verma A.K. and Bhalla O.P., 1985. Prediction of the cabbage aphid, *Brevicoryne brassicae* (L.) peaks on cauliflower seed crop from thermal unit accumulations. Entomon. 10(3): 219-222.

Sharma K.C. and Bhalla O.P., 1992. Studies on the life of *Metasyrphus corollae* (Fab.) a predator of the cabbage aphid (*Brevicoryne brassicae* L.) on cauliflower seed crop. Entomon. 17(1;2): 49-53.

Thakur J.N., Rawat U.S., Pawar A.D. and Sidhu S.S., 1989. Natural enemy complex of the cabbage aphid *Brevicoryne brassicae* L. (Homoptera: Aphididae) in Kullu Valley, Himachal Pradesh. J. Biol. Contrl. 3(1): 69.

Okra

Okra Shoot and Fruit Borer [*Earias vittella* (F.)]

Ahmad S., Gupta S.C., Prakash N. and Mandal S.K., 2000. Weather factors and larval population of *Earias vittella* Fabr. in summer sown okra (*Abelmoschus esculentus* L.. J. Res. Birsa Agril. Univ. 12(2): 215-217.

Chakraborti S., 2001. An alternate approach in the management of okra fruit borer. J. Appl. Zool. Res. 12(1): 47-50.

Dubey V.K. and Ganguli R.N., 1998. Fruit loss in okra due to *Earias vitella* (Fab.. Insect Environ. 4(1): 25.

Gupta A., 1988. Effect of food on the life-processes of *Earias fabia* Stoll with references to three varieties of okra. Bull. Entomol. New Delhi. 29(2): 190-198.

Krishnakumar N.K. and Srinivasan K., 1985. Evaluation of extended spray intervals with synthetic pyrethroids for the control of shoot and fruit borer, *Earias vittella* on okra. Progr. Hort. 17(3): 254-258.

Pawar D.B., Lawande K.E. and Warade S.D., 1996. Effects of different sowing dates on the incidence of leafhopper, mite and fruit borer of okra. J. Maharashtra Agril. Univ. 21(3): 375-377.

Rai S. and Satpathy S., 1999. Influence of sowing date and insecticides on the incidence of jassid and fruit borer on okra. Veg. Sci. 26(1): 74-77.

Ramachander P.R., Biswas S.R., Jaganmohan N. and Krishnaiah K., 1983. Determination of optimum plot size for pesticide trials on okra fruitborer, *Earias vittella* Fabr. Indian J. Entomol. 45(3): 218-221.

Sardana H.R. and Tewari G.C., 1990. Spatial distribution pattern of eggs of *Earias vittella* Fabricius in okra field. Entomon. 15(1-2): 53-58.

Satpathy S and Rai S., 1998. Influence of sowing date and crop phenology on pest infestation in okra. Veg. Sci. 25(2): 178-180.

Srinivasaperumal S., Samuthiravelu P. and Muthukrishnan J., 1992. Life tables and energetics of *Earias vittella* (Fab.) (Noctuidae: Lepidoptera) reared on three different hosts. Insect Sci. Appl. 13(6): 749-754.

Tamhankar A.J., Shantharam K., Harwalkar M.R. and Rahalkar G.W., 1992. Development of an artificial diet for the laboratory rearing of the spotted bollworm, *Earias vittella* (Fabricius) (Lepidoptera: Noctuidae. Insect Sci. Appl. 13(2): 241-243.

Leafhopper (*Amrasca biguttula biguttula* Ishida)

Faleiro JR, Rai S, 1985. Determination of vulnerable stage of crop growth to leafhopper attack in okra. Indian Journal of Entomology, 47(2):238-239

Faleiro JR, Rai S, 1988. Yield-infestation relationship and economic injury level for okra leafhopper management in India. Tropical Pest Management, 34(1):27-30, 114-115, 119-120

Hooda VS, Dhankhar BS, 1992. Evaluation of wild taxa of okra against the leafhopper *Amrasca biguttula biguttula* (Ishida. Tests of Agrochemicals and Cultivars, No. 13:108-109

Lal H, Mahal MS, Rajwant Singh, Balraj Singh, 1990. Influence of rainfall on population build-up of *Amrasca biguttula biguttula* (Ishida) on okra. Journal of Insect Science, 3(2):169-171

Uthamasamy S., 1988. Seasonal incidence and damage potential of leafhopper (*Amrasca biguttula biguttula*) on okra. Indian J. Ecol. 15(1): 117-118.

Uthamasamy S., Subramaniam T.R. and Jayaraj S., 1973. Studies on the varietal resistance of bhindi (*Abelmoschus esculentus*) (L.) Moench) to the leafhopper, *Amrasca* (=*Empoasca*) *devastans* (Dist.) (Homoptera: Jassidae) I. Screening of varieties under field conditions. Madras Agril. J. 60(1): 27-31.

Onion

Onion Thrips (*Thrips tabaci* Lind.)

Bhardwaj B.S., Srivastava P.K. and Gupta R.P., 1995. Economic threshold value of thrips population on the yield of kharif onion crop. News-Letter Nat. Hort. Res. Develop. Found. 15(3): 10-12.

Brar K.S., Sidhu A.S. and Chadha M.L., 1993. Screening onion varieties for resistance to *Thrips tabaci* Lind. and *Helicoverpa armigera* (Hubner. J. Insect Sci. 6(1): 123-124.

Kumar N.K.K. and Rawal R.D., 1999. Onion thrips, *Thrips tabaci*, a vector of onion tospovirus. Insect Environ. 5(2): 52.

Sinha A.K., Sinha R.B.P. and Kumar A., 1993. Reaction of onion cultivars to thrips (*Thrips tabaci*. J. Appl. Biol. 3(1-2): 104-105.

Leguminaceous Vegetables

Two Spotted Spider Mite (*Tetranychus telarius* Linn.)

Bhanderi G.P., Rai A.B. and Patel C.B., 1995. Efficacy of certain Neem based products (Azadiractin) in comparison to conventional chemicals against *Tetranychus macfarlanei* (Acari: Tetranychidae) on okra and their safety to predatory mites. *Indian J. Environ. Toxicol.* 5(1): 21-24.

Desai H.R., Patel J.R., Rai A.B. and Patel J.R., 2003. Studies on the resurgence of mite, *Tetranychus macfarlanei* on okra. *Indian J. Appl. Entomol.* 17(1): 4-8.

Krishnamoorthy A. and Mani M., 1989. Effect of releases of *Phytoseiulus persimilis* in the control of two spotted spider mite on French beans. J. Biol. Contrl. 3(1): 33-36.

Patel C.B., Rai A.B., Patel M.B., Patel A.J. and Shah A.H., 1993. Acaricidal tests of botanical pesticides in comparison to conventional acaricides/pesticides against red spider mites (Acari: Tetranychidae) on okra, brinjal and Indian bean. *Indian J. Ent.* 55 (2): 184-190.

Patel C.B., Shah A.H. and Patel S.H., 1982. Effect of insecticides on the resurgence of two spotted spider mites, *Tetranychus telarius* Linn., in brinjal. Indian J. Agril. Sci. 52(11): 774-776.

Rai A.B., Sejalia A.S., Patel C.B. and Kumar S., 1995. The rate of natural increase of red spider mite when reared on okra. *GAU Res. J.* 21(1): 130-136.

Singh R.N. and Singh J., 1999. Effect of azadirachtin and some conventional acaricides against two-spotted spider mite, *Tetranychus urticae* Koch. Shashpa. 6(1): 85-88.

Bean Fly [*Ophiomyia phaseoli* (Tryon)]

Brar K.S., Dhillon G.S., Singh M. and Dhillon T.S., 1995. A note on the resistance in pea genotypes to leaf miner, *Chromatomyia horticola* (Goureau. Haryana J. Hort. Sci. 24(2): 156-158.

Gunjal S.B. and Mote U.N., 1994. Determination of economic threshold level of stem fly in french bean. J. Maharashtra Agril. Univ. 19(3): 427-430.

Moorthy P.N.K. and Srinivasan K., 1989. Petiole-mining thresholds and potassium chloride sprays for management of bean fly (Diptera: Agromyzidae) in snap beans and cowpea. J. Econ. Entomol. 82(1): 246-250.

Ramasubramanian G.V. and Babu P.C.S., 1991. Field evaluation of insecticides and botanicals against the spotted pod borer, *Maruca testulalis* (Geyer) on lablab. Madras Agril. J. 78(1-4): 17-19.

Sharma K.K., Yadav H.S. and Chandra A., 1999. Reaction of field bean varieties to pod borer complex. JNKVV Res. J. 33(1-2): 78-79.

Singh S., 1982. Ecology of the Agromyzidae (Diptera) associated with leguminous crops in India. Memoirs of the School of Entomology, St. John's College, Agra. No. 8, pp.126.

Srivastava A.S. and Singh Y.P., 1974. Survey, damage, life history and control of pea pod borer, *Etiella zinckenella* Tr. (Pyralidae: Lepidoptera. Sci. and Cult. 40(8): 370-372.

Cucurbits

Caterpillar, *Diaphania indica* (Sanders)

Pareek B.L. and Bhargava M.C., 2003. Estimation of avoidable losses in vegetable crops caused by borers under semi-arid conditions of Rajasthan. Insect Environ. 9(2): 59-60.

Patel R.C. and Kulkarny H.L., 1956. Bionomics of the pumpkin caterpillar *Margaronia indica* Saund. (Pyralidae: Lepidoptera. J. Bombay Natural Hist. Soc. 54: 118-127.

Peter C. and David B.V., 1991. Natural enemies of the pumpkin caterpillar *Diaphania indica* (Lepidoptera: Pyralidae) in Tamil Nadu. Entomophaga, 36(3): 391-394.

Peter C. and David B.V., 1991. Population dynamics of the pumpkin caterpillar, *Diaphania indica* (Saunders) (Lepidoptera: Pyralidae. Trop. Pest Mangt. 37(1): 75-79.

Peter C. and David B.V., 1992. Studies on the thermal requirement for development of *Diaphania indica* (Saunders) (Lepidoptera: Pyralidae. J. Insect Sci. 5(2): 172-174.

Red Pumpkin Beetle, *Aulacophora foveicollis* Lucas

Chandravadana M.V. and Pal A.B., 1983. Triterpenoid feeding deterrent of *Raphidopalpa foveicollis* L. (red pumpkin beetles) from *Momordica charantia* L. Curr. Sci. 52(2): 87-88.

Chaudhary R.N., 1995. Management of red pumpkin beetle, *Aulacophora foveicollis* (Lucas) using polyethylene cages on cucumber. Pest Mangt Hort. Ecosys. 1(1): 55-57.

Chaudry Z.A. and Alikhan M.A., 1990. Effects of temperature and relative humidity on the development and the fecundity of the red pumpkin beetle *Aulacophora foveicollis* Lucas (Chrysomelidae Coleoptera. Indian J. Entomol. 52(2): 274-278.

Dhillon N.P.S. and Wehner T.C., 1991. Host-plant resistance to insects in cucurbits–germplasm resources, genetics and breeding. Tropical Pest Management, 37(4): 421-428.

Mehta P.K. and Sandhu G.S., 1990. Monitoring of red pumpkin beetle (*Aulacophora foveicollis*) by cucurbitacins as kairomones through poison baits. Indian J. Agril. Sci. 60(3): 225-226.

Singh D. and Gill C.K., 1982. Estimation of losses in growth and yield of muskmelon due to *Aulacophora foveicollis* Lucas. Indian J. Entomol. 44(3): 294-295.

Sinha S.N. and Chakrabarti A.K., 1983. Effect of seed treatment with carbofuran on the incidence of red pumpkin beetle, *Raphidopalpa foveicollis* Lucas on cucurbits. Indian J. Entomol. 45(2): 145-151.

Fruit Fly, *Bactrocera cucurbitae* Coq.

Batra H.N., 1964. The population, behaviour, host specificity and development potential of fruit flies bred at constant temperature in winter. Indian J. Entomol. 26: 195-206.

Borah R.K., 1996. Influence of sowing seasons and varieties on the infestation of fruitfly *Bactrocera cucurbitae* (*Dacus cucurbitae*) in cucumber in the hill zone of Assam. Indian J. Entomol. 58(4): 382-383.

Singh S. and Singh R.P., 1998. Neem (*Azadirachta indica*) seed kernel extracts and azadirachtin as oviposition deterrents against the melon fly (*Bactrocera cucurbitae*) and the oriental fruit fly (*Bactrocera dorsalis*. Phytoparasitica, 26(3): 191-197.

Stonehouse, J.M., Singh, H.S., Patel, R.K., Satpathy, S., Shivalingaswamy, T.M., Rai, S., Verghese, A., Mumford, J.D., 2007. The measurement and modeling of losses of cucurbits to tephritid fruit flies. Comm. Biometry Crop Sci. 2(1): 17-25.

Stonehouse, J.M., Patel, R.K., Verghese, A., Mumford, J.D., Shukla, R.P., Satpathy, S., Singh, H.S., Jiji, T., Thomas, J., Patel, Z.P., Jhala, R.C., Patel, V.M., Manzar, A., Shivalingaswamy, T.M., Mohantha, A.K., Nair, B., Vidya, C.V., Jagadale, V.S., Sisodiya, D.B., Joshi, B.K., 2007. Village-level area-wide fruit fly suppression in India: bait application and male annihilation at village level and farm level. Crop Protection.26: 788-793.

Tewatia AS, Dhankhar BS, 1996. Inheritance of resistance to melon fruitfly (*Bactrocera cucurbitae*) in bitter gourd (*Momordica charantia*. Indian Journal of Agricultural Sciences, 66(10):617-620.

Thakur J.C., Khattra A.S. and Brar K.S., 1992. Comparative resistance to fruit fly in bitter gourd. Haryana J. Hort. Sci. 21(3-4): 285-288.

Chapter 32

Red Rot Menace and its Sustainable Management in Sugarcane

Narendra Singh*
Advisor, Cane Development
Indian Sugar Mills Association, New Delhi

Introduction

Sugarcane crop in India is grown in about 4.6 million hectares of land, providing, livelihood to about 35 million farmers. Indian Sugar industry is one of the main industries. Diseases are one of the major constraints in the profitable cultivation of sugarcane. More than 60 diseases of sugarcane caused by fungi; bacteria; viruses, and mycoplasma have been reported from different parts of India. The most important diseases that hit the Indian Sugar industry are red rot, wilt, smut, pineapple disease, rust, grassy shoot, mosaic, ratoon stunting, red strip, and leaf scald. The sugarcane crop during its stand of 10 to 16 months in the field is attacked by about more than 60 diseases in India and other countries. The most important diseases that hit the Indian sugar industry are red rot, wilt, smut, ratoon stunting disease, leaf scald, grassy shoot disease and mosaic.

It has been estimated that diseases and pests cumulatively cause a loss of nearly 15 to 20 per cent cane yield and quality under moderate levels of incidence (David and Alexander 1977). However, disease like red rot under epidemic conditions cause nearly the total loss (Alexander 1986). The sugarcane diseases hit the farmers on one hand and the millers on the other (Flow Chart 32.1). In India crores of rupees are lost annually on account of diseases and pests of sugarcane. The annual loss due to diseases is of the order of Rs. 225 crores (Agnihotri 1990).

Red Rot Disease

Red rot is one of the oldest known diseases of sugarcane and has been called "cancer" of sugarcane. In India the disease was reported in 1901 at Godavari delta of Andhra Pradesh. Bulter (1906) conducted comprehensive studies on the disease and proposed "Red Rot" as the common name.

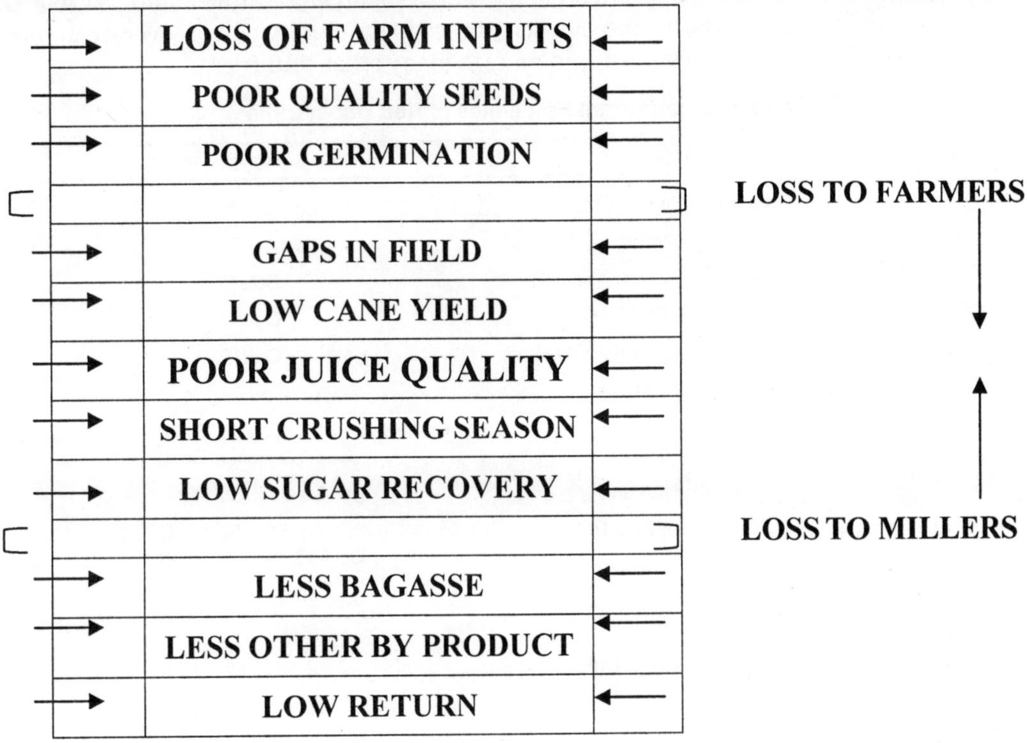

Flow Chart 32.1: Impact of Diseases and Insect Pests on Sugarcane Industry

Distribution, Prevalence and Economic Importance

Red rot has been widely distributed from all the major sugarcane growing countries of the world. However, it is regarded as a disease of major importance in Southern U.S.A., Queensland, Pakistan, Bangladesh, Myanmar and India. In India, the disease is most destructive in sub-tropical parts of India and coastal areas of Andhra Pradesh and Tamil Nadu. However, the picture is fast changing and the disease has now spread to all the sugarcane growing states. Recently during 1997-98, the occurrence of the disease was reported for the first time from northern Karnataka and southern Maharasthra (Anonymous, 1998). The disease is fast becoming a limiting factor in sugarcane production, particularly in Uttar Pradesh, Uttarakhand, Punjab, Haryana and Bihar in subtropical, Andhra Pradesh, Tamil Nadu and Gujarat in tropical India (Table 32.1).

Table 32.1: Status of Red Rot in India

Status	*State*
Major Problem	Uttar Pradesh, Haryana, Punjab, Bihar, Andhra Pradesh
High incidence	Tamil Nadu, Orissa, Gujarat, West Bengal
Medium/Low incidence	Assam, Kerala
Rare	Maharashtra, Karnataka

Epidemics of red rot have occurred as a regular event in sugarcane culture in India (Table 32.2). In the subtropical and tropical parts, the disease has resulted in the elimination of several commercial varieties. There have been reports of red rot epidemics in many other countries.

Table 32.2: Important Epidemics of Red Rot in India

Year	*Area*	*Major Genotype Affected*
1903	Godavari, Andhra Pradesh	Local varieties
1906	Champaran, Muzaffarpur, Bihar	Local varieties
1932	Pusa, Bihar	Co210
1936	Intiathoka, Gonda, U.P.	Co210
1937	Nagina, Golagokarannath, U.P.	Co210
1938-40	Bihar, Eastern U.P.	Co213, Co 299, Co 331
1946-47	U.P., Bihar	Co 299, Co 313, Co 395, Co 453, Co 421, Co 312, Bo 11, Bo 17
1950-51	Punjab, Haryana	Co 313, Co 313
1952-54	Bihar, Eastern U.P.	Co 453, Co 513, Bo 11, Co 331, Co 443, Co 419,Co 421
1957-58	Bihar	Bo 17, Co 313
1957-58	Delhi	Co 312
1959-62	Punjab, Western U.P.	Co 312, Co 313
1970-72	Tamil Nadu	Co 419, Co 658
1980-81	Mawana, Meerut, U.P.	Co 1148
1980-81	Andhra Pradesh	Co C 671, Co 419, Co 997, Co A 7701
1982-84	Kerala	Co 997, Co 419, Co 785
1990-93	Western U.P., Eastern U.P., Haryana, Punjab, Bihar	Co 1148, Bo 70, Bo 54, Co 7717
1992-97	Gujarat	CoC 671
1998-99	Punjab, Haryana	CoJ 64, CoJ 82, CoJ 84

* Reasons for epidemics of red rot.

Significance of incipient infection in disease spread.

The epidemics of red rot in India can mainly be attributed to the use of infected seed materials containing incipient/dormant infections in the nodal region which normally remain undetected unless otherwise insepected by trained eyes. Many workers have demonstrated the importance of incipient infections in the dissemination of the pathogen to a larger area (Srinivasan and Alexander, 1966 and Singh, 1997). Cane development workers were totally unaware to the presence of incipient infections in the nodal region of cane. The nodes, which externally appear to be normal to the naked eyes can be detected by trained eyes and appear as brown lesion in the place of root primordial, leaf scar and growth ring upon scrapping the region. Such diseased setts cause reduction in germination, crop growth, cane yield and Juice quality.

The incipient infections express themselves in the form of disease during the month of May to July. The symptoms produced are dark brown to straw colored continuous lesions with heavy sporulations on both the surfaces of midrib of the spindle (unfolded) leaves or leaf shath. With the advent more favourable environment (humidity and temperature) during July and thereafter, the disease bursts out suddenly into an epidemic form. Virk *et al.* (1996) showed that presence of spindle infections is a positive sign towards crop susceptibility.

*Abiotic Stress

Moisture Stress

Dry spells, particularly during the pre-monsoon period and sometimes also in between monsoon rains, predispose the plants to red rot attack. This also leads to the formation of more dormant/resting structures of the pathogens (Singh *et al.*, 1985). These surviving structures present on the crop or debris of the previous crop sporulate and produce numerous acervuli loaded with conidia upon receiving favourable conditions due to recurrent rains during monsoon.

The water stress leads to the formations of hallow pith in some varieties. This may be one of the reasons fir predisposition of sugarcane. Papplis and Katsanos (1965) observed presence of more dead parenchymatous cells in canes to facilitate the red rot attack.

Water Logging

Sugarcane in many areas is subjected to water logging for varind periods during the rainy monsoon season (July to September) which also tend to predispose the plant to red rot attack (Butler 1918). This results from more infections in the nodal regions, particularly those which are present at the air water interface, as the spores, acervuli or other fungal structures floating on water easily come in contact with root primoida, leaf scars, buds and growth rings under ideal conditions for their germination and entry into the host tissue. Apart from this, under water logged conditions, conidia of *C.falcatum* in the acervili starts germinating and fuse with each other thereby forming aggregates of fused conidia. After fusion, the resultant conidia have shown higher virulence resulting in the break down of resistance of sugarcane genotypes (Dattamajumdar *et al.*, 1990)

Variability

C. *falcatum* is a facultative saprophyte and keeps on changing in nature due to mutation, heterokaryosis and adaptation as a result new races of the pathogen are formed. The information generated on the behaviour of various isolates

To date, there is no variety, which is immune to this disease. The presence of red rot in seed cane is known to reduce germination (Agnihotri *et al.*, 1979; Singh and Singh 1994). The loss in cane yield due to red rot has been estimated to be 44.1 per cent in Co5430 (Kirtikar and Verma 1962), 12 to 41.5 per cent in variety Co313, and 35 to 86 per cent in cane yield of ratoon over preceding plant crop (CoLK 7701) having different disease levels (Singh *et al*,1991).

Reduction in sucrose content in red rot affected canes has been reported by many workers (Kirtikar and Verma 1962, Sandhu *et al.*, 1969; Singh 1997). It reduces brix (7.03 to 32.9 per cent), pol percent (7.43 to 38.7 per cent), purity coefficient (0.45 to 8.26 per cent), extraction (4.97 to 40.79) and commercial cane sugar (7.84 to 43.93 per cent) depending on the variety and intensity of the infection (Tables 32.3a–32.3b). On the other hand, it increases the amount of reducing sugars (19.23 to 30.93 per cent), depending on variety and intensity of the infection (Beniwal *et al.*, 1989).

Table 32.3a: Effect of Red Rot on Juice Quality

Variety	Juice Quality Parameters					
	Brix %			Pol %		
	Healthy	Diseased	Reduction %	Healthy	Diseased	Reduction %
Co 7314	17.08	15.88	7.03	13.72	12.70	7.43
Co 1148	16.58	13.50	18.58	13.31	10:63	20.14
CoS 802	18.43	15.78	14.38	15.42	12.58	18.42
CoJ 64	18.68	15.50	17.02	15.47	11.89	23.14
CoL 9	19.83	13.38	32.53	16.62	10.62	36.10

Table 32.3b: Effect of Red Rot on Juice Quality

Variety	Juice Quality Parameters					
	Extraction %			Commercial Cane Sugar %		
	Healthy	Diseased	Reduction %	Healthy	Diseased	Reduction %
Co 7314	72.4	68.8	4.97	8.55	7.88	7.84
Co 1158	74.0	59.2	20.00	9.01	6.50	27.85
CoS 802	68.0	60.4	11.18	9.94	7.78	21.73
CoJ 64	74.0	61.2	17.30	9.89	7.10	28.21
CoL 9	76.0	45.0	40.79	10.73	6.55	38.96

Data after Beniwal *et al.* (1989), Indian Sugar 39:403-405.

Diagnostic Symptoms

Expression of the disease varies depending on the nature of infection and prevailing environment. If sufficient inoculum is present in the sett in active form, it causes both pre and post emergence death of the sprouts. Incipient infections on seed sets are present on nodal region of cane in form of very minute red brown to dark lesions, usually detected after scrapping of nodal region. Such infections express during the months of April to July as dark brown colored continuous lesions with heavy sporulation on both the surfaces of the midrib of unfolded leaf (spindle) and old leaf sheaths (Singh 2000). Typical symptoms in standing crops are seen during monsoon season. Main diagnostic symptoms of red rot can be observed in the internodes of stalk by splitting it longitudinally. Such canes emit peculiar sour smell of fermentation. Pith of splitted cane displays reddish brown discoloration with white transverse bands (Figure 32.1). In advanced stage, the red discoloration turns grayish brown, stalk becomes hollow and the fungus grows and sporulate in its lumen.

The Pathogen

The causal organism, *Ceolletotrichum falcatum* Went imperfect stage has been extensively studied. The fungus grows and sporulates luxuriantly on oat meal agar medium. The mycelium on the fungus in host tissues is mostly intracellular and grows in the parenchymatous cells of the pith. The hyphae are thin, hyaline, branched, spetate and contians oil droplets. Conidia are produced on short conidiophores packed inside the acervulus.

Glomerella tucumanensis (speg) Arx and Muller is a ascigeral or sexual stage of red rot pathogen. Parithenia of *G. acumumanensis* are inconspicuous and mostly embeded in leaf sheath and blades. These are dark brown to black and measure to 62 to 250 mm in diameter. Numerious hyaline, spetate, filiform paraphyses having oil globules are present in the slightly papillate ostiole. Asci are clavate, unitunicate, eight ascospores are elliposoid, hyaline, straight and measure 12-20 mm x 5-11mm (Singh and Singh 1989).

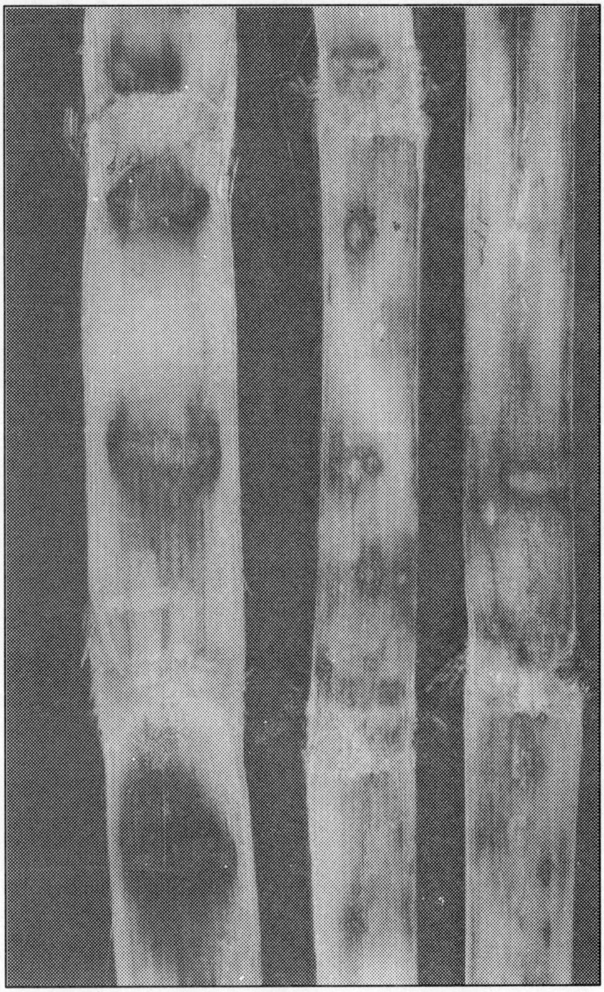

Figure 32.1: Red Rot Symptoms

Variability in *Colletotrichum faleatum* Went

C. falcatum is a facultative saprophyte and keeps on changing in nature due to mutation, heterokaryosis and adaptation and as a result new races of the pathogen are formed. Studies on variability in *C. falcatum* carried out at U.P. Council of Sugarcane Research indicated appearance of several new strains based on cultural, morphological variations and relative pathogenic behaviour on sugarcane varieties (Kirtikar *et al.,* 1964; Singh and Rana 1968; Pandey and Sakal 1974). In the last few years, efforts are being made to identify standard differentials for designation of physiologic races of red rot pathogen.

Few studies were correlated to identify specific pathotypes/races based on differential reactions on host differentials. Alexander *et al.* (1979) classifeid *C. falcatum* isolate of India in five major pathotypes *viz.* MP-1 (Tamil Nadu area), MP-2 (coastal Andhra Pradesh), MP-3 (eastern U.P. and Bihar), MP-4 (western U.P., Haryana and Punjab) and MP-5 (Kerala). Khirbat *et al.* (1980) and Beniwal *et al.* (1989) reported the existence of 3 pathotypes in Haryana. Singh *et al.* (1993) reported 4 distinct races from U.P.

The major limitations of pathogenic variability studies in *C. falcatum* was non-availability of common set of differentials. Keeping this in view new experiment was initiated under All India Coordinated Research Project on Sugarcane to identify existence of physiologic races on a common set of host differentials at different locations in India. For this 13 differentials *viz. Baragua, Khakai,* SES-594, Bo 91, CoS 767, CoC 671, CoJ 64, Co 419, Co 975, Co 997, Co 1148, Co 7717 and Co 62399 were used. Information generated on the behaviour of various isolates on the differentials demonstrated existence of at least 6 races (Anonymous 1992). These races were designated as CF01 (Co 1148), CF02

(Co 7717), CF03 (CoJ 64), CF04 (Co 419), CF05 (Co 997) and CF06 (CoC 671). Subsequent variability studies conducted revealed the existence of 4 new races as CF07 (CoJ 64), CF08 (CoJ 64/CoJ 84), CF09 (CoS 767) and CF010 (85A261) (Table 32.4).

Annual Recurrence

Annual recurrence of red rot disease is through infected seed materials, diseases stubbles and by resting propagules in the soil (Singh and Singh 1983). Sett transmission is the most common method of spread of the disease. The setts carry infection which may be internal, external both. In nature, red rot pathogen enters through the leaf scar, root primordia, and bud (Singh 1988).

Nature of Resistance

The various factors determining resistance against red rot pathogen have not been fully understood. Resistance to red rot is morphological (mechanical), physiological and protoplasmic.

Morphological

It refers to structures or modifications in the plant tissue which mechanically restrict or prevent the entry or spread of the pathogen of epidermis, cuticle, bud scales, rind and the relative abundance of vascular bundles under the rind and the presence of septa in vascular bundles thus preventing the rapid migration of spores (Jaglan *et al.*, 1995; Verma *et al.*, 1971).

Physiological Resistance

Physiological resistance in which the living cells of the plant suppress pathogen development. In the resistant varieties, gums are formed in the intercellular spaces of the parenchyma, in response to the presence of red rot fungal hyphae in the vicinity of the infection which adversely affect the progress of the infection (Srinivasan, 1965). Investigations also revealed no positive correlation between resistance and phenolic content in sugarcane (Verma *et al.*, 1971).

Genetics of Resistance

Most of the evidence available indicates that either there is no association between the degree of resistance of the progenies and parents or that crosses involving resistant parents tend to have more resistant off-springs than crosses involving susceptible parents (Mehla *et al.*, 1997). Chaudhary *et al.*,(1988) observed that resistance was controlled by more than one gene and their effect may be cumulative.

Ram *et al.* (1998) assessed the inheritance of resistance with seedlings of six sugarcane crosses. High juice quality was found to restrict nodal transgression and lesion width and each in a cross. More stalk diameter helped in disease development vertically and horizontally through nodal transgression and transverse white spot formation. Resistance selection should be based on the component traits like nodal transgression and lesion width. The distribution of seedlings of each cross in different resistant groups indicated that the use of resistant parents resulted in more number of resistant seedlings.

Sources of Resistance

Basic species of *Saccharum* assume considerable importance as sources of genes for meeting the challenges of the rapidly changing race picture of red rot pathogen. The information on screening of these clones for red for resistance is presented in Table 32.5 (Srinivasan, 1995).

Table 32.4: Current Status of Physiological Races of *C. falcatum* in India

Race No.	*Location/ Zone*	*Source*	*Co997*	*Co 1148*	*Khakai*	*CoJ64*	*Co 62399*	*Co419*	*Co975*	*Co 7717*	*CoC 671*	*Baragua*	*CoS 767*	*BO91*	*SES 594*
Cf01	North-West	Co1148	S	S	S	S	S	R	S	R	S	R	R	R	R
Cf02	North-West	Co7717	S	R	S	I	I	I	R	S	S	R	R	R	R
Cf03	North-West	CoJ 64	S	R	S	S	R	R	R	R	I	R	R	R	R
Cf04	East Coast	Co 419	I	S	S	S	R	S	S	S	S	R	R	R	R
Cf05	East Coast	Co 997	S	I	I	I	R	R	R	R	S	R	R	R	R
Cf06	East Coast	CoC 671	S	I	I	S	R	R	I	R	S	S	R	R	R
Cf07	North-West	CoJ 64	S	S	S	S	R	I	R	R	R	R	R	I	R
Cf08	North-West	CoJ 64, CoJ 84	S	S	S	S	S	I	S	S	S	R	I	R	R
Cf09	North-West	CoS 767	S	S	S	S	R	I	I	R	I	R	S	R	R
Cf10	East Coast	85A261	S	R	R	-	-	S	S	S	S	S	R	-	R

S: Susceptible; R: Resistant; I: Intermediate.

Table 32.5: Red Rot Reaction of *Saccharum* Species

Species	*Red Rot Reaction*					
	R	*MR*	*MS*	*S*	*HS*	*Total*
S. officinarum	7	15	14	33	210	279
S. barberi	1	3	3	6	5	18
S. sinense	2	6	3	1	9	21
S. robustum	2	8	9	9	37	65
S. spontaneum	109	114	26	54	12	315

Results in Table 32.5 clearly indicate the relatively very low number of clones of *S. officinatum*, *S. barberi*, *S. sinense* and *S. robustum* showed resistance as compared to *S. spontaneum*. Twenty eight of 67 clones evaluated against 3 races of north western zone showed resistance against red rot. *S. spontaneum* and Indo-American (IA) clones exhibited greater degree of resistance to red rot pathogen (Singh *et al.* 2001).

Morphological Factors in Relation to Red Rot Resistance

Generally three types of structure modifications are mostly limited with the mechanical resistance *i.e.*

- ☆ The nodal tissues consisting of thick walled slerenchyma which operate against transnodal growth of the pathogen (Edgerton, 1959).
- ☆ Presence of septa in the trachcheal tubes at the nodes which block the free movement of Conidia of fungus upwards in the vascular stream and consequently the development of secondary lesions.
- ☆ Continuity of vessels in internodes of sugarcane clones also influence the resistance to red rot. Resistant clones Co 7317 and CoS 767 recorded least continuity where as susceptible varieties recorded higher continuity.

Integrated Management System

Following steps would be effective and helpful to minimize the spread of red rot and losses in cane and sugar production.

Domestic Quarantine

In India lot of seed cane often move from one state to another state and no regulatory measures are imposed during such an inter state or inter zonal movement of canes. Introduction of red rot into new areas (Gujarat, Karnataka and Maharashtra) are glaring examples. Therefore, implementation of domestic quarantine 'Disease and Insect pest Act' at state or zonal level is absolutely necessary.

Healthy Seed Production Programme

The key factor in sugarcane production is the use of disease free high quality seed. The following aspects need to be taken care of in the seed programme.

Foundation Seed

The material used for raising foundation seed (Stage I) nursery should be free from any infection of diseases. After cutting the setts, further treatment with bavistin or mercurial fungicides would provide protection against soil borne pathogens. In order to produce healthy seeds, following steps should be taken into account.

- ☆ Easily approachable seeds
- ☆ Fertile field with assured water supply and free from water logging
- ☆ Cane affected with borers should be avoided
- ☆ Nucleus seed for planting should be taken from 10 months old or late planted crop
- ☆ The seed material should be treated with hot water (50°C for 2h) or with moist hot air (MHAT) 54°C for 4 hr

Certified Commercial Seed (Stage II)

The crop raised as foundation seed would provide seed for raising certified seed (stage II) except heat treatment. All other things except heat treatment remain to be followed under foundation seed programme.The certified seed supplied to the farmers need to be changed every five years.

Method of Planting and Seed Certification of Seed Crop

All crops raised for seed purpose should have paired row planting with distance with in pair as 30-60 cm and between the pairs as 90 cm or 120 cm, provides sufficient space for regular inspection of the crop at all stages of growth:

The crop raised for stage I and II seed production should be inspected by the experts; at least 4 times (during April–June), July–September, October-November, December-January). Diseased clumps/ plants as and when detected should be rouged out followed by spot application of 5-10 gm bleaching powder per spot. The permissible limits for the red rot being 0 per cent during the last inspection.

Varietal Balance

At present, generally 1 to 2 varieties occupy as high as up to 80 per cent of the total area under sugarcane in each factory. Even most resistant variety do not escape disease infection when their area becomes very large in a particular region. Prolonged cultivation of a single variety in the area helps in the development of new race of the pathogens and should be avoided. Therefore, a variety should not occupy more than 15-20 per cent area under cane to check epidemics (Singh 2000). Further the varieties from a narrow genetic background should not be cultivated in an area.

Regular Monitoring and Rouging of Sugarcane Crop

Strict vigilance should be kept on the buildup of infection of red rot disease in crop from April–July. Infected shoots/clumps, as when detected, should be rouged out to reduce the secondary spread of the disease. In case of red rot; after rouged out of infected plant, 5-10 g of bleaching powder (which produces chlorine as disinfectant), should be applied to each spot followed by light covering of soil to disinfect site from leaf over inoculum in soil (Singh 2000).

Cropping System

Rotation with rice, linseed, onion, garlic may be helpful to reduce the red rot inoculum in soil. Intercropping with garlic, onion, guar, coriander and linseed showed reduction in red rot incidence (Singh *et al.*, 1983; Singh and Lal, 1993).

Field Hygiene

The crop residues *i.e.* stubbles, dry foliage, dead and dry canes should be burnt after harvesting the crop to reduce the inoculum of the red rot. As the incidence of red rot increases in ratoon crop, rationing of infected crop should be avoided.

Proper Irrigation and Drainage in the Fields

Water logging is also known to predispose the plant to red rot infections. To avoid this, upland fields should be selected as far as possible or good drainage should be ensured. Planting of canes on raised furrows in water logged area may be helpful to reduce red rot (Singh 2000).

Biological Control

It is the best way of combating harmful pathogens with the beneficial micro-organisms. But the problem in this method of control is that in nature the buildup of these antagonists is very slow and stabilization of the pathogen population comes when the crop is badly damaged by the pathogen.

So far few antagonists have been found effective against red rot pathogen. Soil, sett treatment and foliar spray (in last week of June) with *Chactomium globasum,* a fungal bioagent have shown some success against red rot in the field (Singh 1994; Singh 1998). Similarly, sett treatment with *Pseudomonas fluorescence* combined with spray after 60 days to 120 days of planting and soil application at 45 to 75 days reduced the incidence of red rot (Viswanathan and Samiyappan 1999).

Resistant Varieties

Cultivation of a resistant variety is the easiest, economical way to combat the disease. Breeding and selection of resistant varieties against red rot disease of sugarcane is being carried out at different sugarcane research Centres in the country. It has been observed that the resistance of a variety, especially in case of red rot breaks down after some years due to formation of new races of the pathogen. Under such a situation, continuous flow of disease resistant varieties has to be maintained. Following new varieties showed resistance to red rot *i.e.* Co 98014, Co 89029, CoS 96268, CoPant 94211, CoPant 97222, Co 0118, Co 0238, Co 0239, Co 99004, BO 38, CoP 02181, CoSe 03234, CoS 98259, CoS 99259 and CoP 9301 etc.

Management of Red Rot: A Success Story

Work on these lines was started in 1998 at GTB Sugars (presently A.B. Sugars), Dasua (Punjab). Red rot management was carried out in 440 ha with different sugarcane varieties. Following steps were taken for management:

* Strict vigilance was kept to check the build up of initial infection in crop from April to July by

1. Roguing of infected plants/tillers, as and when detected followed by 5-10 g of bleaching powder spot application and covering the spot with moist soil.
2. If the initial infection is more than 2 per cent till June, apply bleaching powder @ 10 kg/acre by mixing it with soil or sand and applying it in the furrow near the root zone followed by light irrigation.

Results of the investigations (Tables 32.6–32.9) clearly indicated that red rot losses effectively reduced by following integrated management practices.

Table 32.6a: Effect of Management Practices on Red Rot Incidence and Cane Yield in GTB Sugars, Dasua (Punjab)

	Number of Villages	*Number of Fields*	*Variety*	*Area (ha)*	*Disease Incidence (%)*		*Cane Yield t/ha*
					Before	*After*	
	6	8	CoJ 64	9.0	18.0	11.0	67.2
	4	5	CoJ 82	4.2	10.2	3.4	73.0
	1	1	CoJ 1148	0.8	14.0	5.5	70.0
	4	6	CoJ 84	4.8	20.0	4.63	72.5
	1	1	CoJ 78	1.0	20.0	2.5	80.0
Total	16	21		17.8			
	Average				16.4	5.4	72.44

Table 32.6b: Disease Incidence and Cane Yields in Unmanaged Fields

	Number of Villages	*Number of Fields*	*Variety*	*Area (ha)*	*Disease Incidence (%)*		*Cane Yield t/ha*
					Before	*After*	
	1	1	CoJ 64	0.8	20.0	100.0	0.0
	1	1	CoJ 82	0.8	15.0	40.0	30.0
	2	2	CoJ 1148	2.0	25.0	70.0	20.0
Total	4	4		3.16			
	Average				20.0	70.0	16.7

Table 32.7a: Effect of Management Practices on Red Rot Incidence and Cane Yield in GTB Sugars, Dasua (Punjab)

	Villages	*Number of Fields*	*Area (ha)*	*Disease Incidence (per cent)*	
				Initial	*Final*
	Maithadi	10	4.05	15.40	4.60
	Raipur Nagli	1	0.47	5.00	0.00
	Khanjapur	5	1.40	5.14	2.40
	Lesoda	10	2.55	11.32	2.62
	Ladpur	5	1.00	3.20	0.50
	Aroda	1	0.30	25.00	10.00
	Mustafabad	1	0.60	30.00	20.00
	Tulsipur	6	2.30	4.70	2.58
Total	16	39	12.67		
	Average			12.47	5.34

Table 7b: Disease incidence in unmanaged field

	Villages	Number of Fields	Area (ha)	Disease Incidence (per cent)	
				Initial	Final
	Basaich	1	0.10	10.00	30.00
	Maithedi	1	0.25	20.00	65.00
Total	16	2	0.35		
	Average			12.47	47.5

Table 32.8a: Effect of Management Practices on Disease Incidence and Cane Yields in Selected Areas of Simbhaoli Sugars, Simbhaoli

Mill area	No. of Infected Fields when Extent of Disease		*No. of Fields Becoming Free of RR	No. of Still Infected Fields After Management (%)
	< 2 per cent	>2 per cent		
Kithor	83	142	105	120(0.1-3.0)
S.loti	0	29	9	20(2.0-5.0)
Garh	1	8	9	0
Derakuti	6	4	10	0
KRC	16	85	60	41(0.5-3.0)
Babugarh	0	2	2	0
Gate	42	161	150	53(0.5-2.0)
Total	148	431	345	234

*Total no. of 579 fields had an area of 400 ha in all.

Table 32.8b: Effect of Management Practices on Disease Incidence and Cane Yields in Selected Areas of Simbhaoli Sugars

Types of Crop	No. of Fields	Area in ha	Initial Incidence %	Final Incidence %	Cane Yields t/ha
Managed ratoon	12	4.13	5.70	1.59	63.70
Unmanaged ratoon	2	0.62	10.50	32.00	18.50
Managed plant crop	7	2.88	2.50	0.50	66.60
Unmanaged plant crop	2	0.75	3.70	19.30	34.70

In order to raise crops of CoJ 64 in a total of 80 ha of land as seed crop, the following practices were used.

- ☆ Selection of seed from disease free fields
- ☆ Two budded setts from upper half of the cane
- ☆ Rejection of such setts showing any kind of reddening at cut ends or nodes
- ☆ Sett treatment with bavistin 0.1 per cent for 5 minutes

- ✩ Avoid planting in previously infected fields
- ✩ Application of organic manure @ 10 q/acre mixed *Trichoderma viridae* and *Chaetomium globosum* at the time of planting
- ✩ Planting of canes in furrows at 70 to 90 cm
- ✩ Roguing and bleaching powder application done as and when required
- ✩ Ten places were selected for detailed data recording

Table 32.9: Effect of Seed Production Techniques

Sl.No.	Farmer's Name	Germination (%)	Red Rot Incidence %		Cane Yield t/ha
			June 2004	Oct 2004	
1.	Satyapal	62.0	0.0	0.0	87.6
2.	Om Prakash	68.0	1.0	0.0	93.7
3.	Devender	57.5	0.0	0.0	90.0
4.	Rambhol	54.5	0.5	0.0	70.0
5.	Virender Singh	50.5	2.5	0.5	63.5
6.	Rish Pal	53.8	20	0.2	80.3
7.	Mukesh	55.5	0.0	0.0	74.5
8.	Omvir	50.2	1.0	0.0	70.6
9.	Rambir	54.3	0.00	0.0	72.0
10.	Ramesh Chand	52.6	0.0	0.0	70.0
Average		55.9			77.24
Control	Virender	34.0	9.7	30.0	37.5

Result (Table 32.6) indicated that use of healthy 2-budded setts, sett treatment with bavistin, application of bioagents with organic manure, timely rogueing of initial infected plants and spot application of bleaching powder are very effective in reduction of red rot, improvement in cane yield and also possible to increase life span of susceptible variety.

References

Abbott, E.V. and Hughes, C. G., 1961. Red rot. In: *Sugarcane diseases of the world*, by Martin *et al.*, Elsevier publishing comp., Amsterdam.

Agnihotri, V.P., 1983. *Diseases of sugarcane*. Oxford and IBH Publishing Co. Pvt. Ltd., New Delhi pp. 373.

Agnihotri, V.P., 1990. *Diseases of sugarcane and sugarbeet*. Oxford and IBH Publishing Co. Pvt. Ltd., New Delhi. pp. 483.

Agnihotri, V.P., 1996. Current sugarcane disease scenario and management strategies. *Indian Phytopath.* 49: 109-126.

Agnihotri, V.P. and Dutta Majumder, S.K., 1994. Diseases management practices in sugarcane. Past, present and future. In: *Current Trends in sugarcane pathology* (Eds. Rao, G.P. *et al.*,) pp. 333-345. Intern Books and Periodicals Supply Service, Pitampura, New Delhi.

Agnihotri, V.P., budhraja, T.R. and Singh, K., 1979. Role of diseased setts and soil, and the annual recurrence of red rot in sugarcane. *Intern. Sugar Jour.* 82: 263.265.

Agnihotri, V.P., Singh, N, Lal, Ramji and Singh, V., 1995. Management of red rot disease of sugarcane. *Proc. Nat. Sem. In: Strategies for research and management of red rot.* (Eds. Agnihotri, V.P. *et. al.*). Indian Institute of Sugarcane Research, Lucknow. pp. 294-213.

Alexander, K.C., 1986. Diseases of sugarcane and their management. Dr. N. Prasad Memorial invited lecture. *Indian J. Mycol. Pl. Path.* 16: 16-19.

Alexander, K.C., Kandasami, P.a., Ramanarao, T.C., Mohan Raj and Rao, M.M., 1979. A new approach to genetic defence against red rot disease of sugarcane. *Indian Sug.* 28: 695-696.

Alexander, K.C 1992. Sugarcane diseases management latest trends in evolving varieties resistant to diseases. *Sugar news*, S.F. Madras: 35: 82.

Anonymous., 1992. Proceedings of the plant pathology group. *All India Coordinated Research Project Workshop* held at IISR, Lucknow. October 14-15, 1992.

Anonymous., 1998. Technical Report–Plant Pathology 1998. All India Coordinated Research Project on sugarcane, VSI, Poona.

Beniwal, M.S., Satyavir and Virk, K.S., 1989. Pathogenic variability in *colleotrichum faleatum* incitant of red rot of sugarcane. *Indian Phytopath.* 42: 95-98.

Beniwal, M.S., Satyavir, and Taneja, A.D., 1989. Effect of red rot on juice quality of sugarcane. *Indian Sugar.* 39: 403-405.

Bhatti, D.S., and Chohan, J.S., 1971. Comparative morphology and nature of pathogenicity of *Cepheclosporium sacchari* Butler and *Fusarium moniliforme* Sheldon, *Jour. Res.* (Punjab Agri. Univ.), 8: 327-330.

Butler, E.J., 1906. Sugarcane diseases in Bengal. *Mem. Dept. Agri. India Bot. Series* 1(B) p 1-52.

Butler, E.J., 1918. *Fungi and diseases in plants.* Thanker spink and Co. Calcutta. 1395-401.

Chaudhary, B.S., Virk, K.S. and Ram, K., 1988. Inheritance of red rot disease in sugarcane. *Agri. Sci. Digest.* 8: 210-212.

David, H. and Alexander, K.C., 1977. Controlling pests and diseases of sugarcane–A means to learn and earn. *Indian Fmg.* 27: 37-38.

Dattamajumdar, S.K, Singh and Agnihotri, V.P., 1990. Behaviour of *Colletotrichum falcatum* under water logged conditions. *Indian Phytopathology*, 43: 227-229.

Edgerton, C.W., 1959. *Sugarcane and its diseases.* Louisiana State Uni. Press. La, USA, pp: 69-100.

Jaglan, B.S., Virk, K.S. and Satyavir., 1995. Studies on the nature of red rot resistance in sugarcane. *Proc. National Seminar on sugarcane production constraints and strategies for research and management of red rot.* IISR, Lucknow. 357-363.

Khirbet, S.K., Satyavir and Beniwal, M.S., 1980. Physiological and pathological variability in sugarcane red rot pathogen (*Colletobricum falcatum*) in Haryana. *Indian Phytopath.* 33: 296-299.

Kirtikar and Verma, H.S., 1962. A review on effect of sugarcane diseases on yield and juice quality in Uttar Pradesh. *Indian Sugar.* 12: 103-108.

Lewis, H.D., Natarajan, S. and Rajan, S.D., 1976. Control of sugarcane red rot (*Physalospora tucumanensis* speg.) by chemotherapy. *Sugarcane Pathol. Newsl.* 17: 17-20.

Mehla, A.S., Virk, K.S., Chaudhary, B.S. and Jaya, R.S., 1997. Inherent gene expression for red rot resistance in sugarcane. *Indian Sugar*. 271-274.

Papplis, A.J., Katsanos, R.A., 1965. An approach to the study of the physiology of senescence and parasitism in sugarcane. *Phytopathology*. 55: 620-622.

Ram, B., Singh, N., Hema Prabhu, G., Sahi, B.K. and Tripathi, B.K., 1998. Studies of red rot resistance and its association with yield and quality traits in sugarcane seedlings. *Proc. 60th Ann. Conv. STAI*, Shimla, Agriculture. 25-33.

Sandhu, S.S., Bhatti, D.S. and Rattan, B.K., 1969. Extent of losses in sugarcane caused by red rot (*Physalospora tucumanensis* speg.) and smut (*Ustilago scitaminea* Syd.). *Jour. Research (PAU, Ludhiana)*, 6: 341-344.

Sarma, M.N., Satyanarayana, Y., Ramarao, M.A. and Ramakrishnarao, S.V., 1980. Soil infection and the spread of red rot disease of sugarcane. *Maharashtra Sugar*. 5: 13-15.

Satyavir, Singh, Virk, K.S., Nagesarwrrao, G.V., Singh, A. and Misra, S.R., 2000. Pathogenic variability of sugarcane red rot system. Pp 109-114. (Eds. S.Nagarajan and D.P. Sigh). In: *Role of resistance in Intensive Agriculture*. Kalyani Pub., Ludhiana.

Singh, G.P. and Rana, O.S., 1968. A new strain of *Glomerella lucumanensis* red rot pathogen in U.P. *Indian Sugar*. 18: 537-540.

Singh, K. and Singh, R.P., 1989. Red rot. In: *Diseases of Sugarcane–Major Diseases* (Eds. Ricaud, C., Egan, B.T., Gillaspie, A.G. Jr. and Hughes, C.G.), Elsevier, Amsterdam. Pp. 169-188.

Singh, N., 1988. Spread of red rot pathogen and infection from primary focus in standing sugarcane crop. *Indian Phytopath*. 41: 253-255.

Singh, N., 1994. *Trichoderma harzianum* and *Chaetomium* sp. potential biocontrol fungi in management of red rot disease in sugarcane. *J. Biol. Control*. 8: 65-67.

Singh, N., 1997. Extent of losses in sugarcane caused by incipient infection of red rot in seed setts. In: *Proc. "Sucrose Synthesis and recovery in Sugarcane–issues and dimensions"*. CCS Haryana Agril. Univ., Reg. Res. Sta., Uchani, Karnal. 142-144.

Singh, N., 1998. Effect of antagonistic fungi on *Colletotrichum falcatum* Went causal agent of red rot of sugarcane. *Indian J. Sugarcane Technol*. 13: 49-56.

Singh, N., 2000. Field diagnosis of red rot of sugarcane and integrated disease management system. *Tech. Bull.*, Sugarcane Breeding Institute, Reg. Centre, Karnal, 4: 1-7.

Singh, N., 2002. Management of red rot disease of sugarcane through biotherapy. Pp 451-463. In: *Sugarcane Crop Management* (eds. S.B. Singh *et al.*) Sci.Tech. Publishing ILC, Houston, Texas, USA.

Singh, N. and Lal, R.J., 1993. Effect of various crops on red rot development in sugarcane. *Pl. Dis. Res.* 8: 129-135.

Singh, N. and Sahi, B.K., 2000. *Annual Report*, Sugarcane Breeding Institute, Coimbatore–1999-2000. pp. 63-64.

Singh, N. and Sahi, B.K., 2003. Impact of sugarcane disease on sugar industry and Integrated disease management system. Pp 153-167. In: *Implications of of plant diseases on produce quality* (Ed D.P. Sigh) Kalyani Publishers, Ludhiana.

Singh, N. and Singh, G.B., 1994. Effect of red rot incipient infections in sugarcane seed setts on germination. *Indian J. Sugarcane Technol.* 9: 149-151.

Singh, N. and Singh, K., 1983. Role of naturally infected debris in red rot development. *Indian Sugar*. 32: 923-926.

Singh, N., Lal, R., Agnihotri, V.P. and Yadav, D.V., 1991. Role of moisture regimes on the development and spread of red rot disease on sugarcane. *53rd Ann. Conv. STAI, Agriculture*. pp. 85-95.

Singh, N., Ram, B. and Sahi, B.K., 2001. Sources of resistance against red rot of sugarcane. *Pl. Dis. Res.* 16: 251-255.

Singh, N., Singh, V., Tewari, S.D. and Verma, K.P., 1993. Occurrence and identification of different physiological races of red rot pathogen of sugarcane in Uttar Pradesh. *Indian J. Sugarcane Technol.* 8: 26-30.

Srinivasan, K.V., 1965. Towards the ideal of red rot resistance–Rndsand means. *Proc. of 12th congr. Inern. Soc. of Sugarcane Technol.* Puerto-Rico. pp. 1108-1117.

Srinivasan, T.V., 1995. Research strategies for breeding for red rot resistance in sugarcane. In: *Proc. Strategies for research and management of red rot*. (Eds. Agnihotri, V.P., Sinha, O.K. and Singh, R.P.). IISR, Lucknow. Part II. pp. 233-243.

Viswanathan, R. and Samiyappan, R., 1999. Induction of systemic resistance by plant growth promoting rhizobacteria against red rot disease in sugarcane. *Sugar. Tech.* 1: 67-76

Chapter 33

Status and Prospects of IPM in Rapeseed-Mustard

Y.P. Singh
Senior Scientist (Entomology),
National Research Centre on Rapeseed-Mustard,
Sewar, Bharatpur – 321 303, Rajasthan

Introduction

Oilseeds contribute an important role in national agricultural economy and Rapeseed–Mustard is one of them as it shared about 26.1 per cent of country's gross oilseed area and contributes about 29.1 per cent of gross oilseed production during the year 2005-06. Among oilseeds, rapeseed-mustard is next to groundnut in both area and production in the country. The production of rapeseed- mustard is low in India in comparison to some other countries of world. This is mainly due to the large area under rain fed conditions, marginal holdings, low and imbalance use of fertilizers and damages due to insect pests and diseases (Singh and Sharma, 2002). Insect pests are causing the damage to crop at different stages. The rapeseed-mustard crops are highly vulnerable to large number of insect pests. About 50 insects are known to damage *Brassica* crops and more than a dozen species as major pests (Bakhetia and Sekhon, 1989). Among these insects, mustard aphid (*Lipaphis erysimi*) is the key pest of the *Brassica* crops while saw fly (*Athalia proxima*), painted bug (*Bagrada hilaris*), pea leaf miner (*Chromatomyia horticola*) and Bihar hairy caterpillar (*Diacrisia obliqua*) are the major pests. The other pests are minor pests, which occasionally cause considerable damage to the crop. On the basis of their economic importance, some of them are grouped as below.

Pest Status	Common Name	Scientific Name
(A) Key Pest	Mustard aphid	*Lipaphis erysimi* (Kalt.)
	Saw fly	*Athalia lugens proxima* (Klug.)
(B) Major Pests	Painted bug	*Bagrada hilaris* (Burm.)
	Pea leaf miner	*Chromatomyia horticola* (Goureau)
	Bihar hairy caterpillar	*Diacrasia obliqua* (Walker)
(C) Minor Pests	Cabbage butter fly	*Pieris brassicae* (Linn.)
	Flea beetle	*Phyllotreta cruciferae* (Goeze)
	Green peach aphid	*Myzus persicae* (Sulzer)
	Diamond back- moth	*Plutella xylostella* (Linn.)
(D) New pests	The larger moth	*Crocidolomia binotalis* (Zeller)
	Cabbage top borer	*Hellula undalis* (Fab.)
	Cabbage aphid	*Brevicoryne brassicae* (Linn.)
	White fly	*Bemisia tabaci* (Gen.)
	Cabbage semilooper	*Plusia orichalsia* (Fab.)
	Turnip moth	*Agrotis segetum* (Dennis & Schiff)
	Gram pod borer	*Helicoverpa armigera* (Hub.)
	Termites	*Odontotermus obesus* (Rambur)

Out of these minor and new pests *Myzus persicae*, *Agrotis segetum*, *Crocidolomia binotalis*, *Plutella xylostella*, *Odontotermus obesus* and *Monomorium* sp. are the new emerging potential threats to rapeseed-mustard in Haryana (Singh and Singh, 1983) and other parts of the country. The cabbage butterfly is also known to cause considerable damage to *B. napus* and *B. carinata* particularly near maturity. However, *L. erysimi* continues to be a key pest on rapeseed-mustard in India.

Integrated Pest Management

Mustard Aphid

The mustard aphid (*L. erysimi*) (Homoptera: Aphididae) is the most serious and the key pest of rapeseed-mustard crops in India and the other associated species of aphids to these crops are *Myzus persicae* Sulzer and *Brevicoryne brassicae* (Linn.). This species has the wide distribution in the world. The mustard aphid (*L. erysimi*) is a proven menace to the mustard crop and other cruciferous plants in India and elsewhere (Agarwala and Bhattacharya, 1999). Bakhetia *et al.* (2002) had the view that this pest is most serious, cosmopolitan and has attend the importance of a national pest and causes field losses up to 96 per cent. The pest suck the cell sap form the plants by thrusting its needle like stylets in to the tissue and has the tremendous reproductive potential to produce the large number of offspring. The infested plants exhibits yellow, curled, wrinkled and withered leaves and the plant growth stunted which fail to produce seeds. Generally the mustard aphid infest the plant above the ground on leaves, stem inflorescence, pods but it has also been seen feeding on roots up to 15 cm. depth, (Singh and Singh, 1988). The insect also pours a good deal of sugary excreta known as honey dew which make the infested plant sticky to touch and facilitate the growth of black fungus on them called "Sooty Mould" that hinders in photosynthesis. The weather parameters have the great impact on the multiplication of mustard aphid and the peak activity of this pest varied in different reasons

(Table 33.1). The mustard aphid is the dreaded pest of rapeseed-mustard causes the damage to the crop ranging from 19 to 96 per cent depending upon its intensity and weather conditions (Table 33.2). The pest causes severe damage to the crop in absence of control measures. The integrated approaches to control the mustard aphid is preferred because the use of chemical pesticides causes the health hazards and environmental pollution.The losses due to insect pests were estimated to be Rupees 6000 crores in 1983 (Rao and Murthy, 1983), 20, 000 crores in 1993 (Jayaraj, 1993) and 29,000 crores in 1996 (Dhaliwal and Arora, 1996). Chemical control tactics are available to combat the mustard aphid effectively. However in addition to the high cost of insecticides, several other drawbacks of the chemical control *viz.* development of resistance to commonly used insecticides, pest resurgence, primary pest out break, build up of insecticide residues in oil and cake beyond the permissible limit and the degradation of the environment. Therefore, it is imperative that available pest management tactics should be such that the situation is not aggravated to a point of no return. Raghunathan (1998) presented the Integrated Pest Management targets for IX plan (Table 33.3).

Table 33.1: Period of Peak Aphid Activity as Influenced by Different Types of Cruciferous Plants

State	*Period of Peak Activity*	*Crop*	*Reference*
Rajasthan	End January	Mustard	Ahuja, 1990
Punjab	Mid February	Mustard	Bakhetia and Sidhu, 1983
	January–February	*B. campestris*	Bakhetia, 1986
	January–March	*B. juncea*	
	March	*B. napus*	
		B. carinata	
Haryana	January–February	*Brassicas*	Singh, 1982; Singh *et al.*, 1986; Bishnoi *et al.*, 1992; Rana *et al.*, 1993.
Delhi	February	*B. campestirs*	Phadke, 1986
Bihar	January–February	Rapeseed-Mustard	Sinha *et al.*, 1990
Uttar Pradesh	January	Mustard	Srivastava and Srivastava, 1972

Source: Bakhetia *et al.*, 2002.

Bio-Ecology of Mustard Aphid

The studies on bio-ecology of mustard aphid has been conducted by several workers like Landing (1982), Sidhu and Singh (1964) to find out the weak links in its life history so that these can be exploited in evolving Integrated Pest Management programme. Many other workers subsequently supplemented these. In relation to the changed spectrum of cultivars, cultural practices and the global environment, there are still many gaps in our knowledge. The mustard aphid survives on some vegetables and wild crucifers during summer months (Sidhu and Singh, 1964; Agarwala and Bhattacharya, 1999), however Sachan and Srivastava (1972) could not locate the pest on cabbage in Rajasthan from July to October, whereas Lal (1977) stated that it was not traceable in the plains of India during summer. On the basis of these contrasting reports, it is utmost necessary to study the off-season biology, colonization and dispersal patterns, migratory system, behaviour etc. The study on alate aphid and its movement in the country will help considerably in developing a forecasting or early warning system of aphid outbreaks and thus ensure an effective and economical management of this serious menace. The alate of *L. erysimi* appears on rapeseed-mustard in October when these crops are just in the seedling stage gradually attain peak population from end-December to mid-March on different host plants in different agro-ecological regions (Table 33.4). It is followed by a decline during February-March, when the proportion of apteroid viviparae increases to help in aphid dispersal (Sachan and Srivastava, 1972; Roy, 1975; Ghosh, 1980; Singh, 1982; Ghosh and Mitra, 1983; Bakhetia *et al.* 1985). The aphid population is mainly regulated by the weather parameters and composition and phenology of the plants.

Table 33.2: Extent of Losses Caused by Mustard Aphid in *Brassica* Crops in India

Per cent loss	*Reference*	*Region/Place*
24-95	Peshwani *et al.*, 1968	Delhi
28-65	Singhvi *et al.*, 1973	Hisar (Haryana)
25-66	Singhvi *et al.*, 1974	Hisar (Haryana)
55-58	Butani, 1974	Delhi
90	Kelkar, 1977	Pantnagar (Uttarakhand)
24-95	Phadke, 1980	National
35-73	Bakhetia, 1983	National
30-81	Bakhetia, 1987	Ludhiana (Punjab)
9-78	Prasad and Phadke, 1983	Delhi
45	Singh *et al.*, 1983	Ludhiana (Punjab)
22.6-52.5	Brar *et al.*, 1987	Ludhiana (Punjab)
69	Bakhetia and Ghorbandi, 1987	Ludhiana (Punjab)
77	Ghorbandi, and Bakhetia, 1987	Ludhiana (Punjab)
19-96	Bakhetia and Sekhon, 1989	Ludhiana (Punjab)

Table 33.3: Integrated Pest Management Targets During the IX Plan

Field releases of bio-agents			9000 million
Area coverage			22.5 lakh hectare
IPM farmers field schools	Central Integrated Pest Management Centres		2,500
	States		30,000
IPM demonstrations			1,50,000
Season long training	By Central Integrated Pest Management Centres–farmers	Farmers (Residential training)	75,000
		Agriculture Extension Officer/Agriculture Officer	60,000
	By States	Agriculture Extension Officer/Agriculture Officer	16,000
		Farmers	10,00,000
		Season long training (SMS)	3,000
State Biological Control Centres			30
Subsidy			Rs. 580 millions

Source: Raghunathan, 1998.

Table 33.4: Economic Threshold of *L. erysimi* Under Different Agro-ecological Situations in India

State/Crop/Variety	*Economic Threshold*	*Reference*
Haryana		
B. campestris	9-19 aphids/central shoot or 20 per cent plant infestation	Singh, 1982
B. napus GSL–1	4 aphids/central shoot or 10 per cent plant infestation	Rohilla, 1990
B. juncea RL 1359	9 aphids/central shoot or 20 per cent plant infestation	Rohilla, 1990
Uttar Pradesh		
Mustard	9–13 aphids/15 cm top terminal	Singh and Misra, 1986
Punjab		
B. juncea RLM 198	50-60 aphids/10 cm shoot or 40-50 plant infestation or 0.5-1.0 cm of aphids colony on central shoot	Bakhetia *et al.*, 1983, 1989; Singh Balraj *et al.*, 1983; Singh *et al.*, 1983; Bakhetia and Ghorbandi, 1986
Radish *Raphanus sativus*	50 aphids/10 cm terminal portion of central shoot terminal portion	Bath and Singh, 1989
Himachal Pradesh		
Brown sarson	4 mm shoot infestation	Suri, 1986
B. campestris Chinese	<20 per cent infestation	Lee, 1988*
B. chinensis		
B. oleracea	< 30 per cent infestation	

* The report pertains to China.

Source: Bakhetia *et al.*, 2002

Effect of Environmental Factors on Mustard Aphid

The multiplication of mustard aphid is favoured by cool, moist and cloudy weather. The prevalence of favourable weather conditions for a longer period can cause severe outbreak of aphids. Various climatic factors like fog; frost, rain, sub-zero temperature and high temperature have been identified as the key factors for the mortality of mustard aphid.

Temperature

Peak incidence of mustard aphid occurs at an average temperature of 17-18°C (Bishnoi *et al.*, 1992). Severe cold during December and high temperature after March deterred its multiplication. Roy (1975) holds the view that temperature had very little effect on its multiplication during the crop season in West Bengal.

Humidity

Relative humidity between 61-83 per cent favours the multiplication of mustard aphid. However, its influence on aphid population during the crop season *i.e.* mid January to mid March was found negligible.

Rainfall

The population of mustard aphid declines after the heavy rainfall but increases within a week during spring season. Bakhetia and Sidhu (1983) found that the continuous rainfall for four to five days towards the end of February kill the mustard aphid population which fail to build up to higher proportions in the subsequent weeks.

Integrated Management

Cultural Control

Time of Sowing

Alteration in crop sowing can be differed so that the susceptible crop is not allowed to coincide with the peak pest infestation. A synchrony can also be achieved through the breeding techniques by incorporating genes for earliness and lateness in *Brassica* species. Flowering period (End of December and first fortnight of January to mid February is the crucial period to have the aphid incidence and crop suffers maximum damage. Therefore, the early sown crops *i.e.* up to 20th October escape aphid infestation because flowering is over and plants become hardy before the peak period of infestation (Singh *et al.*, 1984 and Singh and Bakhetia, 1987). Pal *et al.* (1976) were also of the opinion that the aphid infestation was a major factor in yield reduction in late sown crops. The similar observations were reported by Yein (1985), Phadke and Prasad (1987), Joshi *et al.* (1989) and Bhagat and Singh (1989). Ghosh and Ghosh (1981) found low infestation of mustard aphid in early sown mustard crops. The peak activity period of mustard aphid is presented in Table 33.5.

Table 33.5: Peak Activity Periods of *L. erysimi* on Rapeseed-Mustard in Different States of India

State	*Peak Period*	*Crop*	*Reference*
Rajasthan	End January	Mustard	Ahuja, 1990
Punjab	Mid February	Mustard	Bakhetia and Sidhu, 1983
	Jan.-Feb.	*B.campestris*	
	Jan.-March	*B. juncea*	
	March	*B. napus*	
		B. carinata	Bakhetia *et al.*, 1986
Haryana	Jan.-Feb.	*Brassicas*	Singh, 1982; Singh *et al.*, 1986; Bishnoi *et al.*, 1992; Rana *et al.*, 1993
Delhi	February	B. campestris	Phadke, 1986
Bihar	Jan.-Feb.	Rape/Mustard	Sinha *et al.*, 1989
Orissa	January	Rape/Mustard	Raut and Senapati, 1968
U.P.	January	Mustard	Srivastava and Srivastava, 1972

Empirical Methods

The incidence of mustard aphid appears in the last week of December or first week of January on rapeseed-mustard crops. Initially the north-east border plants or on few plants on border of the fields or in some pockets of the crop found to be infested with mustard aphid. Singh *et al.* (1995) suggested the removal of infested twigs and destroying them help in checking the further development and spread of aphid pest.

Nutrient Application

Use of NPK fertilizers is more important because of their being integral part of Agro-ecosystem. In general heavily fertilized crops are more prone to the incidence of mustard aphid. The population of mustard aphid was 4-8 times more in the mustard crop receiving N at the rate of 40 and 60 kg/ha as compared to the crop under zero level (Rawat *et al.*, 1968, Singh *et al.*, 1990). Bakhetia and Sharma (1979) found that the population of *L. erysimi* was not affected by increasing dose of N up to 80 kg/ha. Bakhetia *et al.* (1982) found that increased sulphur nutrition had a negative correlation with aphid population. Bhat and Sidhu (1983) also observed that host plants grown under increasing supply of K caused adverse effects on the reproduction and excretion of the mustard aphid. These inconclusive results needs more research in this direction.

Irrigation

Mustard crop in want of water results in production of fewer young ones by mustard aphid (Sidhu and Kaur, 1977). Chada and Arora (1982) observed the reduced amino acid concentration in plants kept under dry conditions resulted in imbalance in the excretion of mustard aphid, which was responsible for the reduced fecundity under water stressed conditions. Bakhetia and Brar (1988) reported that the rain fed crop had heavy infestation whereas the irrigated crop maintained a good stand even under heavy aphid attack. Miles *et. al.* (1982) also reported increases in amino acid contents of leaves and phloem sap water stressed rape plants which enhances the development of cabbage aphid, *Brevicoryne brassicae*. Joshan *et al.* (1992) also observed concomitant decrease in aphid's net reproduction rate (R_0) on potted plants with increasing water stress in the three generations.

Source of Resistance

Rai and Sehgal (1975) inferred that tender and thickly packed inflorescence of *B. campestris* varieties were more susceptible than those of *B. juncea* having harder and loose inflorescence. The yellow bright colour of flower, sparse arrangement of the flower buds on inflorescence twig and succulency of the inflorescence twig contribute for the susceptibility of *Brassica* genotypes to aphid incidence. Among the rapeseed-mustard crops *B. juncea* and *B. carinata* genotypes were observed tolerant to aphid attack as compared to the *B. campestris* genotypes (Teotia and Lal, 1970 and Kalra *et al.*, 1987). *B. juncea* strains *i.e.* Laha 101, T 6342, Rai 236, Rai 237, RL 18, RLM 198, RLM 202, B 85, RW 2-2, RH 7846, RH4847, Purple Mutant and RC 199 and *B. carinata* strains DLSC 1and DLSC 2 have been reported fairly resistant to mustard aphid. Laha 101 exhibited all the three types of resistance *i.e.* non-preference, antibiosis and tolerance. Slow development of mustard aphid on T 27 (*Eruca sativa*) and Purple Mutant (*B. juncea*) and more wing production exhibited antibiosis (Rohilla *et al.*, 1999). Non-waxiness of *Brassica* plants has been found to impart resistance to aphids (Srinivasachar and Malik, 1972; Yadava *et al.*, 1985; Angadi *et al.*, 1987; Chatterjee and Sen-Gupta, 1987). The alates of the mustard aphid also got attracted towards yellow colour traps (Dilawari and Atwal, 1989a). It has also been observed that white petal and non-waxy (glossy surface) mutants had lower number of initial settlers and subsequently smaller aphid colonies as against normal (yellow petal) and waxy stems (non-glossy). Rich amount of glucosinolate and phenols have been reported negatively correlated whereas rich amount of sugars and nitrogen content had positive correlation with mustard aphid population (Gill and Bakhetia, 1985 and Singh and Rohilla, 1998). *B. tournefortii* was also found to harbour relatively low population indicating tolerance. Malik (1988), however observed that amino acids in the raceme were relatively higher in tolerant and susceptible ones which ultimately had highest per cent loss due to aphid infestation. *B. napus* strains and colchicines induced tetraploid toria have been known to possess antibiotic factors (Singh *et al.*, 1965; Jarvis, 1970; Gill and Bakhetia, 1985; Kalra *et al.*, 1987).

The importance of host plant resistance as a tool in IPM has been long documented (Snelling, 1941; Painter,1951; Chesnokov, 1953; Russel, 1978; Lara,1979; Panda, 1979; Smith, 1989; Dhaliwal and Dilwari, 1993; Smith *et al.*, 1994; Panda and Khush, 1995; Smith 1999; Dhaliwal and Singh, 2004).

Use of Botanicals

As many as 2121 plant species have been reported to possess pest control properties. Of these plant 1005 have insecticidal, 384 antifeedent, 297 repellents, 27 attractants and 31 growths inhibiting properties (Puri, 1999). India has the vast potential for botanical pesticides particularly neem. An estimated 18 million neem trees with a potential yield of 0.7 million tones of fruits provides about 142–350 tones of azadirachtin (0.2-0.5 per cent) out of 10 per cent seed kernel yield (Narval *et al.*, 1997).

Many plant materials as extracts have been evaluated against mustard aphid, which include nicotine sulphate, rotenone and pyrathrins. All these have shown variable toxicity. Plant extracts of *Azadirachta indica, Lantana camera, Ipomoea carnea, Acorus sp., Solanum xanthocarpum, Swertia chirata, Melia azadirach and Argemone maxicana* proved toxic to mustard aphid (Pandey *et al.*, 1977). Singh (2007) reported neem seed kernel extract (5 per cent) and neem leaves extract (5 per cent) effective against mustard aphid.

Use of Bio-pesticides

Microbial pesticides like B.t., Fungal and baculovirus have been widely used in the management of insect pest and avoiding the development of resistance. Entomopathogenic fungi and bacteria paralyze or kill their hosts by adversely affecting the growth and development of host insects. These pathogens are now commercially available through their marketability is very negligible. The use of *Verticillium lacanii* against the mustard aphid and other aphids has been found encouraging. Despite the fact that many bio-pesticides have been tested, but, their use for control of mustard aphid required much more concentrated efforts at national level for developing a pest management strategy having the plant based biocides as one of its components.

Behavioural Control

Insects do not think and react to stimuli largely chemical from the environment. These reactions are stereotyped and predictable. Chemicals regulating insect behaviour are collectively called semiochemicals having four categories pheromones, allomones, kairomones and Allomone-kairomones. Studies of host insect derived cues have concentrate on parasitoid responses to aphid sex pheromones at IACR, Rothemsted U.K. These pheromones have been successfully isolated and chemically identified for a range of aphid pests. Early field trials with laboratory produced pheromones revealed an apparent response of braconid aphid parasitoid that has been confirmed in the laboratory. This novel pest control strategies based on the field manipulation of parasitoid populations using chemical component of the aphid sex pheromones that can be extracted form certain plants. The use of pheromones for management of aphid pests can be of much significance in the development of pheromone baited traps to attract the winged morphs during their settling on plants early in the season or in dispersal during crop maturity late in the season. Spraying of such chemicals on non-host plants can be very effective in attracting the winged aphids and finally killing them with potent insecticides. It is reported that the alarm pheromone (B-farnesone) is of common occurrence in most of the aphids and thus spraying of formulations based on the B-farnesone on the infested plants or leaves will produce peculiar reactions in the aphids and they will not be able to settle for feeding on the hosts. If, spraying can be synchronized with the first appearance of aphid in the fields, the early settlers can be prevented from alighting on the hosts and thus build-up of their population can be checked to a greater extent. Thus there is a good scope for the use of alarm pheromones in the management of aphid pests. (Bhakhetia *et al.*, 2002).

Biotechnological Approaches

The incorporation of resistance for controlling diseases caused by fungi and bacteria with the help of conventional plant breeding and made some significant achievements. However, the progress in the production insect resistant cultivars has been very limited. Biotechnology provides a potential tool for genetic engineering of crop plants. Weising *et al.* (1988) reviewed the literature on integration and expression of foreign genes in plants through Ti plasmid of *Agrobacterium tumefaciens*, Ti plasmid derived plant gene vectors, transformation techniques, the structure and expression of genes in their transgenic environment and its impact in crop improvements. One of the significant advances of the last decade in the development of plants that are resistant to insect pests arises from the use of one form of biotechnology–the use of markers to identify and track genes of interest for breeders (Yencho *et al.*, 2000). Many workers have reported transgenic plants of *B. napus* obtained by *Agrobactirium* mediated transformation. The advances in the field of biotechnology and molecular genetics can help in over coming the constraints in using the methods of embryo rescue, protoplast fusion, recombinant DNA technology and selection of colony stage in transferring insect pest also for production of biocides. Protoplast fusion has been used to produce genepool with genetic variation through development of amphiploids. To mention, the amphiploid (*E. sativa x B. campetris*) developed recently has proved promising for aphid resistance. Though, there are man reports about the successful transfer of foreign genes into *B. napus* and *B. oleracea* but not field trials have been conducted, so far to show their performance. It is hope that certain desirable foreign genes for resistance will be incorporated in *B. juncea* and *B. campestris* against the most serious pests aphid. Since the introduction of biotechnology for commercialized genetically engineered products in agriculture have been those developed for weed, insect and disease management and until 2005, a cumulative total of over 475 million ha of transgenic crops were grown in 21 countries (James, 2005).

Use of Natural Enemies

The natural enemies play an important role in the reduction of aphid population. In a well studied fauna, such as in the British isles, over 26 per cent species may be parasitoids and another 4 per cent predators making 30 per cent of the insect fauna (Price, 1980). In just one field in England, 25 species of aphids were recorded that were attacked by 18 species of primary parasitoids and several families of predators and the primary parasitoids were attacked by 28 species of secondary parasitoids (Godfray and Muller, 1998). Means of 60 per cent mortality of a cohort caused by natural enemies are observed commonly and in long term population dynamics studies, natural enemies were of major importance in 40 per cent of the cases (Price, 2003). Also when natural enemies were employed as biological control agents, mortality of host may reach 90-100 per cent in numerous cases, and successful biological control is brought about even when only 40 per cent mortality is achieved in some cases (Hawkins, 1994). Bakhetia and Shekhon (1989) reported that mustard aphid has the six Coccinellids, 16 Syrphids, one Chamaemyliids, Hemerobiids as insect predators, 4 species of Hymenopterous parasites, 4 species of Entomogenous fungi and one predatory bird as natural enemies. Coccinellids are the important predator of mustard aphid. Some of the coccinellies *viz.*, *Coccinella septampunctata* L *C. repanda, C. transversalis, Brumoides suturalis, menochilus sexmaculatus and Hippodamia varigeata.* Coccenellids are not the only important factor responsible for the decline of mustard aphid population because flourish more at a temperature below 20°C whereas coccinellids at above 20°C. Coocinellid adults take pollen, and are predatory, but their larvae are predatory only. The use of these floral 'resource subsidies' (Tylianakis *et al.*, 2004) in this way can markedly increase predator or parasitoid fitness (*e.g.* by increases in longevity or fecundity), and can make them more effective biological agents (Barbosa, 1998; Landis *et al.*, 2000; Gurr *et al.*, 2004a,b). Singh and Singh (1983) had the view that predators reduce the already declines population of mustard aphid.

A long list of syrphids have been found to predate on mustard aphid however, their number in the field has been quite low and had a limited scope for the control of mustard aphid. Some of the important species of syrphids are as *Sophaerophoria scutellaris, Eristalis Obscuritarsis, E.tenax, Episyphus belteatus, Syrphus belteatus. Obscuritarsis, E. tenax, Episyphus belteatus, Syrphus belteatus, Metasyrophus adligatus, Metasyrphus corollae, Xanthogramme Scutallare, Syrphus serarius and Syrphus issaci.*

The lacewing *chrysopa scaslastes, chrysoperla carnea* have been found actively predating upon the mustard aphid. However, their scope is very limited in the aphid management.

A small bird (Gray Wag Tail) have also been found as a predator of mustard aphid.

Some of the parasitoids *viz., Diaratus rapae* has been reported to cause 93 per cent parasitization of *L. erysimi* under laboratory conditions. Some other parasitoids *viz. Diaeretiella rapae, Aphiencyrtus sp* and A. *Aphidiorus* have also been observed parasitizing the mustard aphid. The parasitoids also appear at the end of the season *i.e.* somewhere mid February. However, information on their comprehensive role in suppressing the aphid in the field condition is lacking (Bakhetia and Sidhu, 1983).

Agarwala and Bhattacharya (1999) provided a list of parasitoids and predators of *L erysimi* in detail. Raj and Dass (1992) also concluded that coccinellids and syrphids could serve as biological control agents for suppressing the aphid populations. Apart from the studies on population dynamics, biology and predation efficiency of a few natural enemies of mustard aphid, there has been virtually no study on the field efficacy of such bio-control agents. Hence, future research strategies should be aimed at studying the interaction of resistant plant-aphid-natural enemies along with the possible effects of antibiotic factors on them. The studies on the augmentation of these natural enemies with artificial releases, their protection against odd weather and use of comparatively safer insecticides are also to be conducted to make them more fruitful.

Two fungal diseases caused by *Cephalosporium aphidicola* and *Entomophthora* sp. has been reported to take a heavy roll of the already declining population of mustard aphid the month of February and March. The disease is more prevalent when the average temperature was 27.3°C and the relative humidity was 81.0 per cent. The diseased aphids stopped feeding and reproduction, their color changed to dark brown or radish and died. The scope of disease for the control of aphids seems limited but these would certainly supplement the control of aphids.

To understand the incorporation of natural enemies in to IPM systems responses full-or reduced-rate of pesticide use (Poeling, 1989; Van Emden 1990). The concept of 'Integrated Biological Control' (Gurr and Wratten, 1999) captures this idea. With 3 billion kg of pesticide currently used each year with no evidence that this use has led to sustainable pest population reductions (Pimentel, 2004). Biological control can be 'elegant, sustainable, non-polluting and inexpensive' (Gurr *et al.*, 2000b) but its success rate has not been particularly high, especially for biological control of arthropods by arthropods (Gurr *et al.*, 2000a). As pest invasions increase with 'globalization', and pesticide use continues to increase (Pimentel, 2004), greater expectations for natural enemies to reduce pest numbers will result (Loope and Howarth, 2003).

Insecticidal Control

Chemical control would continue to be the first line of defence against mustard aphid particularly under out break situation. More than 90 per cent mortality of aphids is possible by use of systemic insecticides but the population attains similarity under treated and untreated fields within a period of 2 to 3 weeks (due to high rate of multiplication (Singh *et al.* 1984). The economic injury level of mustard mentioned for different states should be followed for the spraying. The use of various insecticides like oxydemeton methyl 0.025, phosphamiden 0.025 per cent, dimethoate 0.03 per cent, Ouinalphos 0.025 per cent, Endosulfan 0.05 per cent and Chlorpyriphis 0.03 per cent have been advocated an all India basis. Oxydemeton methyl 0.025 per cent (2 sprays at 15 days intervals) or 0.05 per cent (one spray) monocrotophos 0.05 per cent (one spray) and phosphamidon 0.05 per cent (one spray) showed quite high cost benefit ratio and proved equally efficient (Singh *et al.*, 1984). Under dry land conditions the use of insecticidal dusts *viz.*, methyl Parathion 2 per cent and endosulfan 4 per cent @ 25 kg/ha were found very effective. Singh *et al.* (1987) mentioned that oxydemeton methyl 0.025 per cent and endosulfan 0.35 per cent are very safe to the pollinators and at the same time very effective against aphids. Arora and Sidhu (1991) and Brar *et al.* (1991) also explained the better efficacy of different insecticides against the mustard aphid under water stressed conditions.

Other Insect Pests

Very limited information is available on other insect pests of economic significance like leaf miner, mustard saw-fly and the painted bug.

Painted Bug

Painted bug is highly polyphagous pest and often infest the rapeseed-mustard crops in India. The pest has been found to cause serious damage in the state of Rajasthan, Punjab, Haryana, Uttar Pradesh and Madhya Pradesh. The pest is more serious on young/seedling stage and both adults and nymphs suck the cell sap from the leaves and shoots. In case of tender two-leaf plant, the infested tender shoot fall down and plant died. The infestation of this pest in the vegetative growth stage results in first whitening of leaves then wilting leading to complete drying of the plant. In both the cases re-sowing of the crop becomes necessary. The pest also attacks the crop at pod formation and maturity stage, which results in curling of the pods and shriveling of grains. Bugs can be seen feeding on the harvested material lying on the threshing floor. Painted bug may lead to losses in yield by 31.1 per cent and in oil content by 3-4 per cent (Singh *et al.*, 1980).

Integrated Pest Management

Singh *et al.* (1993) and Singh and Malik (1993) observed that clean cultivation surely suppresses the build up of population of painted bug in Rapeseed-Mustard crops by weeding, hoeing and burning of debris in and around the field. Deep ploughing in summer and application of first irrigation at 3-4 weeks after sowing of the crop wherever possible is helpful to reduce the pest population. Pandey *et al.*, 1981 reported that leaf extract of *Argemone maxicana* effectively controlled painted bug both in field and laboratory conditions. Dusting of crop with endosulfan 4 per cent or quinalphos 1.5 per cent dust @ 20-25 kg/ha or spraying with malathion 50 EC @ 500 ml or endosulfan 35 EC @ 625 ml in 500 litres of water in one hectare in severe infestation control the pest effectively. Threshing of crop as early as possible, avoids the further losses.

Mustard Saw Fly

The pest is found in all rapeseed-mustard growing areas of the country. The insect is found in the pest form in Uttar Pradesh, Bihar and West Bengal. The insect appears in the pest form in early stage of the crop *i.e.* October and November. The mustard saw-fly, is more serious on the very young seedlings/ plants and some times excessive larval feeding necessitate resowing of the crop. The degree of oviposition was minimum on stoke, Lethbridge, Metapolaska, black mustard, wild mustard and Blaze, while the values were higher on *B. juncea* cv. RC 781. Srivastava (1957) regarded importance of certain olfactory/ gustatory stimuli in host food selection while sinigrin acted as phago-stimulant for the larvae (Bogawat and Srivastava, 1968). In another study, Sehgal and Ujagir (1977) observed that certain unrelated plant species such as *E. sativa, lberis amagra, Cheiranthus cheiri* and *Tropaeohummajus* have evolved marked degree of natural defence and juices of such plants suppressed the feeding of grubs on preferred hosts. Similar studies by Ahuja and Sehgal (1982a) also indicated feeding deterrency of seed extracts containing glucosinolates and isothiocyanates. Besides, the extract of isothiocyanate significantly reduced the adult survival, larval weight and fecundity of females emerged from treated larvae. Ahuja and Sehgal (1982b) further observed that the extracts from such plants also reduced the ovipositional preference of female.

Integrated Pest Management

Crop sanitation practices reduce the pest population. Timely irrigation helps in killing the larvae through drowning in water. Spray the crop with malathion 50 EC @ 500 ml or endosulfan 35 EC @ 625 ml in 500 litres of water in one hectare to control the pest Bakhetia and Sachan (1997).

Pea Leaf Miner

The pea leaf miner is a highly polyphagous pest and found in all the mustard growing areas of the country but attains the status of a serious pest by mining the leaves of rapeseed and mustard suffer severe damage during their vegetative and flowering phases. The damage leaves present zig-zag silvery lines with black pupae at the end of the mines (Singh *et al.*, 1985).

Integrated Pest Management

Plucking the infested leaves and burry them in soil helps in the reduction of the pest. Foliar spray of systemic insecticide such as oxydemeton methyl 25 EC or dimethoate 30 EC @ 1.0 litre in 600-800 litres of water per hectare controls the pest effectively (Bakhetia and Sachan, 1997).

References

Agarwala, B. K. and Bhattacharya, S., 1999. Effective bio-control agents and their use in IPM strategy of the mustard aphid. *IPM System in Agriculture.* Rajeev K. Upadhyay, K. G. Mukerji and R. L. Rajak, Aditya Books Pvt. Ltd., New Delhi, India. Vol. 5 Oilseeds, pp 77-89

Ahuja, D.B., 1990. Population dynamics of mustard aphid, *L. erysimi* on Indian mustard, *Brassica juncea., Indian J. Plant Prot.*, 18: 233-235.

Ahuja, D.B. and Sehgal, V.K., 1982a. Effect of seed extracts containing mustard oil glucodides on feeding responses of the mustard saw fly, *Athalia proxima* Klug. *Indian J. Ent.*, 44: 43-40.

Ahuja, D.B. and Sehgal, V.K., 1982b. Effect of seed extracts containing mustard oil glucodides on the biology of mustard saw fly, *Athalia proxima* Klug. *Indian J. Ent.*, 44: 264-272.

Angadi, S.P., Singh, J.P. and Anand, I.J., 1987. Inheritance of non-waxiness and tolerance to aphids in Indian Society of Oilseeds Research, Hyderabad.

Arora, R. and Sidhu, H.S., 1991. Efficacy of dimethoate against mustard aphid, *Lipaphis erysimi* (Kalt) in relation to water stress in the host plant. *J. Insect Sci.*, 4: 138-140.

Bakhetia, D. R. C. and Sachan, G. C., 1997. Identification and management strategies for insect-pests of rapeseed and mustard. Directorate of Extentsion, Ministry of Agriculture, Government of India, Krishi Vistar Bhavan, Pusa, New Delhi–110 012.

Bakhetia, D. R. C. and Sharma A. K., 1979. Preliminary observations on aphid infestation on *Eruca sativa* Mil!. *Indian J. Ent.* 81 (3) : 288-289.

Bakhetia, D. R. C. and Sidhu, S. S., 1983. Effect of rainfall and temperature on mustard aphid, *Lipaphis erysimi* (Kalt.) *Indian J. Ent.*, 45 (2) : 202-205.

Bakhetia, D. R. C., Singh, Harvir and Chander, Harish 2002. IPM for Sustainable Production of Oilseeds. Oilseeds and Oils Research and Development Needs. Indian Society of Oilseeds Research, Hyderabad, pp 184-218.

Bakhetia, D.R.C., 1983. Losses in rapeseed / mustard due to *Lipaphis erysimi* (Kaltenback) in India : A literature study. Sixth International Conference, Paris, May 16-22, 1983, p.1142-1147.

Bakhetia, D.R.C., 1986. Pest management in rapeseed and mustard pesticides, May 1986, p. 32-39.

Bakhetia, D.R.C., 1987. Insect pests of rapeseed-mustard and their manatgement. *Plant Protection in field crops.* (Eds. M. Veerabhadra Rao and S. Sithanantham) PPAI, Rajendranagar, Hyderabad, p. 249-259.

Bakhetia, D.R.C. and Brar KS. and Sekhon, B.S., 1986. Seasonal incidence of *Lipaphis erysimi* (Kaltenback) on the Brassica species in the Punjab. (in) Aphidology in India. ed. B.K.Agarwala, A.R. printers, Calcutta, p. 29-36.

Bakhetia, D.R.C. and Brar, K.S., 1988. Water stress in Ethiopian mustard, *B. carinata* and Indian mustard, *B.juncea* on infestation by *L.erysimi. Indian J Agric Sci.*, 58: 638-640.

Bakhetia, D.R.C. and Ghorbandi, A.W., 1986. Economic threshold of *Lipaphis erysimi* (Kaltenhach) on Indian mustard, *Nowl. Aph. Soc.* India, 5 (2) : 5-7.

Bakhetia, D.R.C. and Sekhon, B.S., 1989. Insect pests and their management in rapeseed-mustard. *J. Oilseeds Res.*, 6: 269-299.

Bakhetia, D.R.C., Sushma Rani and Ahuja, K.L., 1982. Effect of sulphur nutrition of some *Brassica* plants on their resistance response to the mustard aphid, *Lipaphis erysimi* (Kaltenbach, *Calicut Uni. Res. J.* Special Conference Number, May 3-5, 1982, P. 36 Abstract).

Barbosa, P., 1998. *Conservation Biological Control.* Academic Press, San Diego, California.

Bath, D.S. and Singh, D., 1989. Studies on the economic threshold of mustard aphid, *Lipaphis erysimi* (Kaltenback) on the radish crop in India. *Trop. Pest. Mgt.* 35 : 145-156.

Bhagat, D.V. and Singh, S 1989. Effect of date of sowing and varieties on the seed quality, yield and aphid infestation on mustard. Bhartiya Krishi Anusandhan Patrika, 4 : 179-183.

Bhat, N.S. and Sidhu, H.S., 1993. Influence of potassium on amino acids of the host plants and reproduction and excretion of the mustard aphid, *Lipaphis erysimi* (Kalt). Proc. International Rapeseed Conference, Paris, France, pp 28.

Bishnoi, O.P., Singh, H. and Singh, R., 1992. Incidence and multiplication of mustard aphid, *L. erysimi* (Kaltenbach) in relation to meteorological variables. *Indian J. agric. Sci.*, 62: 710-712.

Bogawat, J.K. and Srivastava, B.K., 1968. Discovery of sinigrin as a phagostimulant by *Athalia proxima* Klug (Hymenoptera: Tenthridinidae) *Indian J. Ent.*, 30: 89.

Brar, K.S. Bakhetia, D.R.C. and Sekhon,B.S., 1987. Estimation of losses in yield of rapeseed and mustard due to mustard aphid. *J. Oilseeds Res.* 4: 261-246.

Brar, K.S., Singh, I. and Aujla J.S., 1991. Water stress induced effect of raya crop on the efficacy of oxydemeton-methyl for the control of mustard aphid. *J. Insect Sci.*, 4: 141-144.

Butani, D. K., 1974. Effect of various insecticides on the mustard aphid, *Lipaphis erysimi* (Kalt.) and yield of rape,*Brassica campestris* Linn. *Indian J. Ent.* 36: 243-246.

Chada, I.C. and Arora, R., 1982. Influence of water stress in the host plant on the mustard aphid, *Lipaphis erysimi. Entomon*, 7: 75-78.

Chatterjee, S.D. and Sengupta, K., 1987. Observations on reaction of mustard aphid to white petal and glossy plants of Indian mustard. *J.Oilseeds Res.*, 4: 125-127.

Chesnokov, P.G., 1953. *Methods of Investigating Plant Resistance to Pests.* National Science Foundation, Washington, DC.

Dhaliwal, G. S. and Arora, Ramesh 1996. Principles of Insect Pest Management. *National Agriculture Technology Information Centre*, Ludhiana. pp 374.

Dhaliwal, G.S. and Dilwari, V.K., 1993. *Advances in Host Plant Resistance to Insects.* Kalyani Publishers, New Delhi.

Dhaliwal, G.S. and Singh, R.P., 2004. *Host Plant Resistance to insects: Concepts and Applications.* Panima Publishing Corporation, New Delhi.

Dilawari, V.K. and Atwal, A.S., 1989b. Response of mustard aphid, *L. erysimi* (Kalt) to allylisothiocyanate. *J.Insect Sci.*, 2: 103-153.

Ghorbandi, A.W. and Bakhetia, D.R.C., 1987. Assessment of yield losses and determination of economic-injury levels for *Lipaphis erysimi* (Kaltenbak) on Indian mustard. II. Based on per cent plant infestation by the aphid. *11th Intern. Congress of Plant Protection,* Manila (Philippines), October 5-9, 1987 p, 170 (Abstract).

Ghosh, A.K. and Ghosh, M.R., 1981. Effect of time of sowing and insecticidal treatments on the pests of Indian mustard, *Brassica juncea* L. and on seed yield. *Entomon.*, 6 : 357-362.

Ghosh, M.R., 1980. Spread of infestation of *Lipaphis erysimi* (kalt.) in and around Calcutta, West Bengal. *Sci, Cult.* 47 (2) : 58-60.

Ghosh, M.R. and Mitra, A., 1983. Incidence pattern and population composition of *Lipaphis erysimi* (Kaltenbach) on mustard and radish (in) The Aphids ed. B,K, Behaura: The Zoological Society of Orissa, Utkal Universtiy, Bhubaneshwar, India, p. 43-51.

Gill, R.S. and Bakhetia, D.R.C., 1985. Resistance of some *Brassica napus* and *B. campestris* strains to *Lipaphis erysimi* (Kaltenbach) *J. oilseeds Res.* 2 (1): 227-239.

Godfray, H.C.J. and Muller, C.B., 1998. Host-Parasitoid dynamics. In: Dempster, J.P. and McLean, I.F.G. (eds) *Insect population in Theory and in Practice.* Kluwer academic Publishers, Dordrecht, The Netherlands, pp. 135-168.

Gurr, G.M. and Wratten, S.D., 1999. 'Integrated Biological Control': a proposal for enhancing success in biological control. *International Journal of Pest management.* 45: 81-84.

Gurr, G.M., Barlow, N.D., Memmott, J., Wratten, S.D. and Greathead, D.J., 2000a A history of methodological, theoretical and empirical approaches to biological control. In: Gurr, G.M. and Wratten, S.D. (eds) *Biological Control: measures of Success.* Kluwer Academic Publishers, Dordrecht, The Netherlands, pp. 3-37.

Gurr, G.M., Scarratt, S.L., Wratten, S.D., Berndt, L. and Irwin, N., 2004a. Ecological engineering, habitat manipulation and pest management. In: Gurr, G.M., Wratten, S.D. and Altieri, M.A. (eds) *Ecological engineering for pest management. Advances in Habitat Manipulation for Arthropods.* CSIRO Publishing, Collingwood, Victoria, Australia, pp. 1-12

Gurr, G.M., Wratten, S.D. and Altieri, M.A., 2004b. Ecological engineering for enhanced pest management: towards a rigorous Science. In: Gurr, G.M., Wratten, S.D. and Altieri, M.A. (eds) *Ecological engineering for pest management. Advances in Habitat Manipulation for Arthropods.* CSIRO Publishing, Collingwood, Victoria, Australia, pp. 219-225.

Gurr, G.M., Wratten, S.D. and Barbosa, P., 2000b. Success in Conservation biological control of Arthropods. In: Gurr, G.M. and Wratten, S.D. (eds) *Biological Control: measures of Success.* Kluwer Academic Publishers, Dordrecht, The Netherlands, pp. 105-132.

Hawkins, B.A., 1994. *Pattern and Process in Host-parasitoid Interactions.* Cambridge University press, Cambridge.

James, C., 2005. Preview: global status of commercialized transgenic crops: 2003. *ISAAA Briefs No. 34, ISAAA*, Ithaca, New York.

Jarvis, J.L., 1970. Relative injury to some cruciferous oilseeds by the turnip aphid. *J. econ. Ent.* 63 (5): 1498-1502.

Jayaraj, S., 1996. Biopesticides and Integrated pest management for sustainable crop production pp 65- 81 In : *Agrochemicals and sustainable Agriculture.* (Ed.) Roy, N. K. APC Publishers, New Delhi

Joshan, V.S., Narang, D.D. and Dilawari, V.K., 1992. Population build-up of mustard aphid, *Lipaphis erysimi* (Kalt) on *Brassica* crops under water stress conditions. *Proc. Nat. Symp. Recent Adv. IPM.* 12-15 October, PAU, Ludhiana.

Joshi, M.L., Ahuja, D.B. and Mathur, B.N., 1989. Loss in seed yield by insect pests and their occurrence on different dates of sowing in Indian mustard, *B.juncea. Indian J. agric Sci.*, 59: 166-168.

Kalra, V.K., Singh, H. and Rohilla, H.R., 1987. Influence of various genotypes of *Brassica juncea* on biology of mustard aphid, *Lipaphis erysimi* (Kalt). *Indian J. agric. Sci.*, 57: 277-279.

Kelkar, R. V., 1977. Field incidence and spatial distribution of *Athalia proxima* Klug and *Lipaphis erysimi* (Kaltenback) together with estimation of yield losses due to *L. erysimi* (kalt.). M. Sc. thesis, submitted to G. B. Pant University of Agriculture and Technology, Pantnagar (Nainital), 105p.

Lal, O.P., 1977. *Lipaphis erysimi* (Kalt.) p. 335-336. (in) Diseases, Pests and Weeds in Tropical Crops, ed. by J. KRANTZ, H. Schmutterer and W. KOCH, Paul Parey, Berlin and Hamburg.

Landing, J., 1982. Life history, population dynamics and dispersal in the mustard aphid, *Lipaphis erysimi*: A literature study, 18p, Swedish University of Agricultural Research, Research Information Centre, Uppsala (Sweden).

Landis, D.A., Wratten, S.D. and Gurr, G.M., 2000. Habitat management to conserve natural enemies of Arthropod pests in agriculture. *Annual Review of Entomology.* 45, 175-202.

Lara, F.M., 1979. *Principios de Resistancia de plantas a insectos* (in Portugese), Livroceres Ltda.Piracicaba, SP, Brazil.

Lee, H. S., 1988.Action thresholds for turnip aphid and green peach aphid on cabbage and leaf mustard. *J. Agric. Res.* China. 37 : 91-99.

Loope, L.L. and Howarth, F.G., 2003. Globalization and pest invasion: where will we be in five years? *Proceedings of the 1st International Symposium on Biological Control of Arthropods.* USDA Forest Service Health Technology Enterprise Team, FHTET-03-05, pp. 34-39.

Malik, R.S., 1988. Role of amino acids in relation to aphid (L. erysimi Kalt), J. Oilseeds Res., 5: 39-45.

Miles, P.W., Aspinali, D. and Rosenberg, 1982. Performance of cabbage aphid, *Brevicoryne brassicae* on water stressed rape plants in relation to the changes in their chemical composition. *Australian J. Zool.*, 30: 337-345.

Narval, S. S.,Tauro, P. and Bisla, S.S., 1997. Neem in sustainable agriculture. Scientific Publishers, Jodhpur. Pp-266.

Painter, R.H., 1951. *Insect Resistance in Crop Plants.* Macmillan, New York.

Pal, S.R., Nath, D.K. and Saha, G.N., 1976. Effect of time of sowing and aphid infestation on rai (*Brassica juncea* Goss.) *Indian Agriculturist.*, 20 : 27-34.

Panda, N., 1979. *Priciples of Host Plant resistance to Insect Pests.* Allenheld, Osmun and Co. and Universal Books, New York.

Panda, N. and Khush, G.S., 1995. *Host Plants Resistance to Insects.* CAB International, Wallingford, UK.

Pandey, N.D., Singh, M. and Tiwari, G.C., 1977. Antifeedent, repellant and insecticidal properties of some indigenous plant materials against mustard saw fly, *Athalia lugens proxima. Indian J. Ent.*, 39: 62-64.

Pandey, U.K., Pandey, M and Chauhan, S.P., 1981. Insecticidal properties of some plant material extract against the painted bug, *Bagrada cruciferarum. Indian J. Ent.*, 43 : 404-407.

Peshwani, K.M., Kumar, K. and Lal, R., 1968. Delhi farmers need not be afraid of mustard aphid. *Indian fmg.* 18 (7) : 17-19.

Phadke K. G. and Prasad S. K., 1987. Identification of *Brassica* genotypes least susceptible to mustard aphid *Lipaphis erysimi* (Kalt.); *J. Aphidology* 1 93-97.

Phadke, K.G., 1980. Proc, *FAI group discussion on increasing pulses and oilseeds production in India,* New Delhi. Sept. 4-5, 1980 p. 151-158

Phadke, K.G., 1986. Ecological factors influencing aphid, *Lipaphis erysimi* (Kaltenback) Incidence on mustard crop. In: *Aphidology* in India, Agarwala, B.K. (Ed.) pp-37-42. A R Printers. Calcutta.

Pimentel, D., 2004. Foreward. In Gurr, G.M., Wratten, S.D. and Altieri, M.A. (eds) *Ecological engineering for pest management. Advances in Habitat Manipulation for Arthropods.* CSIRO Publishing, Collingwood, Victoria, Australia, pp. vi.

Poeling, H.M. (1989). Selective application strategies for insecticides in agricultural crops. In: Jepson, P.C. (ed.) *Pesticides and Non-target Invertibrates.* Intercept, Wimborne, UK, pp. 151-176.

Prasad, S.K. and Phadke, K.G., 1983. Effect of mustard aphid, *Lipaphis erysimi* (Kalt). infestation on the seed yield of different varieties of *Brassica species.* (In): The aphids, (Ed. B.K.Behura). The Zoological Society of Orissa, Utkal University, Bhubaneshwar, India, p. 283-287.

Price, P.W., 1980. *Evolutionary Biology of Parasites*. Princeton University Press, Princeton, New Jersey.

Price, P.W., 2003. *Macroevolutionary Theory on Macroecological Patterns*. Cambridge University press, Cambridge.

Puri, S.N., 1999. Integrated pest management in India. In Masters trainers training programme on Integrated pest management, 13-17 April, 1999. LBS Building, New Delhi

Raghunathan, V., 1998. Making Farmers Experts. In : National Seminar on Integrated Pest Management. *Pestology* (Special issue), 22 (4) : 10-14

Rai, B. and Sehgal, B.K., 1975. Field resistance of *Brassica* germplasm to mustard aphid, *Lipaphis erysimi* (Kalt). *Sci. and Cult.*, 41: 444-445.

Raj, D. and Dass, G., 1992. Population build-up of aphids and their predators in mid-hill zone of HP. *Proc. Nat. Symp. Recent, Adv. IPM.*, 12-15 October, PAU, Ludhiana.

Rana, J.S., Khokhar, K.S. Singh, H. and Sucheta, K.S., 1993. Influence of abiotic environment on the population dynamics of mustard aphid, *L. erysimi* (Kalt). Crop Res., 6: 116-119.

Rao Krishnamurthy, B. H. and Murthy, K. S. R. K., 1983. Proceeding of National Seminar on Crop losses due to Insect Pests. *Indian J. Ent.* (special issue) Vol. 1 and 2, Hyderabad.

Rawat, R.R., Misra, U.S., Thakar, A.U. and Dhamdhere, S.V., 1968. Preliminary study on the effect of different doses of nitrogen on the incidence of major pests of mustard. Madras agric. J. 55; 363-366.

Rohilla, H R., 1990. Management of mustard aphid. In : Oholsson, I. and Kumar, P. R. (Eds). Proc. Indo-Swedish Symp. Rapeseed-Mustard, Uppasala, Sweden.

Rohilla, H.R., Singh, Harveer and Singh, Ram., 1999. Evaluation of rapeseed mustard germplasm against mustard aphid, *Lipaphis erysimi* (Kalt). *Test of agrochemicals and cultivars* No. 20. *Ann. Appl. Biol.*, 134 : 42-43.

Rout, G. and Senapati, B., 1968. Biology of the mustard aphid, *Lipaphis erysimi* in India. Ann. Ent-Soc. Am. 61(2): 259-261.

Roy, P., 1975. Population dynamics of mustard aphid, *Lipaphis erysimi* (Kaltenbach) (Aphididae: Hemiptera) in West Bengal. *Indian J. Ent.* 37(3): 318-321.

Russel, G.E., 1978. *Plant Breeding For Pest and Disease Resistance*. Butterworths, London.

Sachan, J.N. and Srivastava, B.P., 1972. Studies on the seasonal incidence of insect pests of cabbage. *Indian J. Ent.* 34(1): 123-129.

Sehgal, V.K. and Ujagir, R., 1977. Plants natural defences in Cruciferae and Tropaeolaceae against mustard sawfly, *Athalia proxima* Klug. *Indian J. Ecol.* 4(2): 199-205.

Sidhu, H.S. and Kaur, P., 1977. Influence of nitrogen applicaion to the host plant on fecundity of mustard aphid, *Lipapahis erysimi* (Kalt.). *J. Res. Punjab agric. Univ.* 14(4): 445-448.

Sidhu, H.S. and Singh, S., 1964. Control schedule of mustard aphid in the Punjab. *Indian Oilseeds J.*, 8: 237-256.

Singh H., Singh Z. and Yadava T. P., 1990. Influence of abiotic factors on alatae production in mustard aphid *Lipaphis erysimi* (Kalt.); *J. Aphidology* 4 71-74.

Singh, B. and Bakhetia, D.R.C., 1987. Screening and breeding techniques for aphid resistance in oleiferous brassicae: A review, the oil crops network, International development research centre, Canada., 50p.

Singh, B., Singh, R. and Mahal, M.S., 1983. Assessment of loss in yield of *Brassica juncea* by *Lipaphis erysimi* (Kalt.) II. Economics of aphid control. *Indian J. Ecol.* 10 (2): 279-284.

Singh, Balraj, Singh, R. and Mahal, M.S., 1983. Assessment of loss in yield of *Brassica juncea* by *L. erysimi*. II Economics of aphid control. *Indian J. Ecol.*, 10 : 279-284.

Singh, H., 1982. Studies on the insect-pest complex in *Brassica campestris* L., Var. Sarson Dissertation, Doctor of Philosophy, Haryana Agricultural University, Hisar. Pp., 192.

SIngh, H., 1982a. Studies on insect pest complex in *Brassica campestris* L.var. brown sarson. Ph.D. Thesis, Haryana Agric. Univ. Hissar. 179 p.

Singh, H. and Malik, V.S., 1993. Biology of the painted bug, *Bagrada cruciferarum* (Kirk). *Indian J. agric. Sci.*, 63: 672-674.

Singh, H. and Singh, Z., 1983. New records of insect pests of rapeseed. *Indian J. agric. Sci.* 53(9): 890.

Singh, H. and Singh, Z., 1983. New records of insect pests of rapeseed. *Indian J. agric. Sci.*, 53 (9) : 890.

Singh, H., Kalra, V.K. and Rohilla, H.R., 1986. Studies on the efficacy and economics of sprays of synthetic pyrethroids against mustard aphid, *Lipaphis erysimi* (Kalt). on rapeseed. *Indian J. Plant Prot.*, 10: 1-5.

Singh, H., Rohilla, H.R., Kalra, V.K. and Yadava, T.P., 1964. Response of *Brassica* varieties sown on different dates to the attack of mustard aphid, *Lipaphis erysimi* (Kalt.) *J. Oilseeds Res.* 1(1): 49-56.

Singh, H., Rohilla, H.R., Kalra, V.K. and Yadava, T.P., 1993. Outbreak of painted bug, *Bagrada crucuferarum* (Kirk) on oilseeds brassicas. *Int. J. Trop. Agric.*, 11: 153-154.

Singh, Harveer and Rohilla, H.R., 1998. Plant biochemical contents as parameter of resistance in oilseed *Brassica* to mustard aphid *Lipaphis erysimi* (Kalt). *J. Aphidology*, 12: 81-84.

Singh, Harveer, Rohilla, H.R., Kalra, V.K. and Yadav, T.P., 1984. Response of *Brassica* varieties sown on different dates to the attack of mustard aphid, *Lipaphis erysimi Kalt*. *J. Oilseeds Res.*, 1 (1) : 49-56.

Singh, Harvir, Gupta D. S., Yadava T. P. and Dhawan K., 1980. Post-harvest losses caused by painted bug, *Bagrada cruciferarum* (Kirk.) to mustard. *Journal of Research*, Haryana Agricultural University, Hisar 10 (3) : 407-9.

Singh, Harvir, Rohilla, H.R. and Singh, H., 1995. Emperial approach for the management of mustard aphid, *Lipaphis erysimi* (Kalt.). *J. Insect Sci.*8: 203-204.

Singh, K.J. and Singh, O.P., 1988. *Lipaphis erysimi* (Kalt.) infestation on the roots of mustard in Madhya Pradesh. *J. Aphidology*, 2 : 90-92.

Singh, P., Bakhetia, D. R. C. and Sidhu H. S., 1987. Plant appearancy and chemical defence in *Brassicas* against the leaf miner, *Chromatomyia horticola* Goureau, Proc. Int. Symp. Crop Improvement Soc., 23-27 Feb., 1987, PAU., Ludhiana, pp. 102-103.

Singh, P., Mayi, G.S., Bakhetia, D.R.C. and Sidhu, H.S., 1985. Identification of leaf miner, *Chromatomyia horticola* Goureau punctures in *Brassica* strains. *Annals Biol.*, 1 : 226-228.

Singh, R., 1982b. Estimation of losses in yield of mustard caused by the aphid, *Lipaphis erysimi* (Kalt.) Ph.D. Thesis, Punjab Agricultural University, Ludhiana, 69.

Singh, R.P., 1997. Insect pest vis a vis neem extract/formulations pp-6. In seminar on botanicals and pesticides in plant protection held on Jan. 10, 1997 Chennai. Org. by Fredrick Institute of Plant Protection and Toxicology, Padappai – 601 301.

Singh, S.R., Narain, A., Srivastava, K.P. and Siddique, R.A., 1965. Fecundity of mustard aphid on different rape and mustard species. *Indian Oilseeds J.*, 9: 215-219.

Singh, W. and Misra, D.S., 1986. Economic injury level of mustard aphid, *Lipaphis erysimi* (Kalt.) on mustard. (in) *Aphidology* in India ed. B.K. Agarwala, A.R. Printers, Culcutta, p. 105-112.

Singh, Y.P., 2007. Efficacy of plant extracts against mustard aphid, *Lipaphis erysimi* on mustard. *Indian J. Plant Prot.*, 35 (1): 116-117.

Singh, Y.P. and Sharma, K.C., 2002. Integrated approach to manage the mustard aphid, *Lipaphis erysimi* (Kaltenbach) in oilseed *Brassica* crops–a Review. *J. aphidology*. 16: 77-88.

Singhvi, S. M., Verma, A. N. and Yadava, T. P., 1974. Spraying schedules against rapeseed and mustard pests. *Oilseeds J*.4: 15-19.

Singhvi, S. M., Verma. N. D. and Yadava, T. P., 1973. Estimation of losses in rapeseed (*Brassica compestris* L.var. toria) and mustard (*Brassica juncea* Coss) due to mustard aphid, *Lipaphis erysimi* (Kalt.).*Haryana Agric. Univ. J.* 3(1): 5-7.

Sinha, R.P., Yazdani, S.S. and Verma, S.D., 1990. Population dynamics of mustard aphid, *L. erysimi* (Kalt.) in relation to ecological parameters. *Indian J. Ent.*, 52: 387-392.

Smith, C.M., 1989. *Plant Resistance to insects- A Fundamental Approach.* John Wiley and Sons, New York.

Smith, C.M., 1999. Plant Resistance to Insects. In: Rechcigi, N. (eds) *Biological and Biotechnological Control of Insects*. Lewis Publishers, Boca Raton, Florida, pp. 171-205.

Smith, C.M., Khan, Z.R. and Pathak, M.D., 1994. *Techniques for evaluating Insect Resistance in Crop Plants.* Lewis Publishers, Boca Raton, Florida.

Snelling, R.O., 1941. Resistance to plants to insect attack. *Botany Reviews*. 7: 543-586.

Srinivasachar, D. and Malik, R.S., 1972. An induced aphid-resistant, non-waxy mutant in turnip, *Brassica rapa*. *Curr. Sci.* 41: 820-821.

Srivastava, A.S. and Srivastava, J.L., 1972. Ecological studies on the mustard aphid, painted bug and mustard saw-fly attacking mustard and rape in India. *Plant Prot. Bull.* FAO., 20: 136-140.

Srivastava, P. D., 1957. Studies on the choice of the food plants and certain aspects of digestive physiology of the larvae and adults of *Athalia Lugens proxima* Klug *Bull. Ent. Res.* 48 : 289-91.

Suri, S.M., 1986. Development of pest management programme in *brassica* oilseeds in Kangra velley. Ph.D. thesis. Y.S. Parmar University Hort. Forestry, Nauni (Solan) p 153.

Teotia, T.P.S. and Lal, O.P., 1970. Differential reponse of different varieties and strains of oleiferous *Brassicae* to aphid, *Lipaphis erysimi* kalt. *Labdev J. Sci. Tech.* 8-B(4): 219-226.

Tylianakis, J.M., Didham, R.K. and Wratten, S.D., 2004. Improved fitness of aphid parasitoids receiving resource subsidies. *Ecology*. 85: 658-666.

Van Emden, H.F., 1990. The interaction of host plant resistance to insects with other control measures. *Proceedings of the Brighton Crop ProtectionConferance, Pests and diseases*, 939-948.

Weising K, Schel, l J. and Kall G 1988. Foreign genes in plants: Transfer, structure, expression and applications. *Ann Rev Genet.*, 22: 421-477.

Yadava, A.K., Singh, H. and Yadava, T.P., 1985. Inheritance of non-waxy trait inIndian mustard and its reaction to aphids. *J. Oilseeds Res.*, 2: 339-342.

Yein, B.R., 1985. Effect of dates of sowing on the incidence of insect pests of toria, *B. campestris*. *J. Res. Assam agric. Univ.*, 6(1): 68-70.

Yencho, G.C., Cohen, M.B. and Byrne, P.F., 2000. Applications of tagging and mapping insect resistance loci in plants. *Annual Review of Entomology*. 45, 393-422.

Chapter 34

Termite's Management with Special Reference to Sugarcane

S.K. Pandey
Sugarcane Breeding Institute,
Regional Station, Karnal - 132 001

Introduction

Termites are land dwelling social insects. They live in small or large colonies having been constructed, generally deep down in the earth, in the form of nests. Their common name of termites is derived from the Latin word *tarmes* which was used in ancient times to indicate any small creature, such as termites, wood-worm beetles, the larvae of many kinds of small butterflies and other creatures that bored holes in various materials, especially the dead wood. Today, the name 'termite' is applied to a well defined group of insects, comprising of more than 2,000 known species (though a number of species are still undiscovered).

Termites form an order of their own–Isoptera. Termites being social insects whose largest colony may contain well over a million individuals. Their destructive capacity lies in their numbers. They are polymorphic with various "castes" such as soldiers and workers being non-reproductive with queen and king being the functional females and males respectively. They feed largely on cellulosic materials thereby have a tendency to attack all kinds of cellulosic plant materials, both alive and dead (such as textiles, wood, paper and bamboos, etc.). They live mostly underground but come up for foraging; some species habitually live above ground in dead or living wood and in mounds, carton nests, etc.

Sugarcane is one of the favourite species attacked by termites. In India, about 13 species have so far been recorded to attack sugarcane from germination to harvest, of which the mound forming *Odontotermes obesus* Rambur and the non mound forming *Microtermes obesi* Holmgren are the most important. The other common species of termites attacking on sugarcane are *Odontotermes assmuthi*

Holmgren, *Odontotermes wallonensis* (Wasmann), *Odontotermes gurdaspurensis* (Wasmann), *Odontotermes redemanni* (Wasmann), *Odontotermes taprobanes* (Walker), *M. mycophagus* (Desneux), *Microtermes* beesoni Snyder, *Microtermes* estherae (Desneux), *Coptotermes heimi* (Wasmann), *Eremotermes paradoxalis* Holmgren and *Trinervitermes biformis* (Wasmann).

The Termite Community

The termite community usually comprises of a couple of fertile member and a very large number of sterile individuals, workers, soldiers and their nymphs. The queen is usually the only female in the community which continually lays eggs (Figure 34.2) as egg laying machine (2,000–36,000 eggs per day) thus continually supplying the colony with new members. Striking features of the queen includes her extra ordinary size (6.0cm x 1.3cm when mature), and especially her enormously developed abdomen (Figure 34.1). In adult insects, the front part of the body is the only part that bears legs. The gross disproportion between this part and the enormously built abdomen of the queen renders her incapable of doing anything on her own, except laying eggs. The normal life span of the queen varies from 5–15 years which may be extended up to 50 years in some cases. Indeed, she is always surrounded by a nucleus of workers who feed her 2 and is watched over by a number of soldier guards. Queen

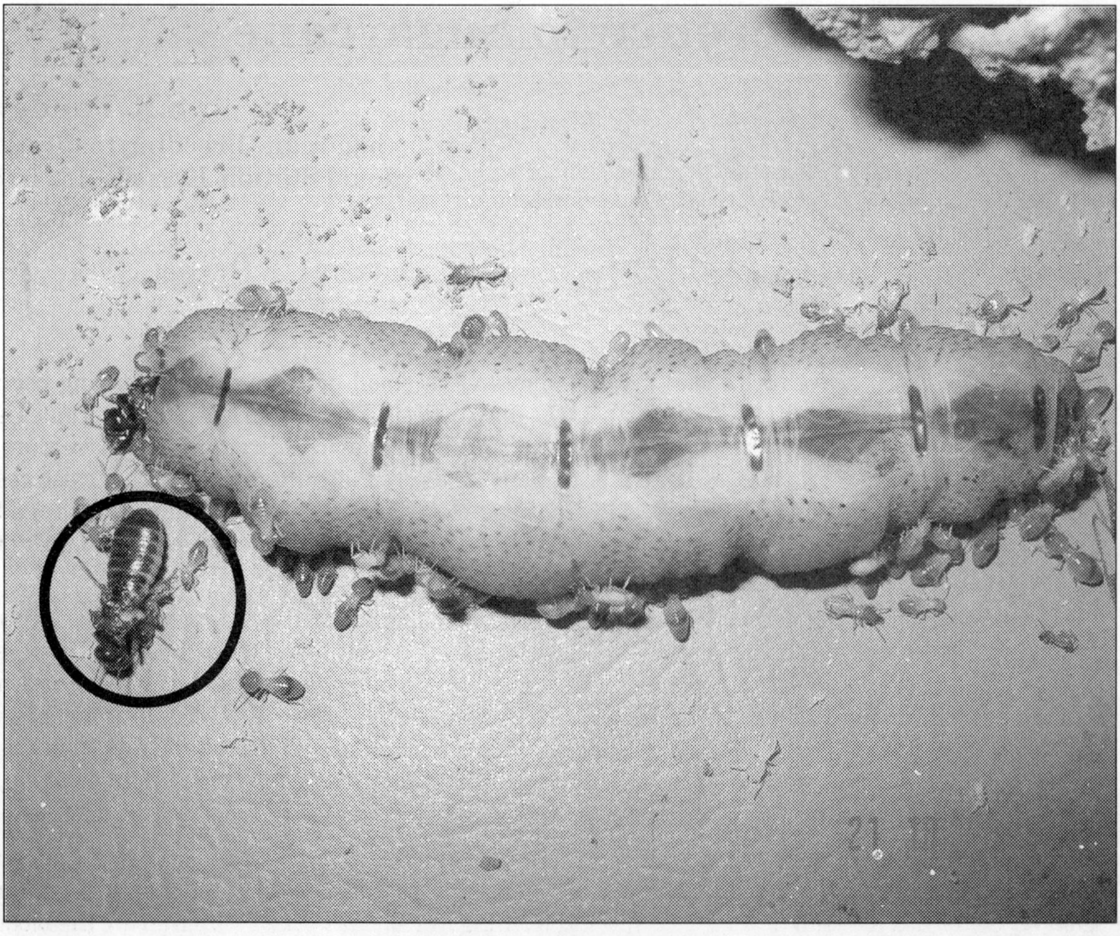

Figure 34.1: King (Circle)

Figure 34.2

secretes a liquid with which she is cleaned by the workers (Figure 34.1) who also frequently move her around and carry away the eggs she continually lays.

The king, the only fertile male in the community, always lives at the queen's side. He can be distinguished from her at a glance by his considerably smaller size, even though his size is larger than those of the workers and the soldiers. His head and thorax may be more or less as big as hers, but his abdomen is normal in size or at most slightly enlarged; his abdomen is never as great as hers.

Formation of Colonies

There is a distinctly marked season of the year during which the nymphs, belonging to a nest, that already has a large number of inhabitants, reach their final stage of development and transform into winged individuals. Before developing into adult, these nymphs gather together and move to the outskirts of the nest. Here they eat more food and empty the alimentary canal of its faecal matter contents. With great efforts, they then transform themselves into winged termites. This process is called swarming and there are approximately equal numbers of males and females for nuptial flight. These large numbers of winged termite (Figure 34.3) are attracted towards light and many of them are attacked by their predators and thus get killed. Those which survive, shed their wings, move in pairs and enter the soil or crevices in the wood, where they copulate and start the formation of a new colony. The female, usually referred to as queen, with the advance in age gradually develops a distended abdomen owing to a large number of eggs with in. The queen is always confined to the royal chamber with the male (king) who performs the sole function of reproduction.

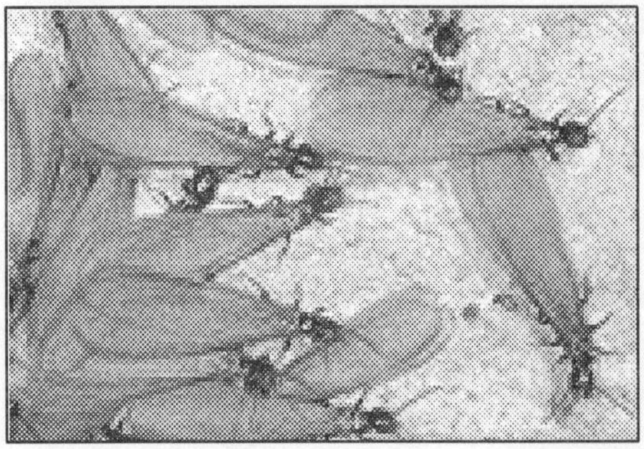

Figure 34.3

Nests

Large proportion of all the termites that inhabit the subsoil, some of the species form mounds whereas the others don't form mounds. The mound building termites build larger structures than any other living being of their size. The king and the queen shift deeper

Figure 34.4

into the soil where the Royal chamber is built by the workers formed from the first batch of eggs laid. Then slowly and slowly, the nest is built with the increase in number of workers to build cathedral–like nest (Figure 34.4). The nest has provides protection to the royal chamber inhabited by the founding king and queen, acts as a refuge and defence for the colony itself and is also an efficient sun–screen, preventing the sun rays from penetrating and upsetting the activities of the workers. During exploration of the nests by the author, big zig-zag, cylindrical inter connected tunnels (whose diameter varies within and amongst the species) which ended up in pockets were observed (Figure 34.5). These pockets were full of fungal gardens and act as nurseries where eggs are stored for hatching and the young ones, the nymphs, are reared by the workers. As we penetrated deeper into the nest, we come across sponge-like structures where fungal crop grows like a real garden. The fungal gardens are dark brown or brownish-yellow structures, spongy in appearance and consistency, housed in special chambers and emitting a characteristic fungus-like smell (Figure 34.6).

The sponge-like structures provide a very favourable environment for the growth of certain species of fungi; have never been found in any other environment. The termites cultivate these fungi with extreme care. The termites seem to eat a negligible amount of fungus, and in practice could only be getting their requirements of vitamins from this source. The author observed the royal apartments to be

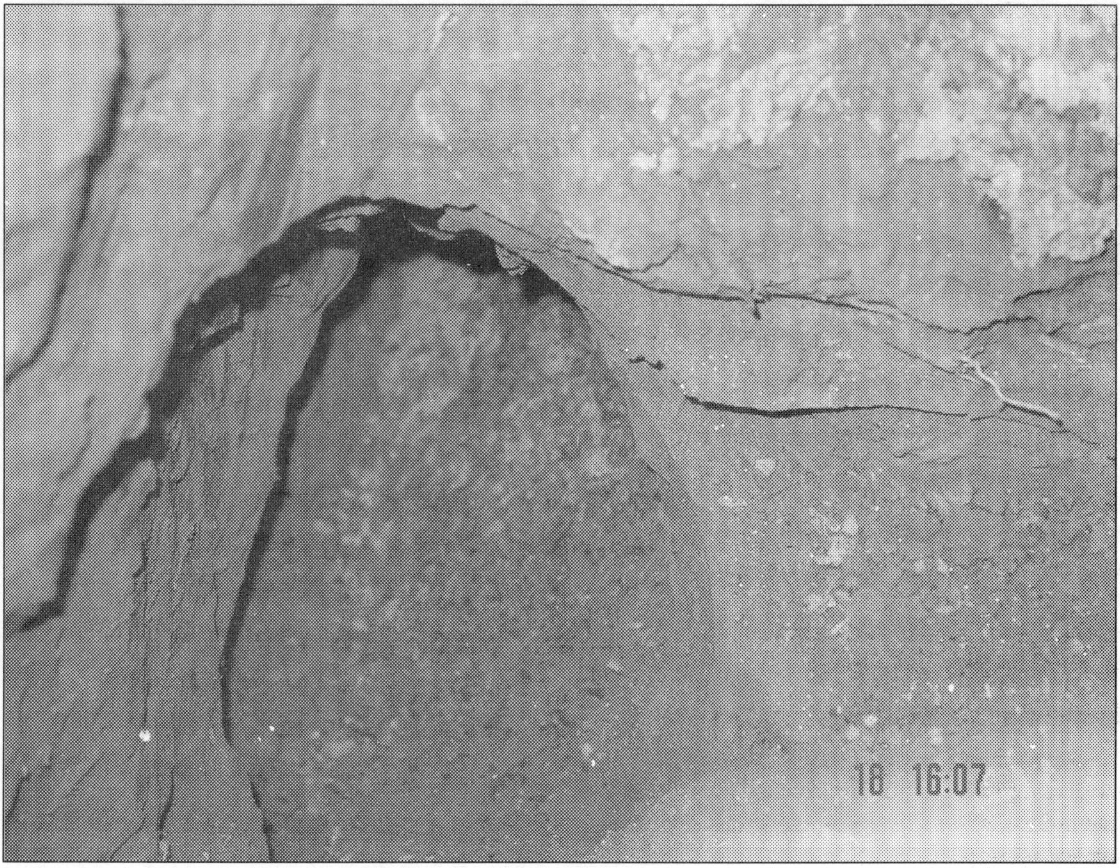

Figure 34.5

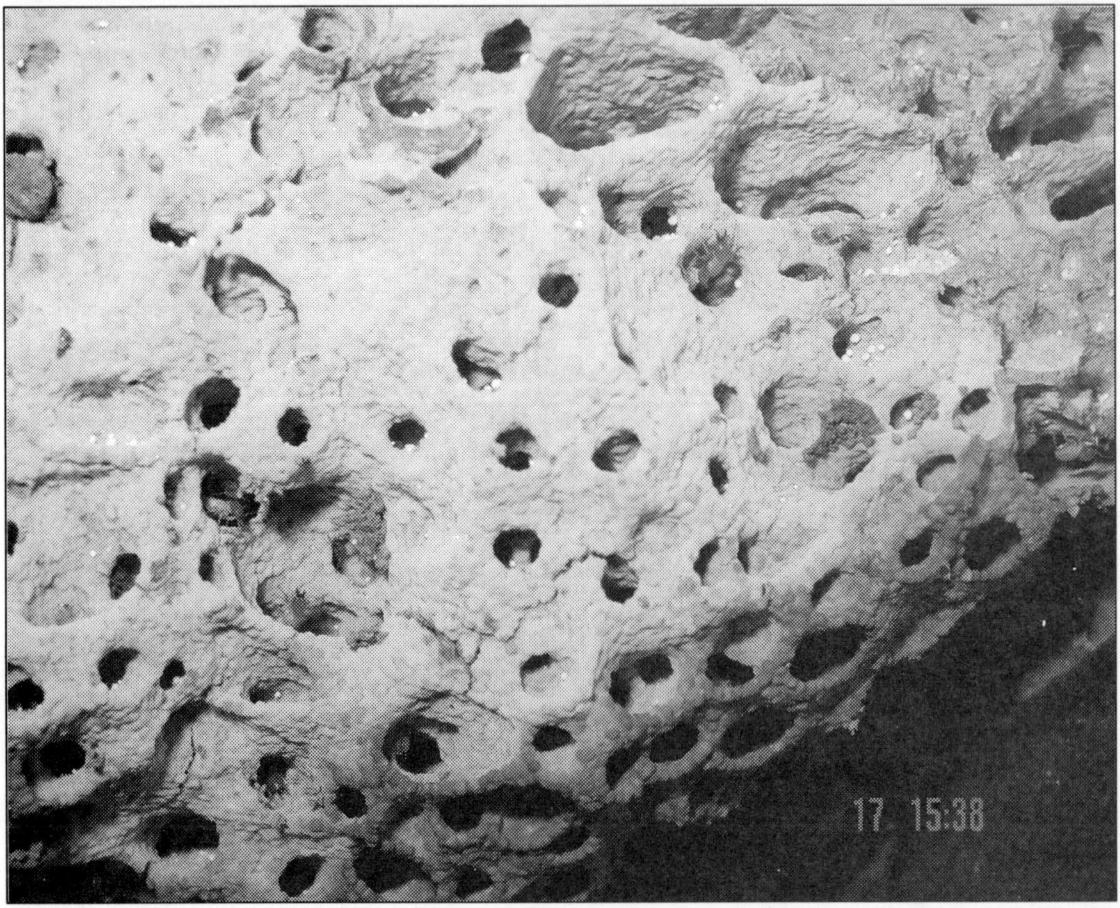

Figure 34.6

present in close vicinity of the fungal gardens. The royal chamber appeared to be built from clay whose wall thickness was around 3 to 5 cm and had passage holes, normally of less that 2 mm diameter. The dimensions of the oval chamber were 10-15 cm broad and 20-30 cm long like a mud clump. The chamber is very smooth and clean from within (Figure 34.8). The abdomen of the queen is lifted by the workers continuously. The eggs are immediately removed after laying and are deposited in the nursery pockets. The royal apartment is a sort of home within a home, inhabited by king and queen; the workers are responsible for cleaning and feeding the queen, and the soldiers responsible for defending the royal couple.

Sterile Castes, the Workers and the Soldiers

Apart from the distinctly marked period of the season during which the nymphs transform into winged reproductive individuals to form swarms, during rest of the period, the eggs laid down by the queen develop in to sterile castes only which are permanently wingless and completely blind. The sterile castes consist of at least one type of worker and in almost all species, at least one type of soldier as well.

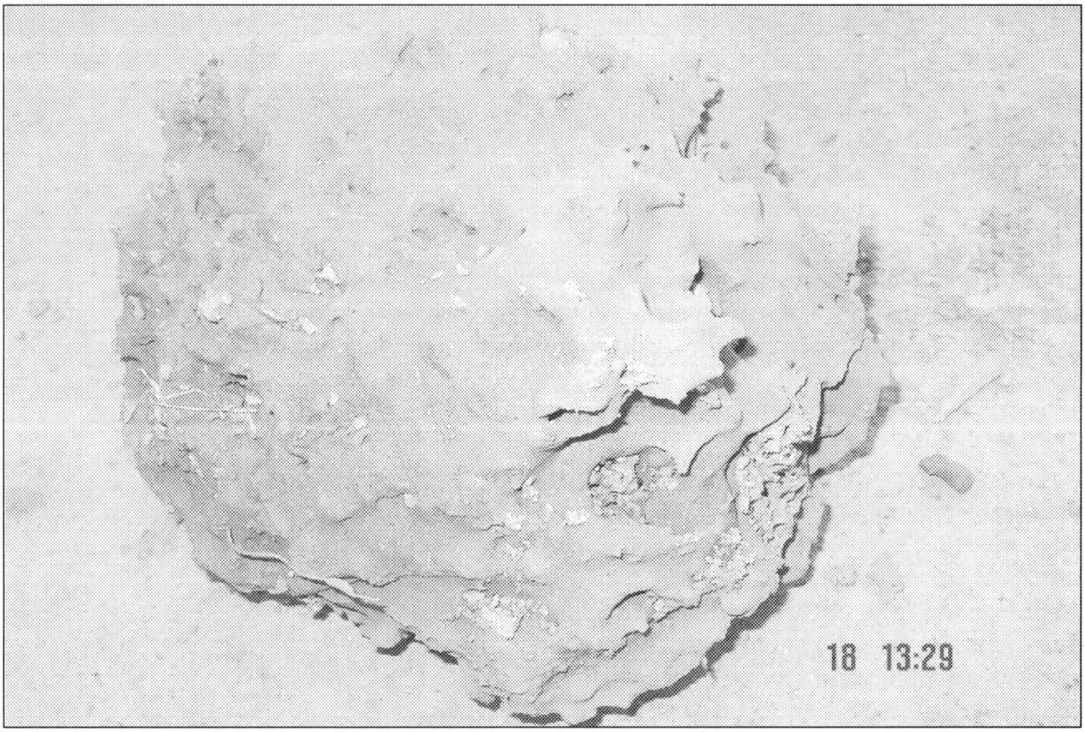

Figure 34.7

Figure 34.8

The workers (Figure 34.9) can be considered as the least specialized members of the community, since they may carry out any number of functions such as the care of the young, from egg to independent adult life. They provide mouth to mouth feeding by giving regurgitated food to both king and the queen and even to the soldiers. They build and maintain the nest and also cultivate fungi to form a fungal garden (details given in a later section).

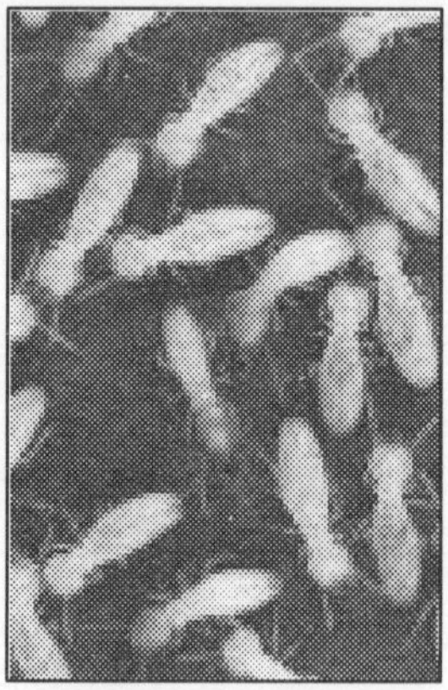

Figure 34.9

Even amongst the workers, however, there is frequently a sub-division of labour works which is often accompanied by the differences in their outer appearance. Many species have at least two classes of workers, "large workers" and "small workers", which are distinguished from one another both by their size and by a number of detailed features, particularly the mouth parts and the limbs, which are adapted for different functions which the two classes perform.

Figure 34.10

The soldiers (Figure 34.10) are structurally the most specialized members of the community. They may be easily distinguished by their large sclerotized heads. Head and thorax are well chitinized whereas the abdomen is delicate and dirty white in color. The soldiers' job is to defend, not only the king and the queen but also the food stocks, the nest itself-especially its entrances and the rows of workers engaged in the collection of food. Soldiers are usually larger than the workers and are immediately recognizable by their exceptionally larger heads. There are four well defined types of soldiers; *Mandibulate* soldiers are the most commonly occurring forms with large powerful mandibles that may assume a strikingly grotesque shape. *Nasute* soldiers in which the mandibles are vestigial, but there is a long median frontal nasus or rostrum at the end of which frontal gland opens and these soldiers occur only in Nasutitermitinae (Termitidae). *Nasutoid* soldiers are found among Rhinotermitidee in which the mandibles vary in size, but the frons, clypeus and labrum are produced into a snout like structure bearing a dorsal channel for the flow of frontal gland sticky liquid secretions and the nasuti can squirt out of the tip of their snout for quite a distance in the direction of attackers in order to envelop them in glue like substance and render them unable to move. *Phragmotic* soldiers are found in Kalotermitidae in which the head is strongly sclerotized, more or less plug like in structure that can be used to block the nest openings. Their mandibles are reduced.

The soldiers defend the colony mainly against other insects like ants and vertebrate predators. Typical soldiers attack with their mandibles, which can pierce or slice and can be used defensively with a snapping action. Other soldiers attack insect intruders with adhesive secretions of the salivary or frontal glands. The other forms of soldiers include *Pseudergates*, which replace the workers in their absence and the *Achrestogonimes*, which have lost their wings and their gonads have undergone atrophy. They do not take part in the maintenance of the colony, and the *intercastes*, which are morphologically intermediate between the different castes.

Damaging Effects of Termites to the Crops, Particularly to Sugarcane

In general, the initial termite attack occurs mostly in patches at the seedling/settling stage when roots are eaten up, resulting in the yellowing of leaves, and dislodgement, finally death of the plant. In sugarcane, the termites even attack the setts soon after planting in the field. The termites enter the setts, either through the injured eye-buds or through both the cut ends of the setts completely eating up the soft and nutritious pith. Thus there is either poor germination which leads to gappy crop stand. In severe cases of infestation, replanting may become a necessity. The post germination attack is also common, particularly in loamy soils and dry areas where proper facilities of irrigation are not usually available. Severely affected sugarcane show yellowing and drying of the older leaves followed by the younger ones). The whole plants may appear yellow followed by withering (Figure 34.11). The plantlets dry up and can be pulled out easily. Such plantlets do not show compensatory tillering in most areas. The severity of termite attack subsides after the onset of monsoon rains. Even during the post monsoon period, young developing tillers as well as mature canes may also be subjected to the attack, and in severe cases, lower portions of the stalk is some times completely eaten away leaving the rind only which is filled with soil which are in fact their foraging mud galleries (Figure 34.12). Thus there is a considerable loss in weight of cane and also in sucrose content. In cases of severe attack up to three-

Figure 34.11

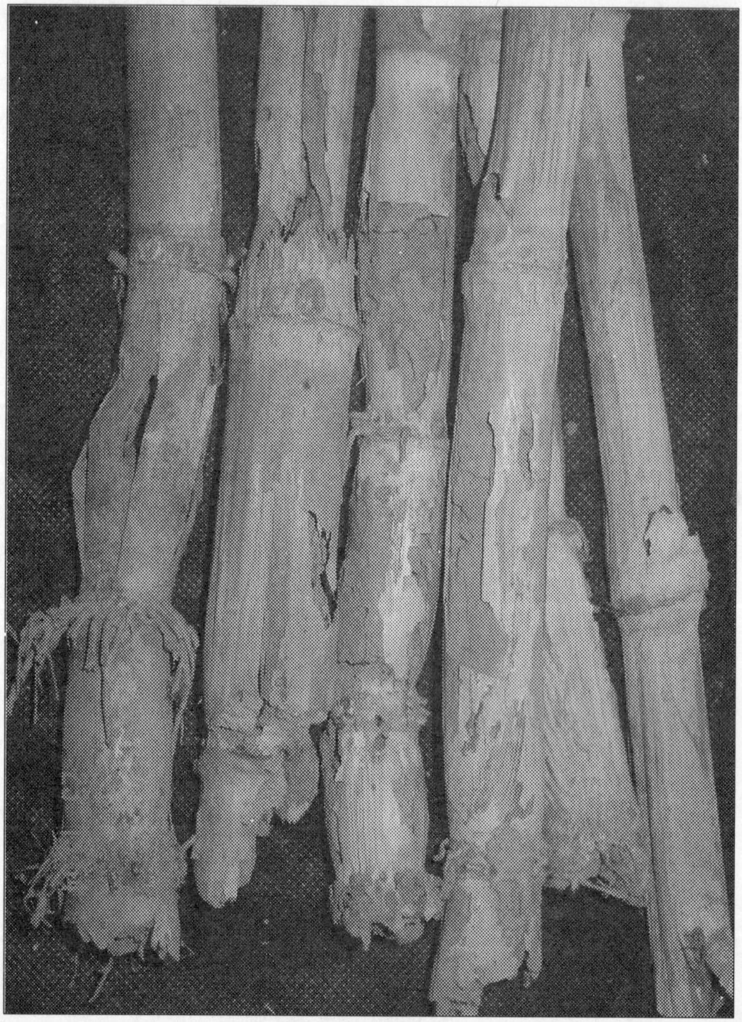

Figure 34.12

fourths of the total cane may be seriously damaged with in a sort period of 6 to 8 weeks; damage is more serious under drought conditions. The leaves and the crown eventually die as a result of severe attack. At our farms, the foraging tunnels were observed on the surface of growing stalk in severely infested fields which even covered whole of the stalk near the ground.

Integrated Approach of Termite's Management

Cultural Control

Frequent irrigation in termite prone or endemic areas help in reducing attack, even though it is of temporary nature. Clearing and burning of crop residues and other debris in the crop ecosystem deprive the subterranean termites of their food and reduce population. Since trash mulching can attract termite infestation, it should not be practiced in areas endemic to termites.

Mechanical Control

Use of light traps during the rainy season in termite affected areas would attract termites swarms and would kill the reproductives going to develop new colonies.

Chemical Control

Chlorpyriphos 20 EC @ 2 litres/acre should applied along with last irrigation (prior to sowing) water by putting a drip system over the channel through which water is entering the field. Drip systems such as used to supply glucose or normal saline to the patients in the Hospitals can be used to distribute chlopyriphos. Old used bottles as well as drippers can be successfully used for such purpose.

In sugarcane, Chlorpyriphos, 20 EC, @ 5ml/litre of water may be applied over the setts placed in furrows using a rose cane prior to covering the setts with soil followed by irrigation. About 2 litres of the formulation diluted in 400 litres of water will be needed per acre. Alternatively, chlorpyriphos can also be applied along with irrigation water after the setts have been covered with soil as has been given in the previous paragraph. Control of termite in the standing crop can be effected by application of chlorpyriphos along with irrigation water as explained in the previous paragraph. In endemic areas, soaking of seed setts with chlorpyriphos affords effective protection until germination. Even though Gamma HCH 20 EC is an effective insecticide, sett treatment with insecticide should be avoided as it has been found to affect germination adversely.

Generally termite attack appears in patches in a cropped area, the affected patches should only be treated as spot treatment by application of insecticidal solutions as collar drenching as well as application of chemical mixed sand. About 2 litres of the formulation of chlorpyriphos 20 EC or Gamma HCH 20EC are sprayed over 50 kg of dry sand and mixed well before broadcasted in the affected area followed by irrigation. Bunding of such spots before application/irrigation would prevent the chemical to flow to the other areas where it is not needed. This will cut down the cost of cultivation and save the environment to the extent possible.

Recently, imidachloprid (Confidor 17.8 per cent SL) has been found to be effective against termites at a lower concentration of 0.4ml/litre of water.

Control through Colony Elimination

The mound forming species can be controlled by digging the colony and killing the queen. Although there are reports that if a queen gets old/becomes non-productive, then it is replaced by another queen which develops by transformation of a nymph. But there are no reports in the literature which indicate that after dismantling a colony, a new couple develops and forms another royal chamber to reactivate the colony.

Chemical control has been reported for colony elimination. Mound building colonies may be eliminated by tossing 2 tablets of aluminum phosphide @ in a mound which fumuigates the colony and is likely to kill the queen and other member s of the colony. Another method to eliminate the colony is to apply 20 to 30 litres (depending upon the size of the mound) suspension of Gamma HCH, 20 EC or Chlorpyriphos, 20 EC @ 2 ml/per litre has been reported.

Bait Station, Colony Elimination Technique

Latest trend in termite management is to install termite bait stations (Figs.) based on "colony elimination Technique", which is reported to eliminate the queen thereby the entire termite colony is destroyed.

In this technique, pieces of wood species, highly susceptible to termites, are impregnated with insect growth regulators (IGR) such as chlorfluazuron, diflubenzuron etc. and kept as bait inside the specially prepared bait stations (Fig). These bait stations are permanently placed below ground level with plain wood pieces or cellulose blocks initially to monitor the termite activities. Once termite activity is noticed during periodical observations, the plane wood is replaced with IGR impregnated cellulose blocks or wood pieces. These treated baits are eaten and carried by the workers to feed the members of the entire colony including queen of the colony. These IGRs block chitin synthesis, a vital process in insects' growth and development thus affecting the very existence of an active colony. IGRs have been reported to even sterilize the queen there by stopping the production of new soldiers and workers. So, in all likely hoods, a colony is bound to become ultimately a non functional one.

Efforts are being made at our Centre to use entomopathogenic fungi to infect the members of the colony which if becomes successful will eliminate the members of the colony itself.

Some Additional Specific Information About the Species of Termites Associated with Sugarcane

The available information on this aspect as reviewed by Chhotani (1980) is given below:

Odontotermes obesus

This is the main mound-building termite in India. The mounds are tall, unilocular with buttresses, or dome-shaped and multilocular with or without turrets or in the form of lumps of earth within the main nest, underground. The mounds may reach a height of three metres and are sub-cylindrical. Generally, only one queen and a king are present, though more than one pair is also reported. In such a case, each pair is present in a separate royal chamber.

Swarming starts with the beginning of monsoon at dusk and may continue during night. The ratio of soldiers, workers and nymphs in the mound is 7.7: 49: 43.3 in the non-mound building months and 5.5: 66.5: 28 in the mound building months.

Workers

Head capsule dark brown, post clypeus pale brown, pronotum and abdomen somewhat paler than head. The former with 'T' shaped pale mark, densely pilose. Head capsule sub circular (width 2.43-2.80 mm); fontanelle minute: eyes large, sub circular, projecting out, maximum diameter 073-0.90 mm; ocelli large, oval. Situated very close to eyes: antennae 19 segmented: post clypeus swollen, about half as long as wide; pronotum sub trapezoidal, somewhat emarginated both interiorly and posteriorly; wings brownish. Length of the body 23-30 mm with wings and 10-14 mm without wings.

Soldiers

Head capsule yellowish to dark brown or dark reddish brown, antennae pale basally and dark distally, body yellowish to yellowish brown; head sparsely and body moderately pilose. Head capsule oval, converging in front (1.20- 1.67 x 1.07-1.37 mm); antennae with 16-17 segments, third segment shortest; labrum tongue-shaped, mandibles thin, sabre-shaped, left mandible with a prominent tooth at the base of distal third, postmentum arched, convex laterally; pronotum saddle-shaped, deeply notched both interiorly and posteriorly. Body length about 4-6 mm.

Odontotermes assmuthi

The mounds of this species appear similar to the buttress type mounds of *O. obesus*. In the underground nests there are long narrow galleries with narrow covered runways at the ground level for foraging. The mass emergence occurs around 4.45 pm followed by swarming during June. The first batch of 100–300 eggs are laid 6-9 days after swarming. After a temporary cessation, egg lying continues.

Workers

Head and body blackish brown, post clypeus pale brownish. Pronotum with 'T' shaped mark in the middle, head and body fairly densely pilose. Head capsule sub circular (1.68-1.70 mm x 2.20–2.30 mm); fontanelle small and circular; eyes moderately large, circular; ocelli oval situated at a distance of more than their length from eyes; antennae 19 segmented, segments four and five are shortest; post clypeus swollen, twice as wide as long: pronotum sub trapezoidal, both anteriorly and posteriorly weakly notched medially; wings dark brown, hairy near distal end. Body length with wings 24.0-26.5 mm and without wings 12.0-14.0 mm.

Soldiers

Head capsule golden yellow, body pale yellow, sparsely hairy. Head capsule, sub- rectangular, slightly converging anteriorly (1.44–1.70 mm x 1.09-1.25 mm antennae 16 segmented, fourth segment shortest; labrum broadly tongue-shaped; mandibles short, stout, sabre-shaped, left mandible with a sharp prominent tooth, a little below distal third, postmentum sub rectangular; pronotum saddle-shaped, deeply notched both at anterior and posterior margins. Body length 4.30- 5.00 mm.

Odontotermes wallonensis

The mounds are made up of low, solid, irregularly shaped structures of earth having a number of open chimneys. Internally the mounds have chambers having fungus colonies or gardens. The fungus combs are arranged in tiers one above the other. The royal chamber lies much below the ground level and may have more than one queen.

Workers

Head capsule dark brown, pale around fontanelle, body slightly pale, pronotum with 'T' shaped median pale mark; head and body densely hairy. Head capsule sub circular, much broader than long (1.88-.95 mm x 2.7-.8mm) with eyes: fontanelle minute. circular; eyes large, sub circular (0.8- 0.85 mm dia): ocelli large (0.3-0.45 mm), oval, situated at a distance of less than half the length from the eyes: antennae 14 segmented, third segment shortest : post clypeus swollen, twice as wide as long: pronotum large, sub trapezoidal. Both anterior and posterior margins notched medially. Body length with wings about 30 mm and without wings 16 mm.

Soldiers

Head capsule yellowish to pale brownish yellow, antennae pale basally, dark apically, abdomen pale yellowish; head sparsely and body fairly densely pilose. Head capsule oval, converging in front (1.53- 1.68 x 1.20-1.28 mm); antennae 17 segmented, third shortest; labrum tongue-shaped: mandibles sabre-shaped, strongly curved at apices, left mandible with a prominent tooth a little above middle; pronotum saddle- shaped, deeply notched anteriorly, sub straight at posterior margin: abdomen elongately oval. Body length 5-7 mm.

Odontotermes gurdaspurensis

This species makes complex underground nests; the swarming of alates is reported during July and August.

Workers

Head and body fuscous, densely hairy, pronotum with a shaped pale mark. Head capsule sub circular, 1.9-2.1 mm long and 2.47-2.70 mm broad; fontanelle minute, lying a little above middle of head dorsum; eyes large, sub circular (0.8-0.78 mm dia): ocelli large (0.35-0.38 x 0.25 mm), oval, situated from eyes at half the distance of their width: antennae with 19 segments, third segment short, sub equal to fourth : post clypeus swollen: almost half as long as wide. pronotum trapezoidal, both anteriorly and posteriorly weakly notched; wings long, pale brown and hairy near distal margin (fore wing 26–27 mm, hind wing 24–25 mm long). Body length with wings 31 or 32mm and without wings 15 or 16 mm.

Soldiers

Head capsule deep golden yellow to dark brown, body yellowish to pale brownish; head sparsely and body moderately hairy. Head capsule sub rectangularly oval (1.40–1.60 mm x l.13–1.40mm); antennae 16-17 segmented: labrum tongue-shaped; mandibles long, slender, left mandible with a sharply pointed tooth near base of distal third and right with a minute tooth in middle third; post mentum sub rectangular: pronotum saddle-shaped, anteriorly slightly to appreciably notched and posterioly distinctly emarginate medially. Body Length 5.0- 6.5 mm.

Odontotermes redemanni

Mounds are large, dome-shaped or conical in shape and have large number of chambers (multilocular type) with fungus gardens interconnected by tunnels, The nest is 1.3 metre deep in the soil, There is one central chamber with fungus combs near which the royal chamber is situated eccentrically, Rarely, the royal chamber (7-12 cm dia, 2.25 cm height) is found above the ground level and generally has a physogastric queen and a king. Sometimes, more than one queen is present. Swarming is reported to occur during September in eastern India.

Workers

Head capsule castaneous brown, body somewhat pale, pronotum brownish with a pale 'T' shaped mark: densely pilose. Head capsule sub- circular, width with eyes 2.67-2.87 mm; fontanelle small, point-like; eyes sub- circular, large (08-0.88 mm dia); ocelli large, oval, situated from eyes at about their short diameter; antennae 19 segmented, segments three and four shortest, sub equal; post clypeus swollen, pronotum trapezoidal, anteriorly slightly notched, posteriorly sub straight or faintly emarginate; wings yellowish brown. Body length with wings 26-31 mm and without wings 11-14 mm.

Soldiers

Head capsule yellowish brown to dark brown, antennae pale basally and dark apically, body straw yellow to yellowish; head sparsely and body moderately or densely hairy. Head capsule oval (length 1.15-1.33, width 0.98-1.05mm), narrowing anteriorly; antennae 16 or 17 segmented, third segment shortest or sub equal to fourth; labrum tongue- shaped; mandibles sabre- shaped, curved at apices, left mandible with a sharp tooth near distal third, right mandible with a small tooth, a little below distal third; pronotum saddle-shaped, deeply notched at middle, both anteriorly and posteriorly; abdomen oblong. Body length 4-6 mm.

Microtermes obesi

It makes its nest in soil or in mounds of other termite species *viz. M. obesus, O. redemanni* and *O. rnicrodentatus* Roonwal and Sen Sarma. The nests have small round chambers, interconnected by thin, long galleries. The chambers contain fungus combs, which are small, rounded structures made up of yellowish, granular pellets. They are extremely fragile and have weak ridges on the surface. Swarming is reported to occur during April to August.

The queen cells are dome shaped with a flat floor and a labyrinth of intercommunicating galleries and cells, with fungus garden in some. The eggs and workers are seen throughout the year, but the potential reproductives are seen only a month or so before the rains. At this stage, nymphs appear with large wing pads.

Workers

Head capsule yellowish brown to dark reddish brown, post clypeus yellowish brown, abdomen slightly paler than head; head fairly but body densely hairy. Head capsule sub circular (length 1.05-1.1 5 mm, width with eyes 1.33-1.60 mm); epicranial suture faintly marked; fontanelle small, circular, whitish spot; eyes round (0.35-0.40 mm dia), bulging, separated from lower margin of head by about one fourth of their diameter oceili oval, medium sized (0.23-0.28 mm long), separated from eyes by about their width; antennae 15 or 16 segmented, segment three and four sub equal and shortest; post clypeus much swollen, length about half the width; pronotum with indistinct median notch anteriorly and with a distinct notch at posterior margin. Body length with wings 13-15 mm. and without wings 7-12 mm.

Soldiers

Head yellowish, body brownish, head and body moderately hairy. Head capsule broadly oval (0.8-1.0 mm x 0.76-0.88 mm); antennae 14 segmented, segment four shortest; labrum lanceolate, reaching up to distal one fourth of mandibles; mandibles thin, slightly incurved distally, postmentum arched, almost as wide as long; pronotum very much saddle-shaped, anteriorly deeply notched, medially and posteriorly sub straight. Body length 3.5-5.0 mm.

Microtermes mycophagus

This species is confined to arid and semi-arid parts of India. It lives in small communities under stones and wooden logs. The chambers are small and inter-connected by galleries. Swarming is reported to occur from July to August.

Workers

Head capsule castaneous brown, abdomen paler than head; head and body densely hairy. Head capsule broad, circular (1.25-1.45 x 1.75-1.93 mm); fontanelle minute, circular, situated on mid dorsum; eyes large, round (0.4-0.5 mm dia); ocelli oval, 0.20–0.26 mm long, separated from eyes by less than half their long diameter; antennae with 17 or 18 segments, segment three as long as two, four shortest; post clypeus swollen, length less than half of width; pronotum sub trapezoidal, anterior margin sub straight and with a faint median notch, posterior margin round and distinctly notched medially. Body length with wings 20-24 mm and without wings 8-12 mm.

Soldiers

Head capsule yellowish brown, body much paler than head, head and body fairly densely hairy Head capsule broadly oval, converging anteriorly (0.85-1.03 x 0.78-0.95 mm); antennae with 15 segments, segments three and five sub equal and shortest; labrum tongue-shaped, extending up to distal third of mandibles; mandibles weak and thin, weakly uncurved at apices; postmentum swollen, arched at sides, almost as long as wide; pronotum strongly saddle-shaped; anterior margin with a deep and posterior margin with a weak median notch. Body length 3.5 mm.

Microtermes beesoni

M. beesoni lives in and under logs of wood, under the bark of trees and in soil. The nests which are made up of carton-type material ate generally found partially buried in the soil and may be spherical, cone-like, sub squarish or flat. The internal structure of the nest is honey-combed and shows two zones, the outer with large irregular communicating passages and the inner one with smaller chambers. This species is highly tolerant to the fluctuations of temperatures and relative humidity. Swarming takes place during May to July.

Workers

Head capsule brown, body paler than head; densely hairy. Head capsule with squarish sides, sub parallel; 'Y' suture fairly prominent; fontanelle plate small, yellowish, situated medially; eyes sub circular (0.20-0,22 mm dia): ocelli oval, situated as far as their long diameter from eyes; antennae with 14 segments, third segment shortest; post clypeus swollen as long as half of width; labrum broad, tongue-shaped; mandibles typically *Microcerotermes* type; pronotum trapezoidal, anterior margin nearly straight, posterior margin emarginate medially; wings pale smoky brown, thickly covered with micrasters and sparsely' hairy. Body length with wings 8-10 mm, without wings 5-6 mm.

Soldiers

Head capsule yellowish, body pale yellow; head sparsely and body densely hairy. Head capsule sub rectangular, sides straight (l.28-1.55 x 0.77- 0.91 mm); fontanelle minute, situated at anterior third of head; antennae 13 segmented, third segment shortest; labrum with parallel sides, anteriorly sharply rounded; mandibles a little longer than half of head, coarsely serrated along its length; postmentum club shaped, widest at anterior third, waist yellow, lying below middle and slightly more than half as wide as maximum width; pronotum saddle–shaped, anteriorly weakly notched, posteriorly slightly emarginate. Body length 4–5 mm.

Microtermes estherae

The nests of *M. estherae* are underground and extend to large areas. The termites collect small heaps of granular soil about 2–8 cm over the ground surface in the vicinity of the nests. The nests are in the form of communicating galleries and chambers running horizontally for a distance of 100 metres or more. The fungus combs are small (5 cm dia) and have cavities like the convolutions of a human ear. The soldiers and workers march in long procession during foraging expeditions. Swarming has been observed after rains in July.

Workers

Head capsule dark reddish brown, pale around fontanelle, body somewhat paler than head; head sparsely but body densely pilose. Head capsule sub circular (2.10x2.93 mm); fontanelle plate large, circular; eyes large, circular (0.83 mm dia); ocelli oval, large (0.35 mm long), situated from eyes at about half of its long diameter; antennae 19 segmented, third segment shortest; post clypeus swollen, twice as wide as long; labrum broad, tongue–shaped; mandibles *Macrotermes* type; pronotum a little narrower than head, anterior margins weakly notched and posterior one straight. Both lengths with wings 32-33 mm and without wings 15-16 mm.

Soldiers

Major: Head capsule pale, reddish brown, body pale brownish: head sparsely hairy, abdomen bare or with a few short hairs. Head capsule large (4.3–6.6 x 3.09–4.1 mm), oval, fontanelle small, circular, situated on mid-dorsum: antennae with 17 segments, third segment larger than second; labrum triangular with a large, prominent hyaline tip; mandibles thick, short and stout, sabre- shaped, left mandible with three crenulations in basal half; pronotum weakly saddle-shaped. anterior margin weakly notched and posterior one distinctly notched medially; body length 11–14 mm.

Minor: Resembles soldier major, except for its generally smaller size; oval and comparatively longer than wide: head (2.2-2.70x 1.88- 2.35 mm) elongate; labrum with narrow and hyaline apex; long and slender mandibles. Body length 7-8 mm.

Coptotermes heimi

The nests of this species are found in soil as well as in dead portions of trees and logs of wood. The nest is dark grayish in colour and is in the form of a porous structure composed of network of strands of hard wood fibers left over from feeding and termite excrement. The soldier population in a colony is fairly high and when disturbed they come out in large numbers, standing guard and waving their antennae. They also exude some sticky, milky white fluid from fontanelle in defence, Swarming has been observed generally at sunset during January to August.

Workers

Head capsule yellowish brown to dark brown, pale in region of fontanelle, thorax and abdomen above somewhat paler than head; wings transparent, head and body fairly hairy, wings densely covered with short hairs. Head capsule sub circular, shallow, depressed in region of fontanelle, with eyes wider than length to base of mandibles (1.04–1.16 mm x 1.30- 1.40 mm); eyes sub circular, medium sized (0.37-0.40 mm dia); ocelli oval, translucent, either touching the eyes or slightly separated; fontaneile minute; antennae 20-22 segmented, third segment shortest: post clypeus flat: mandibles typically of Coptotermes type; pronotum flat, narrower than head with eves, anteriorly weakly incurved and posteriorly prominently notched medially: legs long and hairy; wings thin, transparent and hairy; abdomen elongate, hairy. Body length with wings l0.6-12.3 mm and without wings 5.6-7.0 mm.

Soldiers

Head capsule straw yellow to yellowish, body yellowish white; head sparsely and body fairly hairy, hairs on posterior margin of abdominal terga arranged in a row. Head capsule variable in shape, oval to round but genera]ly oval, longer than wide, rarely wider than long (l.20-1.48x 1.05-1.35 mm); fontanelle large. circular with a sclerotized border, situated at base of clypeus; antennae with a hyaline tip and generally with a pair of long bristles apically; mandibles sabre-shaped, long, left mandible with a basal projection and four crenulations in basal half; postmentum generally long and club shaped, shape and size varying with the shape of head; pronotum flat, interiorly with a prominent median notch and posteriorly with a weak incurving: abdomen oblong. Body length 4.2-6.0 mm.

Eremotermes paradoxalis

Its nests are found in soil under stones. Swarming occurs during March to September.

Workers

Head capsule yellowish to chestnut brown with numerous small rounds pale spots; pronotum slightly paler than head, abdomen pale brownish; head and body densely hairy. Head capsule small. roundish oval (length to the mandible base 0.65-0.75 mm, width 0.7-0.8 mm); fontanelle plate elongate, whitish: antennae with 14 or 15 segments, third segment shortest; eyes round, small (0.2-0.23 mm dia); oceili slightly separated from eyes; post clypeus swollen, length about half of width, sub trapezoidal;

mandibles typically *Eremotermes* type; pronotum flat, sub trapezoidal, very narrow posteriorly, both anterior and posterior margins notched medially; wing membrane smoky and densely covered with micrasters. Body length with wings 8.0-9.8 mm and without wings 4.5-6.7 mm.

Soldiers

Head capsule pale yellowish, abdomen yellowish white; fairly densely hairy. Head capsule sub rectangular, thick and with a prominent frontal protuberance (length to mandible base 0.8-0.95, width 0.68-0.75 mm); fontanelle prominent, lying below protuberance, gland large; antennae with 14 segments, fourth segment shortest; labrum tongue–shaped, narrowed anteriorly; mandible thin, elongate, three fourth as long as head capsule, sharply pointed and weakly incurved at tips, incurved medially at outer margin and with a sharply pointed tooth on inner margin in the middle; postmentum swollen, short, broadest a little in front of middle and narrowest near posterior end; pronotum saddle-shaped, either fairly or not notched anteriorly, sub straight at posterior margin; abdomen elongate to sub globular. Body length 3.5- 4.2 mm.

Trinervitermes biformis

It is a common foraging termite in India. The workers go out on foraging expeditions in thousands, a little before sunset in regular columns, with the soldiers guarding against ants and other predators. The nests of this species are generally subterranean and consist of ramifications of galleries and small chambers which are coated with black excretory matter. The small dome-shaped mounds are covered with grass.

Workers

Head capsule and abdomen brown with reddish tinge above, pronotum pale brown, sternite brownish, pale medially; head and body densely hairy. Head capsule sub circular, broader than long to base of mandibles (length 1.2-1.4 x 1.52-1.82 mm), medially depressed a little; epicranial suture either present or absent; fontanelle pale, Y-shaped or triangular; eyes sub oval, moderately: large (0.5-0.55 mm din): ocelli oval, translucent, length 0.18- 0.25 mm, from eves situated at a distance of one fifth to one third of length: antennae 15 or 16 segmented, segment three longer than two; post clypeus swollen, length a little less than half its width; labrum dome shaped: pronotum trapezoidal, anterior margin straight not incised, posterior margin either straight or weakly convex, with a weak to strong medial emargination; wings hairy and punctate. Body length with wings 18-l9 mm and without wings 912 mm.

Soldiers

Major: Head capsule brown, rostrum reddish brown and pale apically, abdomen pale brown to dark smoky brown dorsally. Head capsule without rostrum, almost round or pear-shaped. not constricted behind antennae (head length with rostrum 1.35–1.60 mm, width 1.25-1.50 mm) in profile with a weak concavity at base of rostrum; occiput greatly arched in profile; rostrum conical, uplifted apically, tip with a few short hairs; antennae 12-15 segmented, segment three variable, depending on number of segments, slightly longer by nearly 2.5 times that of second; mandibles vestigeal without spines; pronotum saddle- shaped, anterior lobe greatly upturned and broadly invaginated at anterior margin, posterior margin weakly convex: abdomen elongate; tergites without any long hairs; sternites with long and short hairs. Bed length 4.5-5,0 mm.

Minor: Head capsule yellowish to dark brown, rostrum dark, abdominal tergites fuscous. Head capsule subrectangulariycal, length about one and a half times that of width (length with rostrum 1.4-

1.7 mm, without rostrum 0, 85-1.10 mm: width 0.55-0.77 mm) in profile, concave; rostrum long, cylindrical, awl-shaped; antennae long. 12-14 segmented, segment three about twice as long as second or a little longer: mandibles vestigeal, without any spines; pronotum small, saddle-shaped, without notched anteriorly and convex posteriorly. Abdomen as in soldier major. Body length 3- 4 mm.

References

Ananathanarayana, K. and David, H., 1986. *Termites. Sugarcane Entomology in India*, 208-231. Sugarcane Breeding Institute, Coimbatore

Chhotani, O.B., 1980. *Termite pests of agriculture in the Indian Region and their control*. Tech. Monograph, Zool. Survey India. 4: 87 pp.

David, H. and Nandgopal, V., 1986. *Pests of Sugarcane: Distribution, Symptomatology of Attack and Identification*. Sugarcane Entomology in India, p. 1-23. Sugarcane Breeding Institute, Coimbatore

Kalra, A.N., 1983. *Sugarcane: Pests and their Control*. 27-28. I.C.A.R. Pub.

Srikanth, J., Singaravelu, B. and Jayanthi, R., 2007. *Management of Termites in Sugarcane*. Exr. Pub. No. 150. Sugarcane Breeding Institute, Coimbatore.

Parthiban, M. and David, Vasantharaj, 2007. *Termites: Manual of Household and Public Health Pests and their Control*. Namrutha Publications, Chennai, pp. 55-64.

Roonwal, M.L., 1983. *Termites of Agricultural Importance. Agril. Ent.* Vol.II: 23-35. All India Scientific Writer Society, New Delhi.

Chapter 35

Changing Scenario of Pests and their Management in Sesame

D.B. Ahuja[1] and T.S. Rajpurohit[2]

[1]National Centre for Integrated Pest Management, New Delhi

[2]Rajasthan Agricultural University, Agricultural Research Station, Mandor

Sesame is mainly grown in kharif season under rainfed farming situation in marginal and submarginal land. It is also grown in rabi and semi rabi farming conditions but in a few thousand hectare area. The productivity of sesame is very low (250 kg/ha) and fluctuating in India. Reasons for it are many but one of the important constraint is its high vulnerability to pests. Earlier 29 insects species (Rai,1976) were reported to damage the crop but the recent survey of literature revealed that pest fauna associated with this crop has greatly changed and new pest problems have emerged. The list now comprises of 65 species of insect and one species of mite feeding and damaging the crop at one or another stage of the crop growth (Ahuja and Bakhetia, 1995). On the basis of avoidable loss caused by different insects, the pests of greater economic importance has been grouped and are presented in Table 35.1. The following account provides information on different aspects of important pests found damaging sesame.

Leaf Webber/Capsule Borer (*Antigastra catalaunalis*)

It is most important pest of sesame in India. It is cosmopolitan in distribution and has been reported from different parts of the world. Biology and bionomics of the pest has been studied in different states of India.The insect has been found to damage all the stages of the crop growth. Its incidence on the crop start soon after the germination when it damage the shoot by webbing the developing leaves. The plants dies without producing any branch or shoot. If attack occurs at a later stage of the crop growth, it rolls the leaves and infested shoot stops growing. At flowering stage, the production of flower is stopped. It also damages the pod. Young larva after hatching bores the pod

Table 35.1: Some of the Important Pests Associated with Sesame

Sl.No.	Common Name	Scientific Name	Pest Status	Stage of the Crop	Nature of Damage	Reference
1.	Leaf webber and capsule borer	*Antigastra catalaunalis* (Dup.)	Key pest	Vegetative, flower, pod	Webs leaves flowers and borers the pods	Fletcher (1921)
2.	Gall fly	*Asphondylia sesami* Felt.	Major pest	Flowering, pod	Malformation of pods	Anonymous (1957)
3.	Jassid	*Orosius albicinctus* Dist.	Major pest	Vegetative	Transmits phyllody and leaf curl disease	Anonymous (1971)
4.	Bihar hairy caterpillar	*Spilosoma obliqua* Walker	Major pest	Vegetative	Defoliate the plants	Fletcher (1921)
5.	Red hairy caterpillar	*Amsacta moorei* (Butter)	Major pest	vegetative	Denude the whole plants	Fletcher (1921)
6.	Mite	*Polyphagotarsonemus latus* (Banks)	Major pest	Vegetative flowering, pod	Suck sap and produce leaf curl type symptoms	Choudhary *et al.* (1986)
7.	Bug	*Cyrtopeltius tenuis* (Reut.)	Minor pests	Vegetative, flowering, pod	Suck sap from leaves, pods	Anonymous (1992)
8.	Bug	*Spilostethus pandurus* (Scopoli)	Minor pests	Vegetative, flowering, pod	Suck sap from leaves, pods	Bhattacharjee (1959)
9.	Hopper	*Hishimonus phycitis* (Dist.)	Minor pests	vegetative	Suck sap from leaves	Anonymous (1985)
10.	Thrips	*Frankliniella schultzei* Tribom	Minor pests	Vegetative, flowering	Malformatian of stamens	Anonymous (1992)
11.	Thrips	*Thrips palmi* Karny	Minor pests	Vegetative, flowering	Malformation of stamens	Anonymous (1971)
12.	Hawk moth	*Acherontia styx* Westw.	Minor pest	Vegetative, flower, pod	Denude the plant	Fletcher (1921)
13.	Tobacco caterpillar	*Spodoptera litura* (Fab.)	Minor pest	Vegetative	Defoliate the leaves	Ahuja (1989)
14.	Black weevil	*Cyrtozemia dispar* Mshl.	Minor pests	Vegetative	Make whole in the leaves	Ahuja (1989)
15.	Ant	*Monomorium destructor* Jordon	Major pests	Sowing and harvesting time	Carry seed from sowing beds and threshing floor	Jakhmola and Kausik (1974)

and feed on newly formed seed. On the basis of some of insecticidal trials it was shown that 25-90 per cent loss in seed yield could be saved by using effective insecticides (Kapadia, 1996; Ahuja, 1990; Srivastava *et.al.*, 1989; Cheema and Singh, 1987; Choudhary and Singh, 1986; Patel and Bhalani,1986).

The pest is known to pass through 6-14 generations in a year and one life cycle takes 18-64 days during different months of the years (Kumar and Goel,1994; Singh *et al.*, 1992; Cheema and Singh,1987; Desai and Patel,1965; Teotia and Hussain 1965; Menon *et al.*, 1960). Female lay up to 223 eggs singly on the tender parts of the plant. The eggs are shinny white to pale green and hatch in 2-9 days depending upon the season. Eggs viability varies from 64 to 100 per cent. Newly emerged larvae measures about 2 mm in length and feed for a while on epidermis or within leaf tissue. Afterwards it start binding the tender leaves with the help of silken threads and continues to feed on the webbed mass. The larval stage lasts for about 8 to 33 days depending upon the season. The grown up larvae pupate either inside the webbed leaf or in the soil in the silken coccons. Pupal development is completed in 3 to 20 days. Adults lives for about 5 to 16 days. The longevity of unmated males and females observed was more than that of mated ones. A polyphagous male mated successfully with 3 different females.

Peak activity of the larvae has been found to occur during September in Madhya Pradesh and Rajasthan, August- September in Maharashtra and July- August in Tamil Nadu. Rabi and semi Rabi crops are known to be infested more than kharif season crop (Bajpai, 1953; Mehta and Purohit,1957; Desai and Patel 1965; Vittal and Saroja, 1966). Effect of various abiotic factors on its abundance has been studied. Ahuja (1989) noted that pest activity continue to rise in July, August and September during year of severe drought but remained at low level in July and August during years of high and normal rainfall at Mandor, Jodhpur. It has been reported that pest hibernate during December and January. Negative correlation was observed between larval population and rainfall. (Singh *et al.*, 1992). But Cheema and Singh (1987) reported that pest is active throughout the year and does not aestivate or hibernate. It continue to breed up to September in case of early maturing varieties and up to November in case of late maturing varieties. Afterwards, it migrated to *Antirhinum majus* and remained on it up to end of March. The period from April to June was passed on stray plants and early sown sesame intercropped with cotton.

Gall Fly (*Asphondylia sesami*)

It is one of the important pest of sesame widely distributed in the state of Maharashtra, Gujrat, Andhra Pradesh, Uttar Pradesh, Bihar, Orissa and Rajasthan. Its incidence on sesame has been greatly influenced by environmental factors. Sachan (1959) reported it as a minor pest from Rajasthan. However severe attack of gallfly was reported in Pali and Udaipur district of Rajasthan in 1970 (Anonymous, 1971) and in Jodhpur district in 1990 when very high rainfall (840 mm) was observed (Anonymous,1990). Its damage has been continuously recorded since 1990 to 1994 at Mandor. Rainfall seems to enhance its damage. The maggot feeds on floral content inside the buds, which as a result of feeding develops into galls and produce no pods and seeds.

Female lay eggs singly within buds, flowers and developing green pods. The larvae hatches in 2-4 days, feed on floral contents. The larval period lasts for 14-21 days. Larvae are small in size and are creamish white. Pupation takes place inside the galls and one life cycle completes in 23-27 days. Pest has been known to pass 2 generations in one season. Maximum infestation occurred when the crop attained the full bloom stage.

Linseed Gall Fly (*Dasineura sesami* Grover and Prasad)

It has been observed to infest sesame in September- October (Grover and Prasad,1966). Recently it was also found to damage sesami in Madhya Pradesh. Average numbers of maggots were highest during Sept. 3-9 and flower damage recorded was to the extent of 27.7 per cent (Anonymous,1998).

Biology of *D. sesami* was studied by Prasad (1967). Clear humid weather is more suitable for oviposition. Eggs are laid in batches of 3 to 7 in newly formed flower buds. A maximum of 20 eggs per bud were observed. The incubation period was 28 to 36 hours and hatching usually occurred from 2 to 6 PM. There are 4 larval instars. The fully grown larva make a hole in the bud, comes out and pupate 5 to 7 cm deep in the soil. Pupation takes place inside the silken cocoon or sometime it may pupate without spinning silken cocoon. Pupal period is 4 to 6 days. At the end of flowering season of sesame, the larva enters the soil and diapause within cocoons. Emergence of the adults taken place in the morning time from 6 A.M. to 8.30 A.M.

Mite (*Polyphagotarsonemus latus*)

It has been reported to attack sesame from Rajasthan (Anonymous, 1995), Punjab (Dhooria, 1996) and Delhi (Mahto,1987). Plants found infested with mite showed severe malformations resulting leaf curl virus symptoms. Both adult and nymphs suck sap from leaves which causes downward curling of leaves and necrosis of young leaves and flowers. Initially the infested leaves give pale to brown colour on lower and middle leaves. Severely infested leaves first turn light brown and then to dark reddish brown. Its incidence on host plant is known to be affected by abiotic factors. Jones and Brown (1983) showed that population increase of mite was greatest at 20° C and 90 per cent relative humidity. Temperature above 30 °C coupled with low humidities were lethal to egg and nymphal stages. Recent studies on its incidence in arid zone of Rajasthan revealed its peak occurrence during mid September on sesame, when the maximum and minimum temperature ranged from 30.1-33.0°C and 22.6-25.2°C, respectively. The relative humidity recorded were 86-92 per cent and 67-76 per cent, respectively. Correlation between relative humidity both morning and evening and number of mite was positive and significant. Maximum temperature showed negative and significant correlation with population of mite. But correlation between minimum temperature and rainfall and number of mite was negative and nonsignificant (Ahuja, 2000).

Adult mites are hyaline in colour. Female lay eggs on the central surface of leaves. The egg look like white and oval in outline, studed with whitish tubercles on the surface. Incubation period ranges from 1-3 days. Hatched out larvae are six legged. Nymphal period last from 3 to 11 days. Nymphs after full fed enters into quiescent stage and later changes into an adult. Male emerges earlier than females and are commonly seen congregating around quiescent female larva.

Leaf Hopper (*Orosius albicinctus*)

Leaf hopper has been identified as the vector of the disease known as phyllody (Bindra and Bakhetia, 1967; Ghauri, 1966). It appears year after year in almost all the sesame growing areas of the country. Incubation period of this disease is.considered to be 30 to 35 days after sowing when initiation of flowering takes place (Sellammal, *et al.*, 1973). Once the pathogen (mycloplasma like organism) is transmitted, further spread is very fast. As a result of infection of the disease, malformation of flowers takes place, the infected plants gives the appearance of bunch of leaf like structures. On such infected plants, no pods are formed and loss in seed yield is ultimately resulted.

Biology of jassid has been studied in great details under North (Bindra and Singh, 1970) and South India (Sundararaju and Jayaraj, 1977) conditions. Mating of the adult insect takes place just after the Sun set which lasts for 3-4 minutes. Maximum eggs are laid from 10.30 hrs to 18.30 hrs in the bright light. The insect requires light stimulus for oviposition and no eggs are laid if the adults are not exposed to light. Eggs are laid singly on lower surface of leaves in the midrib, petioles and tender portion of the stem of young plants. The egg stage was shortest (6.2 days) for the egg laid in June and longest (95.8 days) for the egg laid in December to February. The newly hatched nymphs are creamy white and gradually changed to grey. The nymphal period varied from 11.6 to 16 days for the one hatched during April to September which was prolonged (105 days) for the nymphs hatched in November to February. Adult longevity was highest in December (91 to 105 days) and lowest (19 to 20 days) in June. There were 14.over lapping generations in a year.

Host range studies were also conducted for the jassid (Bindra and Singh, 1970, Sundararaju and Jayaraj, 1977). It is reported that sesame phyllody diseases is prevalent throughout the year. 17 species of plants are known to be host of this insect. Other than sesame, some of wild alternate host plant reported are chicory (*Chicorium intybus* L.), datura (*Datura metel* L.), dhencha (*Sesbania bispinosa*) and cultivated host plants are bengal gram (*Cicer arietinum*), lucerne (*Medicago sativa*), sannhemp (*Crotolaria juncea*) and Indian mustard (*Brassica campestris*).

Sphingid Moth (*Acherorntia styx*)

It is an occasional pest of sesame (Anon, 1962 and Sachan, 1967). Its attack on sesame was reported in the month of September in arid region of Rajasthan in Year 1988. (Anonymous, 1988). It feed voraciously on foliage and sometime entire plant is denuded. In event of severe attack, pods are also eaten). Its damage generally occurs with the initiation of flowering and continue till harvest of the crop.

The moth lay globular eggs singly on the underside of leaves of food plants. The eggs are greenish white turn yellow on hatching and start feeding on leaves. The full grown caterpillar measures about 5 cm in length and 1 cm in width. The larval period is long (11-21 days during March- September) and may last for about 2 months or more (October- November). Larvae undergoes many changes in colour. The changes in coloration of larvae and adults aid them in protective mimicry. The thornlike projections on the hind end of abdomen are conspicuous. The body of larvae is plump and is decorated with pleasant mixture of soft colours. The fully fed larvae stops feeding, rests for 24 hrs and burrows 15 cm deep into the soil and forms an oval cell of earth. The adult is large reddish brown, robust thick set moth with a wing span of about 10 cm and prominent death's head mark on the thorax. The moth are swift flier and often make flight towards the source of light at dust. There are 2-3 generations in a year. The winter is passed as pupal stage in soil.

Bugs

Sixteen species of bugs are known to attack sesame (Ahuja and Bakhetia,1995). Most commonly found species of bugs reported damaging are *Lygaeous militaris* Fab., *Nezara viridula* (Linn.), *Cyrtopeltis tenuis* (Reut.), *Spilostethus pandurus* (Scop.) and *Elasmolomus sordidus* Fab. which causes economic loss. Most of these bugs inflict injury right from the seedling to pod stage. Both nymphs and adults suck the sap from leaves, flowers and pods. As a result of their feeding, the leaf tissue does not grow normally and leaves get distorted, stipulated and discolored.

Available information on the biology of *C. tenuis* (Abraham *et al.*, 1973) indicated that the insect complete development in 15 days. The eggs are thrust inside the tender plant tissues on leaf axils and leaf midribs and nymphs hatch out in 4 days and nymphal period last for 11 days. The incidence of bug has been observed in more numbers on December- March and February- May season crops while on the crop raised in August- October (kharif), the incidence is low.

Table 35.2: Some of the Biocontrol Agents which can be Employed for Management of Important Pests of Sesame

Biocontrol Agents	*Stage of Pest*	*Pest*	*Reference*
Parasitoid			
Apanteles sp.	Larva	*Antigastra catalaunalis*	Anonymous(1971), Kumar and Goel (1994b)
Apanteles acherontiae Camer	Larva	*Acherontia styx*	Bhatnagar (1948)
Bracon hebetor Say	Larva	*Antigastra catalaunalis* *Asphondylia sesami*	Anonymous(1971) Narayanan (1961)
B.brevicornis (Wasm.)	Larva	*Antigastra catalaunalis*	Anonymous(1971)
B.gelechiae Ashm.	Larva	*Antigastra catalaunalis*	Jakhmola (1983)
Campoplex sp.	Larva	*Antigastra catalaunalis*	Krishnamurthy and Usman (1954)
Phanerotoma hendicusisella (Camer)	Larva	*Antigastra catalaunalis*	Bhatnagar and Davies(1979) Kumar and Goel(1994)
Conostigmus sp	Larva	*Antigastra catalaunalis*	Anonymous(1971)
Desmometopa sp.	Larva	*Antigastra catalaunalis*	Anonymous(1971)
Cadura lucous Vilte	Larva	*Antigastra catalaunalis*	Chadha(1974)
Eriborus sp.	Larva	*Antigastra catalaunalis*	Ayyer(1927)
Hockeria sp.	Larva	*Antigastra catalaunalis*	Anonymous(1971)
Nythobia sp.	Larva	*Antigastra catalaunalis*	Anonymous(1971)
Telenomus sp.	Larva	*Antigastra catalaunalis*	Anonymous(1971)
Cremastus (*Trathala*) *flavoorbitalis*	Larva	*Antigastra catalaunalis*	Anonymous(1971) Dutt (1914), Fletcher (1919)
Diadegma sp.	Larva	*Antigastra catalaunalis*	Patel and Bhalani(1986)
Brachymeria sp.	Larva	*Antigastra catalaunalis*	Anonymous(1971) Hallman and Sanchej (1982)
Agathis sp.	Larva	*Antigastra catalaunalis*	Jakhmola (1983)
Euplectrus sp.	Larva	*Antigastra catalaunalis*	Hallman and Sanchej(1982)
Temelucha biguttula Mats.	Larva	*Antigastra catalaunalis*	Bhattacharjee(1960)
Elasmus brevicornis	Larva	*Antigastra catalaunalis*	Kumar and Goel(1994b)
Sarcophaga sp.	Larva	*Acherontia styx*	Anonymous(1971)
Zygobothria ciliata Wulp	Larva	*Acherontia styx*	Anonymous(1971)
Anastatus acherontiae Narayanan *et al.*	Egg	*Acherontia styx*	Narayanan *et al.* (1960)
Eurytoma dentipectus Gahan	Larva	*Asphondylia sesami* Felt.	Narayanan (1961)
Pteromalus fasciatus (Thoms)	Larva	*Dasineura sesami*	Prakash and Caturvedi(1979)
Predator			
Cntheconidea furcellata (Wolff)	Larva	*Antigastra catalaunalis*	Jakhmola (1983)
Cicindella sp.	Larva	*Antigastra catalaunalis*	Jakhmola (1983)
Coccinella septempunctata Linn.		Trips, jassid and bugs	Anonymous(1971)
Menochilus sexmaculatus(*F.*)		Trips, jassid and bugs	Anonymous(1971

Contd...

Table 35.2–Contd...

Biocontrol Agents	Stage of Pest	Pest	Reference
Brumoides suturalis(*F.*)		Trips, jassid and bugs	Anonymous(1971
Laius mallerifer Champ.		Trips, jassid and bugs	Anonymous(1971
Spider	Immature stages	*Antigastra catalaunalis*	Choudhary *et al.* (1986) Anonymous (1994)
Fungal parasite			
Aspergillus flavus Link.		*Orsius albicinctus*	Choudhary *et al.* (1986)
Nematode parasite			
Mermis sp.	Larva	*Antigastra catalaunalis*	Jakhmola and Yadav (1973)

Attack of *E. sordidus* is observed when sesame is harvested, and heaped for threshing (Mohansundaram and Sundara Babu, 1987; Anonymous, 1994). The adult bugs copulate during day and night and the copulated female select shady sandy situations on the sides of the channels for egg laying. They insert their abdomen in the soil and lay eggs and then with their hind legs cover the egg laying site with soil. Eggs are found in batches of 5-10 in a mass, are oblong and light rose in colour. Egg period is 3 days. Nymphs are light brown immediately after hatching and measures about one mm. Nymphal period is 12 days. Adult are dark brown to dark gray in color and always creep into dark situations below heaps and leaves during day time. These are not attracted to light.

S. (L). pandurus has been observed to damage sesame (Bhattacharjee, 1959; Singh, 1966). Female lay 50-60 eggs in clusters underneath the fallen leaves or flowers. Eggs period is 4-6 days. There are five nymphal instars. Nymphal period last for about one month. Adults starts copulating 8-7 days after emergence and the impregnated females commences oviposition from the second day after copulation. Adults may live up to one month. During the rainy season, adults takes shelters underneath the fallen leaves and there is no ovipositon.

Thrips

Six species of thirps have been reported to damage sesame. Most important among these is, *Frankliniella schultzei* Trybom (Anonymous, 1992) which is polyphagous and found to cause economic loss. Ananthakrishnan (1955) reported that attack of *Thrips palmi* Karny results in the malformation of stamens, causing injuring to the ovarian wall and development of dark pigment on the fruit wall, instead of usual green color. The eggs are laid on the anther lobes. 4-8 adults and 8-20 larvae per flower have been observed.

Hairy Caterpillar

Two hairy caterpillar *viz; Spilosoma obliqua* Walk. and *Amsacta moorei* (Butt.) (Fletcher, 1921) have been reported to damage sesame. The incidence of the *A.moorei* takes place at the onset of monsoon in the month of July and other kharif crops mainly pulses. Both the pests are polyphagous and are known to attack a wide range of host plants. As a result of their feeding, the plants get denuded. In case of severe attack, the field gives a deserted look. Female of *S. obliqua* lay eggs on the lower surface of the leaves. The incubation period is 3-4 days. In 2 to 3 weeks, the larvae become full grown. Pupation takes place on the soil under dry foliage and debris. The insect has 6-8 generations per year. According to Singh (1962), the insect hibernates in the pupal stage. Adult emergence takes place towards the end of February. In summer insect aestivate in pupal form.

Tobacco Caterpillar

Spodoptera litura and *S. exigua* are reported to be of little economic importance on sesame (Ahuja, 1989). Biology studies of *S.litura* revealed that sesame is less preferred host plant than castor, chilli and tomato (Ahuja and Noor, 1991). Female lay eggs on the lower surface of leaf covered with mass of hairs. The incubation period is 3-4 days. The newly hatched larvae feed on gregariously for a short period before dispersal. First instar larvae feed only on the leaf surface. Second to fourth instar feed voraciously. Larval period lasts for 20-23 days. There are five instars. Pupal period was observed 7 days. Life cycle completes in 30 days. Only one generation was observed on sesame in kharif season of 1990 at Mandor Jodhpur (Anonymous, 1990).

Leaf Miner

Melanagromyza azawii Spencer (Pillai *et al.*, 1980) and *M. obtusa* (Malloch) (Abraham *et al.*, 1973 and Choudhary *et al.*, 1986) were recorded as pest on sesame. Maggot of both the species bores from tip downwards into the stem of the sesame young plants aged one month and above. Shoots dropped and plants dried up as a result of larval feeding. The eggs are laid singly in the axils of tender leaves. In young seedlings, the larvae tunnels the central core of the shoot from top downward. In older plants, the larvae enters the stem beneath the cortical layer and reach the basal portion. Pupation takes place in tunnels or in soil. The egg, larval and pupal stage lasts for 7, 20 and 6 days, respectively.

Pest Management

As discussed earlier, sesame is largely grown under poor management practices in marginal and submarginal lands under low rainfall conditions. Most of the recently evolved varieties do not yield more than 6-8 q/ha seed yield and the losses due to the pest have been estimated to range between 25 to 90 per cent (Ahuja, 1990, Srivastava *et. al.*, 1989;Choudhary and Singh 1986, Patel and Bhalani, 1986). Farmers are reluctant to spend money on any input except seed. Under these circumstances adoption of pest control measures did not find favour with farmers. Therefore means like augmentation of natural enemies, manipulation of agronomical practices, evolving pest resistant varieties along with less loaded pest control spray schedules may provide solution to otherwise complex problem of raising productivity of sesame. Information gathered from survey of literature had hinted favourably for evolving strategy for evolving practical IPM programme and its implementation is suggested below:

Biological Control

Ahuja and Bakhetia (1995) had collected information regarding various natural enemies found insect pests associated with the insect pests of sesame. Twenty species of larval parasitoid, one predator and one nematode parasite of *A.catalaunalis*, two larval parasitoid of *A. sesami*, one larval parasitoid of *D.sesami*, four parasitoid of *A. styx* and one fungal parasite of *O. albicinctus* have been recorded (Table 35.3). Among various parasitoid of *A. catalaunalis*, *Trathala flavo orbitalis* has been found to attack in various states of India. Murali Baskaran and Thangavelu (1990a) had given detailed account about the relationship between *T. flavo orbitalis* and field incidence *of A. catalaunalis*. Maximum parasitisation went up to 17.1 per cent during January followed by February (14 per cent), March (13.2 per cent) and October (14 per cent). However, 5-10 per cent parasitism was recorded during other months, although incidence of *A. catalaunalis* has been high. Bhattacharjee (1960) observed higher parasitism in the months of July (12.5 per cent), September (13.2 per cent), and December (11.8 per cent) than other months of the year. Cheema and Singh (1987) and Kalra (1986) recorded higher parasitism in the month of October in state of Punjab and Haryana, respectively. Thus *T. flavo orbitalis* along with other parasite if augmented in nature can play a crucial role in keeping a check on the population of *A. catalaunalis*. Spiders can also be utilized in controlling the population of sucking pests of sesame.

Table 35.3: Some of the Recommended Insecticides Against Insect Pest of Sesame

Pest	*Insecticides*	*Dose*	*No of Sprays/Dust (Days Interval)*	*Cost Benefit Ratio*	*References*
A. catalaunalis	Carbaryl 50 WP	0.25%	2 (20, 40 days after germination)	11.9	Srivastava *et al.* (1989)
	Monocrotophos 36 WSC	0.05%	2 (one month interval)	8.7	Patel and Bhalani (1986)
	Endosulfan 35 EC	0.07%	2 (30, 45 days after sowing)	4.4	Ahuja (1991)
	Fenvalerate10 EC	50g, a.i./ha	2 (52, 68 after sowing)	4.3	Singh *et al.*, (1991)
	Quinalphos 1.5D or Methyl parathion dust 2D	20kg/ha	2(45, 60 days after sowing)		Gorpade and Thaku (1995)
A. sesami	Endosulfan 35 EC	0.07%	2 (28, 43 days after sowing)	-	Sundararaju and Jayaraj (1976)
	Acephate 75 WP	0.75 kg a.i/ha	2 (28, 43 days after sowing)	-	Sundararaju and Jayaraj(1976)
	Quinalphos 1.5% dust	20 kg/ha	1 (45 days after sowing)		Sundararaju and Jayaraj (1976)
O. albicintus (Phyllody disease)	Monocrotophos 36 WSC	0.05%	3 (30, 45 and 60 days after sowing)	-	Abaraham *et al.* (1977b)
A. styx	Quinalphos 25 EC	0.025%	1(30–40 days after sowing)	-	Anon (1988)
D. obliqua	Quinalphos 25 EC	1.20 lt./ha	1 (First week of sowing)	-	Singh and Grewall (1982)
P.latus	Quinalphos 25 EC	0.025%		-	Mahto (1987)
	Phosphamidon 85 EC	0.04%		-	Mahto (1987)
	Dicofol 20 EC, or Sulfur 0.16%	0.05%	2 (28, 43 days after sowing)		Sundararaju and Jayaraj (1976)

Resistant Varieties

In integrated pest management programme, resistant varieties plays a crucial role. Even a cultivar not completely resistant but tolerant may require lesser number of sprays and sustains higher number of natural enemies can play a more important role. Recently Bakhetia and Singh (1993) gave a comprehensive account of resistant sources of resistance against *A. catalaunalis*. Early maturing varieties with sparse branching exhibited lower susceptibility than profuse branching types (Battacharjee and Lal, 1962 Selvanarayanan and Baskaran,1996). Hairy varieties were more prone to damage due *to A. catalaunalis*. Kumar *et al.* (1990) and Rath *et al.*, (1997) reported Krishna RAUSS.17-6 and B-67 as resistant to *A. catalaunalis* due to hard pod and another released variety RT-46 suffered less loss in seed yield due to the pest (Ahuja, 1990). Ghauri variety from Andhra Pradesh has also been recorded tolerant to *A.catalaunalis* (Padamja and Savithri 1993). SI-250 and ES-22 germplasm lines have also been reported resistant under both laboratory and field condition (Murali Baskaran and Thangavelu 1990b) can be utilized in breeding programme for evolving resistant varieties against *A. catalaunalis*. Several cultivars has also been identified as resistant to *P.latus* in large scale screening programme.

Cultural Methods

Date of sowing and intercropping has been found to influence the incidence of pests of sesame enormously. This concept can be utilized as component of IPM programme to reduce the pest load. The damage due to *A. catalaunalis* got reduced in the crop sown in May to August (0.6 per cent) as compared to February to April (10.4 per cent to 16.9 per cent) and September to November sowings (19.5 to 28.6 per cent) (Murali Baskaran *et al.*, 1990a). The incidence of phyllody was also low in summer and rabi season sown crop(Abraham *et al.*, 1977a). Damage due to *A. sesami* and *A. catalaunalis* was greatly reduced in sesame sown as mix crop with moong bean or moth bean (Ahuja, 1997) or pearlmillet and groundnut (Murali Baskaran *et al.*, 1991)

Chemical Control

A large number of field trials have been conducted against pests of sesami and insecticides found promising are given in Table 35.3. If these recommended insecticides are used judiciously taking into consideration the economic threshold (ETL) and economic injury levels (EIL), harmful effects of insecticides like residue problems, outbreak of secondary pests, toxicity to non target organism including natural enemies, and health hazards can be avoided or minimized. Choudhary *et al.* (1987) reported 2.87 per cent plant infestation with 1 live larvae in each web as the economic threshold level for *A. catalaunalis*. Recently Ahuja (1999) estimated ETL for RT-46 cultivar of sesame for arid region of Rajasthan as 10 per cent with or without live larvae. Based on such data effective spray schedules have been evolved. Two sprays of 0.05 per cent carbaryl 50 WP, 20 and 40 days after germination (Srivastava *et al.*, 1989) or two sprays of 0.07 per cent endosulfan 35 EC, 30 and 45 days after sowing (Ahuja,1991) or two sprays of 0.05 per cent monocrotophos 36 WSC at vegetative and flowering stage or two sprays of 50 g a.i. fenvalerate 10 EC at appearance of pests and another at flowering stage were found cost effective. Such spray schedules have resulted a return of varying from Rs. 4.3 to 11.9 for each rupee spent on spraying insecticides. Quinalphos 25 EC @ 0.025 per cent and diocfol 20 EC @ 0.05 per cent have been found quite effective for control sesame mite. Good conntrol of phyllody disease was obtained by spraying of monocrotophos @ 0.05 per cent 30, 45 and 60 days after sowing (Abraham *et al.*, 1977a).

Diseases

Root and Stem Rot (*Macrophomina phaseolina*)

Dark cortical lesions occur slightly below or at the soil on stem and root. In advanced stages, the lesions extend upward and down ward. Black dots appear on the infected stem, which are the pycnidia of the fungus. If infected plant is uprooted, black coloured roots are observed having sclerotia of the fungus and looks as charcoal is sprinkled on root. The roots become brittle. In the infected plants, black pods are seen which open prematurely exposing shrivelled and discolored seeds. The seedlings are killed in a group of 5-6.

Seed treatment with either of Bavistin 50 WP, or Carboxin 75 WP, or Captan 75 SD or Thiram D were found highly effective in reducing the disease from 24 to 11.33 per cent and increasing seed yield from 550 to 652 kg/ha of sesame (Rajpurohit, 1997).

Alternaria Leaf Spot (*Alternaria sesami*)

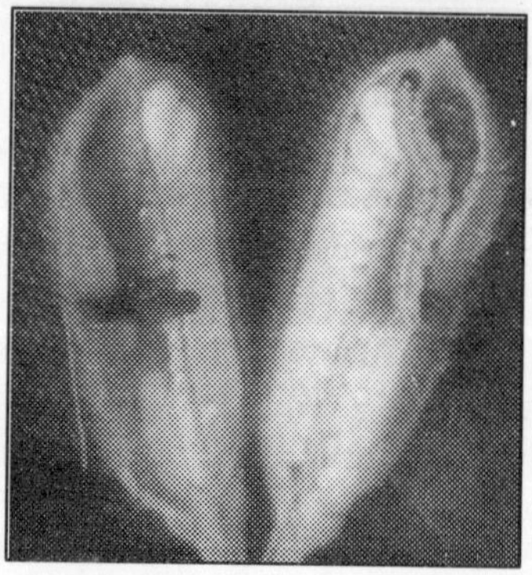

Small, round, brown spots appear on the leaves. Latter these spots increase in size, become irregular and have concentric rings. These lesions coalesce to form large patches covering greater area of the leaf. In severe cases, defoliation occurs. Symptoms also appear on petiole, stem and pods.

Morphology and taxonomy of *Alternaria sesami* was studied and described in detail (Rajpurohit, 1981). The pathogen *A. sesami* was found to be capable of growing on a wide range of temperature. and pH with optimum 28 ºC and 5.5 respectively. Among liquid medium czepeks Dox supported maximum growth and sporulation (Rajpurohit and Prasad, 1982). Effect of carbon and nitrogen on growth and sporulation was studied. Maximum growth was observed in case of glucose followed by sucrose, fructose, lactose and maltose. The fungus is widely adaptable in utilizing various forms of nitrogen for its growth. It has been observed to utilise nitrate nitrogen, ammonical nitrogen and various amino acids (Rajpurohit, *et al.*, 1983). *A. sesami* produced PGTE, PTE and PG enzymes *in vitro* (Rajpurohit and Prasad, 1982). Two entries *viz.* EC-770929 and EC-770932 were found resistant to *Alternaria* leaf spot (Rajpurohit and Solanki, 2002).

Efficacy of plant products and new chemicals was studied against Alternaria blight during kharif 2000 and 2001 under artificial inoculated conditions. The data showed foliar spray of Mancozeb (0.2 per cent) and propicanazole (0.1 per cent) was found highly effective and reduced disease from

AICRPO
SESAME
POWDERY MILDEW OF SESAME

AICRP
SESAM
STEM AND ROOT ROT OF SE

CERCOSPORA LEAF SPOT OF SE

32.7 per cent to 8.4 per cent and increased seed yield from 265 to 401 kg/ha. (Rajpurohit, 2004). *In vitro* studies, different fungicides like Dithane M-45, Dithane Z-78, Auriofungin, Brestan were found very effective in inhibiting the spore germination of *A. sesami* (Rajpurohit, *et al.*, 1984). Effect of Bio agents in comparison to fungicides was studied on Alternaria leaf spot and phyllody disease of sesame, their efficacy was low in comparison to Mancozeb 0.2 per cent (Rajpurohit, 2004).

Cercospora Leaf Spot (*Cercospora sesami*)

On the leaves, minute water soaked lesions, which enlarge to form irregular spots 5-15 cm in diameter on the both surface of leaf. The spotted area is light brown with grey centre and dark margin delimited by veins. The spotted area is light brown to white in beginning latter becoming dark grey. Several spots may occur on a single blade, enlarging to coalesce and blight the leaf. Blighting occurs if the disease appears early and defoliation occurs. Infection also appears on petiole, stem and pods. Spots may occur on a single blade, enlarging to coalesce and blight the leaf. Blighting occurs if the disease appears early and defoliation occurs. Infection also appears on petiole, stem and pods.

Phytophthora Blight (*Phytophthora parasitica* var. *sesami*)

Initially water soaked spots on the cotyledonary leaves of newly emerged seedling and soon the tip is infected. Rotting of tip develops very fast and causes blighting within few hours. In older plants die back symptoms are rare while spots are common on leaves and stem. The spots are brown in beginning later turn to blackish. In humid weather these spots enlarge rapidly coalesce causing death of plant and gives blighted appearance.

Sesame lines No 2-39, F-8 and 19/1/46-2 were found less susceptible to phytophthora blight (Gemawat and Verma, 1971).

Management strategy revealed foliar spray of Ridomil MZ most effective for the control of Phytophthora blight which reduced the disease from 11.7 to 2.2 per cent. Bavistin was found best against Cercospora leaf spot and reduced disease from 18.8 to 2.4 per cent. Mancozeb 0.2 per cent was found highly effective for the control of Alternaria leaf spot and reduced disease from 65.2 to 22.4 per cent.

Occurrence of new powdery mildew (*Erysiphe orontii*) disease was reported. Out of 48 entries screened two entries *viz*; OMT-30 and DORS-101 were found to be tolerant to the disease and RT-225 was highly susceptible. Two sprays of Sulfex (0.2 per cent) were found highly effective in reducing powdery mildew and increasing in yield. This was followed by Bavistin, Karathane and Sulphur dusting. (Rajpurohit, 1992).

Phyllody (MLO Transmitted by *Orosius albicintus*)

The flowers turn into green leafy structures followed by abundant vegetative growth. Infected plant is conspicuous by its stout internodes, abundant abnormal branching which causes top portion of the branches to bend down

It was observed that the incidence of phyllody was more in early sown crop as compared to late sown. Seventeen entries were observed free from phyllody *viz.* ES-29, ES-81-1-84, ES-111-1-84, ES-175-2-84, ES-222-2-84, ES-231-2-84, ES-231-3-84, ES-231-4-84, ES-233-6-84, ES-384-3-84, EC-343402, EC-343406, EC-355662, EC-370663, EC-370716, EC-370721, EC- 370920 under field conditions.

Leaf curl- Virus (*Nicotinia 10 virus*) Symptoms: Curling of leaves and marked thickening of the veins on the under side of the leaf, combined with a reduction in leaf size. Leaves may become leathery possessing dark green colour. Severely affected plants remain stunted and bear few flowers and capsules.

Studies on evaluation of plant products, fungicides and insecticides against Alternaria blight, leaf curl and phyllody diseases conducted during Kharif 2001-03 revealed that the two foliar sprays of Mancozeb @ 0.25 per cent + methyl demeton @1 ml/ha reduced Alternaria blight from 39.35 to 10.1 per cent, Phyllody and leaf curl from 5.24 to 0.83 per cent and increased seed yield from 416 to 721 kg/ ha. (Rajpurohit, 2004)

Integrated Disease Management

Two IDM modules along with farmers practice were tested at ARS, Mandor during Kh-2001 and 2003. Integrated disease management module comprised seed treatment with Thiram 0.2 per cent + carbendazim 50 WP 0.1 per cent + foliar spray of Mancozeb 0.2 per cent + Endosulfan 0.07 per cent at 30-45 DAS and 45-55 DAS reduced Macrophomina stem and root rot from 24.54 to 11.93 per cent, Alternaria leaf spot 41.73 to 10.76 per cent, Phyllody from 3.20 to 0.64 per cent, Leaf curl 3.3 to 0.47 per cent. This module recorded seed yield of 693 kg/ha with 2.92 BC ratio in compared to control plot (376 kg/ha).

References

Abraham, E.V., Natarajan, K. and Muragesan, M., 1977 a. Damage by pests and Phyllody to *Sesamum indicum* in relation to the time of sowing. *Madras Agric. J.* 64: 298-301

Abraham, E.V., Natarajan, K. and Jayaraj, S., 1977b. Investigation on the insecticidal control of phyllody diseases of sesame. *Madras Agric. J.* 64: 379-383.

Abraham, E.V., Thirumurthi, S., Asaf Ali, K and Subramaniam, T.R., 1973. Some New Pests of Sesamum. *Madras Agric. J.* 60: 593.

Ahuja, D.B., 2000. Influence of abiotic factors on the population of mite, *Polyphagotarsonemus latus* (Banks) infesting sesame, *Sesamum indicum* L. in arid region of Rajasthan(India). *J. ent. Res.* 24(1): 1-4.

Ahuja, D.B., 1999. Economic threshold level for leaf webber and capsule borer (*Antigastra catalaunalis*) on sesame (*Sesamum indicum*) in arid region of Rajasthan. *Indian J. Agric. Sci.* 69: 65-66.

Ahuja D.B., 1997. Influence of inter cropping and mix cropping sesame with pearl millet, greengram and mothbean on the incidence of insect and mite pests. *Ramaecos Proceeding. Recent Advances in Management of Arid Ecosystems* (RAM ECOS). March 3-5.

Ahuja, D.B., 1991. Evaluation of spray schedule of endosulfan 35 EC for the control of *Antigastra catalaunalis* (Dup.) in *Sesamum indicum* L. *Pestology* 15: 12-15.

Ahuja, D.B., 1990. Assessment of loss in seed yield due to leaf webber and capsule borer (*Antigastra catalaunalis*) in different varieties of sesame (*Sesamum indicum*). *Indian J. Agric. Sci.* 61: 147-149.

Ahuja, D.B., 1989. Seasonal occurrence of insect pests in *Sesamum indicum* Linn. *J. Ent. Res.* 13: 116-120.

Ahuja, D.B. and Bakhetia, D.R.C., 1995. Bioecology and management of insect pests of sesame. A review. *J. insect Sci.* 8: 1-19.

Ahuja, D.B. and Noor, A., 1991. Effect of different host plants on the development of *Spodoptera litura* (Fab.). *J. insect Sci.* 4: 176-177.

Ananthakrishnan, T.N., 1955. Notes on the thrip *Palmi karny*, attacking *Sesamum indicum*. *J. Bombay and Hist. Soc.* 52: 951-53.

Anonymous., 1998. *Annual Progress Report (Sesame and Niger).* All India Coordinated Research Project on Oilseeds. Directorate of Oilseeds Research, Rajendranagar, Hyderabad.

Anonymous, 1995. *Annual Progress Report (Sesame and Niger).* All India Coordinated Research Project on Oilseeds. Directorate of Oilseeds Research, Rajendranagar, Hyderabad.

Anonymous, 1994. *Annual Progress Report (Sesame and Niger).* All India Coordinated Research Project on Oilseeds. Directorate of Oilseeds Research, Rajendranagar, Hyderabad.

Anonymous 1992. *Annual Progress Report Kharif 1992.* Agricultural Research Station, (Rajasthan Agricultural University) Campus, Mandor, Jodhpur : 379.

Anonymous 1990. *Annual Progress Report Kharif 1990.* Agricultural Research Station, (Rajasthan Agricultural University) Campus, Mandor, Jodhpur.

Anonymous 1988. *Annual Progress Report Kharif 1988.* Agricultural Research Station, (Rajasthan Agricultural University) Campus, Mandor, Jodhpur. P.300.

Anonymous 1971. *Annual Progress Report Kharif 1970-71.* Agricultural Research Station, (University of Udaipur) Campus, Sumerpur pp 65. Distt. Pali, Rajasthan.

Anonymous., 1962. Information collected by Indian Central Oilseeds Committee from Indian States regarding pests of oilseeds, their status, time of appearance and control measures etc. (of Rai, 1976).

Ayyer, T.V.R., 1927. The parasitic Hymenoptera of economoic importance from South India. *Bull. Ent. Res.* 18: 73.

Bajpai, R.N., 1953. *Progress Report on the Scheme of Research on the pests of Oilseeds (Linseed)* for the period from Ist July, 1952 to 30th June, 1953. M.P.

Bakhetia, D.R.C. and Singh, P., 1993. Host plant resistance to insect pests in oilseed crops. P.206-248. In: G.S.Dhaliwal and Y.K.Dilawari (eds) *Advances in Host Plant Resistances to Insect Pests.* Kalyani Publishers, New Delhi.

Bindra, O.S. and Bakhetia, D.R.C., 1967. Note on the natural incidence of sesame phyllody virus in *Brassica* sp. at Ludhiana. *J. Res. Punjab Agric. Univ.* 4: 406-408.

Bindra, O.S. and Singh, Joginder, 1970. Biology and Bionomics of *Orosius albicinctus* Distant, the jassid vector of sesamum- phyllody virus. *Indian J. Agric. Sci.* 40: 340-355.

Bhatnagar, S.P., 1948. Studies on *Apanteles forester* (Vipionidae : Parasitic Hymenoptera) from India. *Indian J. Ent.* 10: 133-203.

Bhatnagar, V.S. and Davies, J.C., 1979. Arthropod endoparasites of insect pests (excluding *Heliothis* sp.) recorded at ICRISAT Centre, Andhra Pradesh, India 1974-1979, *Progress Report* 3: 17

Bhattacharjee, N.S., 1960. Bionomics and economic status of *Cremastus flavo orbitalis* (Cameron) (Ophioninae: Hymenoptera), an endoparasite of til leaf- roller, *Antigastra catalaunalis* (Dup). *Indian Oilseed J.* 4: 176-179.

Bhattacharjee, N.S., 1959. Studies on *Lygaeus pandurus* Scopoli (Heteroptera : Lygaeidae). 1. Bionomics, descrptions of the various stages, biology and control. *Indian J. Ent.* 21: 259-73.

Bhattacharjee, N.S. and Lall, R., 1962. Studies on the varietal susceptibility of til (*Sesamum orientale*) to the attack of *Antigastra catalaunalis* (Duponchel). *Indian J. Ent.* 24: 58-63.

Chadha, S.S., 1974. Effect of some climatic factors on the fluctuation of population of *Antigastra catalaunalis* (Duponchel) (Pyralidae: Lepidoptera), a pest of *Sesamum indicum* L.Samura Mis. Paper No. 48: 23

Cheema, J.S. and Singh, G., 1987. Biology of sesame leaf-webber and capsule borer, *Antigastra catalaunalis* (Duponchel) (Pyralidae: Lepidoptera) in Punjab. *J. Res. Punjab Agric. Univ.* 24: 65-74.

Choudhary, R., Rai, S. and Singh K.M., 1987. Economic injury level of sesame leaf webber, *Antigastra catalaunalis* (Dup.) in Delhi (India). *Indian J. Pl. Prot.* 15: 136-41.

Choudhary, R and Singh, K.M., 1986. Efficiency and economics of different carbaryl spray schedules for the control of *Antigastra catalaunalis* in *Sesamum indicum* Linn. *Indian J. Ent.* 48: 212-215.

Choudhary, R., Singh, K.M. and Singh, R., 1986. Pest complex succession of insect pests in *Sesamum indicum* L. *Indian J. Ent.* 48: 428-434.

Desai, M.T. and Patel, R.M., 1965. Studies on sesame leaf roller, *Antigastra catalaunalis* (Duponchel) in Gujrat. *Indian Oilseed J.* 9: 109-112.

Dhooria, M.S., 1996. Observations on host- range of leaf curl mite, *Polyphagotarsonemus latus* (Banks) (Acari: Tarsonemidae) in Punjab. *J. Insect. Sci.* 9(1): 9-11.

Dutt, G.R., 1914. An annotated list of Ichneumonidae in Pusa collection. *Mem. Dep. Agric. India*, 8: 13-28.

Fletcher, T.B., 1919. *Second hundred notes on Indian Insects.* Bull.No. 89, Agric. Res. Inst. Pusa. pp. 1-102.

Fletcher, T.B., 1921. *An Annotated list of Indian crop pests.* Bull. No. 100., Agric. Res. Inst. Pusa. pp. 1-238.

Gemawat, P. D. and, O. P. Verma 1971. Diseases of *sesamum* in Rajasthan, evaluation of varities of *sesamum* for resistance to phytophthora blight. *Madras Agric. J.* 58: 849-850.

Ghauri, M.S.K., 1966. Revision of genus *Orosius* Distant (Homoptera : Cicadellidae). *Bull. Br. Mus. Nat. Hist.* 8 (7): 231-52.

Ghorpade, S.A. and Thakur, S.G., 1995. Management of major insect pests of sesame in India. *J. Insect Sci.* 8 (1): 43-47.

Grover, P. and Prasad, S.N., 1966. Studies on Indian gall-midges XVIII. Two new species of the genus *Dasineura rondani. Marcellia* 33: 119-32.

Hallman G.J. and Sanchej, G.G., 1974. Possibilities of biological control of *Antigastra catalaunalis* (Dup.) (Lepidoptera : Pyralidae), a new pest of sesme in Western Hemisphere. *Entomophaga* 27: 425-429.

Harvir Singh, Jaglon R.S., Kharub, S.S. and Singh, H., 1990. Antibiosis in some sesame genotypes against shoot webber and pod borer, *Antigastra catalaunalis* (Duponchel). *Journal of Insect Science* 3(2): 174-176.

Jakhmola, S.S., 1983.Natural enemies of til leaf roller, *Antigastra catalaunalis* (Dup.). *Bull. Ent.* 24: 147-148.

Jakhmola, S.S. and Kaushik, U.K., 1974. Occurrence of *Monomortum destructor* Jard. on til (*Sesamum indicum* L.). *Indian J. Ent.* 36: 164-165.

Jakhmola, S.S. and Yadav H.S., 1973. Some observations on *Mermis* sp. as a parasite of til leaf roller, *Antigastra catalaunalis* (Dup.). *Indian J. Ent.* 35: 170-172

Jones, V.P. and Brown, R.D., 1983. Reproductive Responses of the Broad Mite, *Polyphagotarsonemus latus*(Acari: Tarsonemidae), to constant temperature, humidity regimes. *Ann. Ent. Soc. Am.* 76(3): 466-469.

Kalra, V.K., 1986. Record of hymenopterous parasites from sesame leaf roller and pod borer, *Antigastra catalaunalis* (Duponchel) in Haryana. *Haryana Agric. Univ. J. Res.* 16: 86-87.

Kapadia, M.N., 1996. Estimation of losses due to pod borer in oilseed crops. *J. of Oil Seed Res.* 13: 139-140.

Krishnamurthy, B. and Usman, S., 1954. Some insect parasite of economic importance noted in Mysore State. *Indian J. Ent.* 16: 327-344.

Kumar, S.and Goel, S.C., 1994a. Studies on the life history of a pyralid, *Antigastra catalaunalis* (Duponchel.) in western Uttar Pradesh. *Bulletin of Entomology, New Delhi,* 35: 123-128.

Kumar, S.and Goel, S.C., 1994b. Record of new larval parasitoid associated with *Antigastra catalaunalis* (Dup.) (Lepidoptera : Pyralidae). *J. of the Natual History Society.* 91: 331.

Kumar, S., Rohilla, H.R., Harvir Singh and Singh, H., 1990. Evaluation of sesame genotypes against the incidence of shoot webber cum pod borer *Antigastra catalaunalis* (Duponchel). *HAU Journal of Research.* 20(4): 291-294.

Mahto, Y., 1987. Chemical control of *Polyphagotarsonenus latus* (Banks) (Acarina: Tarsonemidae) and malformation in sesame. *Indian J. Ent.* 49: 148-151.

Mehta, T.R. and Purohit, M.L., 1957. Research on diseases and pests of oilseeds. *Gwalior College Agric. J.* 2: 1-16.

Menon, M.G.R., Lal, R and Bhattacharjee, N.S., 1960. Studies on *Antigastra catalaunalis* (Duponchel), the til leaf roller. II. Bionomics and biology. *Indian J. Ent.* 22: 1-6.

Mohansundaram, M. and Sundara Babu, P.C., 1987. *Elasmolomus sordidus* (Hemiptera: Lygaeidae). *Madras Agric.J.* 74(3): 156-157.

Murali Baskaran, R.K., Mahadevan, N.R. and Thangavelu, S., 1991.Influence of intercropping on infestation of shoot webber, *Antigastra catalaunalis* (Dup.) in *Sesamum indicum* L. *Indian J. Agricultural Science,* 61: 440-442

Murali Baskaran, R.K., Mahadevan, N.R. and Thangavelu, S., 1990. Influence of different time of sowing on shoot webber and phyllody incidence and yield of sesame (*Sesamum indicum* L.) . *Sesame Safflower Newsl.* 5: 36-39.

Murali Baskaran, R.K. and Thangavelu, S., 1990b. Germplasm screening against sesame leaf roller and pod borer. *Sesame Safflower Newsl.* 5: 40-45.

Murali Baskaran, R.K. and Thangavelu, S., 1990a. Studies on the incidence of sesame shoot webber, *Antigastra catalaunalis* (Dup.) and its parasitoid, *Trathala flavo orbitatis* Cameron. *Sesame Safflower Newsl.* 5: 29-31.

Narayanan, E.S., 1961. Insect pests of sesamum and methods of control. p.87-91. In: A.B. Joshi(ed.) *Sesamum*. Indian Oilseed Committee, Hyderabad.

Narayanan, E.S., Subha Rao, B.R. and Rao, M.R., 1960. Some new species of chalcids from India. *Proc. Natn. Inst. Sci. India*, 26B: 168-175.

Padamja, S. and Savithri, P., 1993. Field evaluation of sesame varieties against pest complex. *Mysore J. of Agricultural Sciences* 27: 258-261.

Patel, A.A. and Bhalani, P.A., 1986. Chemical control of sesame leafwebber. *Antigastra catalaunalis* (Dup.) (Lepidoptera : Pyralidae). *Pesticide* 20: 23-26

Pillai, K.B., Thomas, M.J. and Nair, N.P., 1980. *Melanagromyza azawii* Spencer (Agromyzidae : Diptera) as a pest of sesame. *Agri. Res. J. Kerala* 18: 144.

Prasad, S.N., 1967. Biology of gingelly blossom midge, *Dasineura sesami* Grover. *Cecidologia Indica* 2: 1-16.

Prakash, A. and Chaturvedi, D.P., 1979. Larval parasite of gingelly blossom midge, *Dasineura sesamii*, *Indisn J. Pl. Prot.* 6: 83.

Rai, B.K., 1976. *Pests of oilseed crops in India and their control*. ICAR, New Delhi.

Rajpurohit, T. S. and N. Prasad 1982. Effect of different media, temperature and levels of pH on the growth and sporulation of *Alternaria sesami.Geobios.* 9(3): 142-144.

Rajpurohit, T. S. and N. Prasad 1982. Production of pectinolytic enzymes by *Alternaria sesami in vitro. Indian. J. Mycol. and Pl.Path.*, 12(2): 220-221.

Rajpurohit, T.S., 1982. Morphology and taxonomy of sesamum *Alternaria. Madras Agric. J.*, 68(10): 696-697.

Rajpurohit, T.S and N. Prasad and P.D.Gemawat1984.Evaluation of fungicides against *Alternaria sesami in vitro. Pesticides Special Issue on Fungicides,* Jan: 57.

Rajpurohit, T.S, .N. Prasad and P.D.Gemawat1983.Effect of carbon and nitrogen sources on growth and sporulation of *Alternaria sesami. Ind.J. of Mycol. and Pl. Path.* 13(2): 234-236.

Rajpurohit, T.S., 1997.Management of Macrophomina stem and root rot of sesame through fungicidal seed treatment and varietal resistance. *Indian. J. Mycol and Pl. Path.* 27(1) P: 98.

Rajpurohit, T.S and Z.S.Solanki., 2002. Identification of resistance sources of *Alternaria* leaf spot of sesame. *Asian Congress of Mycology and Plant Pathology* oct.1-4, University of Mysore.pp: 207.

Rajpurohit, T.S., 2004. Studies on plant products and new chemicals as spray against Alternaria blight of sesame. *Geobios*, 31: 64-65.

Rath, L.K. Mishra, A. and Mukherjee, S.K., 1997. Varietal reaction in sesame to leaf webber and capsule borer, *Antigastra catalaunalis* (Dup.). *Insect Environment* 3: 50-51.

Sachan, J.N., 1959. Common insect pests of crops and vegetable of Jobner (Rajasthan). *Rajasthan Agriculture* 7: 39-44.

Sellammal, M., Ramakrishnan, C., Kandaswamy, T.K. and Murugesan, M., 1973. *Madras Agric. J.* 60: 492-495.

Selvanarayanan, V. and Baskaran, P. Varietal response of sesame to the shoot webber and capsule borer, *Antigastra catalaunalis* (Duponchel). *Advances in Agricultural Research in India.* 2: 131-140.

Singh, C., 1966. Observations on *Lygaeous pandurus* Scopoli in west Rajasthan. *Pl. Prot. Bull.* New Delhi 18: 27-8.

Singh, Gurdeep., 1962. Bionomics and control *Diacrisia obliqua* Walker. *M.S.Thesis* (Unpublished) Punjab Univ. Chandigarh. pp 133.

Singh, G and Grewal, G.S., 1982. Chemical control of the Bihar Hairy Caterpillar, *Diacrisia obliqua* Walker, on castor and sesame. *Ind. J. Ent.* 1(1): 69-71.

Singh, G.P., Sinha, R.P., Singh, S.P. and Hammeed, S.F., 1992. Population dynamics and biology of sesame shoot and leaf webber, *Antigastra catalaunalis* (Dup.). *J. Ent. Res.* 16(4): 305-310

Singh, G., Grewal, G.S. and Parihar, R.S., 1991. Chemical control of sesame leaf webber and capsule borer, *Antigastra catalaunalis* (Dup.) on sesame in Punjab. *J. Insect Sci.* 4: 197-199.

Srivastava, S., Srivastava, N. and Das, S.B., 1989. Comparative efficacy and economics of some of promising insecticides against *Antigastra catalaunalis* (Dup.) (Lepidoptera : Pyralidae) in sesame. *Pestology* 13: 15-17.

Sundararaju, D. and Jayaraj, S., 1977. The bioecology and host range of *Orosius albicinctus* Dist. (Homoptera : Cicadellidae), the vector of sesame phyllody disease. *Madras Agric. J.* 64: 442-446.

Sundararaju, D. and Jayaraj, S., 1976. Efficacy of certain insecticides in the control of gingelly pests. *Madras Agric. J.* 63: 379-381.

Teotia, T.P.S. and Hussain, R., 1965. Bionomics of til leaf and pod borer, *Antigastra catalaunalis* (Duponchel) (Pyralidae: Lepidoptera). Labdev. *J. Sci. Tech.* B-3: 195-198.

Vittal, S.M. and Saroja, R., 1966. A note on the insecticidal control of shoot and leaf webber, *Antigastra catalaunalis* (Dup.) in gingelly. *Madras Agric. J.* 53: 215-216.

Yadav, S.K., 1985. Studies on the susceptibility resistance of sesame genotypes to the infestation of til leaf roller and pod borer, *Antigastra catalaunalis* (Dup.). *M.Sc Thesis* (Unpubl), Haryana Agricultural Univ., Hisar.

Chapter 36

Amino Acid Anlaysis for Biotechnological Crop Improvement and Allied Applications

H.C. Meher and V.T. Gajbhiye
Indian Agricultural Research Institute,
New Delhi - 110 012

Twenty amino acids (Aas) occur in free state or as linear chains in peptides and proteins (Table 37.1). They are the "building Blocks" of the body. Besides building cells and repairing tissue, they form antibodies to combat invading bacteria and viruses; they are part of the enzyme and hormonal system; they build nucleoproteins (RNA and DNA); they carry oxygen throughout the body and participate in muscle activity. Ten are essential (cannot be manufactured by the body) and the rest non-essential (can be manufactured by the body with proper nutrition) (Table 37.1).

Mehods of Amino Acid Analysis (AAA)

Amino acid analysis refers to the methodology used to determine the amino acid composition or content of proteins, peptides, and other pharmaceutical preparations. It is necessary to hydrolyze a protein/peptide to its individual amino acid constituents before amino acid analysis. Following protein/peptide hydrolysis, the amino acid analysis procedure can be the same as that practiced for free amino acids. The amino acid constituents of the test sample are typically derivatized for analysis. Thus, the sample preparation for analysis involves extraction and separation of aas and proteins from the substrate, purification, hydrolysis of protein, derivatization and estimation of aas. Common methods of amino acid analysis include Liquid Chromatography (LC) preferably with fluorescent than UV-visible detector and Gas Chromatography (GC) with flame ionization or nitrogen phosphorous detector. New techniques involve capillary electrophoresis, high performance anion exchange chromatography and mass spectrometry.

Table 37.1: Common Amino Acids

Name	*Symbol*		*Essential for Humans*
	3 Letter	*1 Letter*	
	Acidic amino acids		
Aspartic acid	Asp	D	No
Glutamic acid	Glu	E	No
	Neutral amino acids		
Alanine	Ala	A	No
Asparagine	Asn	N	No
Cysteine	Cys	C	No
Glutamine	Gln	Q	No
Glycine	Gly	G	No
Isoleucine	Ile	I	Yes
Leucine	Leu	L	Yes
Methionine	Met	M	Yes
Phenylalanine	Phe	F	Yes
Serine	Ser	S	No
Threonine	Thr	T	Yes
Tryptophan	Trp	W	Yes
Tyrosine	Tyr	Y	No
Valine	Val	V	Yes
	Basic amino acids		
Argnine	Arg	R	Yes
Histidine	His	H	Yes
Lysine	Lys	K	Yes
	Iminoacid		
Proline	Pro	P	No

Sample Preparation

Extraction

Woo *et al.* (1996) extracted proteins from food samples (soybean powder and whole egg) by hydrolysis using 15 ml of 6MHCl containing 0.1 per cent phenol at 145 °C for 4h under vacuum. Kochhar and Christen (1988) used 4 per cent perchloric acid (PCA) for aaa as substrates or products in enzymic reactions. Siller-Cepada *et al.* (1991) used 10 per cent PCA containing 1mM bathophenanthroline disulfonic acid (BPDS)–prevents thiol oxidation and thiol disulfide interchange, for extraction of free reduced and oxidized glutathione from woody plant tissues. Adamson *et al.* (1999) extracted plant leaves in 5 per cent PCA for determination of organic acids, sugars and alcohols. Zhong *et al.* (2004) extracted proteins from the cells of *Escherichia coli* by using 0.1 per cent trifluoroacetic acid (TFA). For plasma AAA the most widely used method of deproteinization (Deyl *et al.*, 1986; Williams, 1987) was sulfosalicylic acid (SSA) (10 per cent) precipitation (50 μl acid was mixed with

100 µl aqueous sample, left for 30 min at 4°C) followed by centrifugation (10min at 14000g and 4 °C). The technique was adopted for quantification of AAS in plasma/serum and urine (Sherwood *et al.*, 1990) for detecting maple syrup urine disease (primary disorder of aa metabolism and cystinuria–disorders of renal tubular reabsorption). However, SSA deproteinization hindered O-phthalaldehyde/3–mercaptopropionic acid (OPA/3-MPA) and 9-fluorenylmethyl chloroformate (FMOC-Cl) drivatizations and resulted in low aa yield because of its acidic nature inhibiting the derivatization under alkaline condition (Worthen and Liu1992; Liu and Worthen, 1992). SSA also superimposed on peaks of O-phospho-L-serine (OPS), Asp and Glu. Ethanol or methanol deproteinization could not detect OPS, Asp, Glu, AABA and Trp present in low levels in plasma. As against this, ultrafiltration was used for preparation of protein–free sample (Schuster, 1988; Blundell and Brydon, 1987; Feste, 1992). Kataoka *et al.* (1994) extracted free orthophospho amino acids from biological samples by 5 volume of 5 per cent trichloroacetic acid (TCA). For bound orthophospho amino acids, the PCA extract was centrifuged at 2000g for 10 min and the precipitate re-extracted with 5 per cent TCA and the pellet washed with diethyl ether (to remove TCA), warmed to dryness at 60°C and powdered before hydrolysis. Kataoka *et al.* (1993) extracted free sulfur amino acids, glutathione and related amino thiols from animal tissues with 2.4 ml of 0.05M H_2SO_4 and 1.2 ml of 2.5 per cent sodium tungstate (Na_2WO_4) and reduced the extract with 200µl of sodium borohydride ($NaBH_4$) in 0.01M NaOH. For bound forms, they (1993) dissolved the precipitate in 0.01M NaOH. Schofield and Chen (1995) extracted wheat flour (0.2g) by suspending in 2 ml of 5 per cent (w/v) ice-cold PCA and bubbling N_2 at 4°C for 1h for free reduced and free oxidized glutathione. For analysis of bound forms of reduced and oxidized glutathione, Chen and Schofield (1995) re-suspended the pellet in 0.25M of dithiothreitol (DTT)/0.05M 3-(N-morpholino) propanesulfonic acid (MOPS) buffer, pH 8.5 (1.0 ml), incubated at 40°C for 1h with gentle magnetic stirring, cooled to 4°C in an ice bath and reprecipitated with 50 µl of 70 per cent (w/v) PCA and centrifuged at 29000g max at 4°C before derivatization. Weckwerth *et al.* (2004) extracted amino acids, organic acids lipids and sugars from *Arabidopsis* leaves by methanol: chloroform and water (2.5:1:1) at –20°C and proteins by extraction buffer (0.05M Tris, pH7.6, 0.5 per cent sodium dodecyl sulphate and 1 per cent β–mercaptoethanol) and water saturated phenol. Schauer *et al.* (2005) extracted tomato leaf and fruit tissues with methanol for 15 min at 70 °C, mixed vigorously with 1 vol. of water, separated polar and non-polar metabolites by chloroform partitioning, centrifuged at 2200 g, reduced methanol/water phase to dryness *in vacuo*, dissolved in 40 µl of 20 mg ml^{-1} methoxyamine hydrochloride in pyridine before derivatization for analysis of amino acids, organic acids, sugars and sugar alcohols, and phosphorylated intermediates.

Purification

Accurate results from amino acid analysis require purified protein and peptide samples. Sample contaminants contribute to background noise that hampers analysis (Dupont *et al.*, 1989). Excipients to avoid include urea, sugars, and buffers, salts and/detergents. Samples are purified by RP-LC in aqueous TFA/acetonitrile solvents or dialysis into an appropriate buffer or volatile solvent (*e.g.*, water), and concentrated by vacuum concentrator or precipitation using ethanol, acetone, chloroform/methanol and trichloroacetic acid. Acetone precipitation is suitable for recovering proteins from most aqueous solvents and from SDS containing buffers; but not recommended for proteins dissolved in urea or guanidine or for peptides. Weckwerth *et al.* (2004) precipitated protein overnight from SDS containing buffers in phenol phase with ice-cold acetone at –20°C.

Sample conditions for post-column amino acid analyzers are less restrictive than for pre-column analyzers. Post-column analyzers can tolerate a variety of salts and detergents in small amounts (Dupont *et al.*, 1989). Coomassie-stained polyacrylamide gel bands are excised and washed with water and a portion (*e.g.*, 10-20 per cent) dried in a hydrolysis vial for liquid phase HCl hydrolysis and analyzed by ion-exchange post-column amino acid analyzer (Williams and Stone, 1995). Polyvinylidene difluoride (PVDF) immobilized proteins are hydrolyzed in liquid or vapor HCl; amino acids extracted with organic solvent and analyzed by pre- or post-column methods; but average compositional errors is >20 per cent (Mahrenholz *et al.*, 1996). Macchi *et al.* (2000) noted that sample should be pure prior to hydrolysis. Proteins should be desalted to obtain optimal compositional data. Dialysis against 0.1 per cent acetic acid removes salts while keeping proteins in solution, but quantitative estimation requires direct hydrolysis (*i.e.*, without dilution or sample losses introduced during hydrolysis). When proteins must be analysed without desalting, neutral buffers such as 50 mM Tris can be used without comprising the results. However, precipitation methods or RP (reverse phase)–LC result in variable and unacceptable losses of proteins, making quantification difficult. Zhang and Denslow (2003) purified protein using ultramacrospin columns or prosorb sample preparation catridge. Peptides generally can be desalted by RP-LC using 0.1 per cent TFA in water/acetonitrile. Recently, our laboratory could desalt chickpea protein from Tris-glycine buffer (pH 8.3) in a gradient separation on RP_{18} column using similar solvent system. This showed that once the sample is in 0.1 per cent TFA, it can be suitably cleaned by octadecylsilane (ODS) C18 catridge column clean up using a Catridge Vacuum Manifold Processing Station (Agilent P/N 5185-5754) similar to Prosorb sample preparation catridge (Applied Biosystems, Foster City, CA) (unpublished data). Adamson *et al.* (1999) adjusted the pH of PCA extract of plant leaves to 3.5 and purified on C_{18} catridge for simultaneous determination of organic acids, sugars and alcohols. Kataoka *et al.* (1994) adjusted pH by 1MNaOH and cleaned the TCA extract through a Dowex column for analysis of orthophospho amino acids from biological samples. To clean up the analytes from the solvent extract of onion Tsuge *et al.* (2002) subjected the sample solution to a combined solid phase extraction. The eluate from a Sep-Pak C_{18} cartridge was applied to a Bond Elut SCX cartridge (H^+ form), followed by washing with 0.1 M hydrochloric acid and elution with 0.5 M ammonia. Solid-phase extraction was used to analyse free amino acid content of garlic (Lee and Harnly, 2005) and Broccoli (Lee *et al.*, 2005).

Hydrolysis of Proteins

This involves hydrolysis in the tube (Gharahdaghi *et al.*, 1992), vapor phase hydrolysis (West and Crabb, 1992) or microwave hydrolysis (Gilman and Woodward, 1990). Association of Biomolecular Research Facilities (ABRF) demonstrated that hydrolysis step is particularly important (Ozols, 1990; Yiiksel *et al.*, 1994). ABRF studies revealed little difference in results from these methods. HCl hydrolysis of peptides/proteins typically yields 16 amino acids. Many laboratories now satisfactorily perform hydrolysis in 1 h at 150 °C under vacuum. However, Ser and Thr values are greatly reduced. Traditional methods use 6M HCl for 20-24 h at 110 °C under vacuum. This is more convenient from an operational standpoint. Losses of serine, threonine, and a lesser extent, tyrosine were reported to occur under these conditions. Application of adequate vacuum (less than 200 μm of mercury or 26.7 Pa) or introduction of an inert gas (argon) in the headspace of the reaction vessel could reduce the level of oxidative destruction.

Cleavage of Ala-Ala, Ile-Ile, Val-val, Val-Ile, Ile-Val, and Ala-Val linkages required longer hydrolysis time of 48 or 72 h at 110 °C (Ozols, 1990). In vapor phase hydrolysis direct contact with acid is avoided. Though it is advantageous in eliminating the potential source of contamination, Macchi *et al.* (2000) recorded poor hydrolysis of Ile-Ile, Ile-Val, and Val-Val bonds with vapor phase hydrolysis. Acid hydrolysis destroyed tryptophan. Although alkaline hydrolysis with NaOH, $Ba(OH)_2$ or LiOH did not destroy tryptophan (Delhaye and Landry,1986; Cohen *et al.*,1988), acid hydrolysis was required to determine other aas and destruction of tryptophan prevented by additives such as thioglycolic acid

(Yokote *et al.*, 1985), β–mercaptoethanol (Ng *et al.*, 1987) and mercaptoethane sulfonic acid (Yamada *et al.*, 1991). Cysteine and cystine were protected by performic acid oxidation to cysteic acid (which also converts methionine to methionine sulfone) before hydrolysis or by using dithiopropionic acid and dithiobutyric acid (Barkholt and Jensen, 1989) or dithioglycolic acid (Hoogerheide and Campbell, 1992) during acid hydrolysis. The use of dithioglycolic acid and phenylisothiocyanate (PITC) derivatization allowed determination of all aas except tryptophan (Bidlingmeyer *et al.*, 1984). Reduction of disulphide bridges, followed by alkylation of cyseine with iodoacetic acid or 4-vinyl pyridine could determine cysteine plus cystine content of proteins (Inglis, 1983). Carboxymethylation with iodoacetic acid was further extended to determine glutathione (γ-glutamyl cystenyl glycine) in its oxidized and reduced form (Sliier-Cepeda *et al.*, 1991). Clarke *et al.* (1999) followed performic acid oxidation of methionine and cysteine/cystine prior to vapour phase acid hydrolysis in 200 µl of 6M HCl and further employed the same for simultaneous analysis of amino sugars and aas in a single gradient method. As the amide groups of Gln and Asn are cleaved during acid hydrolysis, Glx = Glu+Gln; Asx = Asp+Asn. At 110 °C hydrolysis of phosphoserine and phosphotyrosine required 60 min or less where as phosphothreonine gave better results in 2 h hydrolysis. If the sample is oxidized before acid hydrolysis step, tyrosine content estimation will not be accurate (Bech-Andersen *et al.*, 1990). The sample/hydrolysis mixture or the reagent (for oxidation), additives to HCl solution, nitrogen atmosphere of the reaction vial, hydrolysis time and temperature (classical oven or microwave heating), and treatment of hydrolysate before injection (evaporation or adjustment of pH) must be standardized to obtain comparable results (Wathelet, 1999). A too high ratio of sample/hydrolysis mixture can lead to an incomplete hydrolysis and an underestimation of some amino acid contents. In the oxidation procedure, a minimum ratio of sample:oxidation mixture must also be respected in order to recover all cysteic acid and methionine sulfone. The methionine content is underestimated when the nitrogen content of the sample is significantly above 10 mg. Additives are useful to avoid degradation of some amino acids. An additive such as thioglycolic acid is readily oxidized and minimizes degradation of tyrosine, methionine, threonine and serine. Phenol, which is readily halogenated, serves as a scavenger to prevent degradation of tyrosine (in 3-chlorotyrosine), threonine and serine (Pickering and Newton, 1990). Higher losses of threonine and serine (up to 80 per cent) were observed in absence of phenol (Weiss *et al.*, 1998). Wathelet (1999) observed that a nitrogen atmosphere avoids oxidation and degradation of amino acids, which is important for the estimation of sulfur amino acid contents in acid hydrolysis without previous oxidation. When the hydrolysis time was increased, threonine and serine contents decreased, and hydrophobic amino acids (isoleucine, leucine, valine, etc.) increased. A24 h hydrolysis treatment destroyed about 5 per cent threonine and 10 per cent serine. Peptide bonds involving Leucine-Leucine, Isoleucine-Leucine, Leucine-Isoleucine were difficult to hydrolyse. A hydrolysis time between 48 h and 72 h was necessary to approach the real value of these hydrophobic amino acids. For all these amino acids, extrapolation with different hydrolysis time was necessary to obtain the real values. But this was time and money consuming. Hydrolysis temperature was also critical but the general temperature adopted in oven was 110°C. Hydrolysis with microwave technology seems to give an underestimation of threonine, serine, isoleucine and leucine (Weiss *et al.*, 1998; Chen *et al.*, 1987). Evaporation of the hydrolysate by heating under vacuum after hydrolysis, results sometimes in loss of sensible amino acids. Otherwise, the pH of the sample is adjusted by a NaOH solution. The method recommended by the European Community (EC, 1998) seems adequate if sufficient raw matter is available. This method dedicated to the quantification of all the amino acids except tryptophan involves the oxidation of sulfur amino acids before acid hydrolysis with a hydrogen peroxide/formic acid/phenol solution for 16 h at 0 °C. The excess of reagent is destroyed by a sodium disulfite solution. The acid hydrolysis step is carried out with the utilization of HCl 6N. The open flask is heated during 1 h at 110 °C; after it is closed and heated during 23 h at the same temperature. For analysis by ion exchange chromatography, the pH of the hydrolysate is adjusted to pH 2.2 with NaOH and aminohexanoic acid (norleucine) added as an internal standard. For tryptophan determination in protein hydrolysates, the barium hydroxide digestion and HPLC fluorescence detection (Molnár-Perl, 1997) is recommended by the European Union (Fontaine *et al.*, 1998).

Derivatization and Analysis

Liquid Chromatography (LC)

The automated amino acid (aa) analyzer based on separation of aas on cation exchange resin and quantification using ninhydrin remained popular for about 20 years because of its reliability (Deyl *et al.*, 1986; Williams, 1987; Fekkes, 1996). These analyzes were laborious, costly and time-consuming (Deyl *et al.*, 1986; Uhe *et al.*, 1991; Sarwar and Botting, 1993). Recently, aaa involves LC with ion-exchange separation and post column derivatization with ninhydrin (Samejima *et al.*, 1971), fluorescamine (Stein *et al.*, 1973) or O-phthalaldehyde (OPA) (Roth, 1971) or RP–LC separation after pre column derivatization forming dansyl (Tapuhi *et al.*, 1981), phenylisothiocyanate (PITC) (Bid1ingmeyer *et al.*, 1984), 9-fluorenylmethyl chloroformate (FMOC) (Haynes *et al.*, 1991), and 6–aminoquinolyl-N-hydroxysuccinimyl carbamate (AQC) (Strydom and Cohen, 1993) and OPA (Hill *et al.*, 1979) with the detection level at fmol in fluorscent and pmol in UV (ultra violet) detections (*viz.*, PITC derivatives). AQC derivatives were found more stable than PITC derivatives (Strydom and Cohen, 1994). Over the past two decades, the precolumn derivatization has been an alternative to post column derivatization due to greater sensitivity, faster analysis, less expensive and more flexible LC instrumentation compared to the dedicated amino acid analyzer (Sarwar and Botting, 1993; Ersser and Davey, 1991) (Table 37.2).

Table 37.2: Comparison of Different Derivatizations of Amino Acid

	Ninhydrin	*OPA*	*OPA*	*PITC*	*FMOC*	*AQC*
Derivatizations[1]	postc	postc	prec	prec	prec	prec
Detection mode[2]	c	f	f	UV	f	f
Sensitivity	pmol	fmol	fmol	pmol	fmol	Fmol
Chromatography[3]	*i.e.*	*i.e.*	r.p.	r.p.	r.p.	r.p.

OPA: Orthophthalaldehyde; PITC: Phenyl isothiocyanate; FMOC: 9-fluorenylmethyl chloroformate; AQC: 6-aminoquinolyl-N-hydroxysuccinimidyl carbamate.

1: Postc: Post column; Prec: Precolumn.

2: c: Colorimetry; f: Fluorescence; UV: Ultraviolet.

3: *i.e.*: Ion exchange; r.p: Reversed phase HPLC.

OPA (Hill *et al.*, 1979) and PITC (Edman's reagent) (Koop *et al.*, 1982; Bidlingmeyer *et al.*, 1984; Heinrikson and Meredith 1984; Cohen and Strydom, 1988) were widely employed. OPA reacted rapidly with primary aa at room temperature in presence of a thiol compound at alkaline pH in aqueous solution and formed highly fluorscent isoindole derivatives which were less polar than aas and could be well resolved in RP-LC (Jones and Gilligan, 1983; Schuster, 1988). OPA being not fluorescent did not interfere in the analysis. However, OPA did not react with secondary (imino acids), Gly and Lys (Lindroth and Mopper 1979; Turnell and Cooper 1982). Moreover, OPA adducts lacked stability when 2–mercaptoethanol (2-ME) was used as sulfhydryl reagent (Jones and Gilligan, 1983; Schuster, 1988; Turnell and Cooper 1982) because of displacement of sulphur by oxygen from ethanolic portion (Turnell and Cooper 1982). Worthen and Liu (1992) used less toxic and non-volatile 3–mercaptopropionic acid (3-MPA) in place of 2-ME to improve stability in automatic on-line OPA derivatization for profiling plasma aas. Liu and Worthen (1992) and Valles *et al.* (2005) incorporated 9-fluorenylmethyl-chloroformate (FMOC-Cl) into on-line OPA derivatization with 3-MPA and used both photodiode array (PDA) and programmable fluorescence detectors to quantify both primary and secondary amino acids. However, complete automation was required for satisfactory results. PITC although reactive with secondary amino acids, produced unstable derivatives of Asp and Glu and its excess amount required to be removed using high vacuum and extraction with toulene (Cohen and Strydom, 1988) to avoid column contamination and poor resolution. PITC derivatives were suitable for RP-LC and UV detection at 254 or 340 nm (Cohen, 1990) but could not detect cysteine and cystine. This problem was overcome by AQC (Cohen and Michaud, 1993). AQC reacted with both primary and secondary amines to form highly fluorscent derivatives and its excess was not required to be removed (Cohen and De Antonis, 1994; Liu, 1994; Strydom and Cohen, 1994; Strydom and Cohen, 1993; Van Wandelen and Cohen, 1997). This could detect 50-300 fmol of aas and required as little as 50 ng protein hydrolysate for compositional analysis or 100-200 ng proteins from 2-D gel electrophoresis (Grant and Fieno, 1998), using a ternary gradient LC. Gln, Val and Trp could not be analysed. However, using in situ AQC derivatization and ion pair chromatography, it was feasible to identify a multidrug resistant 160 kD protein 1 (MDR 1) of mouse leukemia P388 cells, separated by SDS-PAGE and blotted on to a PVDF membrane, with the aid of AACompIdent tool in ExPASy and PROPSEARCH in EMBL via WWW (Shindo *et al.*, 1997; 1998).

Dansyl (Tapuhi *et al.*, 1981; DeJong *et al.*, 1982) dabsyl (Lin and Chang 1975; Knecht and Chang, 1986) and 4-nitrophenylthiocarbamyl (NPTC) (Cohen and Strydom, 1988; Cohen, 1990) derivatives were also used. Dansyl derivatives were unstable (DeJong *et al.*, 1982) whereas dabsyl derivatives were interfered by reaction byproducts (Lin and Chang 1975; Knecht and Chang, 1986). Butylisothiocyanate (BITC) could be successfully utilized to derivatize 22-standard protein AAS and analysed via C18 RP-LC and UV detection at 250 nm (Woo and Lee 1994; Woo *et al.*, 1996) except that aspargine and serine peaks overlapped and BTC derivatives remained stable for 8h only at room temperature. Benzylisothiocyanate (BZITC) (an analog of PITC) derivatives of all 22 aas could be resolved on a Nova-Pak C18 column better than PTC derivatives (Woo and Ahan, 1996).

With Marfey's reagent (1-fluoro-2, 4-dinitrophenyl-5-alanine amide) and UV-LC without column heating, 19 aas except glutamine and tryptophan could be quantified as diastereomeric stable derivatives at 340 nm insensitive to solvent impurities and successfully employed to detect substrates and products in enzymatic reactions following hydrolysis at 110 °C for 24h or 150 °C for 2 h in a dry block heater (Kochhar and Christen, 1988; 1989; Szokan *et al.*, 1988; Martinez del Pozo *et al.*, 1989; Szokan *et al.*, 1989; Adamson *et al.*, 1992; Goodlett *et al.*, 1995; Goodnough *et al.*, 1995; Graber *et al.*, 1995; Tran *et al.*, 1990; Vacca *et al.*, 1997; Wu and Furlanut, 1997).

Rowley *et al.* (1995) described a procedure for precolumn derivatization with OPA and LC–electrochemical detection (ECD) of GABA, glutamate and arginine in brain microdialysates. Clarke *et al.* (1999) used anion exchange chromatography and integrated amperometric detection of AAS from soybean powder.

Orthophosphoamino acids–phosphoserine (Meyer *et al.*, 1987; Ringer, 1991), phosphothreonine (Ringer, 1991), and phosphotyrosine (Ringer, 1991) could be better analyzed by precolumn methods. Analysis using precolumn LC-UV derivatization with PITC (Meyer *et al*, 1987; Murthy and Iqbal,

1991; Aebersold *et al.*, 1991), dabsyl chloride (Malencik *et al.*, 1990), 1-fluoro-2, 4-dinitrophenyl-5-alanine amide (Goodnough *et al.*, 1995) or capillary zone electrophoresis with PITC or dabsyl chloride (Herber *et al.*, 1992; Meyer *et al.*, 1993) and post column fluorescence derivatization with FMOC (Niedbalski and Ringer, 1986; Ringer, 1991) were sensitive, but some lacked specificity and many required lengthy separation and sample clean-up (Aitken and Learmonth, 1997).

For Sulfur aas, glutathione and related aminothiols the reported methods included LC with UV (Reed *et al.*, 1980; Di Pietra *et al.*, 1994; Asensi *et al.*, 1994; Yoshida, 1996) and post column derivatization with 4,4′-dithiopyridine (Andersson *et al.*, 1993) and 5,5-dithiobis (2–nitrobenzoic acid) (Nozal *et al.*, 1997), the precolumn fluorescence derivatization with 4-(amino-sulfonyl)–or ammonium–7–fluoro-2, 1,3-benzoxadiazole-4-sulfonate (Ubbink *et al.*, 1991; Cornwell *et al.*, 1993) methyl 4–(6-methoxynapthale-2-yl)-4-oxo-2-butenoatene (Gotti *et al.*, 1994), N–(1-pyrenyl) maleimide (Winters *et al.*, 1995), monobromobimane (Newton *et al.*, 1981; Jacobsen *et al.*, 1989; Mansoor *et al.*, 1992; Fiskerstrand *et al.*, 1993; Yan and Huxtable, 1995), O-phthaldialdehyde (Fermo *et al.*, 1992; Paroni *et al.*, 1995), 9-fluorenylmethyloxycarbonyl chloride (Staffeldt *et al.*, 1991), and 2-chloro-1-methyl pyridinium (Sypniewski and Bald, 1994), and electrochemical detection (Richie and Lang, 1987; Carvalho *et al.*, 1994; Yang *et al.*, 1995; Kleinman and Richie, 1995; Lakritz *et al.*, 1997). Many were highly sensitive, but required extensive sample clean up and some lacked specificity. Nitrophenylation with DNFB (dinitro fluorobenzene) under alkaline conditions following carboxymethylation of PCA extract with iodoacetic acid, was successfully employed for analysis of free and bound forms of reduced and oxidized glutathione form wheat (Schofield and Chen 1995; Chen and Schofield 1995) and woody plant tissues (Siller-Cepeda *et al.*, 1991).

Precolumn derivatization with RP-LC may not be suitable for detection of trace levels of minor amino acids because the peak resolutions are diminished when sample loads are increased, where as resolution is maintained with higher loads using the cation exchange systems. In addition, precolumn accuracy may be limited if derivatization is incomplete, a problem that does not occur with post column derivatization systems. Similarly, cation-exchange systems are more tolerant of salts and residual HCl than precolumn systems (Macchi *et al.*, 2000).

Gas Chromatography (GC)

GC with FID (Flame Ionization Detector) and NPD (Nitrogen Phosphorus Detector) and of late flame photometric detector has been used for aaa. Capillary gas chromatography has been widely used in aaa because of its high resolving power, sensitivity, simplicity and low cost. The major problem has been deivatization to less polar and more volatile derivatives. Ester derivatives–N–trifluoroacetyl methyl (Darbre and Islam, 1968), propyl (Gamerith, 1983), n-butyl (Gehrke *et al.*, 1987; Zumwalt *et al.*, 1987); N-pentafluoropropionyl isopropyl (Singh and Ashraf, 1988), N-heptafluorobutyryl isobutyl (Zumwalt *et al.*, 1987; Schneider *et al.*, 1985; Labadarios *et al.*, 1986; MacKenzie, 1987; Philpott *et al.*, 1991); N-pentafluorobenzoyl isobutyl (Yeung *et al.*, 1986) and silyl derivatives including trimethyl (Gehrke and Leimer, 1971) and N,(O)–ter-butyldimethyl (ter-BDMS) (Biermann *et al.*,1986; Mawhinney *et al.*,1986; MacKenzie *et al.*,1987; Early *et al.*, 1987; Goh *et al.*, 1987; Woo and Lee, 1995) have been studied extensively and some successfully employed for quantification of aas in biological samples. Requirements of anhydrous conditions, high reaction temperatures, long time, low solubility of some aas in higher alcohols, multiple peaks of some aas and decomposition on long capillary columns limited these derivatizations. This problem was overcome by development of N (O,S)-isobutoxylcarbonyl (isoBOC) tert-BDMS derivatives (Oh *et al.*, 1994) by reaction of aas with isobutyl chloroformate (isoBCF) in aqueous alkaline condition and esterification with diazomethane. However, these methods required dual columns for complete resolution of aas and conversion of arginine to ornithine by arginase. Alternatively, N (O, S)-ethoxycarbonyl ethyl esters (Husek, 1991; Cao and Moini, 1997) were simpler, but reaction with fatty acids (Oh *et al.*, 1994), amines (Husek, 1993) and other organic acids also occurred. Subsequently, N (O, S)-isoBOC methyl esters of aas could be successfully employed to resolve and quantify protein and non-protein aas in biological samples by GC equipped with DB-17 capillary column and FID (Matsumura *et al.*, 1995; Matsumura *et al.*, 1996; Kataoka *et al.*, 1997) or NPD (Kataoka, 1997; Kataoka *et al.*, 1997) up to a detection limit of 5-25 µg in FID and 0.2-0.4 µg in NPD. In aqueous solution, the isoBOC group substituted selectively amino, imino, imidazole, phenolic hydroxyl and sylfhydral groups without heating and not the alcoholic hydroxyl groups, indole ring of nitrogen and amides. The derivatives were stable and could be quantitatively reproducible and resolved in capillary column. However, this did not react with guanidine group of arginine and hence, the method required conversion of arginine to ornithine by arginase before derivatization. Lee and Harnly (2005) used alkylchloroformate derivatization and GC-FID, methods to resolve amino acids in garlic. Schauer *et al.* (2005) iccubated the residues in methoxyamine hydrochloride in pyridine at 37 °C for 90 min. and derivatized with MSTFA at 37 °C for 30 min for analysis of amino acids, organic acids, sugar and sugar alcohols, and phosphorylated intermediates. Earlier Weckwerth *et al.* (2004) also used MSTFA derivatization for analysis of amino acids, organic acids, lipids and sugars.

FPD-GC of phosphoamino acids as their N-isobutoxylcarbonyl (isoBOC) methyl esters (Kataoka *et al.*, 1988; 1990; 1991a; 1991b; 1992a; 1992b; 1993a; 1993b; 1994) was successfully employed for analysis of their free (Kataoka *et al.*, 1990; 1991a; 1993b) and protein-bound (Kataoka *et al.*, 1991b; 1992b) forms in tissues and urine. Phosphoamino acids were shaken with isobutyl chloroformate (isoBCF) in alkaline medium (pH 10-12) for 10 min and then esterified by bubbling with diazomethane in a well-ventilated hood. However, interferences were observed due to diazomethane generating reagents.

The problem was solved by the use of hydrogen chloride-methanol (HCl-MeOH) instead of diazomethane. N (S)–isoPOC methyl esters derivatives of sulfur amino acids could be easily and rapidly prepared by the reaction with isopropyl chloroformate (isoPCF) by shaking in aqueous alkaline media, and methylation with HCl-MeOH with in 5 min. This was employed for selective and sensitive determination of sulfur aas (Kataoka *et al.*, 1994) using long DB-17 column (15m x0.53mm id x 1.0 µm film thickness) and of glutathione and related aminothiols such as cysteine (Cys), cysteinyl glycine(CysGly) and γ-glutamyl cysteine (γ-GluCys) using short DB-1 column (5m x 0.53mm id x 1.5 µm film thickness) (Kataoka *et al.*, 1995) from urine and plasma (Kataoka *et al.*, 1995) as well as blood and tissues (Takagi *et al.*, 1996) as against trimethylsilyl (Kataoka *et al.*,1986), neopentylidine (Jellum *et al.*,1969) and n-trifluoroacetyl n-butyl esters (Mee, 1973) derivatives lacking sensitivity and requiring anhydrous conditions. Sulfur and non-sulfur aas could be detected by FID and sulfur aas by both FID and FPD. FPD-GC was 20 times sensitive than FID-GC and could detect as low as 0.2-5 pmol of GSH and related aminothiols (except GSSG–disulfide form of GHS). However, the oxidized and protein bound forms of these compounds were required to be reduced to free thiols by addition of sodium borohydride ($NaBH_4$) before derivatization.

Mass Spectrometry

Current techniques are not efficient in determining protein modifications. Protein modifications, such as post-translational modifications, can be characterized only by examining a protein directly. At present, reading the amino acid sequence of a protein is performed by the Edman degradation method and increasingly by tandem mass spectrometry (MS/MS) (Hunt *et al.*, 1992; Kinter and Sherman, 2000; Kelleher *et al.*, 1999; Sze *et al.*, 2002). Other MS-based techniques, such as ladder sequencing (Patterson *et al.*, 1995; Chait *et al.*, 1993; Tsugita *et al.*, 1992; Vorm and Roepstorff, 1994; Zubarev *et al.*, 1994; Gobom *et al.*, 1999; Shevchenko *et al.*, 2000; Lin *et al.*, 2001) in-source fragmentation (Lin *et al.*, 2001; Reiber *et al.*, 1998) and chemical derivatization (Lennon and Walsh, 1997; Keough *et al.*, 1999) have been reported for sequencing peptides with varying degrees of success. Compared to the Edman method, the MS approach has the advantages of having high sensitivity and generating structural information on modifications. However, to map the sequence of an entire protein for examining all possible modifications, one often needs to produce, detect and sequence many short, partially overlapping peptides, which can be very difficult. The previous attempts using enzymes (Kelleher *et al.*, 1999), modified Edman chemistry (Chait *et al.*, 1993) and acids (Tsugita *et al.*, 1992; Gobom *et al.*, 1999; Vorm and Roepstorff, 1994; Zubarev *et al.*, 1994; Shevchenko *et al.*, 2000; Lin *et al.*, 2001) to produce peptide ladders were not generally applicable to sequencing peptides because of a lack of specificity (that is, the methods generated internal peptides along with some peptide ladders) and universality (that is, most peptides failed to generate any ordered peptide ladders for sequence reading). Zhong *et al.* (2004) reported an approach for sequencing proteins and determining modifications with high speed, sensitivity and specificity by mass analysis of polypeptide (MAP) of a protein. They (2004) discovered that a protein could be readily acid-hydrolyzed within 1 min by exposure to microwave irradiation to form, predominantly, two series of polypeptide ladders containing either the N–or C-terminal amino acid of the protein, respectively. The protein was purified by cation exchange LC and then fractionated by RP-LC. This was followed by mass spectrometric analysis. A simple mass spectrum consisting of peaks exclusively from these polypeptide ladders was produced. This allowed direct reading of amino acid sequence and modifications of the protein. The technique was successfully applied to determine protein phosphorylation sites as well as the sequences and several previously unknown modifications of 28 small proteins isolated from *Escherichia coli* K12 cells. This technique can potentially be automated for large-scale protein annotation. Among the twenty common amino acids, leucine and isoleucine have the same mass and, therefore, cannot be distinguished by this method. Glutamine and lysine have similar masses and cannot be readily distinguished within the molecular mass measurement accuracy of TOF MS; but it should be entirely possible with MALDI Fourier Transform-ion cyclotron resonance (FT-ICR) MS (Jones *et al.*, 2003). However, distinguishing these pairs of amino acids is only required for *de novo* sequencing of an unknown protein from a species with no or little genome or proteome database. Fortunately, with the rapid expansion of genome databases as well as the possibility of performing cross-species database searching, *de novo* sequencing is becoming an increasingly rare practice. Thus the MAP sequencing technique should be very powerful for reexamining protein sequences translated from their genome and determining their modifications. Zhou *et al.* (2001) described an approach to protein phosphorylation analysis consistsing of three steps: (1) selectivephosphopeptide isolation from a peptide mixture via a sequence of chemical reactions, (2) phosphopeptide analysis by automated liquid chromatography–tandem mass spectrometry (LC-MS/MS), and (3) identification of the phosphoprotein and the phosphorylated residue(s) by correlation of tandem mass spectrometric data with sequence databases. By utilizing various phosphoprotein standards and a whole yeast cell lysate, they demonstrated that the method was equally applicable to serine-, threonine–and tyrosinephosphorylated proteins, and was capable of selectively isolating and identifying phosphopeptides present in a highly complex peptide mixture. Beck *et al.* (2005) characterized monoclonal anti-IGF-1 receptor antibodies produced in CHO and NS0 cells by analyzing the primary structures and disulfide bridge pairing by microsequencing (Edman degradation) and LC-MS (MALDI-TOF, ES-TOF) and peptide mapping after enzymatic digestion.

Hasegawa *et al.* (2005) determined D–and L-enantiomers of methionine and [2H3] methionine in plasma by (GC-MS) with selected-ion monitoring (SIM).

GC-MS and LC-MS methods for analysis of ter-butyldimethylsilyl (t-BDMS) (Stabler *et al.*, 1987) and N (O,S)-propoxycarbonyl propyl ester (Sass and Endres, 1997) derivatives of sulfur amino acids, glutathione and related amino thiols were highly sensitive, specific, but very expensive. GC-MS of t-BDMS derivatives was employed to determine *S*-Alk(en)yl-L-cysteine sulfoxide from onion up to a detection limit of 0.3-1.8 ng (Tsuge *et al.*, 2002) and for the analysis of glycated and ascorbylated proteins (Hasenkopf *et al.*, 2002). Lee and Harnly (2005) used alkylchloroformate derivatization and GC-MS methods to resolve 18 free amino acids in garlic within 10 min. Füzfai *et al.* (2004) could simultaneously identified and quantified sugar, sugar alcohol, and carboxylic acid contents of sour cherry, apple, and ber fruits, as their trimethylsilyl derivatives, by GC-MS. GC-MS following MSTFA derivatization was used for simultaneous quantification of amino acids, organic acids, lipids and sugars from *Arabidopsis* (Weckwerth *et al.*, 2004) and tomato (Schauer *et al.*, 2005).

Recently, LC-MS was used to determine early glycation products and protein cross links due to Millard reaction (non-enzymatic glycosylation–NEG) or glycation (Smith *et al.*, 1992; Lapolla *et al.*, 1994; 1994; Prabhakaram and Ortwerth, 1994) *in vivo* between aldose or ketose sugars and protein-bounds free amino groups (Monnier, 1989), implicated in diabetic or age-related complications (Baynes, 1996), Alzheimer's disease (Markesbery, 1997) and in cataract formation (Monnier and Cerami, 1983).

GC-MS (Sass and Endres, 1997) and LC-MS (Fermo *et al.*, 1998) was used to measure total plasma homocysteine (tHcy)–free and protein bound homocysteine, homocystine and homocysteine-cystine mixed disulfides, implicated in atherosclerosis. Reduction of disulfide bonds by $NaBH_4$ and fluorescence derivatization with SBD-F (ammonium-7-fluorobenzo-2-oxa-1,3-diazole-4-sulfonate) formed clean and stable adduct for LC-MS (Fermo *et al.*, 1998) and was superior to LC with ninhydrin (Andersson *et al.*, 1989), electrochemical (Malinow *et al.* (1989) or precolumn derivatization with OPA (Fermo *et al.*, 1992).

Application

Amino acid analysis is widely used in biotechnology, biomedical and food analysis laboratories. The importance and utility of amino acid analysis lies in characterization of protein modifications essential for the study of protein function using functional genomic, proteomic and metabonomic approaches. Amino acid analysis remains the best method for protein/peptide quantification and identification (Hobohm *et al.*, 1994; Schegg *et al.*, 1997) as the AA composition reflects the highly conserved primary structure of protein/peptide as against N-terminal sequencing, peptide mapping and Western blotting based on partial information of protein structure. It is a prerequisite for Edman degradation and mass spectrometry. It is essential for maintenance of correct nutrient levels in cell culture broths in recombinant technology (Clarke *et al.*, 1999). In biotechnology, amino acid analysis is used to detect substitution of Nle (norleucine) in Met (methionine) positions (Tsai *et al.*, 1988; Bogosian *et al.*, 1989), screen hydroxylysine (Hyl) in recombinant proteins (Kivirikko *et al.*, 1992; Molony *et al.*, 1995), resolve Gln and Asn in analysis of carboxypeptidase and quantify Asn-linked glycosylation (Tarentino *et al.*, 1985) and to monitor cysteine alkylation levels for reduced proteins (Garnick *et al.*, 1988; Lundell and Schreitmiiller, 1999). Amino acid analysis is used to study the protein composition of food and feed (Woo *et al.*, 1996; Woo and Ahan, 1996). It is also utilized to study protein function in post-translational modifications, including glycosylated aas (Packer *et al*, 1998) and glycated aas (Walton and McPherson 1987; Cayot and Tainturler 1997). Amino acid analysis of body fluids (plasma, urine) and tissues serves as the diagnostics of diseases linked with AA metabolism–maple syrup urine disease (accumulation of branched-chain L-aas), phenylketonuria (high concentration of phenylalanine), atherosclerosis (elevated level of homocysteine) (Masser *et al*., 1994; Stampfer and Malinow, 1995) and galactosemia (high methionine), diabetic complications and Alzheimer's disease (glycated lysine) (Baynes, 1996; Markesbery, 1997).

References

Adamson, J. G., Hoang, T., Crivici, A., and Lajoie, G. A., 1992. Use of Marfey's reagent to quantitate racemization upon anchoring of amino acids to solid supports for peptide synthesis. *Anal. Biochem.* 202, 210-214.

Aebersold, R., Watts, I. D., Morrison, H. D., and Bures, E. J., 1991. Determination of the site of tyrosine phosphorylation at the low picomole level by automated solid-phase sequence analysis. *Anal. Biochem.* 199, 51—60.

Aitken, A. and Learmonth, M., 1997. Analysis of sites of protein phosphorylation. *Methods Mol. BioI.* 64, 293-306.

Andersson, A., Brattstrom, L., Isaksson, B., and Israelsson, B., 1989. Determination of homocysteine in plasma by ion-exchange chromatography. *Scand. J. Clin. Invest.* 49, 445-449.

Andersson, A., Isaksson, A., Brattstrom, L., and Hultberg, H., 1993. Homocysteine and other thiols determined in plasma by HPLC and thiol-specific postcolumn derivatization. *Clin. Chem.* 39, 1590-1597.

Asensi, M., Sastre, J., Pallardo, F. V., de la Asuncion, J. G., Estrela, J. M., and Vina, J., 1994. A high-perfonnance liquid chromatography method for measurement of oxidized glutathione in biological samples. *Anal. Biochem.* 217, 323-328.

Barkholt, V. and Jensen, A.L., 1989. Amino acid analysis: determination of cysteine plus half-cystine in proteins after hydrochloric acid hydrolysis with a disulphide compound as additive. *Anal. Biochem.* 177, 318-322.

Baynes, J. W., 1996. The role of oxidation in the Maillard reaction *in vivo*, in *The Maillard Reaction: Consequences for the Chemical and Life Sciences* (Ikan, R., ed.), Wiley, New York.

Bech-Andersen, S., Mason V., and Dhanoa, M., 1990. Hydrolysate preparation for amino acid determinations in feed constituents. *J. Anim. Physiol. Anim. Nutr.*6 3, 188–197.

Beck, A., Bussat, M. C., Zorn, N., Robillard, V., Klinguer-Hamour, C., Chenu, S., Goetsch, L., Corvaia, N., Van Dorsselaer, A. and Haeuw, J. F., 2005). Characterization by liquid chromatography combined with mass spectrometry of monoclonal anti-IGF-1 receptor antibodies produced in CHO and NS0 cells. *J. Chromatogr. B. Analyt. Technol. Biomed. Life Sci.* 819, 203-218.

Bid1ingmeyer, B. A., Cohen, S. A., and Tarvin, T., 1984. Rapid analysis of amino acids using pre-column derivatisation. *J. Chromatog.* 336, 93-104.

Biermann, C. J., Kinoshita, C. M., Marlett, J. A., and Steele, R. D., 1986. Analysis of amino acids as tert.-butyldimethylsilyl derivatives by gas chromatography. *J. Chromatog.* 357, 330-334.

Blundell, G. and Brydon, W. G., 1987. High performance liquid chromatography of plasma amino acids using orthophthalaldehyde derivatisation. *Clin. Chem. Acta.* 170, 79-83.

Bogosian, G., Violanci, B. N., Dorward-King, E. J., Workman, W. E., Jung, P. E., and Kane, J. F., 1989. Biosynthesis and incorporation into protein of norleucine by *Escherichia coli*. *J. Biol. Chem.* 264, 531-539.

Cao, P. and Moini, M., 1997. Quantitative analysis of fluorinated ethylchloformate derivatives of protein amino acids and hydrolysis products of small peptides using chemical ionization gas chromatography-mass spectrometry. *J. Chromatog.* A, 759, 111-117.

Carvalho, F. D., Remiao, F., Vale, P., Timbrell, J.A., Bastos, M. L., and Ferreira, M. A., 1994. Glutathione and cysteine measurement in biological samples by HPLC with a glassy carbon working detector. *Biomed. Chromatolog:* 8, 134-136.

Cayot, P. and Tainturier, G., 1997. The quantitation of protein amino groups by the trinitrobenzenesulfonic acid method: a reexamination. *Anal. Biochem.* 249, 184–200.

Chait, B. T., Wang, R., Beavis, R. C. and Kent, S. B. H., 1993. Protein ladder sequencing. *Science* 262, 89?92.

Chen, S. T., Chiou, S. H., Chu, Y. H., and Wang K. T., 1987. Rapid hydrolysis of proteins and peptides by means of microwave technology and its application to amino acid analysis. *Int. J. Peptide Protein Res.* 30, p. 572–576.

Chen, X. and Schofield, J. D., 1995. Determination of protein-glutathione mixed disulfides in wheat flour. *J. Agric. Food. Chem.* 43, 2362-2368.

Clarke, A. P., Jandik, P., Rocklin, R. D., Liu, Y., and Avdalovic, N., 1999. An integrated amperometry waveform for the direct, sensitive detection of amino acids and amino sugars following anion-exchange chromatography. *Anal. Chem.* 71, 2774–2781.

Cohen, S. A., 1990. Analysis of amino acids by liquid chromatography after precolumn derivatization with 4-nitrophenylisothiocyanate. *J. Chromatog.* 512, 283-290.

Cohen, S. A. and De Antonis, K. M., 1994. Applications of amino acid analysis derivatization with 6-aminoquinolyl-N-hydroxysuccinimidyl carbamate: Analysis of feed grains, intravenous solutions and glycoproteins. *J. Chromatog.* 661, 25-34.

Cohen, S. A. and Michaud, D. P., 1993. Synthesis of a fluorescent derivatizing reagent, 6-aminoquinolyl-N-hydroxysuccinimidyl carbamate, and its application for the analysis of hydrolysate amino acids via high performance liquid chromatography. *Anal. Biochem.* 211, 279-287.

Cohen, S. A. and Strydom, D. J., 1988. Amino acid analysis utilizing phenylisothiocyanate derivatives. *Anal. Biochem.* 174, 1-16.

Cohen, S. A., Meys, M., and Tarvin, T. L., 1988. *The PicoTag Method.* A Manual of Advanced Techniques for Amino acid analysis. Water Chromatography Division, Millipore Corp., Milford, MA.

Cornwell, P. E., Morgan, S. L., and Vaughn, W. H., 1993. Modification of a high–performance liquid chromatographic method for assay of homocycteine in human plasma. *J. Chromatog.* 617, 136-139.

Darbre, A. and Islam, A., 1968. Gas-liquid chromatography of trifluoroacetylated amino acid methyl esters. *Biochem. J.* 106, 923-925.

DeJong, C., Hughes, G. J., Wieringen, E. V., and Wilson, K. J., 1982. Amino acid analysis by high-performance liquid chromatography, An evolution of usefulness of pre-column Dns derivatization. *J. Chromatog.* 241, 345-359.

Delhaye, S. and Landry, J., 1986. High-performance liquid chromatography and ultraviolet spectrophotometry for quantitation of tryptophan in barytic hydrolysates. *Anal. Biochem.* 159, 175-178.

Deyl, Z., Hyanek, J., and Horakova, M., 1986. Profiling of amino acids in body fluids and tissues by means of liquid chromatography. *J. Chromatog.* 379, 177–250.

Di Pietra, A. M., Gotti, R., Bonazzi, D., Andrisano, V., and Cavrini, V., 1994. HPLC determination of glutathione and L-cysteine in pharmaceuticals after derivatization with ethacrynic acid. *J. Pharm. Biomed. Anal.* 12, 91-98.

Dupont, D. R., Keim, P. S., Chui, A. H., Bello, R., Bozzini, M., and Wilson, K. J., 1989. A comprehensive approach to amino acid analysis, in *Techniques in Protein Chemistry* (Hugli, T. E., ed.). Academic, London and New York, pp. 284-294.

Early, R. J., Thompson, J. R., Sedgwick, G. W., Kelly, J. M., and Christopheason, R. J., 1987. Capillary gas chromatographic analysis of amino acids in blood and protein hydrolysates as tert.-butyldimethylsilyl derivatives. *J. Chromatog.* 416, 15-23.

EC (1998). Establishing Community methods of analysis for the determination of amino-acids, crude oils and fats, and olaquindox in feedingstuffs, and amending Directive 71/393/EEC. Commission Directive 98/64/EC, *Off. J. Eur. Communities* L257, 19.9.98, p. 14–23.

Ersser, R. S. and Davey, J. F., 1991. Liquid chromatographic analysis of amino acids in physiological fluids: recent advances. *Med. Lab. Sci.* 48, 59-71.

Fekkes, D., 1996. State-of-the-art of high-performance liquid chromatographic analysis of amino acids in physiological samples. *J. Chromatog. B. Biomed. Appl.* 682, 3-22.

Fermo, I., Arcelloni, C, Mazzola, G., D' Angelo, A., and Paroni, R., 1998. HPLC method for measuring total plasma homocysteine levels. *J. Chromatog. B* 719, 31–36.

Fermo, I., Arcelloni, C., De Vecchi, E.; Vigano', S., and Paroni, R., 1992. HPLC method with fluorescence detection for the determination of total homocysteine in plasma. *J. Chromatog. (Symp.* 593, 171-176.

Feste, A. S., 1992. Reversed-phase chromatography of phenylthiocarbamyl amino acid derivatives of physiological amino acids: an evaluation and a comparison with analysis by ion-exchange chromatography. *J. Chromatog.* 574, 23-34.

Fiskerstrand, T., Refsum, H., Kvalheim, G., and Ueland, P. M., 1993. Homocysteine and other thiols in plasma and urine: Automated determination and sample stability. *Clin. Chem.* 39, 263-271.

Fontaine, J., Bech-Andersen, S., Bütikofer, U., and de Froidmont-Görtz, I., 1998. Determination of tryptophan in feed by HPLC-Development of an optimal hydrolysis and extraction procedure by the EU Commission DGXII in three international collaborative studies. *Agrobiol. Res. Z. Agrarbiol.-Agrikulturchem.-Ökol.* 51, 97–108.

Füzfai, Z., Katona, Z. F., Kovács, E. and Perl, I. M., 2004. Simultaneous identification and quantification of the sugar, sugar alcohol, and carboxylic acid contents of sour cherry, apple, and ber fruits, as their trimethylsilyl derivatives, by gas chromatography-mass spectrometry. *J. Agric. Food Chem.* 52, 7444–7452.

Gamerith, G., 1983. Gas-liquid chromatographic determination of N (O, S)–trifluoroacetyl n-propyl esters of protein and non-protein amino acids. *J. Chromatog.* 256, 267-281.

Garnick, R. L., Solli, N. J., and Papa, P. A., 1988. The role of quality control in biotechnology: an analytical perspective. *Anal. Chem.* 60, 2546-2557.

Gehrke, C. W. and Leimer, K., 1971. Trimethylsilylation of amino acids. Derivatization and chromatography. *J. Chromatog.* 57, 219-238.

Gehrke, C. W., Kuo, K. C., Kaiser, F. E., and Zumwalt, R. W., 1987. Analysis of amino acids by gas chromatography as the N-trifluoroacetyl n-butyl esters. *J. Assoc. Off. Anal. Chem.* 70, 160-170.

Gharahdaghi, F., Atberton, D., DeMott, M., Mische, S. M., 1992. Amino acid analyis of PVDF bound proteins, in *Techniques in Protein Chemistry III* (Angeletti, R. H., ed.), Academic, London and New York, pp. 249-260.

Gilman, L. B. and Woodward, C., 1990. Hydrolysis methods in current research, in *Protein Chemistry* (Villafranca, J. J., ed.), Academic, London and New York, pp. 23-26. Gobom, J., Mirgorodskaya, E., Nordhoff, E., Hojrup, P. and Roepstorff, P., 1999). Use of vapor-phase acid hydrolysis for mass spectrometric peptide mapping and protein identification. *Anal. Chem.* 71, 919?927.

Goh, C. J., Craven, K. G., Lepock, J. R., and Dumbroff, E. B., 1987. Analysis of all protein amino acids as their tert.-butyldimethylsilyl derivatives by gas-liquid chromatography. *Anal. Biochem.* 163, 175-181.

Goodlett, D. R., Abuaf, P. A., Savage, P. A., Kowalski, K. A., Mukherjee, T. K., Tolan, J. W., *et al.*, 1995. Peptide chiral purity determination: hydrolysis in deuterated acid, derivatization with Marfey's reagent and analysis using high-performance liquid chromatography-electrospray ionization-mass spectrometry. *J. Chromatog.* A 707, 233-244.

Goodnough, D. B., Lutz, M. P., and Wood, P. L., 1995. Separation and quantification of D–and L-phosphoserine in rat brain using N-alpha–(2, 4-dinitro-5–fluorophenyl)-L-alaninamide (Marfey's reagent. by high-performanceliquid chromatography with ultraviolet detection. *J. Chromatog. B. Biomed. Appl.* 672, 290-294.

Gotti, R., Andrisano, V., Gotti, R., Cavrini, V., and Candeletti, S., 1994. Determination of glutathione in biological samples by high performance liquid chromatography with fluorescence detection. *Biomed. Chromatog.* 8, 306-308.

Graber, R., Kasper, P., Malashkevich, V. N., Sandmeier, E., Berger, P., Gehring, H., *et al.*, 1995. Changing the reaction specificity of a pyridoxal-5 '-phosphate-dependent enzyme. *Eur. J. Biochem.* 232, 686-690.

Grant, R. A. and Fieno, A. M., 1998). Use of the AQC pre-column method for determination of protein composition and identification from PVDF blots. Presented at ABRF'98: From genomes to Function–Technical Challenges of the Post Genome Era, htpp://www.Abrf.org/ABRFMeetings/abrf98/grant.pdf.

Hasegawa, H., Shinohara, Y., Akahane, K. and Hashimoto, T., 2005. Determination of D–and L-enantiomers of methionine and [2H3]methionine in plasma by gas chromatography-mass spectrometry. *J. Chromatogr. B. Analyt. Technol. Biomed. Life Sci.* 823, 203-208.

Hasenkopf, K., Rönner, B., Hiller, H. and Pischetsrieder, M., 2002. Analysis of glycated and ascorbylated proteins by gas chromatography-mass spectrometry. *J. Agric. Food Chem.* 50, 5697–5703.

Haynes, P. A., Sheumack, D., Kibby, J., and Redmond, J. W., 1991. Amino acid analysis using derivatization with 9-fluorenylmethyl chloroformate and reversed phase high performance liquid chromatography. *J. Chromatog.* 540, 177-185.

Heinrikson, R. L. and Meredith, S. C., 1984. Amino acid analysis by reversed phase high-performance liquid chromatography: pre-column derivatization with phenylisothiocyanate. *Anal. Biochem.* 136, 65-74.

Herber, M., Liedtke, C., Korte, H., Hoffmann-Posorske, E., Donella-Deana, A., Pinna, L. A. *et al.*, 1992. Non-radioactive determination of PTH and dabsyl phosphoamino acids by capillary electrophoresis. *Chromatographia* 33, 347-350.

Hill, D. W., Walters, F. H., Wilson, T. D., and Stuart, J. D., 1979. High performance liquid chromatographic determination of amino acids in the picomole range. *Anal. Chem.* 51, 1338-1341.

Hobohm, U., Houthaeve, T., and Sander, C., 1994. Amino acid analysis and protein database compositional search as a rapid and inexpensive method to identify proteins. *Anal. Biochem.* 222, 202-209.

Hoogerheide, J. G. and Campbell, C. M., 1992. Determination of cysteine plus half-cystine in protein and peptide hydrolysates: use of dithiodiglycolic acid and phenylisothiocyanate derivatisation. *Anal. Biochem.* 201, 146-151.

Hunt, D.F. *et al.*, 1992). Characterization of peptides bound to the class I MHC molecule HLA-A2.1 by mass spectrometry. *Science* 255, 1261?1263.

Husek, P., 1991. Rapid derivatization and gas chromatographic determination of amino acids. *J. Chromatog.* 552, 289-299.

Husek, P., 1993. Capillary GC analysis of biogenic amines, their precursors and catabolytes after fast derivatization with ethyl chloroformate. *J. Microcol. Sep.* 5, 101-103.

Inglis, A. S., 1983. Single hydrolysis method for all amino acids, including cysteine and tryptophan, in *Methods in Enzymology*, vol. 91. Academic, San Diego, CA, pp. 26-36.

Jacobsen, D. W., Gatautis, V. J., and Green, R., 1989. Determination of plasma homocysteine by high-perfonnance liquid chromatography with fluorescence detection. *Anal. Biochem.* 178, 208-214.

Jellum, E., Bacon, V. A., Patton, W., Pereira, W., Jr., and Halpern, B. Quantitative determination of biologically important thiols and disulfides by gas-liquid chromatography. *Anal. Biochem.* 31, 339-347.

Jones, B. N. and Gilligan, J. P., 1983. o-Phthaldialdehyde precolumn derivatization and reversed-phase high–performance liquid chromatography of polypeptide hydrolysates and physiological fluids. *J. Chromatog.* 266, 471-482.

Jones, J. J., Stump, M. J., Fleming, R. C., Lay, J. O. and Wilkins, C. L., 2003. Investigation of MALDI-TOF and FT-MS techniques for analysis of *Escherichia coli* whole cells. *Anal. Chem.* 75, 1340?1347.

Kataoka, H., 1997. Selective and sensitive determination of protein and non-protein amino acids by capillary gas chromatography with nitrogen-phosphorus selective detection. *Biomed. Chromatogr.* 11, 154-159.

Kataoka, H., Matsumura, S., *and* Makita, M., 1997. Determination of amino acids in biological fluids by capillary gas chromatography with nitrogen-phosphorus selective detection. *J. Pharm. Biomed. Anal.* 15, 1271-1279.

Kataoka, H., Matsumura, S., Koizumi, H., and Makita, M., 1997. Rapid and simultaneous analysis of protein and non-protein amino acids as N(O, *S*)–isobutoxycarbonyl methyl ester derivatives by capillary gas chromatography. *J. Chromatog.* A, 758, 167-173.

Kataoka, H., Nakai, K., and Makita, M., 1993a. Identification of O-phosphoamino acids in urine hydrolysate by gas chromatography-mass spectrometry. *J. Chromatog.* 615, 136-141.

Kataoka, H., Nakai, K., and Makita, M., 1994. Increase of phosphotyrosine levels in mouse urine and liver during liver regeneration after partial hepatectomy. *Biochem. Biophys. Res. Commun.* 201, 909-916.

Kataoka, H., Nakai, K., Katagiri, Y., and Makita, M., 1993b. Analysis of free and bound O-phosphoamino acids in urine by gas chromatography with flame photometric detection. *Biomed. Chromatogr.* 7, 184-188.

Kataoka, H., Nakai, K., Ueno, Y., and Makita, M., 1992a. Analysis of O–phosphoamino acids in the protein fractions of mouse tissue by gas chromatography. *Biosci. Biotech. Biochem.* 56, 1300-1301.

Kataoka, H., Ohnishi, K., and Makita, M., 1986. Gas chromatographic determination of cysteic acid. *J. Chromatog.* 354, 482-485.

Kataoka, H., Sakiyama, N., and Makita, M., 1988. Gas chromatographic analysis of aminoalkylphosphonic acids and aminoalkyl phosphates. *J. Chromatog.* 436, 67-72.

Kataoka, H., Sakiyama, N., and Makita, M., 1990. Occurrence of free O–phosphoserine and O-phosphothreonine in porcine liver. *Agric. Biol. Chem.* 54, 1731-1733.

Kataoka, H., Sakiyama, N., and Makita, M., 1991a. Distribution and contents of free O-phosphoamino acids in animal tissues. *J. Biochem.* 109, 577-580.

Kataoka, H., Takagi, K., and Makita, M., 1995. Determination of glutathione and and related aminothiols by gas chromatography with flame photometric detection. *J. Biomed. Chromatogr.* 9, 85-89.

Kataoka, H., Takagi, K., and Makita, M., 1995. Determination of total plasma homocysteine and related aminothiols by gas chromatography with flame photometric detection. *J. Chromatog.* B 664, 421-425.

Kataoka, H., Ueno, Y., and Makita, M., 1991b. Analysis of O-phosphoamino acids in proteins by gas chromatography with flame photometric detection. *Agric. Biol. Chem.* 55, 1587-1592.

Kataoka, H., Ueno, Y., and Makita, M., 1992b. O-Phosphoamino acid analysis of phosphorylated proteins by gas chromatography with flame photometric detection. *J. Pharm. Biomed. Anal.* 10, 365-369.

Kelleher, N. L. *et al.*, 1999). Top down versus bottom up protein characterization by tandem high resolution mass spectrometry. *J. Am. Chem. Soc.* 121, 806?812.

Keough, T., Youngquist, R. S., and Lacey, M. P., 1999. A method for high-sensitivity peptide sequencing using postsource decay matrix-assisted laser desorption ionization mass spectrometry. *Proc. Natl. Acad. Sci. USA* 96, 7131?7136.

Kinter, M., and Sherman, N.E., 2000). Protein Sequencing and Identification Using Tandem Mass Spectrometry. (Wiley, New York).

Kivirikko, K. I., Myllyla, R., and Pihlajaniemi, T., 1992. Hydroxylation of proline and lysine residues in collagens and other animal and plant proteins, in *Post-translational Modifications of Proteins* (Harding, J. J. and Crabbe., M. J., eds.), CRC, Boca Raton, FL, pp. 1-51.

Kleinman, W. A. and Richie, I. P., Jr., 1995. Determination of thiols and disulfides using high-performance liquid chromatography with electrochemical detection. *J. Chromatog. B* 672, 73-80.

Knecht, R. and Chang, J. Y., 1986. Liquid chromatographic determination of amino acids after gas-phase hydrolysis and derivatization with (dimethylamino)azobenzene-sulfonyl chloride. *Anal. Chem.* 58, 2375-2378.

Kochhar, S. and Christen, P., 1988. The enantiomeric error frequency of aspartate aminotransferase. *Eur. J. Biochem.* 175, 433-438.

Kochhar, S. and Christen, P., 1989. Amino acid analysis by high-performance liquid chromatography after derivatization with 1-fluoro-2, 4-dinitrophenyl-5-L-ala–nine amide. *Anal. Biochem.* 178, 17-21.

Koop, D. R., Morgan, E. T., Tarr, G. E., and Coon, M. J., 1982. Purification and characterization of a unique isozyme of cytochrome P-450 from liver microsomes of ethanol-treated rabbits. *J. Biol. Chem.* 257, 8472-8480.

Labadarios, D., Shephard, G. S., Botha, E., Jackson, L., Moodie, I. M., and Burger, J. A., 1986. Determination of plasma amino acids by gas chromatography. *J. Chromatog.* 383, 281-295.

Lakritz, J., Plopper, C. G., and Buckpitt, A. R., 1997. Validated high-performance liquid chromatography-electrochemical method for determination of glutathione and glutathione disulfide in small tissue samples. *Anal. Biochem.* 247, 63-68.

Lapolla, A., Baldo, L., Aronica, R., Gerhardinger, C., Fedele, D., Elli, G., Seraglia, R., Catinella, S., and Traldi, P., 1994. Matrix-assisted laser desorption ionization mass spectrometric studies on protein glycation 2. The reaction of ribonuclease with hexoses. *Biol. Mass. Spectr.* 23, 241-248.

Lapolla, A., Fedele, D., Seraglia, R., Catinella, S., and Traldi, P., 1994. Matrix–assisted laser desorption/ ionization capabilities in study of non-enzymatic protein glycation. *Rapid Commn. Mass Spectr.* 8, 645-52.

Lee, J. and Harnly, J.M., 2005. Free amino acid and cysteine sulfoxide composition of 11 Garlic (*Allium sativum* L.. cultivars by gas Chromatography with flame Ionization and mass selective detection. *J. Agric. Food Chem.* 53, 9100-9104.

Lee, J., Finley, J. W. and Harnly J. M., 2005. Effect of selenium fertilizer on free amino acid composition of broccoli (*Brassica oleracea* Cv. Majestic. determined by gas chromatography with flame ionization and mass selective detection *J. Agric. Food Chem.* 53, 9105–9111.

Lennon, J. J. and Walsh, K. A., 1997). Direct sequence analysis of proteins by in-source fragmentation during delayed ion extraction. *Protein Sci.* 6, 2446?2453.

Lin, J. K. and Chang, J. Y., 1975. Chromophoric labeling of amino acids with 4-dimethylaminoazobenzene-4'-sulfonyl chloride. *Anal. Chem.* 47, 1634-1638.

Lin, S. H., Tornatore, P., and Weinberger, S.R., 2001). Limited acid hydrolysis as a means of fragmenting proteins isolated upon Protein Chip Array surfaces. *Eur. J. Mass Spectrom.* 7, 131?141.

Lindroth, P. and Mopper, K., 1979. High performance liquid chromatographic determination of subpicomole amounts of amino acids by precolumn fluorescence derivatization with o-Phthaldialdehyde. *Anal. Chem.* 51, 1667-1674.

Liu, H. and Worthen, H. G., 1992. Measurement of free amino acids levels in ultra filtrates of blood plasma by high-performance liquid chromatography with automatic pre-column derivatization. *J. Chromatog.* 579, 215-224.

Liu, H. J., 1994. Determination of amino acids by pre-column derivatization with 6-aminoquinolyl-N-hydroxysuccininidyl carbamate and high performance liquid chromatography with ultraviolet detection. *J. Chromatog.* 670, 59-66.

Lundell, N. and Schreitmuller, T., 1999. Sample preparation for peptide mapping–a pharmaceutical quality-control perspective. *Anal. Biochem.* 266, 31–47.

Macchi, F. D., Shen, F. J., Keck, R. G. and Harris, R. J., 2000. Amino acid analysis, using post column ninhydrin detection in a biotechnology laboratory. In: *Methods in Molecular Biology*, vol.159: Amino Acid Analysis protocols (Eds. Cooper, C., Packer, N. and Williams, K), Humana Press Inc., Totowa, NJ, pp 9-30.

MacKenzie, S. L., 1987. Gas chromatographic analysis of amino acids as the N–heptafluorobutyryl isobutyl esters. *J. Assoc. Off. Anal. Chem.* 70, 151-160.

MacKenzie, S. L., Tenaschuk, D., and Fortier, G., 1987. Analysis of amino acids by gas-liquid chromatography as tert.-butyldimethylsilyl derivatives. Preparation of derivatives in a single reaction. *J. Chromatog.* 387, 241–253.

Mahrenholz, A.M. *et al.*, 1996. In: *Techniques in Protein Chemistry VII* (D.R. Marshak, ed.. Academic Press, in press.

Malencik, D. A., Zhao, Z., and Anderson, S. R., 1990. Determination of dityrosine, phosphotyrosine, phosphothreonine, and phosphoserine by high-performance liquid chromatography. *Anal. Biochem.* 184, 353-359.

Malinow, M. R., Kang, S. S., and Taylor, L. M., 1989. Prevalence of hyperhomocysteinemia in patients with peripheral arterial occlusive disease. *Circ. Res.* 79, 1180-1188.

Mansoor, M. A., Svardal, A. M., and Ueland, P. M., 1992. Determination of the *in vivo* redox status of cysteine, cysteinylglycine, homocysteine, and glutathione in human plasma. *Anal. Biochem.*, 200, 218-229.

Markesbery, W. R., 1997. Oxidative stress hypothesis in Alzheimer's disease. *Free Rad. Biol. Med.* 23, 134-147.

Martinez del Pozo, A., Merola, M., Ueno, H., Manning, J. M., Tanizawa, K., Nishimura, K., *et al.*, 1989. Stereo specificity of reactions catalyzed by bacterial D-amino acid transaminase. *J. BioI. Chem.* 264, 17784-17789.

Masser, P. A., Taylor, L. M. D., and Porter, J. M., 1994. Importance of elevated plasma homocysteine levels as a risk factor for atherosclerosis. *Ann. Thorac. Surg. 58*, 1240-1246.

Matsumura, S., Kataoka, H., and Makita, M., 1995. Capillary gas chromatographic analysis of protein amino acids as their N (O, S)-isobutoxycarbonyl methyl ester derivatives. *Biomed. Chromatogr.* 9, 205-210.

Matsumura, S., Kataoka, H., and Makita, M., 1996. Determination of amino acids in human serum by capillary gas chromatography. *J. Chromatog.* B681, 375-380.

Mawhinney, T. P., Robinett, R. S. R., Atalay, A., and Madson, M. A., 1986. Analysis of amino acids as their tert-butyldimethylsilyl derivatives by gas chromatography and mass spectrometry. *J. Chromatog.* 358, 231-242.

Mee, J. M. L., 1973). Gas-liquid chromatographic analysis of glutathione. *J. Chromatog.* 87, 258-262.

Meyer, H. E., Eisermann, B., Heber, M., Hoffmann-Posorske, E., Korte, H., Weigt, C., *et al.*, 1993. Strategies for nonradioactive methods in the localization of phosphorylated amino acids in proteins. *FASEB J.* 7, 776-782.

Meyer, H. E., Swiderek, K., Hoffmann-Posorske, E., Korte, H., and Heilmeyer, L. M. G., Ir., 1987. Quantitative determination of phosphoserine by high-performance liquid chromatography as the phenylthiocarbamyl-S-ethylcycteine: application to picomolar amounts of peptides and proteins. *J. Chromatog.* 397, 113-121.

Molnár-Perl, I., 1997. Tryptophan analysis in peptides and proteins, mainly by liquid chromatography. *J. Chromatogr. A* 763, 1–10.

Molony, M. S., Wu, S. L., Keyt, L., and Hams, R. J., 1995. The unexpected presence of hydroxylysine in non-collagenous proteins, in *Techniques in Protein Chemistry VI* (Crabbe, J., ed.), Academic, San Diego, CA, pp. 91-98.

Monnier, V. M., 1989. *Toward a MaillardReaction Theory of Aging* (Baynes, J. W. and Monnier, V. M., eds.). Alan R. Liss, New York.

Monnier, V. M. and Cerami, A., 1983. Detection of nonenzymatic browning products in the human lens. *Biochem. Biophys. Acta* 760, 97-103.

Murthy, L. R. and Iqbal, K, (1991. Measurement of picomoles of phosphoamino acids by high-performance liquid chromatography. *Anal. Biochem.* 193, 299-305.

Newton, G. L., Dorian, R., and Fahey, R. C., 1981. Analysis of biological thiols: Derivatization with monobromobimane and separation by reversed-phase high performance liquid chromatography. *Anal. Biochem.* 114, 383-387.

Ng, L. T., Pascaud, A., and Pascaud, M., 1987. Hydrochloric acid hydrolysis of proteins and determination of tryptophan by reversed-phase high-performance liquid chromatography. *Anal. Biochem.* 167, 47-52.

Niedbalski, J. S. and Ringer, D. P., 1986. Separation and quantitative analysis of O-linked phosphoamino acids by isocratic high-performance liquid chromatography of the 9-fluorenylmethyl chloroformate derivatives. *Anal. Biochem.* 158, 138-145.

Nozal, M. J., Bernal, J. L., Toribio, L., Marinero, P., Moral, O., Manzanas, L., and Rodriguez, E., 1997. Detennination of glutathione, cysteine and N-acetylcysteine in rabbit eye tissues using high-performance liquid chromatography and postcolumn derivatization with 5, 5-dithiobis (2-nitrobenzoic acid). *J. Chromatog.* A, 778, 347-353.

Oh, C.–H., Kim, J.–H., Kim, K.–R., Brownson, D. M., and Mabry, T. J., 1994. Simultaneous gas chromatographic analysis of non-protein and protein amino acids as N (O, S)-isobutoxycarbonyl tert.-butyldimethylsilyl derivatives. *J. Chromatog.* 669, 125-137.

Ozols, J., 1990. Amino acid analysis, in *Methods in Enzymology,* vol. 182. Academic, San Diego, CA, pp. 587-601.

Packer, N. H., Lawson, M. A., Jardine, D. R., Sanchez, J. C., and Gooley, A. A., 1998. Analyzing glycoproteins separated by two-dimensiona1 gel electrophoresis. *Electrophoresis* 19, 981-988.

Paroni, R., DeVecchi, E., Cighetli, G., Arecelloni, C., Fermo, I., Grossi, A., and Bonini, P., 1995. HPLC with o-phthalaldehyde precolumn derivatization to measure total, oxidized, and protein-bound glutathione in blood, plasma and tissue. *Clin. Chem.* 41, 448-454.

Patterson, D. H., Tarr, G. E., Regnier, F. E. and Martin, S. A., 1995). C-terminal ladder sequencing via matrix-assisted laser desorption mass spectrometry coupled with carboxypeptidase time-dependent and concentration-dependent digestions. *Anal. Chem.* 67, 3971?3978.

Philpott, M. F. and Van, M. J., der Merwe (1991. Non-linear calibration of a nitrogen-phosphorus detector for the GC determination of amino acids. *Chromatographia* 31, 500-504.

Pickering, M., and Newton, P (1990. Amino acid hydrolysis: old problems, new solutions. *LC-GC Intl.* 3, 22–26.

Prabhakaram, M. and Ortwerth, B. J., 1994. Determination of glycation cross linking by the sugar-dependent incorporation of [14C] lysine into protein. *Anal. Biochem.* 216, 305-312.

Reed, D. J., Babson, J. R., Beatty, P.W., Brodie, A. E., Ellis, W. W., and Potter, D. W., 1980. High-performance liquid chromatography analysis of nanomole levels of glutathione, glutathione disulfide, and related thiols and disulfides. *Anal. Biochem.* 106, 55-62.

Reiber, D. C., Brown, R. S., Weinberger, S., Kenny, J., and Bailey, J., 1998). Unknown peptide sequencing using matrix-assisted laser desorption/ionization and in-source decay. *Anal. Chem.* 70, 1214?1222.

Richie, J. P. and Lang, C. A., 1987. The determination of glutathione, cysteine, and other thiols and disulfides in biological samples using high-performance liquid chromatography with dual electrochemical detection. *Anal. Biochem.* 163, 9-15.

Ringer, D. P., 1991. Separation of phosphotyrosine, phosphoserine, and phosphothreonine by high-performance liquid chromatography. *Methods Enzymol.* 201, 3-10.

Roth, M., 1971. Fluorescence reaction for amino acids. *Anal. Chem.* 43, 880-882.

Rowley, H. L., Martin K. F., and Marsden C. A., 1995. Determination of *in vivo* amino acid neurotransmitters by high-performance liquid chromatography with o-phthalaldehyde-sulphite derivatisation and electrochemical detection. *J. Neurosci. Methods* 57, 93-99.

Samejima, K., Dairman, W., and Udenfriend, S., 1971. Condensation of ninhydrin with aldehydes and primary amines to yield highly fluorescent ternary products. 1. Studies on the mechanism of the reaction and some characteristics of the condensation product. *Anal. Biochem.* 42, 222-236.

Sarwar, G. and Botting, H. G., 1993. Evaluation of liquid chromatographic analysis of nutritionally important amino acids in food and physiological samples. *J. Chromatog.* 615, 1-22.

Sass, J. O. and Endres, W., 1997. Quantitation of total homocysteine in human plasma by derivatization to its N (O, S)-propoxycarbonyl propyl ester and gas chromatography-mass spectrometry. *J. Chromatog. A* 776, 342-347.

Schauer, N, Zamir D. and Fernie, A.R., 2005. Metabolic profiling of leaves and fruit of a wild species tomato: a survey of the *Solanum lycopersicum* complex. *J. Exp. Bot.* 56, 298–307.

Schegg, K. M., Denslow, N. D., Andersen, T. T., Bao, Y. A., Cohen, S. A., Mahrenholz, A. M., and Mann, K., 1997. Quantitation and identification of proteins by amino acid analysis: ABRF-96 collaborative trial, in *Techniques in Protein Chemistry VIII* (Marshak, D., ed.), Academic, San Diego, CA, pp. 207-216.

Schneider, K., Neupert, M., Spiteller, G., Henning, H. V., Matthaei, D., and Scheler, F., 1985. Gas chromatography of amino acids in urine and haemofiltrate. *J. Chromatog.* 345, 19-31.

Schofield, J. D. and Chen, X., 1995. Analysis of free reduced and free oxidized glutathione in wheat flour. *J. Cereal Sci.* 21, 127-136.

Schuster, R., 1988. Determination of amino acids in biological, pharmaceutical, plant and food samples by automated precolumn derivatization and high-performance liquid chromatography. *J. Chromatog.* 431, 271-284.

Sherwood, R. A., Ritheradge, A. C., and Richards, D. A., 1990. Measurement of plasma and urine amino acids by high-performance liquid chromatography with electrochemical detection using phenylisothiocyanate derivatization. *J. Chromatog.* 528, 293-303.

Shevchenko, A. *et al.*, 1997. Rapid *de novo* peptide sequencing by a combination of nanoelectrospray, isotopic labeling and a quadrupole/time-of-flight mass spectrometer. *Rapid Commun. Mass Spectrom.* 11, 1015?1024.

Shevchenko, A., Loboda, A., Shevchenko, A., Ens, W., and Standing, K. G., 2000. MALDI quadrupole time-of-flight mass spectrometry: a powerful tool for proteomic research. *Anal. Chem.* 72, 2132?2141.

Shindo, N., Fujimura, T., Kazuno, N. S., Mineki, R., Furusawa, S., Sasaki, K., and Murayama, K., 1998. Identification of multidrug resistance protein 1 of mouse leukemia P388 cells on a PVDF membrane using 6-aminoquinolyl-carbamyl (AQC)–amino acid analysis and World Wide Web (WWW)-accessible tools. *Anal. Biochem.* 264, 251-258.

Shindo, N., Nojima, S., Fujimura, T., Taka, H., Mineki, R., and Murayama, K., 1997. Separation of 18 6-aminoquinolyl-carbamyl-amino acids by ion-pair chromatography. *Anal. Biochem.* 249, 79-82.

Siller-Cepeda, J. H., Chen, T. H. H and Fuchigami, L. H., 1991. High performance liquid chromatography analysis of reduced and oxidized glutathione in woody plant tissues. *Plant Cell Physiol.* 32, 1178-1185.

Singh, A. K. and Ashraf, M., 1988. Analysis of amino acids in brain and plasma samples by sensitive gas chromatography-mass spectrometry. *J. Chromatog.* 425, 245-255.

Smith, J. B., Sun, Y., Smith, D. L., and Green, B., 1992. Identification of the post–translational modifications of bovine lens alpha B-crystallin by mass spectrometry. *Protein Sci.* 1, 601-608.

Stabler, S. P., Marcell, P. D., Podell, E. R., and Allen, R. H., 1987. Quantitation of total homocysteine, total cysteine, and methionine in normal serum and urine using capillary gas chromatography-mass spectrometry. *Anal. Biochem.* 162, 185-196.

Staffeldt, B., Brockmoller, J., and Roots, I., 1991. Determination of S-carboxymethyl-L-cysteine and some of its metabolites in urine and serum by high performance liquid chromatography using fluorescence pre-column labelling. *J. Chromatog.* 571, 133-147.

Stampfer, M. J. and Malinow, M. R., 1995. Can lowering homocysteine levels reduce cardiovascular risk? *N. Engl. J. Med.* 332, 328-329.

Stein, S., Bohlen, P., Stone, J., Dairman, W., and Udenfriend, S., 1973. Amino acid analysis with fluorescamine at the picomole level. *Arch. Biochem. Biophys.* 155, 202-212.

Strydom, D. J. and Cohen, S. A., 1994. Comparison of amino acid analyses by phenylisothiocyanate and 6-aminoquinolyl-N-hydroxysuccinimyl carbamate precolumn derivatisation. *Anal. Biochem.* 222, 19-28.

Strydom, D.J. and Cohen, S. A., 1993. Sensitive analysis of cystine/cysteine using 6-aminoquinolyl-N-hydroxysuccinimidyl carbamate derivatives, in *Techniques in Protein Chemistry IV* (Angeletti, R. H. ed.), Academic, San Diego, CA, pp. 299–306.

Sypniewski, S. and Bald, E., 1994. Ion-pair high-performance liquid chromatography of cysteine and metabolically related compounds in the form of their S–pyridinium derivatives. *J. Chromatog. A* 676, 321-330.

Sze, S. K., Ge, Y., Oh, H., and McLafferty, F. W., 2002. From the cover: Top-down mass spectrometry of a 29-kDa protein for characterization of any post-translational modification to within one residue. *Proc. Natl. Acad. Sci. USA* 99, 1774?1779.

Szokan, G., Mezo, G., Hudecz, F (1988. Application of Marfey's reagent in racemization studies of amino acids and peptides. *J. Chromatog.* 444, 115-122.

Szokan, G., Mezo, G., Hudecz, F., Majer, Z., Schon, I., Nyeki, O., *et al.*, 1989. Racemization analyses of peptides and amino acid derivatives by chromatography with pre-column derivatization. *J. Liq. Chromatogr.* 12, 2855-2875.

Takagi, K., Kataoka, H., and Makita, M., 1995. Determination of glutathione and and related aminothiols in mouse tissues by gas chromatography with flame photometric detection. *Biosci. Biotech. Biochem.* 60, 729-731.

Tapuhi, Y., Schmidt, D. E., Lindner, W., and Karger, B. L., 1981. Dansylation of amino acids for high-performance liquid chromatography analysis. *Anal. Biochem.* 115, 123-129.

Tarentino, A. L., Gomez, C. M., and Plummer, T.H., 1985. Deglycosylation of asparagine-linked glycans by peptide: N-glycosidase. *Biochemistry* 24, 4665–4671.

Tran, A. D., Blanc, T., and Leopold, E. J., 1990. Free solution capillary electrophoresis and micellar electrokinetic resolution of amino acid enantiomers and peptide isomers with L–and D-Marfey's reagents. *J. Chromatog.* 516, 241-249.

Tsai, L. B., Lu, H. S., Kenney, W. C., Curless. C. C., Klein, M. L., Lai, P.H., *et al.*, 1988. Control of misincorporation of *de novo* synthesized norleucine into recombinant interleukin-2 in *E. coli. Biochem. Biophys. Res. Commun.* 156, 733-739.

Tsuge, K., Kataoka, M. and Seto, Y., 2002 . Determination of *s*-methyl-, *s*-propyl-, and *s*-propenyl-L-cysteine sulfoxides by gas chromatography-mass Spectrometry after *tert* Butyldimethylsilylation. *J. Agric. Food Chem.* 50, 4445–4451.

Tsugita, A., Takamoto, K., Kamo, M., and Iwadate, H., 1992). C-terminal sequencing of protein. *Eur. J. Biochem.* 206, 691?696.

Turnell, D. C. and Cooper, J. D., 1982. Rapid assay for amino acids in serum or urine by pre-column derivatization and reversed-phase liquid chromatography. *Clin. Chem.* 28, 527-531.

Ubbink, J. B., Vermaak, W. I. H., and Bissbort, S., 1991. Rapid high–performance liquid chromatographic assay for total homocysteine levels in human serum. *J. Chromatog.* 565, 441-446.

Uhe, A. M., Collier, G. R., McLennan, E. A., Tucker, D. J and O'Dea, K., 1991. Quantitation of tryptophan and other plasma amino acids by automated pre-column o-phthaldialdehyde derivatization high-performance liquid chromatography: improved sample preparation. *J. Chromatog.* 564, 81-91.

Vacca, R. A., Giannattasio, S., Graber, R., Sandmeier, E., Marra, E., and Christen, P., 1997. Active-site Arg?Lys substitutions alter reaction and substrate specificity of aspartate aminotransferase. *J. Biol. Chem.* 272, 21932-21937.

Valles, B. S., García, N. P., Madrera, R. R. and Lobo, A. P., 2005. Influence of yeast strain and aging time on free Amino acid changes in sparkling ciders. *J. Agric. Food Chem.*, 53, 6408-6413.

Van Wandelen, C. and Cohen, S. A., 1997. Using quaternary high-performance liquid chromatography eluent systems for separating 6-aminoquinolyl-N-hydroxysuccinimidyl carbamate-derivatized amino acid mixtures. *J. Chromatog.* 763, 11-22.

Vorm, O. and Roepstorff, P., 1994). Peptide sequence information derived by partial acid hydrolysis and matrix-assisted laser desorption/ionization mass spectrometry. *Biol. Mass Spectrom.* 23, 734–740.

Walton, D. J. and McPherson, J. D., 1987. Analysis of glycated amino acids by high-performance liquid chromatography of phenylthiocarbamyl derivatives. *Anal. Biochem.* 164, 547-553.

Wathelet, B (1999. Nutritional analyses for proteins and amino acids in beans (*Phaseolus* sp.. *Biotechnol. Agron. Soc. Environ.* 3, 197–200

Weckwerth, W., Wenzel, K. and Fiehn, O., 2004. Process for the integrated extraction, identification and quantification of metabolites, proteins and RNA to reveal their co-regulation in biochemical networks. *Proteomics* 4, 78-83.

Weiss, M., Mannenberg, M., Juranville J. F., Lahm, H W., and Fountoulakis, M., 1998). Effect of the hydrolysis method on the determination of the amino acid composition of proteins. *J. Chromatogr. A* 795, 263-275.

West, K. A. and Crabb, J. W., 1992. Application of automatic PTC amino acid analysis, in *Techniques in Protein Chemistry III* (Angeletti, R. H., ed.), Academic, London and New York, pp. 233-242.

Williams, A. P., 1987. General problems associated with the analysis of amino acids by automated ion-exchange chromatography. *J. Chromatog.* 373, 175-190.

Williams, K. R. and Stone, K. L., 1995. In: *Techniques in Protein Chemistry VI* (J.W. Crabb, ed.. Academic Press, pp. 143-152.

Winters, R. A., Zukowski, J., Ercal, N., Mattews, R. H., and Spitz, D. R., 1995. Analysis of glutathione, glutathione disulfide, cysteine, homocysteine, and other biological thiols by high-performance liquid chromatography following derivatization by N–(l-pyrenyl)maleimide. *Anal. Biochem.* 227, 14-21.

Woo, K. L. and Ahan, Y. K., 1996. Determination of protein amino acid as benzylthiocarbamyl derivatives compared with phenylthiocarbamyl derivatives by reversed-phase high-performance liquid chromatography, ultraviolet detection and precolumn derivatization. *J. Chromatog.* 740, 41-50.

Woo, K. L. and Lee, D. S., 1995. Capillary gas chromatographic determination of proteins and biological amino acids as N (o)-tert-butyldimethysilyl derivatives. *J. Chromatogr. B.* 665, 15-25.

Woo, K. L. and Lee, S. H., 1994. Determination of protein amino acids as butyl–thiocarbamyl derivatives by reversed-phase high-performance liquid chromatography with precolumn derivatization and UV detection. *J. Chromatog A.* 667, 105-111.

Woo, K. L. Hwang, Q. C., and Kim, H. S., 1996. Determination of amino acids in the foods by reversed-phase high-preformance liquid chromatography with a new precolumn derivative, butylthiocarbamyl amino acid, compared to the conventional phenylthiocarbamyl derivatives and ion-exchange chromatography. *J. Chromatog A.* 740, 31-40.

Worthen, H. G. and Liu, H., 1992. Automatic pre-column derivatization and reversed-phase high performance liquid chromatography of primary and secondary amino acids in plasma with photo-diode array and fluorescence detection. *J. Liq. Chromatogr:* 15, 3323-3341.

Wu, G. and Furlanut, M., 1997. Separation of DL-dopa by means of micellar electrokinetic capillary chromatography after derivatization with Marfey's reagent. *Pharmacol. Res.* 35, 553-556.

Yamada, H, Moriya, H., and Tsugita, A., 1991. Development of an acid hydrolysis method with high recoveries of tryptophan and cysteine for microquantities of protein. *Anal. Biochem.*198, 1-5.

Yan, C. C. and Huxtable, R. J., 1995. Fluorimetric determination of monobromobimane and o-phthalaldehyde adducts of g-glutamylcysteine and glutathione: application to assay of y-glutamylcysteine synthetase activity and glutathione concentration in liver. *J. Chromatog. B* 672, 217-224.

Yang, C.S., Tsai, P.J., Chen, W. Y., Liu, L., and Kuo, J. S., 1995. Determination of extracellular glutathione in livers of anaesthetized rats by microdialysis with online high-performance liquid chromatography. *J. Chromatog.* 667, 41-48.

Yeung, J. M., Baker, G. B., and Coutts, R. T., 1986. Simple automated gas chromatographic analysis of amino acids and its application to brain tissue and urine. *J. Chtomatog.* 378, 293-304.

Yiiksel, K. U., Andersen, T. T., Apostol, I., Fox, J. W., Crabb, J. W., Paxton, R. J., and Strydom, D. J., 1994. The hydrolysis process and the quality of amino acid analysis: ABRF-94AAA collaborative trial, in *Techniques in Protein Chemistry VI* (Crabb, J.W., ed.), Academic, San Diego, CA, pp. 185-192.

Yokote, Y., Murayama, A., and Akahane, K., 1985. Recovery of tryptophan from 25-minute acid hydrolysates of protein. *Anal. Biochem.* 152, 245-249.

Yoshida, T., 1996. Determination of reduced and oxidized glutathione in erythrocytes by high-performance liquid chromatography with ultraviolet absorbance detection. *J. Chromatog. B* 678, 157-164.

Zhang, L. and Denslow N., 2000. Purification of proteins using ultramacrospin columns or prosorb sample preparation cartridges for amino acid analysis. In: *Methods in Molecular Biology*, vol.159: Amino Acid Analysis protocols (Eds. Cooper, C., Packer, N. and Williams, K), Humana Press Inc., Totowa, NJ, pp 31-37.

Zhong, H, Zhang, Y, Wen, Z and Li, L., 2004). Protein sequencing by mass analysis of polypeptide ladders after controlled protein hydrolysis. *Nat. Biotechnol.* 22, 1291–1296.

Zhou, H., Watts, J. D., and Aebersold, R., 2001. A systematic approach to the analysis of protein phosphorylation. © 2001 Nature Publishing Group http://biotech.nature.com

Zubarev, R. A., Chivanov, V. D., Hakansson, P. and Sundqvist, B. U. R., 1994). Peptide sequencing by partial acid hydrolysis and high-resolution plasma desorption mass spectrometry. *Rapid Commun. Mass Spectrom.* 8, 906?912.

Zumwalt, R. W., Desgres, J., Kuo, K. C., Pautz, J. E., and Gehrke, C. W., 1987. Amino acid analysis by capillary gas chromatography. *J. Assoc. Off. Anal. Chem.* 70, 253-262.

Chapter 37

An Existing and Futuristic Studies of Neem in 21st Century

Ashutosh Isaiah and Manoj S. Paul
Department of Botany, St. John's College, Agra – 282 002

There are more than 18 million Neem trees in India with yield potential of about 414000 metric tones of seeds per year, which can yield 83000 metric tones of oil and 3,30,000 metric tones of cake. Both oil and cake possess chemical which have diverse biological effects (antifeedant, growth disruption (IGR) oviposition deterrent, sterilant, repellent, direct toxicity etc.) on insect-pest. Thus, enough material is available for utilization.

Neem is a deciduous tree, grows faster, and often reaches a height of 25–40 meters. The leaves are crowded at the end of each branch. The flowers are small having honey like scent that attracts many bees. Its fruit is one seeded drupe.

The very special properties of the neem tree were recognized in the Indian subcontinent from ancient times, and western scientists through various reports knew these properties in this century. There was a botanical link recognized, too between neem (*Azadirachta indica*, A juice or media azadirachta) and Persian lilac or Chinaberry (*Melia azedarach*). This latter plant was well known in north Africa to be resistant to locust attack, as neem was in India.

Neem tree is a part of Indian history, tradition and culture. Widely used and recommended in folklore medicines sicne time immemorial, modem research has rediscovered it to be a rich mine of natural pesticides. Neem is one of such versatile trees which is a natural harmless pesticide and has found multiple uses. The importance of this tree is further supported by the fact that at least four international Neem conferences have been held at Roltach Egem, Germany (1980), at Rouischholzhausen, Germany (1983), at Nairobi, Kenya (1986) and at Banglore India, 1993 and two more have been held in the United States.

Margosa tree, Indian Lilac or Azadirachta from the Persian Azad-darakhat-e-hind the later name, the earliest mention of which was Abu Mansour in 970 A.D., "The free tree of India" in Persian, Urdu and Hindustani. The Neem-a native of Indo-Pakistan region, is a fast growing, evergreen and hardy tree, thrives were under semi-arid to sub humid conditions, and is also introduces into warm and dry areas where annual rain fall is less than 500 mm.

Neem By-Products

The history of commercial uses of neem tree is shrouded in the mystery and lore of vedic period of India, which began about more than 4000 years B.C. In Ayurvedic medicines neem bark and leaves were used in skin troubles, flowers as tonic stomachic and fruits as purgative and emollient.

Neem Bark

The bark is a good bitter tonic, astringent and anti periodic. It is said to contain a resinous bitter principle and is usually prescribed in the form of a tincture or an infusion. It is also regarded as beneficial in malarial fever the bark is consider useful in skin diseases too.

Some toothpaste in Europe and India contains extract of Neem bark.

Neem Leaves

The leaves are bitter and have a faint, but characteristic unpleasant smell. Leaves are dried in shade before use as moth repellant. The leaves are applied to boils in the form of poultice, and a decoction is recommended in ulcers and eczema. The dry flowers are considered tonic and stomachic. The berries are regarded as purgative, emollient and anthelmintic. It has been shown that the juice of leaves and the oil do not have any anthelmintic properties.

Neem leaves are used as a component of various cosmetics and extracts of leaves for hair and skin lotions in (Germany).

Neem Oil

The seeds of neem are reported to contain upto 45 per cent oil. Neem oil is used against stomach ulcers, worm infection and rheumatism. The neem oil used in manufacture of soap. Medicmall medicated soap (Margo) with the odour of neem is available in the market. In India, Sri Lanka, Pakistan people use Neem oil in hair to kill lice.

Neem Cake

It is the residue of kernel after extraction of oil. It is rich in nutrients than other cakes, contents being N 5 per cent, P2 Os 2 per cent and potash 10 per cent.

Neem Gum

The gum, which exuded from the bark, occurs in the form of clear bright amber coloured tears or fragments. It is found mixed with gum ghatti. It contents: water 1315 per cent, ash 3 per cent, glactants 12 per cent, pentosans 26 per cent, some albumins and oxidase.

Neem Today

A sweet liquid neem today is occasionally obtained as an exudation from the upper parts of some trees. It has the unpleasant odour of neem and contains sugars 6.5 and albuminous and gummy matter, about 6.5 per cent.

Neem Wood

The wood is moderately heavy, medium textured and narrowly interlocked-grained (8p. gr. 0.68: air dry wt., 44 lb/ft³). The sapped is grayish-white. The heartwood is red when first exposed, and later turns reddish brown. It is not very lustrous and is slightly aromatic. Anatomically, it is featured by distinct growth rings, medium sized vessels occluded by reddish brown gum parenchyma cells with deposits of gum, and fine rays.

Neem Timber

The timber seasons well and are durable even in exposed situation. It is not attract by white ants, perhaps due to its bitter character. It is not difficult to saw. It can be worked both by hand and machine and is suitable for carving but it does not take polish well. The timber used in house building and for making boards, panels, toys and ploughs.

Mode of Action

Neem causes extensive deleterious effects on target pests. Azadirachtins affect insect-pests chiefly in four modes antifeedents action, insect growth regulation, sterilant action and incapicitation action. The insect growth regulatory action amongst them is universal to all insect-pests affected by neem and dosages are uniformly reproducible. It is also noteworthy that development of pest resistance against neem has not been reported so far. The apparent reason due to its multi-pronged attack.

Another highlight of neem pesticides is their selectivity and safety to non-target beneficial organism like biological nitrogen fixer in soil, honeybee, insect predator and parasitoids etc. These product are also compatible with well known biopesticides such as NPV, Bt etc.

Neem Research

In India, research on neem has been both sporadic and scattered. Through pioneering work was done on commercial exploitation of neem oil and neem cake, the center of neem research has since shifted to other countries like Japan, UK, USA and Germany.

The institution involved in this work are Indian Institute of Science, Bangalore, Indian Agricultural research Institute (IARI), New Delhi, Central Tobacco Research Institute (CTRI) Rajahmundary (Andhra Pradesh), National Institute of Immunology, and Indian School of mines, Dhanbad (Bihar). The scientists at National Institute of Immunology have found a potent contraceptive in neem oil, called "Praneem". This seems to have promising future. At the Indian Institute of Mines, Dhanbad studies on corrosion inhibiting characteristics of alcoholic neem extract on mild steel have been done. The inhibiting efficiency of the extract is found to be 93 to 97 per cent at pH 1.0.

In Indian several companies have entered into trade of neem based pesticides while big business houses like TOMCO and SPIC are cashing in on export market, companies like ITC (Wellgro), West Coast Herbochem Ltd (Neemark), Gharda Chemicals (Neemguard), T-Staine's (Nimbicidin); Sri Bio Multitech Pvt. Ltd. (Limonol) and others are trying to capture local markets for their products. It is estimated that they have sold neem based pesticide worth rupees five crores in 1992.. Some companies are manufacturing pesticides by the name of Bioneem, Azatin Turplex and Margosam from Margosa oil. Some of the recent commercial preparations are:

Nimbola

A preparation based on neem oil in gelation capsules is begin manufactured by M/s Keepharma, B-34, Mayapuri Industrial Area, Phase-I, New Delhi – 110 064, Indian, as an oral treatment for diabetes control. No side effect are so far reported.

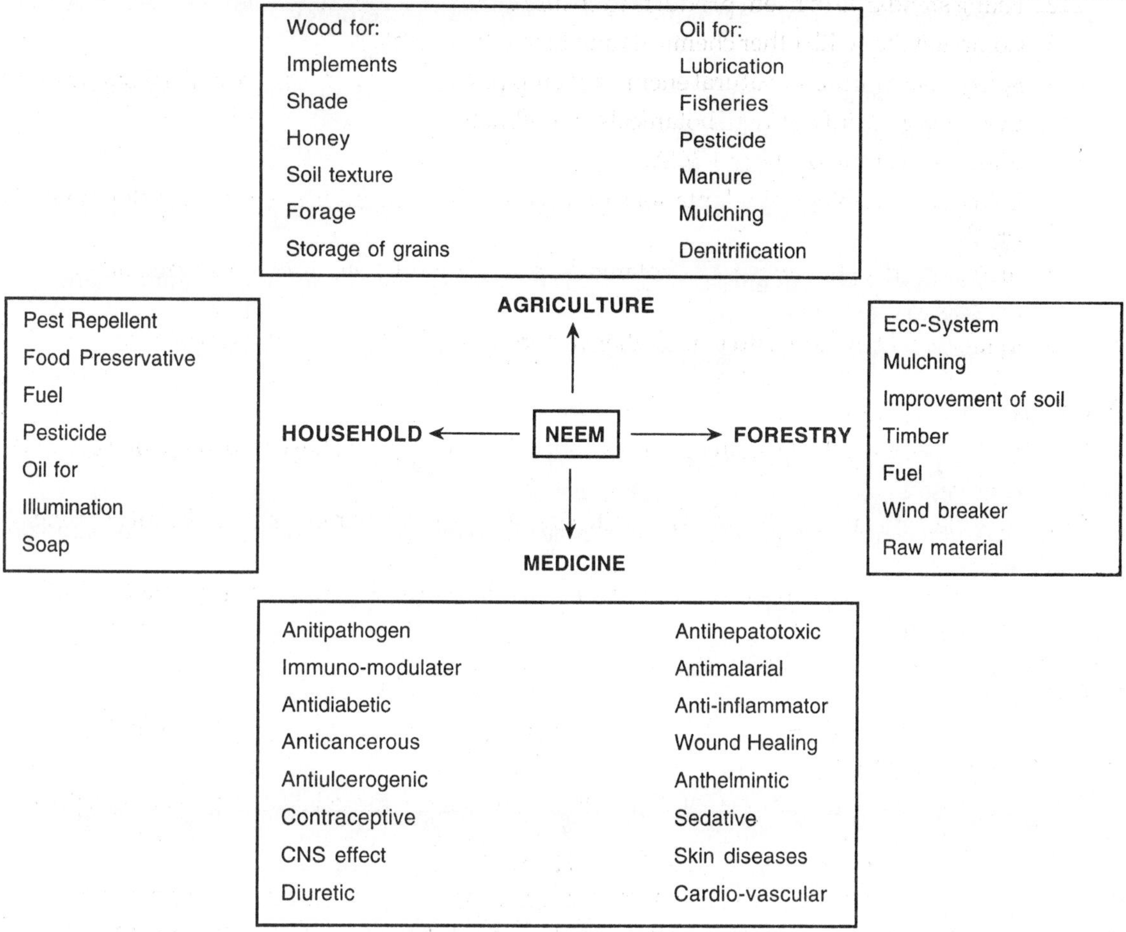

Figure 37.1: Activity of Neem

Sensal

A herbal, medicine, liquid vaginal contraceptive based on neem oil is being manufacture by M/s Excelsior Enterprises, B-12, A.M. Jaipuria Road Cantt., Kanpur – 208 004, U.P., India. It is claimed to be an antiseptic and also tones up the vaginal. No contra-indications are reported.

Future Research

Neem seed kernel extract and neem oil checked stem fly, white fly and pod borers of black gram. It was decided to take up research on following aspects:

Insect and Mite Pests and Vectors

1. Improving stability and development of stable formulations with emphasis on cost and accessibility.

2. Fixing standards in neem products to ensure quality and efficacy.
3. Compatability with other chemicals and biocontrol agents.
4. Safety of botanicals to natural enemies of crop pests, fishes and other beneficial organisms.
5. Evaluating the influence of botanicals on pollinators.
6. Effect on pests of azolla and BGA.
7. Effect on physiological adaptations of insects to neem products and on development of resistance.
8. Increasing the effectiveness of botanicals through emulsifier, adjuvants, antioxidant, UV protectants etc.
9. Management of mosquitoes and other vectors in rice fields with botanicals.

Diseases

1. Use of neem products, managements of rice disease and post harvest disease of fruits and vegetables.
2. Role of neem cake in slow release of nitrogen and consequent reduction in severity of fungal diseases.
3. Role of neem cake in prevention of germination failure and as a food base for multiplication of antagonists.

Nematodes

1. Evaluation of neem products on nematodes infesting rice, cotton, vegetables, black gram, banana, turmeric.
2. Influence of neem cake in promotion antagonistic fungi and predators of parastic nematodes.

Table 37.1: Board Effects of Neem Products on Insects

Neem Products	*Type of Bioactivity*
Seed /Kernel/oil extracts	Insect antifeedant, metamorphosis and growth disruption, oviposition deterrent, inhibition of chitin synthesis, nematicidal, insecticidal, nitrification, inhibition, antifungal to certain human pathogens, antiviral, blocking the molting of larvae of nymphs, spermicidal and antiembryonic implantation agents etc.
Leave/bark/twigs/flowers	Nematicidal, antimicrobial, pathogen inhibitors, repellent.

Neem and Environments

The neem trees is an excellent air purifier. Various parts of this tree are said to be magical due to their wonderful health remedies. The scope of neem has been expanded to cover its application in Ayurvedic medicine, Agriculture, Poultry and Animal husbandry. Some of the medicine values of the neem products with reference to its parts are as follows:

The seeds have emetic, laxative and anithelminitic properties. A Pungent, bitter, and yellow coloured oil comprising mainly the bitter components obtained from the seeds is applied as an antiseptic dressing in leprosy and in chronic skin diseases. Neem soaps are prepared from this oil.

Scientific investigations on the immunomodulatory contraceptive and antimicrobial activities of neem oil have been undertaken. Purified extract of neem oil (Praneem) along with to other active components of herbal origin have been formulated as "Praneem" polyherbal herbal cream which is under clinic trials for contraception and in the treatment of cervicities/vaginitis caused by various genital pathogens. The seed extract are known to activate the local cell mediated immune reactions after a single intrauterine administration and this leads to a long term reversible block of male fertility. Scientist at India's Defence Institute of Physiology an Allied Sciences (DIPAS) have isolated a neem oil extract (Nim-76) which can be refined in to new birth control with strong spermicidal activity. The volatile constituents identified from neem oil mainly comprise the sulphurous compounds.

The leaves are carminative, expectorant, anthelminic, antidotal, diuretic and insecticidal. The fresh juice of leaves is given in case of intestinal worms and with honey the juice is prescribed for jaundice and skin disease. An infusion of fresh leaves is a bitter vegetable tonic and alternative especially in chronic malarial fevers, because of its action on liver. Externally the leaves are applied over the skin diseases as a disutient, stimulant, antiseptic and are especially used for boils, chronic ulcers, eruptions of small pox, syphilitic sores, glandular swellings, wounds etc. Aqueous extract of leaves administered intravenous to dogs decreased blood sugar leaves and prevented adrenaline as well as glucose hyperglycemia.

The neem twigs are used as toothbrush and supposed to be excellent for protection of teeth and gums. The bark is also a bitter tonic, astringent alternatively is has anthelmintic, antispasomodic and stimulant properties. The root bark has the same properties as the stem bark. The neem-bark is a stimulant and demulcent tonic. The fermented sap that exudes from the trunk is a refrigerant nutrient, stomachic and alternative to tonic. Glycoproteins containing carbohydrate and protein (19 : 18) have been isolated from "neem-gum". It contains manaose, glucosamine, arabinose, galactose, fructose, xylose and glucose in molar ratio of 4 : 3, 3 : 2, 2 : 1, 1 : 1. Two related arabinofucoglucans Gla and Gla are isolated from the bark, both composed of main chain of repeating units (1-4 linked glucopyranose units with side chain of L-arabinofuranosyl units). Clinical activity of nimbidin for the pain in epigastric region due to ulcer in duodenum was taken up and it was fount that out of 13 patients ulcer was healed in three cases and five had fair response and the remaining five had only poor response. "Nimbatiktam"–a drug containing terpenic ester. "Nimbidin" (1.2 per cent w/w) showed significant ulcer healing activity with no side effects. The neem is supposed to be useful in the treatment of AIDS and Cancer.

Not only in India but in United states also neem products are used commercially. At specific departmental stores one can look for neem brush and neem soap. The Environmental Protection Agency (EPA) approved commercial neem based insecticide Margosan-OR and Align(R) for use on crops. Two related neem formulations, Bio Neem(R) for the consumer market for lawn and turf care are available. The other products Azatin(R) and Turplex(R) have also entered the United States insecticide market.

Conclusion

Now-a-days, when ecological and toxicological aspects are of steadily increasing interest in integrated pest management (IPM), it is of special importance that azdirachtin and related compounds from Halic. *Indica* have very low toxicity or are practically non-toxic to warm blooded organism including man. Further development of resistance of insects to insecticides is today a serious and steadily worsening problem of pest control when using synthetic products and certain biopesticides.

Neem-sounds like a designer tree, the brain chilled of genetic engineer for developing countries. Almost every part of tree-leaves, flowers fruits, bark and seeds is utilized as a pesticides, insecticides, medicine, diabetic food, mosquito repellant; animal food, and also to produce soap, lubricants, paper, briquettes, fuelwood, gums, plough, tooth paste, tooth brush, and even a contractive. The neem is eco-friendly bio pesticides compatible in IPM. One day, nature's bitter tasting neem will be a boon to farmers in developing and developed world.

References

Attri, B. S. and Ravi, P. (1980). Neem oil extractive - an effective larvaecide. *Indian J. Ent.*, **42**, : 371.

Devakumar, C., Saxena, U. S. and Mukerjee, S. K. (1986). Evaluation of neem (*Azadirachta indica* A. Juss.) limmonoids and azadirachtin against safflower aphid (*Dactynotus carthami* HRL). *Indian J. Ent.*, **48**: 467-470.

Krishnaiaha, N. V., Kalode, M.V. and Passalu, I. C. (1990). New approaches in utilization of botanicals in rice insect pest control. *In* : Botanical Pesticides in Integrated Pest management, (*Eds*. M. S. Chari and G. Ramprasad), Indian Society of Tobbaco Science, Rajahmundry, pp.203-216.

Sardana, H. R., Bambawale, O. M., Singh, D. K. and Kadu, L. N. (2005). A novel repellent action of neem cake against rabbit foraging on brinjal (Solanum melongena l.) crop. *Pestic. Res. J.*, 17 (1)23-24.

Sardana, H. R., Bambawale, O. M., Singh, D.K. and Kadu, L. N. (2008). Larger area validation of adaptable integrated pest management technology in okra (*Abelmoschus esculentus* Moench) through farmers' participation. *Indian J. Agric. Sci.*, **78** (12): 1063-66.

Srivastava, C. P., Singh, O. P.and Singh, N. K. (1994).Testing of neem derivatives with some commonly used insecticides to control pod borer in chickpea. Pestology, **18** (8)36-39.

Chapter 38

Pests of Irrigated Oil Palm and their Management

P. Kalidas
National Research Centre for Oil Palm,
Pedavegi – 534 450, Andhra Pradesh

Introduction

Oil palm (*Elaeis guineensis* Jacquin), which is the best source for high productivity of vegetable oil throughout the world, is grown in large areas of India to overcome the edible oil shortage. Commercial cultivation of oil palm was started in Kerala during 1971, but could pick up momentum only from 1989. About 0.796 million ha of area in eleven states in the first instance in 1986 followed by 10.00 million ha in 2006 has been found potential for oil palm cultivation in India (Anon., 1988 and Chadha, 2006). The incidence of pests and diseases is of low to very low in most of the areas, except in few pockets where the areas are reported as hot zones for certain pests and diseases of other palmaceae palms.

Although oil palm is an imported crop to India, the pest incidence on various parts of palm has been found increasing as the age of the palms is increased. As the chances of introduction of foreign pests is remote due to the strict quarantine measures that are being carried out at various ports of import of seed sprouts, the only chance for the pest out break is the migration of pests from other crops that are adjacent to the oil palm plantations mainly due to the continuous availability of food material (Kochu Babu and Kalidas, 2004). The present spacing of 9 meters flowed by flood irrigation were also found to enhance the pest incidence in the aged gardens. The leaves of the adjacent palms are overlapping each other arresting the sunrays falling down to the soil creating the microclimate suited for pest growth. With tne flood irrigation the humidity is also increased making congenial conditions for the survival of pest population. Similarly the cultural practices that are being followed in farmers' fields are also found one of the reasons for the pest out breaks.

In this paper a brief account on pests of nursery, young palms of juvenile stage and adult palms is presented based on the survey carried out during the last 12 years in different cultivated areas. The pests are further classified as root feeders, bunch feeders, defoliators and dry matter foragers. Of these, insect pests are found to be indirect damagers compared to avian and mammalians, which cause direct damage by feeding on the fruit, bunches. Mortality of the palm due to these pests is very rare and losses occurred is very negligible.

Pests of Nursery

On the nursery plants, spindle bug, *Carvalhoia arecae* Miller and China (Hemiptera: Miriidae) (Dhileepan, 1992; Nair and Daniel, 1982), Tussock caterpillar, *Dasychira mendosa* Hb. (Lepidoptera: Lyrnantridae) (Wood, 1968; Dhileepan, 1992), *Sesamia inferens* Walker (Lepidoptera: Noctuidae) (Jacob and Kochu Babu, 1995), Psychid, *Metisa plana* Walker (Psychidae: Lepidoptera), Tobacco caterpillar, *Spodoptera litura* (Lepidoptera: Noctuidae), Black slug: *Laevicaulus alte*, Termites: *Odontotermes* sp. (Isoptera: Termitidae) (Kalidas *et al.*, 2002, Kalidas *et al.*, 2006) and grass hoppers were reported in various nurseries as defoliators causing stunted growth and ultimately long duration for readiness for transplantation. Cockchafer beetles (Root grub), *Apogonia/Adoretus* spp. (Scarabaeidae: Coleoptera), wild boar, *Sus scrofa* and Indian Squirrel were the major root feeders on nursery plants causing mortality of the plants. The economic losses due the root feeders are very heavy compared to defoliators as they cause mortality to the plants.

Among the various foliage feeders, spindle bug whose damage symptoms are seen on the tender spear leaves with typical linear brown lesions that turn to brown, dry up forming numerous holes on the leaves and fails to open the spindle leaves fully in the advance stage of infestation, is mostly observed in cool climate areas of Kerala and Karnataka migrating from areca nut which is cultivated in large areas. While the tussock caterpillar, *Dasychira mendosa* Hb. (Lepidoptera: Lyrnantridae) whose identity is peculiar with dense tufts of hairs on the body makes significant damage to nursery seedlings in Kerala, Karnataka, Andhra Pradesh and Tamil Nadu. In recent years, the sorghum shoot borer, *Sesamia inferens* Walker (Lepidoptera: Noctuidae) is gaining importance with its incidence observed in both primary and secondary nurseries causing holes on the leaves of primary nursery and dead heart and little leaf symptoms on secondary nursery plants of Andhra Pradesh, Karnataka and Kerala. Incidence of Psychid, *Metisa plana* Walker (Psychidae: Lepidoptera) with brown patches and holes on leaves is a common occurrence in the nursery seedlings where management practices are not followed at regular intervals. While Tobacco caterpillar, *Spodoptera litura* (Lepidoptera: Noctuidae) incidence is seen in the nurseries that are raised adjacent to maize / tobacco fields. While the above may cause low to moderate incidence, Black slug, *Laevicaulus alte* could cause significant infestation and mortality to the primary nursery in Andhra Pradesh by feeding on plumule region of the one day old planted sprout. Nurseries that are raised in termite inhabited areas coupled with poor maintenance and lack of irrigation were subjected to severe incidence of termite attack in Visakhapatnam district of Andhra Pradesh even though termites in general don't feed on the green tissues.

Of the important root feeders reported on oil palm nursery plants, Cockchafer beetles (Root grub) *Apogonia/Adoretus* spp. (Scarabaeidae: Coleoptera), Black rat, *Rattus rattus* Wroughtoni, Wild boar, *Sus scrofa*: and Indian squirrel are the important ones that cause low to moderate incidence. While the incidence of root grubs was observed more in red soils, the incidence of rats was observed in black soils and poorly managed nurseries. The problem of wild boar was reported mostly from nurseries that are adjacent to forest areas whereas Indian squirrel was observed in all most all the areas.

Apart from above Leaf hopper, *proutista moesta* Westwood (Derbidae: Homoptera) was also reported feeding on nursery plants in Kerala. Even though it has not been reported causing any physical damage, it has been proved as carrier of many viral/MLO diseases particularly the Spear rot (Kochu Babu, 1993; Kochu Babu, 2002; Shanta *et al.*, 1970).

Management of the nursery pests is very easy provided the prophylactic control measure are followed at regular intervals. To control the root feeders, it is advisable to mix Phorate 10G granules with the soil while filling the bag. This will effectively control all the root feeders as the granules emit the phosphene gas by reacting with the water. Where as for the control of defoliators, regular application of contact insecticides and biopesticides like *Beauveria bassiana* and *Metarhizium anisopliae* will give effective check. For the control of rats and squirrels all the possible measures instead of one practice should be carried as it is very difficult to manage them.

Pests of Juvenile Palms

During the juvenile period of fist three years after planting, irrigated oil palm is more susceptible to the attack of pests like rhinoceros, psychids, cockchafer beetles, rodents, wild boars and porcupines. The last three are more attracted to the boll region of the palm, which is tender and preferred. However the attack is such that it leads to mortality of the palms. Infestation caused by rhinoceros is sometimes dangerous as it may lead to death of the plant. The adult beetle enters thru the base of the stem that gives an entry to all other pests like red palms weevil and bud rot fungus. Incidence was observed heavy when the plantation is nearer to the coconut gardens and palmyrah palms surrounded by dead logs of the same palms. As the pest is migrating from coconut and palmyrah palm (*Borasses flabedrifer*), which are collateral hosts of this pest, it is essential to take control measures on these palms also. Incidence of cockchafer beetles is observed increasing in all the oil palm growing areas with the defoliation of leaves on the margins as well as in the lamina by two different species. Severity of the incidence was observed in Andhra Pradesh and Karnataka as during the monsoon period as the pest emerges out with the onset of monsoon showers. Incidence of psychids was however observed at low levels on the juvenile period. All these pests were found specific to oil palm and not having any relation with the intercrops that are being grown during the juvenile period. Since no congestion is seen for the competition of light or other climatic parameters as the palms are young and occupy very little space (25 per cent in the first year, 50 per cent in the second and 75 per cent in third year) the spacing between palms is sufficiently high and as such climate has no role in increasing incidence. As the palms are more susceptible to the pest attack during this period and hence the incidence is observed (Wood, 1968; Dhileepan, 1992; Ponnamma *et al.*, 2001)

Pests of Adult Palms

In the yielding irrigated oil palm gardens, the pest outbreaks of leaf web worm, *Acria* sp. and slug caterpillar, *Darna catenatus* were found to occur in many areas in Andhra Pradesh during the last 12 years (Kalidas, 2003; Kalidas, 2004; Ponnamma and Biju, 1998). Of these, leaf web worm is a new pest that is reported only in Thailand and Papua New Guinea only. The pest was not reported from any crop in the country and hence its immigration is suspected. The probable migration is thru the seed sprouts that were imported from PNG during 1990-2000. The slug caterpillar that causes severe infestation occasionally leading to yield losses during the next two to three years of attack is probably migrating from coconut where it was a minor pest. Incidence is observed more in flood-irrigated gardens compared to micro irrigated ones. Availability of abundant food material (with the production of more than two leaves per month) combined with the existence of congenial conditions like low temperatures and high humidity due to overlapping of leaves of adjacent palms and flood irrigation practices are the few reasons for the severity of the pest. The present spacing of 9 mtrs that was adopted from Malaysian practices need to be reviewed as the cultivation practices for irrigated oil palm are entirely different with that of Malaysia. It is clearly visible in case of leaf length, weight, girth, topography of the garden as well as FFB production. This is one of the reasons for the increasing incidence of psychids in the aged gardens. However in the recent years the pest was observed shifting from oil palm to cocoa, which is grown as intercrop in adult oil palm gardens. This is mainly because of availability of more leaves in cocoa compared to oil palm followed by the congenial conditions for pest growth.

Rhinoceros beetle was observed feeding on the spindle leaves causing permanently marked hold on the petiole apart from v shaped cutting on the leaves. Apart form its feeding, the hold made by this may allow the bud rot causing fungi to enter and finally lead to death of the palm. Incidence of red palm weevil was however restricted to bud portions that too in the palms that were infested with rhinoceros beetle. The hole made by rhinoceros is giving entry to the red palm weevil but it is not going deep into the stem portion, which is its base for attack in coconut. Where as it is unable to enter into the stem portions as is seen in case of coconut. Presence of leaf buts all around the stem is the main hindrance for the pest to enter into stem portion is advantage for the palm, which is lacking in case of coconut (Pillai *et al.*, 1993; Sunderababu *et al.*, 1983; Mohan and Pillai, 1982; Kalidas, 2000; Kalidas *et al.* 2001; Kalidas and Ramprasad, 2001; Kalidas, 2004, Kalidas, 2004a))

Few more insect pests were observed feeding on adult palms but are of minor importance. These include scales and mealy bugs and bunch borers that feed directly on the fruits causing yield losses (Dhileepan and Jacob, 1992; Nair *et al.*, 1997; Wood, 1968). However, the quantification of these losses has not been done any wherein the world. These were observed as endemic pests with their continuous presence in the gardens of infestation. This may be due to the poor sanitation practices that are being followed in such gardens.

Termites were also observed as important insects in the oil palm gardens. However their role as pests needs serious discussion (Dhileepan, 1992). These insects feed on dried material of leaf buts as well as male inflorescence by covering the area with earthen sheathings. Since these earthen sheathings are formed with sub soil that contains more of potassium hence it is an advantage to the garden with the increased amounts of potassium. At the same time the dried leaf buts as well as the male inflorescence that are eaten by them have no economic value. Rather it is advantage if they are destroyed otherwise they may pose problem by harboring disease causing fungi. This is particularly so in case of male inflorescence and hence crown cleaning is recommended at frequent intervals to avoid the same. However in Theni of Tamil Nadu, which is, a dry tract containing red soils, heavy incidence of termites causing mortality of the palms by invading the inner tissue of the stem portion was a common scenario. Lack of irrigation, poor sanitation conditions and complete negligence of palms without destroying the termite mounds are the few reasons for such a heavy incidence. Where as in Andhra Pradesh the incidence of termites that was observed feeding on the harvested leaves that are piled at few places was an advantage to the farmers as it completely degrade the material very quickly thus making the gardens more fertile as well as free from reptiles.

Apart form insects pests, avians and mammals cause moderate to heavy incidence making considerable losses on the yields. As these directly feed on the FFB, hence the loss is found on the yield levels compared to insects that cause indirect loss in the form of loss in photosynthetic areas and thereby reduction in the yield.

Crows, *Convus splendens protegatus and Corvus macrorhynchus cuiminatus*, Mynah, *Acridotheres tristis*, Parrots, *Psitticula krameri manillensis* among avians, Indian rat among mammals were observed as serious pests. Incidence of crows and mynahs was seen more in the isolated gardens where as the parakeet was observed in the gardens adjacent to forests. Crows were found to cause partial damage compared to mynahs as the latter causes 100 per cent damage to the fruits in the FFB. These were observed causing heavy damage during the summer months when no other crop exists in the fields. Indian rat was observed causing damage to all parts of the palms during all stages including young seedlings, juvenile and adult palms. In case of young seedlings the pest is attracted towards the nut portion where as in the juvenile plants, it is the boll region while in case of adult palms its incidence is observed on both FFB and in the stem portion. In all the stages the incidence ends with the mortality of the palms except the feeding on FFB where the pest feeds on the immature fruits of FFB for the sake of kernel, which is in semi solid condition. Other pests that are found feeding on adult palms are bear which feeds on the ripened FFB, porcupines that feed on the boll region, eagles, jackals, dogs and vultures on harvested FFB (Kalidas, 2006).

Management

An integrated pest management approach incorporating all possible and feasible practices like cultural, mechanical, sanitation, chemical and biological aspects is required to combat the pest menace in an effective manner. In case of rhinoceros it is better to use bioagent, *Metarhizium anisopliae* in the breeding sites at frequent intervals. On the young and juvenile palms the incidence can be arrested by applying Lambda cyhalothrin, which is a second generation synthetic pyrethroid. The pheromone lure and castor cake baits though attract the beetles to some extent, may not of much use due to climatic abnormalities. Similarly mechanical practices may also be of least useful.

In this connection, it is worth mentioning the technology developed at NRCOP under NATP project on biopesticides namely *Metarhizium anisopliae*, that causes green muscardine disease to all the stages of the pest and *Trichoderma viride*, that controls many root diseases like Ganoderma in a combined capsule with magnesium carbonate ($MgCO_3$) as filler material. Since the application is recommended in the FYM pits, the green muscardine fungus will take care of the rhinoceros beetle, while the *Trichoderma viride* will develop on the FYM continuously. When the FYM is taken out of the pit for application in the fields, the fungus will start its action on the Ganoderma using helix and coil theory. The filler material will be useful to the palms as elemental magnesium is needed for the better growth of the crop. Thus with one module three uses can be achieved. In case of psychids, biological agent *Apanteles* sp. is the most common larval parasite. Since the pest is embedded in a cocoon, it is essential to use a strong contact insecticide like Quinalphos (0.05 per cent). The research work carried out at NRC revealed that the lepidopteron pests like psychids, leaf web worms and slug caterpillars were effectively controlled using white muscardine fungus *Beauveria bassiana*. Apart form using these pesticides, it is better to follow good cultural practices that can arrest the build up of the pest population. For this purpose, the flood irrigation should be avoided followed by allowing the sunrays down to the ground. If need be pruning can also be carried out leaving minimum number of leaves that are required for the optimum yields.

For the successful management of bird menace, it is better to use novel technologies rather routinely practiced ones like covering the bunches with oil palm leaves or with wire mesh or hanging the dead crop or eagle in front of the garden to scare the birds. As these methods did not prove effective, NRCOP has come out with a technology of hanging fishnets in between two palms like volleyball net. Different colours were also tried and found that green and violet ones were effective in scaring away the birds. Birds, which come for feeding on FFB in the oil palm gardens, are getting entangled into these nets making scaring sounds that ultimately scare the other birds from feeding. Since the green and violet are not visible or poorly visible more number of birds are trapped in such nets. This was found giving good results reducing the infestation by 70 per cent. However it needs to be shifted to different places within the garden instead of keeping at one place for weeks together. The scaring sound technology developed by Punjab Agricultural University was also found effective in scaring away the birds to some extent. The alarm sounds that were recorded on tapes were played using amplifiers system proved effective in confusing the birds in the beginning followed by scaring them. This can also be used as one of the practices for the menace of birds (Kalidas,2006; Jacob, 1993; Nair *et al.*, 1997).

Since mammals are the most problematic and difficult to manage pests in oil palm it is better to take management practices from the beginning of the plantation in those areas where the problem is serious. For this purpose, it is advisable to cover the root portion with a used chicken wire mesh while plating it self as it was found effective in arresting the rat damage in young palms. Apart from this it is necessary to take all the possible methods of rat control like weed removal, engaging the rat catchers, using the traditional traps, use of zinc phosphide with bait and anticoagulants like brodifacum and bromadiolone. Inspiteof carrying all these practices, it was not found possible to manage the incidence 100 per cent in many cases. As the rat control needs a unified effort in the entire village at a time, it is suggested to carryout the above operations at community level rather than individual. Even though the above methods are effective for the control of other mammals like porcupine and wild boar, however these can be managed using scaring devices that were developed at NRCOP, Palode.

References

Anon., 1988. *Potentialities of oil palm cultivation in India.* Report of the Working Group constituted by Department of Agriculture and Cooperation, Ministry of Agriculture, Govt. of India, 72p.

Chadha, K. L., 2006. Progress and potential of oil palm in India. Report of the committee constituted to reassess fresh/potential areas of oil palm in India. Deapartment of Agriculutre and Cooperation, Govt. of India, New Delhi. 216p.

Dhileepan, K., 1992. Insect Pests of oil palm (*Elaeis guineensis*) in India. *The Planter,* Kuala Lumpur, 68: 183-191.

Dhileepan, K. and Jacob, S.A., 1992. Pests. In : *Oil Palm Production Technology,* Central Plantation Crops Research Institute, Research Centre, Palode, Kerala, India, p. 49-57.

Jacob, S.A., 1993. A simple device for scaring away wild boar (*Sus scrofa*) in newly planted oil palm fields. *ThePlanter,* 69(811):475-477.

Jacob, S.A. and Kochu Babu, M., 1995. *Sesamia inferens,* a new pest of oil palm seedlings. *The Planter,* Kuala Lumpur. 71 (831): 265-266.

Kalidas Potineni, K.V.Ramprasad and K. Rammohan, 2001, Artificial multiplication of *Baculovirus* on rhinoceros beetle affecting oil palm. Proceedings of Symposium on Biocontrol based pest management for quality crop production in the current millennium, July 18-19, 2001, PAU, Ludhiana. pp.86.

Kalidas Potineni and K.V.Ramprasad, 2001, Efficacy of green muscardine fungus *Metarhizium anisopliae* and its commercial multiplication on maize grains. International journal of Oil Palm, 2(2):51-53

Kalidas, P., 2001, Integrated management practices for rhinoceros beetle affecting Oil palm. Proceedings of Second National Symposium on Integrated Pest management (IPM) in Horticultural crops: New molecules, biopesticides and environment. October 17-19, 2001, Bangalore. Pp. 110.

Kalidas, P., Ramprasad, K.V. and Rammohan, K., 2002. Pest status in irrigated oil palm orchards of coastal areas of India. Journal of Indian Soc. Coastal agric. Res., 20(1). 41-50

Kalidas, P., 2003, Studies on slug caterpillar, *Darna catenatus* of Oil Palm and its management. Proceedings of the National Symposium on Frontier areas of entomological Research, IARI, New Delhi. PP 88-89.

Kalidas, P., 2004, Effect of abiotic factors on the efficiency of rhinoceros beetle pheromone oryctalure in the oil palm growing areas of Andhra Pradesh. The planter, 80(935)103-115.

Kalidas, P., 2004, Stress management of insect pests on Oil Palm, *Elaeis guineensis* Jacq. Journal of Oilseeds Research, 21(1):.220-223.

Kochu Babu, M. and P. Kalidas, 2004, Key pests and diseases of Oil Palm in India–Their biology, epidemiolcgy and method of control. Proceedings of the *International Conference on the pests and diseases of importance to the Oil Palm Industry* held in Malaysia during 18-19th May, 2004. pp.184-206.

Kalidas, P[a]., 2004, Susceptibility of Oil palm (Tenera hybrid) planting material for rhinoceros beetle, *Oryctes rhinoceros* L. Journal of Plantation Crops, 32 (Suppl.): 385-387.

Kalidas, P.,Ch.Venkateswara Rao, Nasim Ali and M. Kochu Babu., 2006. *New pest incidence on Oil Palm seedlings in India- A study of black slug (Laevicaulus alte). The Planter, Kuala Lumpur 82(960):181-186.*

Kalidas, P., 2006. Avian pest menace in oil palm and the impact of novel technologies on its management. Journal of plantation crops. 34: 172- 178

Kochu Babu, M., 1993. Investigations on spear rot complex of oil palm (*Elaeis guineensis* Jacq.) *Ph. D. Thesis,* Mangalore University, Mangalore, Karnataka, India, 183p.

Kochu Babu, M., 2002. *Final Report,* Cess Fund Scheme "Determining the vectorial role of putative insect vectors with reference to spear rot disease of oil palm, NRC for Oil Palm, Palode, 19p.

Mohan, K.S. and Pillai, G.B., 1982, A method for laboratory scale mass cultivation of *Metarrhizlum anisopliae. Foila Microbiol.,* 27 : 281–283.

Nair, C.P.R. and Daniel, M., 1982. Pests. In: *The Arecanut Palm, A monograph* (Eds. Bavappa, K. V.A., Nair, M.K. and Premkumar, T.) Central Plantation Crops Research Institute, Kasaragod, India.

Nair, C.P.R., Daniel, M. and Ponnamma, K.N., 1997. Integrated Pest Management in Palms (Eds. Nambiar, K, K. N. and Nair, M. K.), Coconut Development Board; Kochi, 30p.

Pillai, G.B., Sathiamma, B.and Danger, T.K., 1993. Integrated control of Rhinoceros beetle. In: *Advances in coconut research and development* (Eds Nair, M.K., Khan.H.H., Gopalasundaram, P. and Rao, E.V.V.B.) Central Plantation Crops Research Institute, Kasaragod, p. 58-65.

Ponnamma, K. N. and Biju, B., 1998. Record of *Hatictophagus* sp. {Strepsiptera: Halictophagidae) a parasitoid of *Proutista moesta* Westwood (Homoptera;. Derbidae), a vector of Mycopiasma disease of palms. *The Planter,* Kuala Lumpur, 74 (862): 37-40.

Ponnamma, K.N., Lalitha, N and Sajeeb Khan, A., 2001. Oil Palm mesocarp waste a potential breeding medium for Rhinoceros beetle, *Oryctes rhinoceros* L, *International Journal of oil palm,* 2(1): 37-40

Shanta, P., Gopinathan Pillai. N. and Lal, S.B., 1970. Oil palm (*Elaeis guineensis* Jacq.) A natural host of root (wilt) pathogen of coconut. *Curr, Sci.*, 39 : 260-261.

Sundarababu, PC, Balasubramanian, M. and Jayaraj, S., 1983. Studies on the pathogenicity of *Metarhizium anisopliae* (Metsehnikoff) Sorokin var major Tulloch on *Oryctes rhinoceros* (L.) In: *Research Publication No. 7*, Tamil Nadu Agricultural University, Coimbatore, T.N.

Wood, B.J., 1968. *Pests of oil palms in Malaysia and their control. The Incorporated Society of Planters*, Kuala Lumpur, Malaysia, 222p.

Index